D0850412

Chemistry and Chemical Taxonomy of the Rutales

Annual Proceedings of the Phytochemical Society of Europe

1. **Biosynthetic Pathways in Higher Plants**
   Edited by J. B. PRIDHAM AND T. SWAIN

2. **Comparative Phytochemistry**
   Edited by T. SWAIN

3. **Terpenoids in Plants**
   Edited by J. B. PRIDHAM

4. **Plant Cell Organelles**
   Edited by J. B. PRIDHAM

5. **Perspectives in Phytochemistry**
   Edited by J. B. HARBORNE AND T. SWAIN

6. **Phytochemical Phylogeny**
   Edited by J. B. HARBORNE

7. **Aspects of Terpenoid Chemistry and Biochemistry**
   Edited by T. W. GOODWIN

8. **Phytochemical Ecology**
   Edited by J. B. HARBORNE

9. **Biosynthesis and its Control in Plants**
   Edited by B. V. MILBORROW

10. **Plant Carbohydrate Biochemistry**
    Edited by J. B. PRIDHAM

11. **The Chemistry and Biochemistry of Plant Proteins**
    Edited by J. B. HARBORNE AND C. F. VAN SUMERE

12. **Recent Advances in the Chemistry and Biochemistry of Plant Lipids**
    Edited by T. GALLIARD AND E. I. MERCER

13. **Biochemical Aspects of Plant-Parasite Relationships**
    Edited by J. FRIEND AND D. R. THRELFALL

14. **Regulation of Enzyme Synthesis and Activity in Higher Plants**
    Edited by H. SMITH

15. **Biochemical Aspects of Plant and Animal Coevolution**
    Edited by J. B. HARBORNE

16. **Recent Advances in the Biochemistry of Cereals**
    Edited by D. L. LAIDMAN AND R. G. WYN JONES

17. **Indole and Biogenetically Related Alkaloids**
    Edited by J. D. PHILLIPSON AND M. H. ZENK

18. **Nitrogen Fixation**
    Edited by W. D. P. STEWART AND J. R. GALLON

19. **Recent Advances in the Biochemistry of Fruits and Vegetables**
    Edited by J. FRIEND AND M. J. C. RHODES

20. **Seed Proteins**
    Edited by J. DAUSSANT, J. MOSSÉ AND J. VAUGHAN

21. **Metals and Micronutrients**
    Edited by D. A. ROBB

22. **Chemistry and Chemical Taxonomy of the Rutales**
    Edited by PETER G. WATERMAN AND MICHAEL F. GRUNDON

ANNUAL PROCEEDINGS OF THE PHYTOCHEMICAL
SOCIETY OF EUROPE NUMBER 22

Chemistry and Chemical Taxonomy of the Rutales

Edited by

PETER G. WATERMAN

Department of Pharmaceutical Chemistry, University of Strathclyde,
Royal College, Glasgow, Scotland

and

MICHAEL F. GRUNDON

Department of Chemistry, The New University of Ulster, Coleraine, Co.
Londonderry, Northern Ireland

1983

. ACADEMIC PRESS

A Subsidiary of Harcourt Brace Jovanovich, Publishers

London · New York
Paris · San Diego · San Francisco · São Paulo
Sydney · Tokyo · Toronto

ACADEMIC PRESS INC. (LONDON) LTD.
24–28 Oval Road
London NW1 7DX

U.S. Edition published by
ACADEMIC PRESS INC.
111 Fifth Avenue
New York, New York 10003

British Library Cataloguing in Publication Data

Chemistry and chemical taxonomy of the Rutales.—
(Symposia/Phytochemical Society of Europe,
ISSN 0309-9393)
1. Botany—Classification—Congresses
2. Biological chemistry—Congresses
I. Waterman, P. G.   II. Grundon, M. F.
III. Series
581′.012   QK95

ISBN 0-12-737680-1
LCCCN 83-70297

Typeset by Speedlith Photo Litho Limited, Longford Trading Estate, Manchester M32 0JT.

Printed in Great Britain by Thomson Litho Limited, East Kilbride, Scotland.

Contributors

J. D. CONNOLLY, *Department of Chemistry, University of Glasgow, Glasgow G12 8QQ, UK*

D. L. DREYER, *USDA-SEA, Western Regional Research Center, Berkeley, California 94710, USA*

A. I. GRAY, *Department of Pharmacognosy, School of Pharmacy, Trinity College, 18 Shrewsbury Road, Dublin 4, Eire.*

M. F. GRUNDON, *Department of Chemistry, The New University of Ulster, Coleraine BT52 1SA, Co. Londonderry, Northern Ireland.*

J. B. HARBORNE, *Plant Science Laboratories, University of Reading, Reading RG6 2AH, Berkshire. UK*

R. HEGNAUER, *Laboratorium voor Experimentele Plantensystematiek, 2313 AV Leiden, The Netherlands*

S. A. KHALID, *Phytochemistry Research Laboratory, Department of Pharmaceutical Chemistry, University of Strathclyde, Glasgow G1 1XW, UK (Present address: School of Pharmacy, University of Khartoum, P.O. Box 321, Khartoum, Sudan)*

J. KUMAMOTO, *Department of Botany and Plant Science, University of California at Riverside, California 92521, USA*

J. R. LEWIS, *Department of Chemistry, University of Aberdeen, Aberdeen AB9 1FX, UK*

V. P. MAIER, *Fruit and Vegetable Chemistry Laboratory, USDA, Pasadena, California 92215, USA*

A. D. J. MEEUSE, *Hugo de Vries Laboratorium, University of Amsterdam, SPUI 21 Amsterdam, The Netherlands*

I. MESTER, *Institut für Pharmazeutische Chemie der Universität, 4400 Münster, West Germany*

J. POLONSKY, *Institut de Chemie des Substances Naturelles, CNRS, 91190 Gif-sur-Yvette, France*

J. O'SULLIVAN, *Department of Pharmacognosy, School of Pharmacy, Trinity College, 18 Shrewsbury Road, Dublin 4, Eire.*

R. W. SCORA, *Department of Botany and Plant Science, University of California at Riverside, California 92521, USA*

D. A. H. TAYLOR, *Department of Chemistry, University of Natal, King George V Avenue, Durban, South Africa*

P. G. WATERMAN, *Phytochemistry Research Laboratory, Department of Pharmaceutical Chemistry, University of Strathclyde, Royal College, 204 George Street, Glasgow G1 1XW, UK*

Preface

The Rutales, in the strict sense in which it is interpreted in this volume, represents only a medium-sized order of flowering plants, probably containing no more than 3000–3500 species distributed in only five or six families: Rutaceae, Meliaceae, Ptaeroxylaceae, Simaroubaceae, Cneoraceae, and, more uncertainly, Burseraceae. Within this group there has developed an extremely rich and diverse range of secondary metabolites, comparable to that of much larger taxa such as the Leguminosae and Compositae. Many of these metabolites are unique to the order, such as the limonoids, quassinoids, and some of the alkaloid groups (acridones, carbazoles, and hemiterpenoid quinolines), and exhibit major levels of structural diversity. In view of the richness and diversity it is hardly surprising that the Rutales have been the focus of a considerable amount of effort by scientists interested in the identification of novel secondary metabolites and in understanding their biogenesis.

From an economic standpoint the genus *Citrus* is overwhelmingly the most important in the order. The biological activity of the limonoids and flavonoids of *Citrus* and their influence on the palatability of citrus products has for long been the subject of considerable research effort and is now largely understood. However, biological activity is widespread among the compounds produced by rutalean plants, and investigations of these have been responsible for an increasing volume of literature in recent years. Subject to particular scrutiny have been the pronounced cytotoxicity exhibited by some of the quassinoids of the Simaroubaceae and acridone and benzophenanthridine alkaloids of the Rutaceae, and the strong anti-feedant properties of some limonoids of the Meliaceae.

Studies on the distribution of secondary metabolites within the order have proved fertile ground for the speculations of the chemotaxonomist, and yield much information suggestive of relationships between taxa; for example, they strongly support the retention of the sub-family Flindersioideae within the Rutaceae and the separation of the Ptaeroxylaceae from the Meliaceae. Just as the "uniqueness" of many types of rutalean compounds points to the distinctiveness of the Rutales as a group, so the co-occurrence of some metabolites in other parts of the plant kingdom has given rise to considerable speculation as to their importance in pointing out phylogenetic relationships for the Rutales, most notably the co-occurrence of 1-benzyltetrahydroisoquinoline alkaloids in the supposedly primitive Polycarpicae (Ranales) and of allied, hemiterpenoid, coumarins in the Umbelliferae.

This book is based on the review lectures given at the symposium of the Phytochemical Society of Europe held at the University of Strathclyde in Glasgow in April 1982 on the subject of the Chemistry and Chemical Taxonomy of the Rutales. This meeting brought together leading authorities on the various classes of compounds found in the Rutales, especially coumarins, quinoline and acridone alkaloids, flavonoids, limonoids, quassinoids, and lignans. Discussion centred on aspects of the chemistry, distribution, biological activity, and biosynthesis of these compounds, and involved the participation of pharmaceutical chemists, organic chemists, botanists, and chemical taxonomists. Chapter 1 deals briefly with the concept of the Rutales and its botanic delimitation. Chapters 2–10 are concerned primarily with the chemistry of the order, covering aspects of structure elucidation, structural diversity, biogenesis, and distribution for all the more important groups of compounds. In several of these chapters note is also taken of the biological activity of some of the metabolites under discussion, and this aspect is pursued as the central theme of Chapters 11 and 12, one of which deals with some of the potentially useful compounds to have been isolated from the Rutaceae whilst the other deals in detail with the problems caused by some of the limonoids and flavonoids found in *Citrus* fruits. Finally, Chapters 13–16 deal mainly with chemotaxonomic implications of the distribution of compounds. Three of these chapters are concerned with problems within the order whilst in the final chapter the chemotaxonomic relationship between the Rutales and other taxa is the focus of attention.

In conclusion the editors would like to thank all those who helped make the meeting a success, particularly the contributors for the prompt submission of their manuscripts. They also gratefully acknowledge financial assistance for the meeting from the Royal Society and the Nestles Foundation. Finally Academic Press must be acknowledged for their expert assistance in preparing this volume for publication.

July 1983   *P. G. Waterman*
                                                               M. F. Grundon

Contents

CHAPTER 1

The Concept of the Rutales

A. D. J. Meeuse

CHAPTER 2

Aspects of the Biosynthesis of Coumarins and Quinoline Alkaloids in the Rutales

M. F. Grundon

CHAPTER 3

Structural Diversity and Distribution of Alkaloids in the Rutales

I. Mester

ix

CHAPTER 4

Structural Diversity and Distribution of Coumarins and Chromones in the Rutales

A. I. Gray

CHAPTER 5

The Flavonoids of the Rutales

J. B. Harborne

CHAPTER 6

Chemistry of the Limonoids of the Meliaceae and Cneoraceae

J. D. Connolly

CHAPTER 7

Limonoids of the Rutaceae

D. L. Dreyer

CHAPTER 8

Chemistry and Biological Activity of the Quassinoids

J. Polonsky

CHAPTER 9

Structural Diversity and Distribution of Lignans in the Rutales

J. O'Sullivan

CHAPTER 10

Chemistry of the Burseraceae

S. A. Khalid

CHAPTER 11

Biological Activity of Some Rutaceous Compounds

J. R. Lewis

CHAPTER 12

Chemistry and Significance of Selected *Citrus* Limonoids and Flavonoids

V. P. Maier

CHAPTER 16

Chemical Characters and the Classification of the Rutales

R. Hegnauer

CHAPTER 1

The Concept of the Rutales

A. D. J. MEEUSE

Hugo de Vries-Laboratorium, University of Amsterdam, Netherlands

I. INTRODUCTION

A taxonomist is double handicapped when he has to discuss a systematic subject before a mixed audience mainly consisting of uninitiated persons. Firstly, he has to touch upon some generalities and basic principles before he can get down to brass tacks, and, secondly, he has to avoid as much as possible the use of the pertaining scientific jargon.

Taxonomists compile evidence in order to unravel certain relationships between organisms, these connections being, at least theoretically, expressions of a lesser or greater amount of shared genetical factors. The assessment of the nature of such relationships consists in actual practice of a comparison of phenetic characteristics (characters, features, peculiarities) which are the discernible expressions of the genetical make-up of the individuals under taxonomic investigation.

The principal purpose of most taxonomic studies is to arrive at a surveyable arrangement or "system" of classification for easy reference. Such "systems" are *hierarchic*, which means that several superposed categories are distinguished starting from the species category (see Fig. 1). For present purposes the following practical but not quite adequate definition will suffice: A species is the assembly of individuals or populations of individuals which do not only share the great majority of their (hereditary) genetical factors but can

1

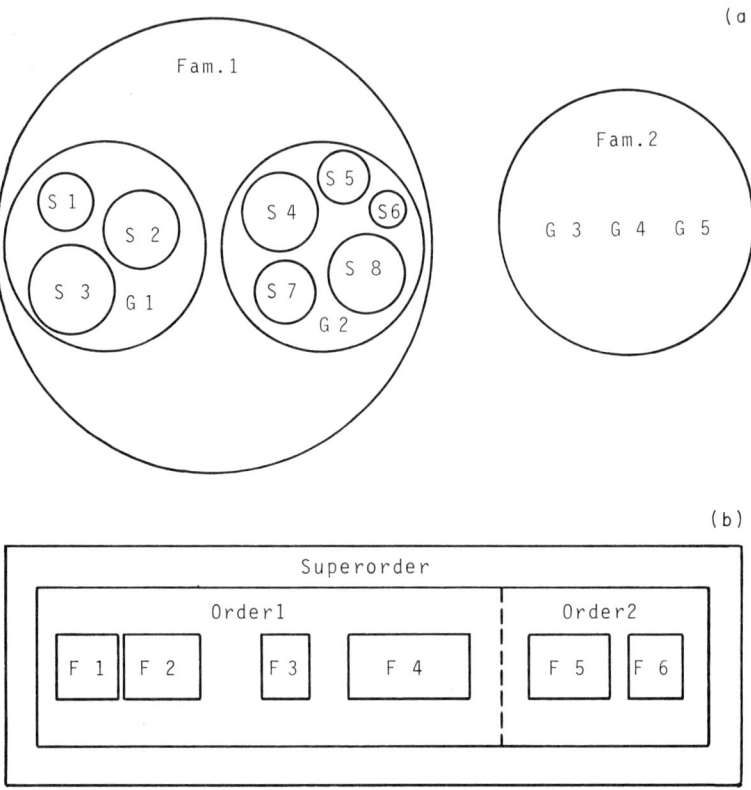

FIG. 1. (a) Left: diagram of family (1) comprising two genera (G1, G2) each containing a number of species; right: another family (2) consisting of three genera. (b) Illustration of an arbitrary classification of two orders (1, 2) together constituting a superorder; by deleting the dashes between 1 and 2 one may consider them a single order; F1 to F6 represent families, distances symbolizing the degree of relationship between them.

normally also mate among themselves thus producing progenies with approximately the same set of phenetic features as the parent generation. The "higher" categories are more or less arbitrarily based on the principles that (1) the systematic units constituting one of the hierarchic levels or "ranks" of the "system" above the species level—genus, family, order, superorder, class, etc., in this sequence—include all representatives of the lower rank or ranks included as subordinate taxa in the taxon of a higher rank, and (2) within each hierarchic category the individual units are recognised by discontinuities in their phenetic features. (In other words, each "higher" unit contains units of a lower rank which share more characters than they do with all other assemblies of such taxonomic units immediately subordinate to another unit of the next higher rank.)

In the present context I shall deal principally with the categories of the family and of the order. A family contains a number of genera (usually two to several, sometimes many, rarely a single one), and an order consists of an aggregate of families which are supposed to share more features than they do with all other families (which, therefore, belong to different and separate orders). A superorder is a conglomerate of two or more orders which share so many features that they are, in taxonomic terms, "more closely related" than they are to any other family or order and seem to hang together to form a "natural" assembly. As intimated before, these concepts are arbitrarily applied and interpreted, with the result that there may be differences of opinion among taxonomists attempting to classify the same taxonomic group (taxon) in a hierarchic system. This may seem to be a bit strange, but the disagreements can be easily understood if one takes into account that systematists do not all attribute the same "value" (significance, weight) to a given character; in their terminology, a certain feature is supposed to have a greater (or lesser, as the case may be) taxonomic significance than others, and therefore a greater (or lesser, respectively) discriminative power.

Since modern systematists tend to exploit all kinds of characters, including embryological, structural (e.g. xylotomic, i.e. anatomical), ultrastructural, palynological, ecological, and (bio)chemical, it stands to reason that such additional information may lead to readjustments of the classification and that new evidence may even result in "shifts" of a taxon from one major group to another. More conservative workers do not always accept the cogency of such unconventional characteristics, so that even the most recent systems of classification of the Flowering Plants show discrepancies depending on, for example, the lesser or greater confrontation with meaningful phytochemical data and other untraditional evidence.

Now that the present author proceeds to a survey of the concept of the Rutales, one must bear in mind that the older systematists based their conclusions almost exclusively on the traditional "macromorphological" features and simply could not attain such a detailed overview as present-day taxonomists are able to do. Another restriction is the stipulation that phytochemical features would have to be omitted from the evidence, because they will be treated by more competent speakers (see the contributions by Hegnauer and Waterman). This is not such a great drawback, however, because the introduction of phytochemical evidence in plant systematics only began in earnest a few decades ago and most systems of classification or their earlier versions date back before that time. The proposers of more recent systems have often ignored the chemical pointers, for that matter.

The inclusions by different taxonomists of unequal numbers of families in an assembly bearing a given name is another evil partly dealt with here by adding *s.l.* (in a broader sense, i.e. containing a greater number of subordinate units) or *s.s.* (more restricted, containing fewer such units).

II. THE OLDER CLASSIFICATIONS

The concept and nomenclatural status of the "order" as a taxonomic unit (and level = rank) in a hierarchic system of classification only rather recently became standardized and used in a consistent way. The idea of a "family" as a "natural group" took root much earlier, but a more or less coherent aggregate of families appears in older works under such names as "series" (Reihen) or "cohors" or even as a nameless category, and under names with inconsistently formed endings (the normalized and now standard ending "-ales" was accepted by the decision of a botanical congress, presumably not before 1900). Each such assembly of families immediately above the family level will be treated here as equivalent to a modern "order" irrespective of its older name and given the ending "-ales", so as to avoid confusion.

From a nomenclatural point of view each assembly of this kind must contain the (nomenclaturally) "typical" family; for instance, the family of the Rutaceae is the "type family" of the order Rutales and must always be retained in this order to which it also gives its name, irrespective of the comprehensiveness ("size") of that order.

Many of the following details are borrowed from the thesis of Boesewinkel,[3] who kindly provided some additional information and to whom I owe the idea of the survey diagram (see Table 1). As a starting point the classification of Bentham and Hooker[2] was chosen.

Apart from the Rutaceae these British systematists also included other families nowadays usually referred to the Rutales *s.s.* and the Geraniales in the current modern sense in an assembly here referred to by the name of Geraniales (*s.l.*). Most subsequent authors recognized a smaller order Geraniales and sometimes included the Rutales in the Sapindales (or vice versa), although the last two are usually kept separate as "twin" orders. This suggests a close relationship between them, and a lesser affinity to the Geraniales (or to segregates of this order such as the Linales and Polygalales).

Several other older classifications not included in Table 1 differ appreciably from this general idea and a detailed discussion is not warranted here (see Chapter 16).

TABLE 1

Various interpretations of the limits of the Rutales, Geraniales, and Sapindales.

Family	Bentham and Hooker[2]	Melchior[8] (Engler)	Takhtajan[9]	Thorne[10]	Dahlgren[5]	Cronquist[4]
Aceraceae						
Anacardiaceae						
Hippocastaneaceae		Sapindales				
Melianthaceae						
Sapindaceae						
Burseraceae						
Cneoraceae						
Meliaceae		Rutales				
Rutaceae						
Simaroubaceae						
Balsaminaceae					Balsaminales	
Geraniaceae		Geraniales				
Oxalidaceae						
Erythroxylaceae						Linales
Linaceae						
Malpighiaceae			Polygalales			
Polygalaceae						

III. RECENT SYSTEMS

Nearly all recent systems are similar, as mentioned above, but one need not accept the general viewpoints for this reason alone.

Most contemporary systematists would accept (or at least imply) a more or less direct descent of early rutalean forms from a magnolioid-ranunculoid group of ancestors, but only a few suggest links with certain more advanced groups (cf. ref. 6). The Rutaceae almost certainly constitute the most primitive member of an aggregate to which, apart from the other families of the Rutales s.s., belong *perhaps* the Sapindales and *perhaps* the Geraniales, and from which most probably the basic stock of the families Araliaceae and Apiaceae (= Umbelliferae) are derivatives and from the latter, in turn, the forebears of the Campanulales and Asterales. The most cogent arguments pleading in favour of such relationships are, as far as can be ascertained, of a phytochemical nature and will for that reason certainly be amply dealt with in a number of other chapters in this volume. However, there are some other unconventional, relevant data. A characteristic feature continuously linking "ranalean" groups with Rutales and the latter with Apiales/Araliales and, ultimately, Asterales is the relation between representatives of the plant taxa concerned and the caterpillars of the butterfly family or superfamily Papilionidae (swallowtails). Such a selective dependence of a herbivore on its host plant almost always has some chemical background,[7] but the interesting part is the gradual "shift" of phytochemical characteristics (from benzylisoquinoline alkaloids in the basic stock to coumarins and other groups of compounds in the most advanced ones) which did not disrupt the host–parasite relationships.

IV. VARIOUS OTHER ASPECTS

The above-mentioned characteristic of a host–parasite relationship which is not broken up by changes in the chemical set-up strongly implies the diagnostic significance of this feature. It may be very meaningful that the larvae of papilionid Lepidoptera are not (at least not at present) found on other rutalean families as far as can be ascertained. They have not been recorded from Sapindales *s.s.* or Geraniales either. The corollary is that the Rutaceae are more primitive than the other Rutales *s.s.* (which is not at variance with certain other characteristics of this family such as apocarpy of the gynoecia in some of its subordinate groups), but it does not support the idea of a close link between Rutales and Sapindales and between Rutales and Geraniales.

The studies of ovule and seed coat development in representatives of these orders[3] do not throw much light on relations between them, unless one accepts the negative or inconclusive results of investigations as cogent evidence of the unrelatedness of Rutales on the one side and the Geraniales/Polygalales/Malpighiales on the other.

The painstaking and thorough work of Behnke and some of his pupils[1] on the structure of microplastids found in the sieve tubes of angiosperms has resulted in a fairly complete survey of the distribution of the various kinds of plastids distinguished over the various families, but does not throw much light on the relationships of the Rutales (compare Table 2). All representatives of the Rutales *s.s.* have the S-type which is of very common occurrence among the dicotyledons. Some Geraniales have other and most probably more advanced types, but this does not necessarily preclude a close affinity because the S-type is also found in this group. The situation in the Sapindales is possibly of greater interest in connection with presumed relationships between the Sapindales and the Connaraceae and with the Leguminosae (considered to belong to the major group of the Rosidae by many taxonomists). The Sapindales have the S-type, but the Leguminosae or Fabales (sometimes) have a rare type of microplastid (the P-type), which is not at all suggestive of a link Sapindales–Leguminosae or of the rosalean affinities of the Sapindales; the Connaraceae have the P1c type and also differ in this respect from the Sapindales.

The outcome of this brief survey of characteristics is that there is no conclusive and convincing evidence from various subdisciplines supporting the *close* affinities of the Rutales with the Geraniales *s.l.* or with the Sapindales, but not altogether denying it either, so that, for instance, one may consider the

TABLE 2

Sieve element characters (data from Behnke[1]).

Taxa	Type	Taxa	Type
Rutales *s.s.*		Geraniales	
Rutaceae	S (P 1c)	Geraniaceae	S
Simaroubaceae	S	Oxalidaceae	P 1c'
Cneoraceae	S	Erythroxylaceae	P 5c
Meliaceae	S	Balsaminaceae	S
Burseraceae	S	Miscellaneous	
Sapindales *s.s.*		Linaceae	S
Sapindaceae	S	Polygalaceae	S
Aceraceae	S	Malpighiaceae	S o
Hippocastaneaceae	S	Krameriaceae	S
Anacardiaceae	S	Connaraceae	P 1c
Rosales	S, P 5 and P 6	Fabales	S (sometimes P 4 types)

possibility of a magnolioid rather than a rosoid phylogenetic origin of the Sapindales *s.s.* but this presumably does not hold true for the Connaraceae and/or the Fabales. These rather disappointing conclusions strengthen the taxonomic significance of biochemical and phytochemical pointers purposely left out of consideration here. However, the "ranalean" affinities of the Rutales can almost certainly be taken for granted, which at least provides taxonomists and phytochemists with a sound starting point for evolutionary deductions.

REFERENCES

1. Behnke, H.-D. (1981). *Nordsk. J. Bot.* **1**, 381.
2. Bentham, G. and Hooker, J. D. (1862). "Genera Plantarum". London.
3. Boesewinkel, F. D. (1980). "Development of ovule and seed-coat in the Rutales-Geraniales assembly". Ph.D. Thesis, University of Amsterdam.
4. Cronquist, A. (1981). "An Integrated System of Classification of Flowering Plants". Columbia University Press, New York.
5. Dahlgren, R. M. T. (1980). *Bot. J. Linn. Soc.* **80**, 91.
6. Fish, F. and Waterman, P. G. (1973). *Taxon* **22**, 177.
7. Meeuse, A. D. J. (1979). *Symb. Bot. Uppsala* **22**, 32.
8. Melchior, H. (1964). *In* "Syllabus der Pflanzenfamilien", 12th Edn, Vol. 2, pp. 262–277. Borntraeger, Berlin.
9. Takhtajan, A. (1973). "Evolution und Ausbreitung der Blütenpflanzen". Stuttgart.
10. Thorne, R. F. (1976). *In* "Evolutionary Biology", Vol. 9, pp. 35–106. Plenum Press, New York.

Note added in proof. After the manuscript had been submitted for publication, a paper appeared on the co-evolution of the butterfly family Papilionidae and groups belonging to, or associated with, the Magnoliales [Richard, D. and Guédès, M. (1983). *Phyton* (Austria) **23**, 117]. The basic idea is not new, but confirms previously made suggestions regarding the position of the Rutales (and some associated groups), the Umbellales (= Apiales or Araliales s.s.) and the Compositae including those expressed by Hegnauer and by myself in the present volume.

CHAPTER 2

Aspects of the Biosynthesis of Coumarins and Quinoline Alkaloids in the Rutaceae

MICHAEL F. GRUNDON

School of Physical Sciences, New University of Ulster, Coleraine, UK

I. INTRODUCTION

During the last few years a number of accounts have been published of the biosynthesis of coumarins[7,20,21] and of quinoline alkaloids.[23] Many of these natural products are hemiterpenes, containing C_5 prenyl groups attached to the carbon atoms of aromatic and heterocyclic systems or to oxygen or nitrogen functions; coumarins and quinolines with C_{10} monoterpenoid groups are encountered less frequently.[21,23,41] In another recent article[22] the remark was made that compartmentation is a problem not only for biosynthetic research but also for a reviewer, and a discussion of the biosynthesis of coumarin, chromone, acetophenone, and indole hemiterpenes occurring principally in species of the Rutaceae and Umbelliferae and in microorganisms emphasized biosynthetic pathways common to these secondary metabolites. The same general approach is adopted here, except that biosynthetic differences between coumarins and quinolines of the Rutaceae as well as similarities are identified. A summary of our present knowledge of the biosynthesis of furocoumarins and furoquinoline alkaloids is followed by more detailed accounts of dihydropyrano derivatives and of coumarins and quinolines that contain side chains that may originate from sigmatropic rearrangement reactions. In the final sections the biosynthesis of 2,2-dimethylpyrano derivatives and of dimeric coumarins and quinolines, topics that have been neglected in recent reviews, is discussed.

II. Biosynthesis of Furocoumarins and Furoquinolines

A. FUROQUINOLINES

Two types of hemiterpenoid quinoline alkaloids occurring widely in the Rutaceae are the furoquinolines, which are derivatives of the simplest member dictamnine (4), and the hydroxyisopropyldihydrofuroquinolines, exemplified by platydesmine (6). Full accounts are available of the occurrence, structure, and biosynthesis of these compounds.[22,23,39] The biosynthetic pathway illustrated in Scheme 1 is supported by tracer feeding experiments with *Skimmia japonica* and *Choisya ternata* in which 4-hydroxy-2-quinolone (1), a 3-prenylquinolone (2), and platydesmine (6) were shown to be good precursors of dictamnine and the *N*-methylplatydesminium cation (9).[14] Dictamnine is a specific precursor of skimmianine (7) and other furoquinoline alkaloids oxygenated in the aromatic ring in these species;[25] similar results were obtained with cell cultures of *Ruta graveolens*.[4]

The pathway to platydesmine given in Scheme 1 occurs readily *in vitro*, and chemical analogy is the only evidence available to indicate that an epoxide (3) is an intermediate.[6] The 3-position of 4-hydroxy-2-quinolone is a powerful nucleophilic centre and a prenyl group should be readily introduced at this point; methylation of the acidic 4-hydroxyl group then occurs preferentially and ensures that subsequent oxidative cyclization gives compounds with linear annelation. Irradiation of platydesmine in the presence of lead tetra-acetate and iodine furnishes dictamnine.[16] Angular furoquinoline alkaloids, although more stable than the linear compounds,[15] have only once been obtained from natural sources; this may be a consequence of chemical reactivity in this series of compounds. Although extensive oxygenation of the

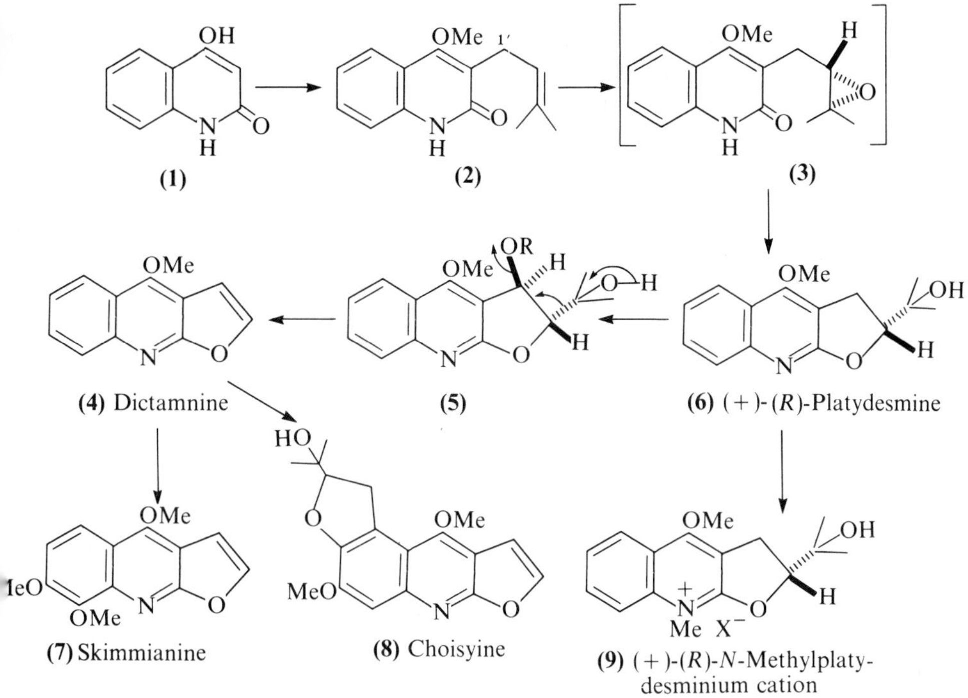

SCHEME 1. Biosynthesis of quinoline alkaloids.

aromatic ring is characteristic of quinoline alkaloids, C-prenylation of that ring is rare and only occurs in alkaloids in which prenylation has also taken place in the 3-position, e.g. choisyine (**8**); again this may be attributed to the high nucleophilicity of the 3-position of 4-hydroxy-2-quinolones.

B. FUROCOUMARINS

Evidence for the biosynthetic pathway to hemiterpenoid coumarins has been reviewed in considerable detail.[7,20,22] The route is exemplified in Scheme 2 for the linear hydroxyisopropyldihydrofurocoumarin marmesin (**11**) and the linear furocoumarin psoralen (**10**), which co-occur in *Ruta graveolens*. Tracer feeding experiments with *R. graveolens* showed that 7-hydroxycoumarin (**12**), a prenyl derivative (**13**), and (+)-(S)-marmesin (**11**) were incorporated into psoralen,[9,50] and there is evidence that the latter compound is a precursor of linear furocoumarins containing additional oxygen functions in the aromatic ring.[1]

Angular furocoumarins (cf. **17** and **18**) are also constituents of rutaceous plants, and their biosynthesis, demonstrated mainly with plants of the Umbelliferae, follows a route from 7-hydroxycoumarin (Scheme 2) parallel to that of the linear compounds.[50] It is of considerable interest that the prenylase

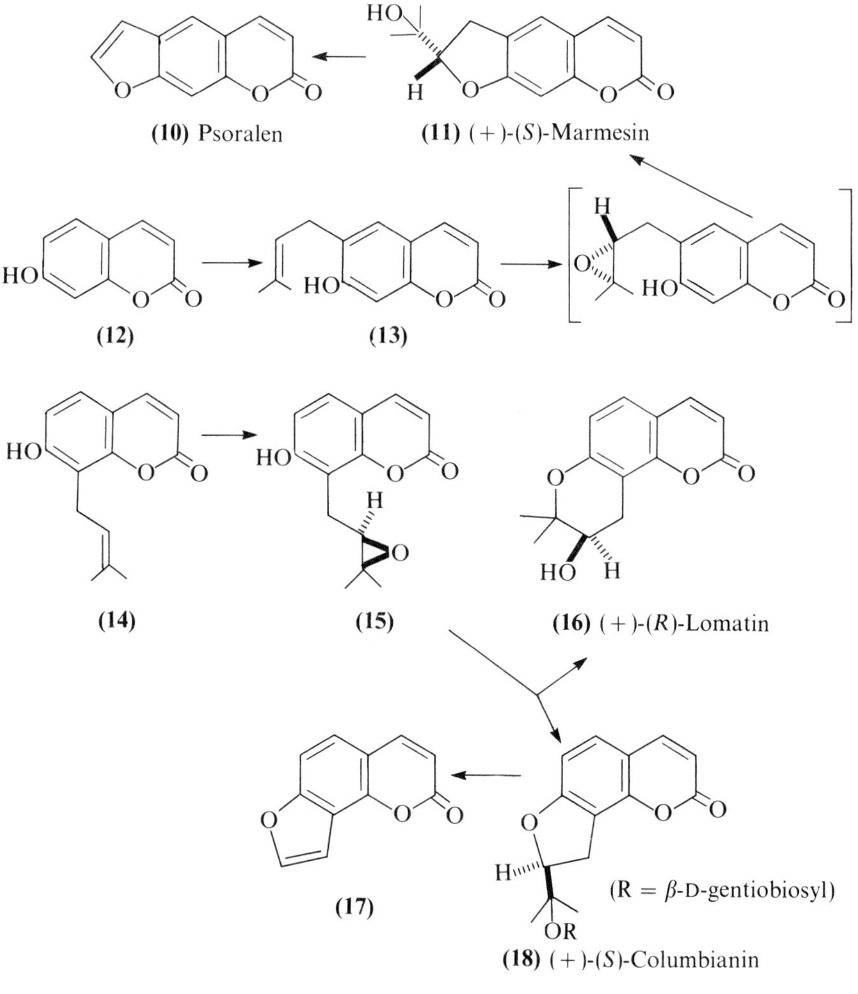

(**10**) Psoralen   (**11**) (+)-(S)-Marmesin

(**12**)   (**13**)

(**14**)   (**15**)   (**16**) (+)-(R)-Lomatin

(**17**)   (R = β-D-gentiobiosyl)

(**18**) (+)-(S)-Columbianin

SCHEME 2. Biosynthesis of coumarins.

responsible for the formation of 6-prenylcoumarin (**13**) from 7-hydroxy-coumarin and 3,3-dimethylallyl pyrophosphate has been isolated from *R. graveolens* and was shown not to affect the transfer of a prenyl group to C-8 of 7-hydroxycoumarin (cf. **14**).[18] Although there is the need for much more experimental evidence, the pathways to furoquinolines and furocoumarins appear to be similar; the contrasts in structural variations between these two groups of compounds seem to be dictated by the heterocyclic starting compounds, 7-hydroxycoumarin and 4-hydroxy-2-quinolone, and the different possibilities for prenylation that these molecules provide.

C. FORMATION OF FURAN RINGS

The biomechanism of the loss of a three-carbon fragment, for example in the transformation of platydesmine (**6**) into dictamnine (**4**) and in the conversion of (+)-(S)-marmesin (**11**) into psoralen (**10**), has been the subject of considerable speculation; the only experimental evidence available so far was obtained for quinoline alkaloids. Platydesmine, in the form of the specifically labelled racemate, was incorporated (18.8 %) into dictamnine in *S. japonica*, and on the assumption that the reaction is enantiomer specific this represents an incorporation of more than 37 %; since platydesmine is also a precursor of the (+)-(R)-N-methylplatydesminium cation (**9**), it appears that (+)-(R)-platydesmine is the bio-active enantiomer.[14] 4-Methoxy-3-prenylquinolone (**2**) and dictamnine (**4**) are known to be precursors of skimmianine in *C. ternata*, and feeding doubly labelled prenylquinoline (**2**) showed that only one hydrogen at C-1′ was lost in the conversion of platydesmine into skimmianine.[26] The most likely mechanism involves stereospecific oxidation of platydesmine to an alcohol derivative (**5**) which then undergoes fragmentation (Scheme 1). The configuration of the proton lost during conversion into the furan ring is not known, and formation of an intermediate with an electron-deficient centre at C-1′, rather than an oxygen function, cannot be ruled out. The isolation of myrtopsine (**5**; R = H, unknown stereochemistry) from *Myrtopsis sellingii*[34] provides chemotaxonomic evidence for its role as a biosynthetic precursor of dictamnine.

III. BIOSYNTHESIS OF HYDROXYDIMETHYLDIHYDROPYRANS

Hydroxydimethyldihydropyranoquinolines, for example ribalinine (**20**) and isobalfourodine (**26**), have been obtained from several rutaceous species. Hydroxydimethyldihydropyranocoumarins (cf. **16**) have been found less frequently in the Rutaceae but are better known in umbelliferous species. Although tracer feeding experiments have not been reported, the co-occurrence of furo and pyrano isomers in the quinoline and coumarin series suggested that the biosynthesis of both groups of compounds involved oxidative cyclization of a common prenyl precursor; knowledge of the absolute stereochemistry of furo and pyrano isomers found in a single species was used to test this proposal.[5,27] Thus the isolation of (+)-(S)-columbianin (18) and (+)-(R)-lomatin (16) from *Lomatium nuttalia* (Umbelliferae) suggests that the biosynthesis of these compounds via an (R)-epoxide (15) occurs by cyclization with stereochemical inversion furnishing the furocoumarin (18; R = H) and reaction at the tertiary carbon atom of the epoxide ring giving the pyranocoumarin (16) without affecting the chiral centre (Scheme 2). *Araliopsis soyauxii* contains the quinoline alkaloids (+)-(R)-isoplatydesmine (21) and (−)-(S)-ribalinine (20),[51] and on the basis of these configurations a biosynthetic pathway via an epoxide (19) similar to the coumarin case can be proposed (Scheme 3). Furo- and pyrano-quinoline alkaloids present in other rutaceous species do not have the same stereochemical relationship. Thus, isoplatydesmine and ribalinine isolated from *A. tabouensis* were both (R)-enantiomers[19] as were the methoxy analogues (+)-(R)-balfourodine (24) and (+)-(R)-isobalfourodine (26) of *Balfourodendron riedelianum*;[46] it seems therefore that the biosynthetic route given in Scheme 3 does not apply in these cases.

A possible rationalization of this situation resulted from a study of the stereochemical consequences of an *in vitro* furoquinoline–pyranoquinoline rearrangement. Heating the furoquinolone (+)-(R)-isoplatydesmine (21) with acetic anhydride in pyridine gave (+)-(S)-ribalinine acetate (22) which was hydrolysed to (−)-(S)-ribalinine (20).[32] The rearrangement must occur by inversion of configuration at the chiral centre, and mechanism A in accord with this observation is shown in Scheme 3. A similar rearrangement of (+)-(R)-balfourodine (24), differing from isoplatydesmine only in the presence of a methoxy group, gave the pyranoquinolone (R)-isobalfourodine acetate (25) hydrolysed by base to (R)-isobalfourodine (26).[35,46] The reaction must involve overall retention of configuration at the chiral centre and an appropriate mechanism (B) has been suggested (Scheme 4).[5] The rearrangements of balfourodine and isoplatydesmine result in considerable reduction of optical purity, and it was first suggested in the case of balfourodine that some racemization occurs.[5,46] A recent proposal is that a dual mechanism is involved, mechanism A (inversion of configuration) predominating with isoplatydesmine and mechanism B (retention) predominating with balfourodine.[32]

If the biosynthesis of hydroxydimethyldihydropyranoquinolone alkaloids

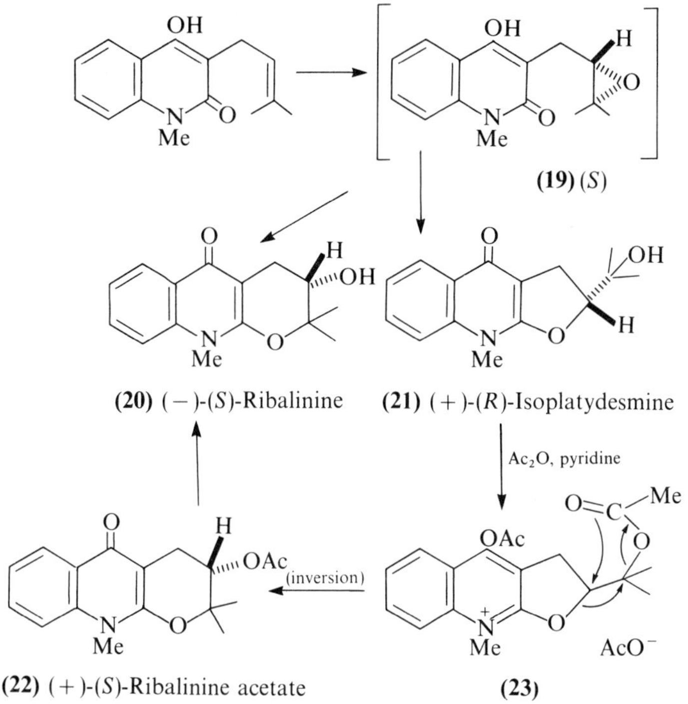

SCHEME 3. Biosynthesis of isoplatydesmine and ribalinine in *Araliopsis soyauxii*, and rearrangement by mechanism A.

occurs by rearrangement of the hydroxyisopropyldihydrofuroquinolone isomers, as proposed earlier,[5] a fine balance between competing reactions A and B might explain the stereochemical differences in the alkaloids of the two *Araliopsis* species.[32] The variation in specific optical rotation of ribalinine from various sources[23] ($0°$, $+8°$, $-10°$, $+14°$) is consistent with this suggestion. An experimental study of the biosynthesis of these pyrano-quinolines is clearly overdue.

Hydroxyisopropyldihydrofurocoumarins do not seem to undergo the rearrangements characteristic of the furoquinolone analogues; this is not unexpected since the latter reactions appear to depend on the formation of quaternary ammonium salt intermediates (Schemes 3 and 4). The biosynthesis of pyranocoumarins may then occur by direct cyclization of an epoxide (Scheme 2). The contrast between natural pyranocoumarins and pyrano-quinolines thus appears to arise from prenylation of the nitrogen-containing ring in the latter group.

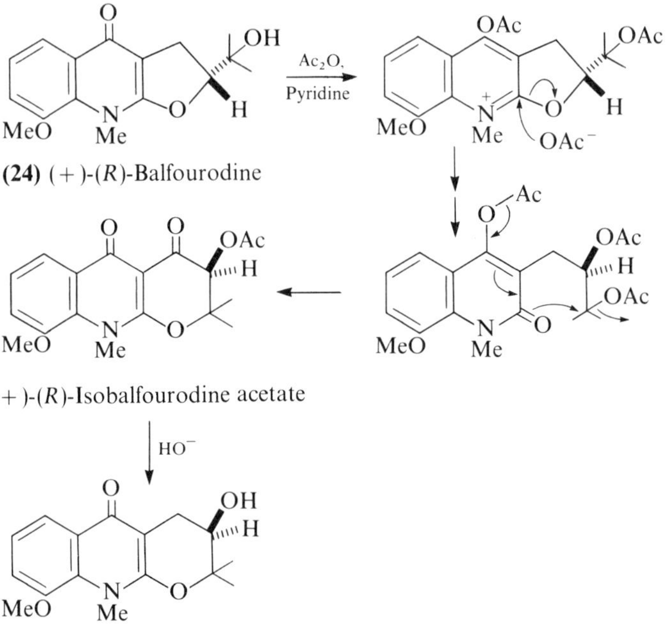

(24) (+)-(R)-Balfourodine

(25) (+)-(R)-Isobalfourodine acetate

(26) (+)-(R)-Isobalfourodine

SCHEME 4. Rearrangement of balfourodine; mechanism B.

IV. SIGMATROPIC REARRANGEMENTS IN BIOSYNTHESIS

A. 3-(1,1-DIMETHYLALLYL)COUMARINS

While quinoline hemiterpenes originate by prenylation in the heterocyclic ring, coumarin hemiterpenes, as has been emphasized, usually contain prenyl groups only in the homocyclic ring. The 1,1-dimethylallyl coumarins in which a reversed prenyl group is substituted at C-3 are exceptions to this general rule. A large number of these compounds, for example coumarins (**27**; R^1, R^2, R^3 = H, OH, or OMe), chalepensin (**28**), and rutamarin (**34**), have been

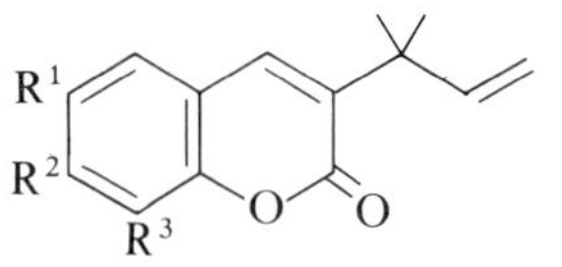

(27) 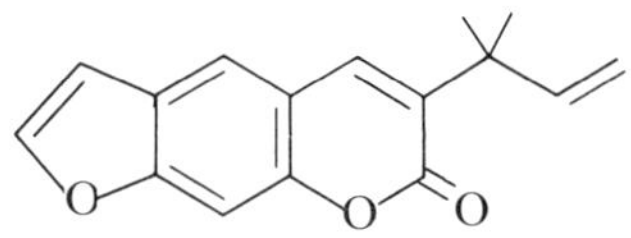   **(28)**

isolated, mainly from *Ruta graveolens* by Reisch *et al.* and from *R. pinnata* by González *et al.* (listed in ref. 22) but also from other members of the Rutaceae.[44] Their biosynthesis may involve direct substitution of a 1,1-dimethylallyl group by reaction between a coumarin and dimethylallyl pyrophosphate, but chemical instincts are against this route, first because coumarins are insufficiently nucleophilic at C-3 for that reaction to occur readily, and secondly because steric effects would discourage direct substitution at the tertiary centre of dimethylallyl pyrophosphate.

A more attractive alternative is the involvement of sigmatropic rearrangements of the Claisen and Cope types,[12] already shown to occur in the biosynthesis of prephenic acid. This general pathway to 1,1-dimethylallyl derivatives and related compounds will now be discussed in relation to coumarin and quinoline derivatives found in the Rutaceae.

An experimental study has been carried out of the biosynthesis of the 1,1-dimethylallyl coumarin rutamarin (**34**) in *Ruta graveolens*.[17] Although the 7-(3,3-dimethylallyloxy)coumarin **29** is converted into the 3-(1,1-dimethylallyl)-coumarin **30** *in vitro*,[2] no evidence was found for a similar biosynthetic route

(29)   (30)

to rutamarin. Another pathway (Scheme 5) involving rearrangement of a 4-(3,3-dimethylallyloxy)coumarin, however, received some experimental support. Thus, specifically labelled 7-hydroxycoumarin (**31**), 4,7-dihydroxy-coumarin (**32**), and $[1'-^3H_2]$-4-(3,3-dimethylallyloxy)-7-hydroxycoumarin (**33**) were incorporated into rutamarin; in the latter case, degradation showed that the label was confined to the terminal carbon of the prenyl group of rutamarin as the proposal requires. This cannot yet be regarded as the main pathway to the 1,1-dimethylallylcoumarins since incorporations are low and 4,7-dihydroxycoumarin and the prenyl ether (**33**) have not yet been detected in *R. graveolens*; proof is also required that the prenyl group of the ether remains intact in its conversion into rutamarin.

Coumarins containing 1,1-dimethylallyl groups adjacent to oxygen functions in the homocyclic ring have been isolated from *Ruta* species, for example pinnarin (**35**) and furopinnarin (**36**); the biosynthesis of these compounds may also involve sigmatropic rearrangements.

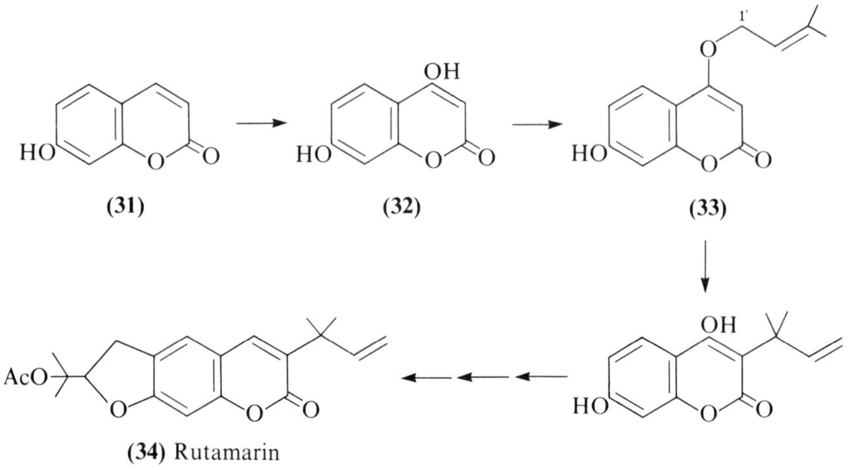

(31)　　　　　　　　(32)　　　　　　　　(33)

(34) Rutamarin

Scheme 5. Biosynthesis of rutamarin in *Ruta graveolens*.

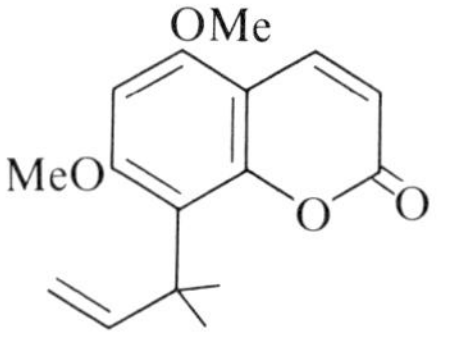

(35) Pinnarin 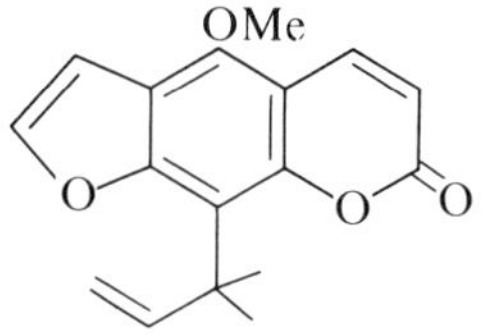　　　(36) Furopinnarin

B. 1,1- AND 1,2-DIMETHYLALLYLQUINOLINE DERIVATIVES

The study of rutamarin biosynthesis showed that the presence of a hydroxyl group at C-4 in a coumarin provides an opportunity for the introduction of a prenyl group into the heterocyclic ring. 4-Hydroxy-2-quinolone appears to exercise a similar function in the biosynthesis of the small group of 1,1- and 1,2-dimethylallyl quinolines and the corresponding geranyl derivatives that occur in rutaceous species and will now be described.

The 1,1-dimethylallyl derivative ifflaiamine (41) is a constituent of *Flindersia ifflaiana* and one proposal for its biosynthesis involves Claisen rearrangement of the 4-(3,3-dimethylallyloxy)-2-quinolone 37 followed by cyclization of the 1,1-dimethylallyl derivative 38 (Scheme 6).[12,13] The ether 37 (ravenine) was isolated from *Ravenia spectabilis* (*Lemonia spectabilis*) together with the 1,2-dimethylallylquinolone ravenoline (40) and its cyclization product lemobiline* (39).[45] Lemobiline is also a constituent of *F. ifflaiana*[13] and of *Euxylophora paraensis*.[36] Ravenoline (40) appears to be a product of abnormal Claisen rearrangement of ravenine (37), a reaction that has been shown *in vitro* to involve initial formation of a normal Claisen product; the co-occurrence of 1,1- and 1,2-dimethylallyl derivatives in *F. ifflaiana* and of 1,2-dimethylallylquinolones and a prenyl ether in *R. spectabilis* and the facile synthesis of these alkaloids by means of the rearrangement reactions[13] provide strong circumstantial support for the biosynthetic pathway summarized in Scheme 6. The optical activity of ravenoline and ifflaiamine shows that the

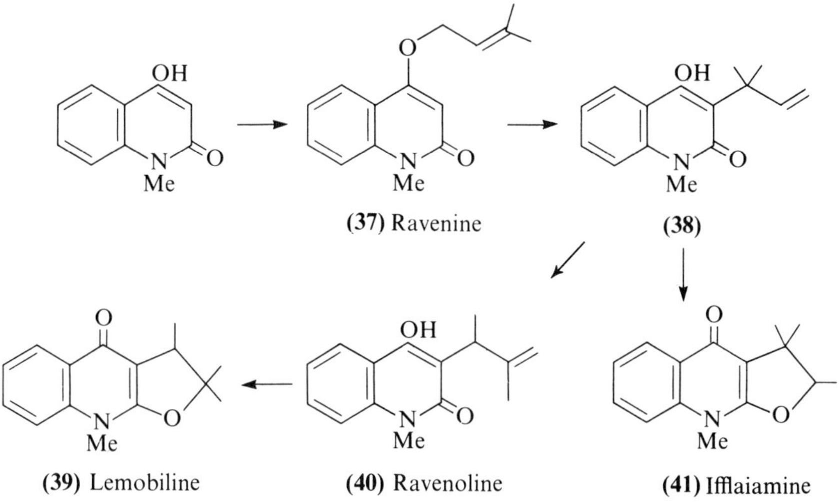

SCHEME 6. Biosynthesis of ifflaiamine, ravenoline, and lemobiline.

alkaloids are not artefacts formed from ravenine during isolation. Lemobiline was obtained as an enantiomer from the three plant sources; since an elevated temperature is required for the conversion of ravenoline into lemobiline it appears that the latter alkaloid is also formed under enzymic control. Some preliminary experimental evidence for the biosynthesis of 1,2-dimethylallyl-quinolines by sigmatropic rearrangement is available. Thus, specifically labelled ravenine (37) was incorporated into ravenoline (40) in *R. spectabilis*, although the site of the label in the latter alkaloid has not yet been determined.[12]

---

* First called spectabiline, but to avoid confusion with a compound of the same name this was changed to lemobiline.

C. GERANYLQUINOLINE ALKALOIDS

The geranylquinoline alkaloids bucharaine (**43**), bucharamine (**46**), and bucharidine (**47**) were isolated from *Haplophyllum bucharicum* by Yunusov *et al.* and the work has been reviewed.[23] Although only the structure of bucharaine has been established with certainty,[28] the three compounds appear to be the geranyl analogues of the *Ravenia* alkaloids. A reasonable biosynthetic pathway, based on the co-occurrence of the three alkaloids, involves Claisen rearrangement of the geranyl ether **42** to give bucharamine (**46**) and abnormal Claisen rearrangement furnishing bucharidine (**47**), in each case after appropriate cyclization (Scheme 7). An experimental study has not been attempted but the observations that bucharaine (**43**) undergoes Claisen and abnormal Claisen rearrangements under electron impact and gives bucharidine (**47**) on heating, and that the geranyl ether **42** is readily converted into compound **44**,[29] supplement the chemotaxonomic evidence for the proposed biosynthetic scheme.

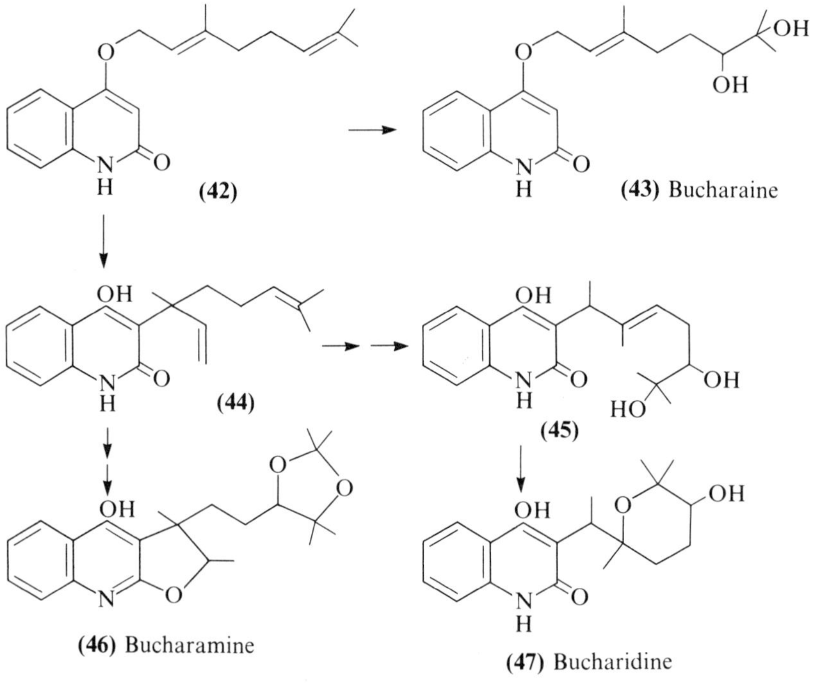

Scheme 7. Biosynthesis of bucharaine, bucharamine, and bucharidine.

V. Biosynthesis of 2,2-Dimethylpyrans

2,2-Dimethylpyranocoumarins are widely distributed in the Rutaceae, xanthyletin (**48**) and seselin (**49**) being typical examples. In the quinoline alkaloids the corresponding compounds usually contain a pyran ring incorporating the oxygen atom at C-4, thus resulting in angular annelation. Flindersine (**51**), the simplest member of the group, has been known for many years, but N-methylflindersine (**52**), which was first recognized as a natural product in 1973, has now been obtained from at least ten species of the Rutaceae.

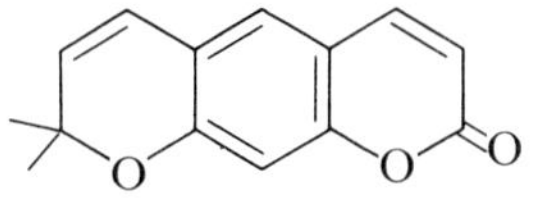

(48) Xanthyletin

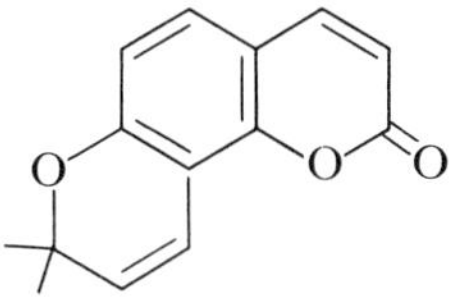

(49) Seselin

No studies of the biosynthesis of the 2,2-dimethylpyrano ring in coumarins or quinolines have apparently been reported, although an *ortho*-prenyl phenol is surely the most probable precursor. Transformation of an *ortho*-prenyl phenol to a 2,2-dimethylpyran requires conversion into a higher oxidation state, and an attractive route is benzylic oxidation to a quinone methide and then to a hydroxy-diene (Scheme 8); either species can cyclize to a dimethylpyran. There are many *in vitro* examples of this route, illustrated by a synthesis of N-methylflindersine (**52**) in which the prenylquinolone **50** was oxidized with 2,3-dichloro-5,6-dicyano-1,4-benzoquinone (DDQ) (Scheme 8).[24]

VI. Biosynthesis of Dimeric Coumarins and Quinolines

Until recently the only known dimeric hemiterpenoid coumarin or quinoline was thamnosin (cyclobisuberodiene) (**53**) obtained from *Thamnosma montana*[38] and independently from *Zanthoxylum ovalifolium*;[33] it was isolated subsequently from *Ruta pinnata*.[48a] The situation has now been altered dramatically by the isolation of twelve dimeric quinoline alkaloids from rutaceous species and merits a review of these interesting compounds.

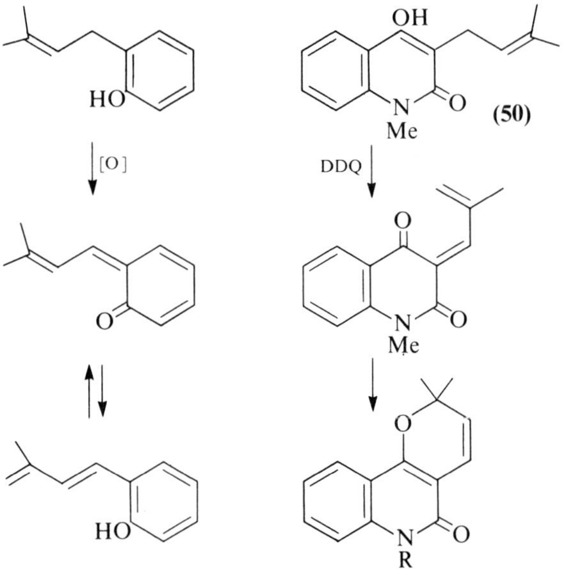

(50)

(51) Flindersine (R = H)
(52) N-Methylflindersine (R = Me)

SCHEME 8. Biosynthesis and synthesis of 2,2-dimethylpyrans.

A. DIMERIC COUMARINS

The constitution of thamnosin (53)[33,38] indicates that the molecule consists of two identically substituted coumarin groups linked at C-6 through a C_{10} unit. The isomeric phebalin (54) from *Phebalum nudum*[8] and isothamnosin A isolated from a *Ruta* species[48b] differ from thamnosin only in the attachment of the terpenoid portions of the molecules to C-8 of the coumarin rings in phebalin and to C-6 and C-8 in the case of isothamnosin A. A third coumarin dimer, mexolide (55), isolated from *Murraya exotica* is a methoxy phebalin.[11]

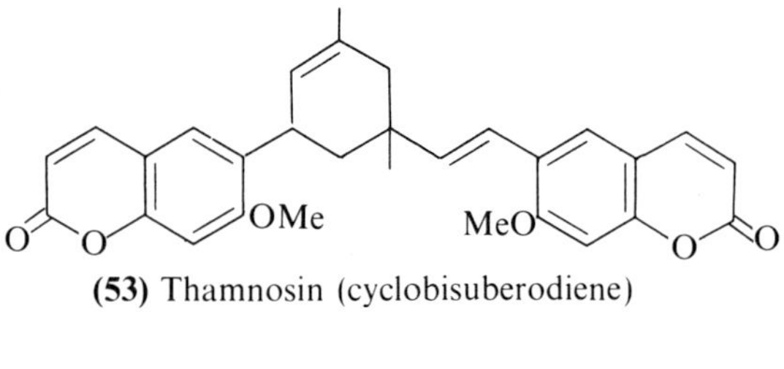

(53) Thamnosin (cyclobisuberodiene)

Toddasin, obtained from *Toddalia asiatica*, was assigned the same structure as mexolide.[49] The melting points of mexolide and toddasin differ considerably, as do the melting points of their dihydro derivatives, so proof of identity must await the opportunity to compare samples. It should be noted that configurations at the two chiral centres of the coumarin dimers are unknown, suggesting the possibility that mexolide and toddasin are stereoisomers.

A reasonable proposal for the biosynthesis of thamnosin involves Diels–Alder addition of two molecules of a coumarin diene (**56**), one molecule acting as a diene and the terminal double bond of the other serving as dienophile.[33,38] The pathway is applicable to the other coumarin dimers and is illustrated for mexolide and phebalin in Scheme 9. The synthesis of mexolide by dehydration of the diol mexoticin (**57**) (also a constituent of *M. exotica*) supports this mechanism.[11] The biosynthetic role of coumarin dienes (cf. **56**) is indicated by the co-occurrence of xanthyletin (**48**) and thamnosin (**53**) in *R. pinnata*, the presence of an OH group adjacent to the diene substituent leading to the formation of the 2,2-dimethylpyran and protection of the OH group by methylation resulting in dimerization to thamnosin. In spite of the presence of chiral centres, the coumarin dimers are optically inactive; this suggests that dimerization of a coumarin diene takes place during isolation or that dimerization occurs non-enzymically in the plant tissue.

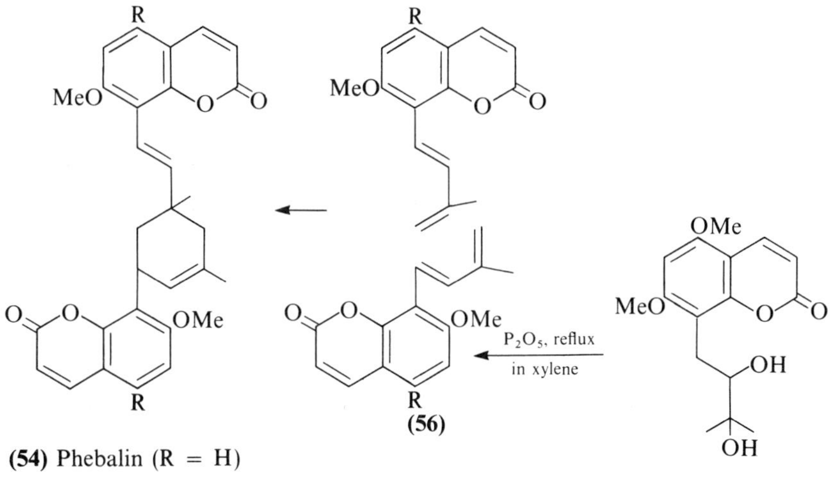

(**54**) Phebalin (R = H)

(**55**) Mexolide (toddasin) (R = OMe)

(**56**)

(**57**) Mexoticin

SCHEME 9. Coumarin dimers.

1. *Vepridimerines*

An investigation of the barks of *Vepris louisii* and *Oricia renieri* resulted in the isolation of four isomeric quinolone dimers, vepridimerines A–C from *V. louisii* and vepridimerines B–D from *O. renieri*. Structures **58**–**61** (relative stereochemistry shown) were assigned on the basis of spectroscopic evidence, mainly ^{1}H and ^{13}C n.m.r. data.[43] The vepridimerines differ in the nature of the heterocyclic system and in stereochemistry at one of the ring junctions. Vepridimerines A and B are 2-quinolones, the relationship of H_d and H_e being *cis* in vepridimerine A (**60**) and *trans* in the B isomer (**61**); vepridimerines C and D each have 2- and 4-quinolone groups but differ in stereochemistry with H_d and H_e *cis* in the case of vepridimerine C (**58**) and *trans* in the D isomer (**59**).

The C_{10} alicyclic moiety of the vepridimerines is similar to that present in isoalfileramine (**67**), the acid-catalysed cyclization product of the alkaloid alfileramine from *Zanthoxylum punctatum*,[10] and found in the 2,2-dimethylbenzopyran dimer **69**. The latter compound was prepared from the allylic alcohol **68** and is believed to originate by dimerization of an *ortho*-hydroxy-diene followed by cyclization.[3] The biosynthesis of the vepridimerines may involve a similar route (Scheme 10). Diels–Alder addition of one molecule of the diene **65** to the internal double bond of a second molecule would be expected to furnish the dimer **63** with a *cis* relationship of H_d and H_e; addition of the oxygen functions at position 2 or 4 of the quinolone portions of the molecule to the residual double bonds then furnishes vepridimerine A and vepridimerine C. The biosynthesis of the *trans* isomers is less easy to rationalize but may involve isomerization of the intermediate to the cyclohexene **62**, epimerization of H_e by a homolytic or heterolytic mechanism, and then ring closure to vepridimerines B and D. A major alkaloid of *V. louisii*[42] and *O. renieri*[37] is the 2,2-dimethylpyranoquinolone veprisine (**64**); on the assumption that this alkaloid also originates from the *ortho*-hydroxy-diene **65** (cf. previous section) it appears that dimerization to the vepridimerines competes with cyclization to veprisine. The observation that the vepridimerines are racemic is consistent with their formation from the diene **65**. It is of interest to note that the vepridimerines and the coumarin dimers appear to represent two modes of Diels–Alder dimerization of prenyl dienes involving in the dienophile an internal double bond and a terminal double bond, respectively.

2. *Paraensidimerine D*

A recent investigation of the heartwood constituents of *Euxylophora paraensis* resulted in the isolation of a number of quinoline alkaloids including

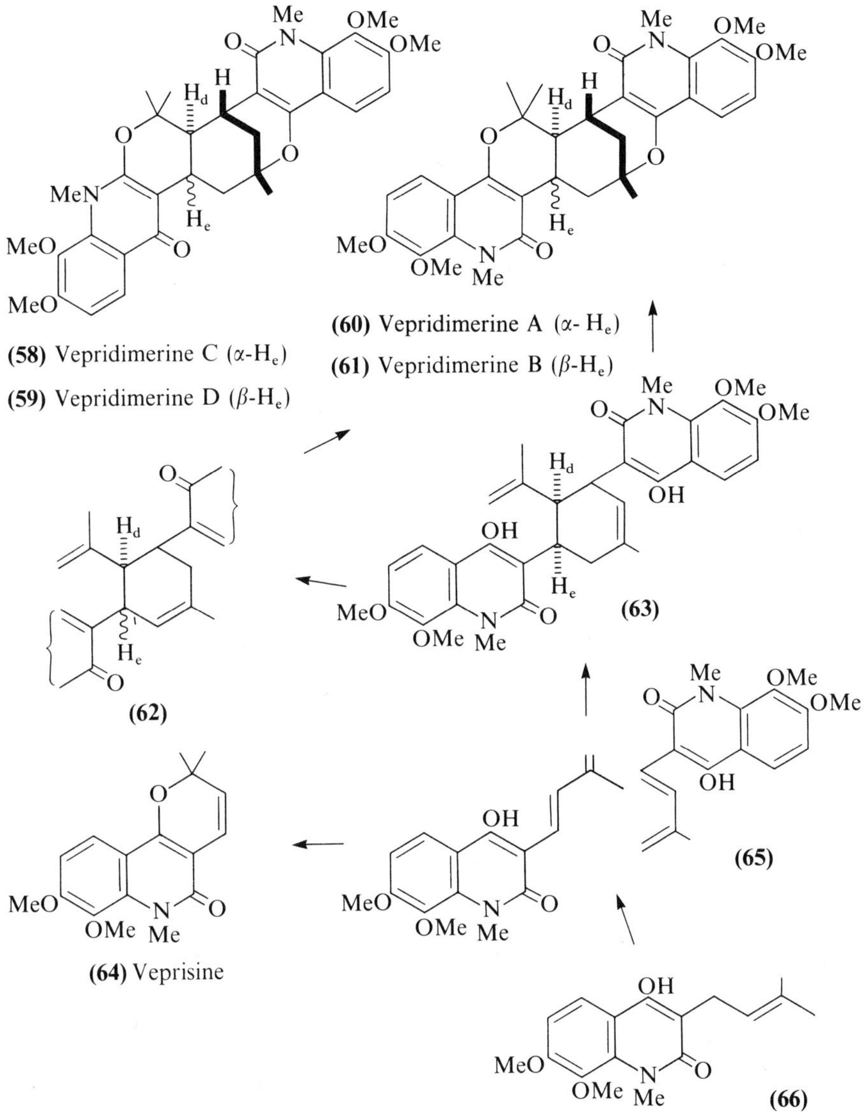

(58) Vepridimerine C (α-H$_e$)

(59) Vepridimerine D (β-H$_e$)

(60) Vepridimerine A (α- H$_e$)

(61) Vepridimerine B (β-H$_e$)

(62)

(63)

(64) Veprisine

(65)

(66)

SCHEME 10. Biosynthesis of the vepridimerines.

N-methylflindersine (72) and five isomeric dimers, $C_{30}H_{30}N_2O_4$. The structure of one of the dimers, paraensidimerine D (73), was established by X-ray crystallography.[36] The alkaloid is thus a new type of dimer and may originate by Diels–Alder addition of a quinoline quinone methide (70) to its tautomeric diene (71) followed by cyclization or by addition of the quinone

(67) Isoalfileramine

(68)   (69)

methide to N-methylflindersine (Scheme 11). *In vitro* support for the latter route is provided by the observation that the electron-deficient enedione **75**, generated *in situ* from the 2-quinolone **74**, reacts readily and specifically with N-methylflindersine (**72**) to give the Diels–Alder adduct (**76**) containing the same ring system as paraensidimerine D (Scheme 12).[30]

3. *Pteledimerine and Pteledimeridine*

A third type of quinolone dimer was obtained from *Ptelea trifoliata*. Two of these alkaloids, pteledimerine (**78**)[47] and pteledimeridine (**79**),[40] are isomeric with the paraensidimerines and their structures were determined by [1]H n.m.r. and mass spectroscopy; the third dimer from *P. trifoliata* is of unknown constitution. *Ptelea trifoliata* is the only known source of dihydrofuro-quinoline alkaloids containing terminal double bonds, and a possible biosynthetic route to pteledimerine involves an ene reaction of the olefin **77** with N-methylflindersine (**72**) (also found in *P. trifoliata*) or, more likely, acid-catalysed reaction of the two components (Scheme 13). The ability of N-methylflindersine to behave in this way is shown by its reaction with acid to give a dimer of probable structure **80**.[31]

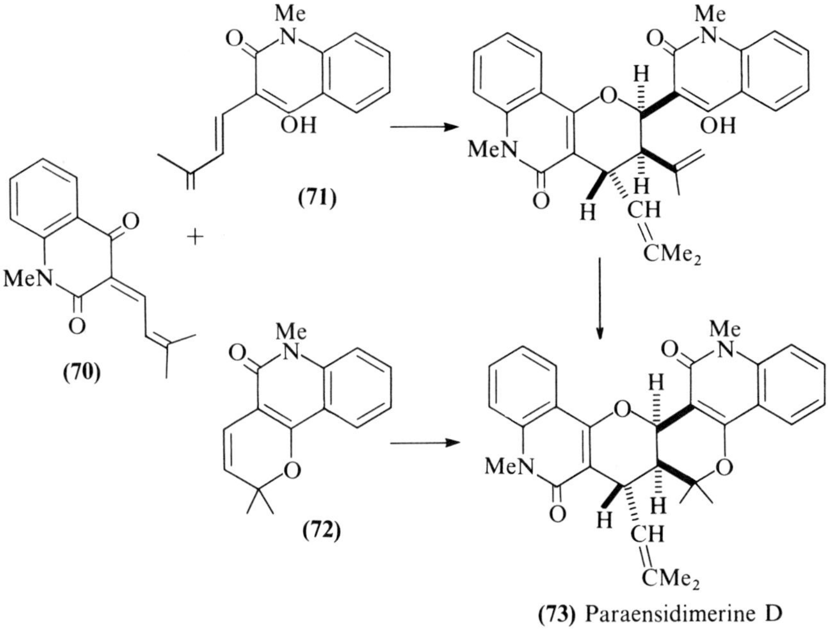

SCHEME 11. Biosynthesis of paraensidimerine D.

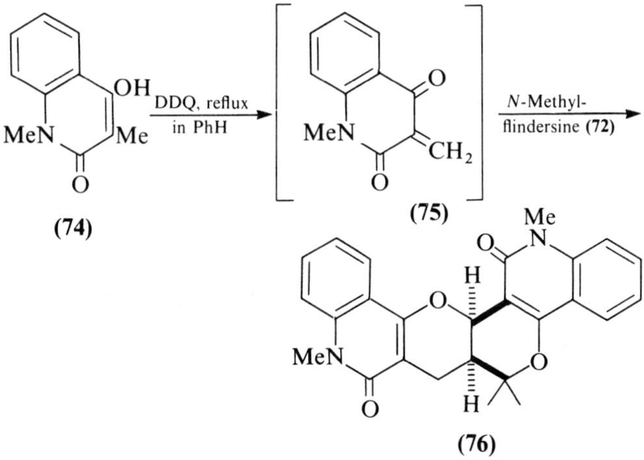

SCHEME 12. Synthesis of a quinolone dimer.

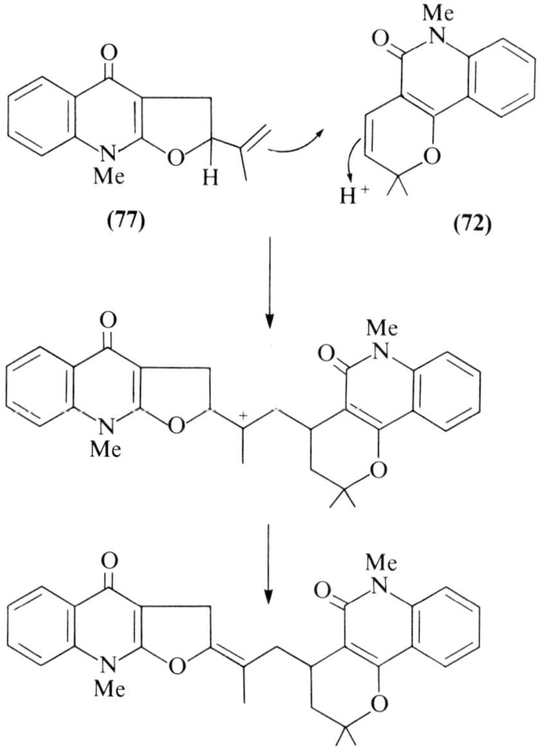

(77)   (72)

(78) Pteledimerine

SCHEME 13. Biosynthesis of pteledimerine.

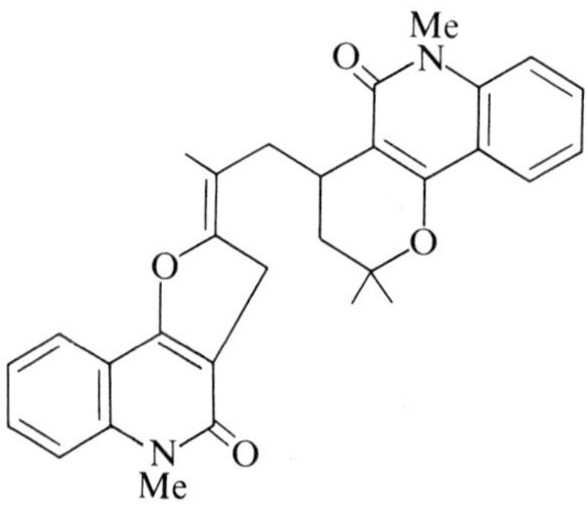

(79) Pteledimeridine 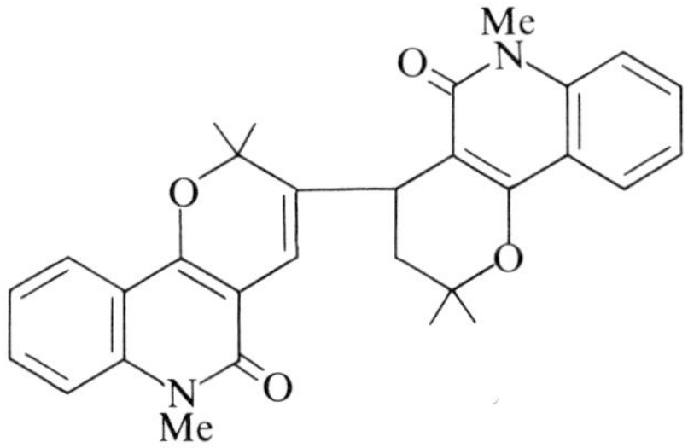   (80)

VII. Conclusion

Although the origin of the main groups of hemiterpenoid coumarins and quinolines is now broadly established, a great deal more detailed work remains to be done; the topics included in this chapter have been selected to emphasize the gaps in our knowledge and to stimulate further experimental biosynthetic research in the area.

REFERENCES

1. Austin, D. J. and Brown, S. A. (1973). *Phytochemistry* **12**, 1657.
2. Ballantyne, M. M., McCabe, P. H. and Murray, R. D. H. (1971). *Tetrahedron* **27**, 871.
3. Barnes, C. S., Strong, M. I. and Occolowitz, J. L. (1963). *Tetrahedron* **19**, 839.
4. Boulanger, D., Bailey, B. K. and Steck, W. (1973). *Phytochemistry* **12**, 2399.
5. Bowman, R. M., Collins, J. F. and Grundon, M. F. (1973). *J. Chem. Soc. Perkin Trans. 1*, 626.
6. Bowman, R. M. and Grundon, M. F. (1966). *J. Chem. Soc. (C)*, 1504.
7. Brown, S. A. (1979). *Planta Med.* **36**, 299.
8. Brown, K. L., Burfitt, A. I. R., Cambie, R. C., Hall, D. and Mathai, K. P. (1975). *Aust. J. Chem.* **28**, 1327.
9. Brown, S. A. and Steck, W. (1973). *Phytochemistry* **12**, 1315.
10. Caolo, M. A. and Stermitz, F. R. (1979). *Tetrahedron* **35**, 1487.
11. Chakraborty, D. P., Roy, S., Chakraborty, A., Mandal, A. K. and Chowdhury, B. K: (1980). *Tetrahedron* **36**, 3563; Mock, J. R. *et al.* (1980). *Aust. J. Chem.* **33**, 395.
12. Chamberlain, T. R., Collins, J. F. and Grundon, M. F. (1969). *J. Chem. Soc. Chem. Commun.*, 1269.
13. Chamberlain, T. R. and Grundon, M. F. (1971). *J. Chem. Soc. (C)*, 910.
14. Collins, J. F., Donnelly, W. J., Grundon, M. F. and James, K. J. (1974). *J. Chem. Soc. Perkin Trans. 1*, 2177.
15. Collins, J. F., Gray, G. A., Grundon, M. F., Harrison, D. M. and Spyropoulos, C. G. (1973). *J. Chem. Soc. Perkin Trans. 1*, 94.
16. Diment, T. A., Ritchie, E. and Taylor, W. C. (1969). *Aust. J. Chem.* **22**, 1797.
17. Donnelly, W. J., Grundon, M. F. and Ramachandran, V. N. (1977). *Proc. Royal Irish Acad.*, 443.
18. Ellis, B. E. and Brown, S. A. (1974). *Can. J. Biochem.* **52**, 734.
19. Fish, F., Meshal, I. A. and Waterman, P. G. (1976). *Planta Med.* **29**, 310.
20. Floss, H. G. (1972). *Rec. Adv. Phytochem.* **4**, 143.
21. Gray, A. I. and Waterman, P. G. (1978). *Phytochemistry* **17**, 845.
22. Grundon, M. F. (1978). *Tetrahedron* **34**, 143.
23. Grundon, M. F. (1979) *In* "The Alkaloids" (Manske, R. H. F. and Rodrigo, R. G. A. eds.), Vol. 17, p. 105. Academic Press, New York.
24. Grundon, M. F., Harrison, D. M., Magee, M. G. and Rutherford, M. J. (1983). *Proc. Royal Irish Acad.*, 103.

25. Grundon, M. F., Harrison, D. M. and Spyropoulos, C. G. (1974). *J. Chem. Soc. Perkin Trans. 1*, 2181.
26. Grundon, M. F., Harrison, D. M. and Spyropoulos, C. G. (1975). *J. Chem. Soc. Perkin Trans. 1*, 302.
27. Grundon, M. F. and McColl, I. S. (1975). *Phytochemistry* **14**, 143.
28. Grundon, M. F. and Ramachandran, V. N. (1981). *J. Chem. Soc. Perkin Trans. 1*, 633.
29. Grundon, M. F. and Ramachandran, V. N., unpublished work.
30. Grundon, M. F., Ramachandran, V. N. and Sloan, B. M. (1981). *Tetrahedron Lett.*, 3105.
31. Grundon, M. F. and Rutherford, M. J., unpublished work.
32. Grundon, M. F. and Surgenor, S. A. (1978). *J. Chem. Soc. Chem. Commun.*, 624.
33. Guise, G. B., Ritchie, E., Senior, R. G. and Taylor, W. C. (1967). *Aust. J. Chem.* **20**, 2429.
34. Hifnawy, M. S., Vaquette, J., Sévenet, T., Pousset, J.-L. and Cavé, A. (1976). *Planta Med.* **29**, 346.
35. James, K. J. and Grundon, M. F. (1979). *J. Chem. Soc. Perkin Trans. 1*, 1467.
36. Jurd, L. and Wong, R. Y. (1981). *Aust. J. Chem.* **34**, 1625.
37. Khalid, S. A. and Waterman, P. G. (1981) *Phytochemistry* **20**, 2761.
38. Kutney, J. P., Inaba, T. and Dreyer, D. L. (1970). *Tetrahedron* **26**, 3171.
39. Mester, I. (1973). *Fitoterapia* **44**, 123; (1977) **48**, 268.
40. Mester, I., Reisch, J., Szendrei, K. and Körösi, J. (1979). *Liebigs Ann. Chem.*, 1785.
41. Murray, R. D. H. (1978). *Fortschr. Chem.* **35**, 199.
42. Ayafor, J. F., Sondengam, B. L. and Ngadjui, B. (1980). *Tetrahedron Lett.*, 3293.
43. Ngadjui, T. B., Ayafor, J. F., Sondengam, B. L., Connolly, J. D., Rycroft, D. S., Khalid, S. A., Waterman, P. G. and (in part) Brown, N. M. D., Grundon, M. F. and Ramachandran, V. N. (1982). *Tetrahedron Lett.*, 2041.
44. Najar, M. N. S., Bhan, M. K. and George, V. (1973). *Phytochemistry* **12**, 2073; Rao, A. V. R., Bhide, K. S. and Mujumdar, R. B. (1980). *Indian J. Chem. (B)* **19**, 1046.
45. Paul, B. D. and Bose, P. K. (1969). *J. Indian Chem. Soc.* **7**, 678; Talapatra, S. K., Maiti, B. C., Talapatra, B. and Das, B. C. (1969). *Tetrahedron Lett.*, 4789.
46. Rapoport, H. and Holden, K. G. (1960). *J. Am. Chem. Soc.* **82**, 4395.
47. Reisch, J., Mester, I., Szendrei, K. and Korosi, J. (1978). *Tetrahedron Lett.*, 3681.
48. (a) Reyes, E. R. and González, A. G. (1970) *Phytochemistry* **9**, 833. (b) González, A. G., Cardona, R. J., Diaz-Chico, E., Lopez-Dorta, H. and Rodriguez-Luis, F. (1977). *An. Quim.* **73**, 1510; *Chem. Abs.* **89**, 160105.
49. Sharma, P. N., Shoeb, A., Kapil, R. S. and Popli, S. R. (1980). *Phytochemistry* **19**, 1258.
50. Steck, W. and Brown, S. A. (1971). *Can J. Biochem.* **49**, 1213.
51. Vaquette, J., Hifnawy, M. S., Pousset, J. L., Fournet, A., Bouquet, A. and Cavé, A. (1976). *Phytochemistry* **15**, 743.

CHAPTER 3

Structural Diversity and Distribution of Alkaloids in the Rutales

I. MESTER

Institut für Pharmazeutische Chemie der Universität, Münster, West Germany

I. INTRODUCTION

According to the botanical systematics of Dahlgren[51] the order Rutales consists of the families Rutaceae, Cneoraceae, Surianaceae, Simaroubaceae, Kirkiaceae, Burseraceae, and Meliaceae. No alkaloids have yet been reported from any of these families except from the Rutaceae, Simaroubaceae, and Meliaceae. Of these three families the Rutaceae is the most examined for alkaloidal constituents. Already in 1826 Chevalier and Pelletan reported the isolation of the alkaloid xanthopicrit from the West Indian medicinal plant *Zanthoxylum clava-herculis*, and in 1862 Perrins identified xanthopicrit as berberine (23-2).[109] Only a few years later, in 1875, the isolation of pilocarpine (27-4) from another drug, the Jaborandi leaves originating from Brazil, was reported.[109] Today more than 470 structurally defined alkaloids are known from species belonging to the Rutaceae.

Despite the intensive chemical exploration of many species from Simaroubaceae and Meliaceae, the first alkaloid, 4,5-dimethoxycanthin-6-one

31

(**18**-20), from a simaroubaceous plant[81] was not reported until 1961, and the isolation of a simple amide, tiglamide (**32**-3), from an unidentified *Aglaia* species[89] (Meliaceae) as late as 1969.

While some review articles are available concerning the alkaloids isolated from rutaceous species, they are either 5–10 years old,[57,109,110,188] or they treat only some distinct alkaloid types, e.g. anthranilic acid derived quinoline alkaloids[74] or the carbazole alkaloids[36] etc. On the other hand, no review has appeared on the alkaloids from Simaroubaceae and Meliaceae. The intention of this chapter is to review the literature reports on the alkaloids obtained up to now from species belonging to the above mentioned three families.

II. ALKALOID TYPES OCCURRING IN THE RUTALES

Due to the structural diversity of the alkaloids the most practical classification is in accordance with their known or hypothetical biogenesis. Thus, the reported alkaloids from the Rutales are derivatives of anthranilic acid, tryptophan, phenylalanine and/or tyrosine, histidine, nicotinic acid, ornithine, and lysine, or in some cases they belong at the same time to two amino acid precursors.

A. ALKALOIDS DERIVED FROM ANTHRANILIC ACID

1. *Simple Quinoline Derivatives* (Tables 1–3)

Quinoline (**1**-1) itself occurs in *Citrus aurantium* and in *Galipea officinalis*.[109] Regarding the substitution pattern of quinolines, 2-alkylated derivatives are known, such as 2-amylquinoline (**1**-3), 2-arylated compounds such as dubamine (**1**-6), and arylalkylated compounds such as galipine (**1**-11). Some quinolines are 4-methoxylated, e.g. galipine (**1**-11), or 2,4-dimethoxylated like orixine (**1**-12). Cuspareine (**1**-9) has a hydrogenated pyridine nucleus.

2-Quinolone (carbostyril) is not a natural product, but its *N*-methyl derivative (**2**-1) was reported from *Galipea officinalis*.[109] All the other known 2-quinolones are oxygenated at position 4. The oxygen containing substituent can be a free hydroxy, methoxy, prenyloxy, or geranyloxy group. Very rarely a methoxy substituent occurs in position 3, as in swietinidine A (**2**-10), but frequently this position is substituted with a prenyl group such as in atanine (**2**-12) or glycosolone (**2**-17). In the benzene nucleus methoxy, methylenedioxy, or prenyloxy substituents appear in positions 6, 7, and 8, but phenolic derivatives are also known, e.g. glycosolone (2-17). A few representatives contain two prenyl substituents, such as 3-prenyl-4-prenyloxy-2-quinolone (2-22) found in *Haplophyllum tuberculatum*[109] and 3,3-diprenyl-1-methyl-1,2,3,4-tetrahydro-quinolin-2,4-dione (2-28) recently reported from *Esenbeckia flava*.[58]

The prenyl substituents in a series of 2-quinolones isolated from the root bark of *Ptelea trifoliata* are worth mentioning, due to the unusual terminal double bond, one example being isoptelefoline (2-43). In pteleoline (2-46) the last carbon atom of the prenyl side-chain is oxidized to carboxy.

In the series of 4-quinolones, 2-alkyl and 2-aryl substitutions are frequent (3-1, 3-3), and a 3-prenylated compound, pilokeanine (3-10), is also known. Japonine (3-15) and 5-hydroxy-1-methyl-2-phenyl-4-quinolone (3-5) are examples of the very rare 3- and 5-oxygenated substitution pattern of quinolines.

2. *Furo- and Pyrano-quinoline Derivatives* (Tables 4–7)

Cyclization of 3-prenylated 4-hydroxy-2-quinolones leads to the isopropyl-dihydrofuroquinoline alkaloids. The anellation of the three-ring system is generally linear, such as in platydesmine (4-4), but recently a few representatives with angular anellation have been found, e.g. compound 4-1 isolated from *Almeidea guyanensis*.[155] Not only quinoline (4-4) and quinolone structures (4-13, folisine) are known, but also quinolinium compounds like veprisinium (4-32). If the precursor of a dihydrofuroquinoline was a 4-prenyloxy- or a 4-geranyloxy-2-quinolone, then the cyclized product of the corresponding Claisen rearrangement occurs, examples being lemobiline (4-3) and bucharaminol (4-31). Oxygenated substituents appear in positions 6, 7, and 8, and very rarely in position 3 of the dihydrofuran ring. The only example of hydroxylation at position 3, apart from two dimeric acridone alkaloids, is myrtopsine (4-11).

The furoquinoline alkaloids of the dictamnine group are the most widespread in the Rutaceae. The simplest representative is dictamnine (5-1) and the most common is skimmianine (5-17) which has been found in 137 out of the 286 species from which alkaloids have been reported. Regarding the anellation and the substitution pattern of these alkaloids, the angular anellation of the three-ring system and furan ring substitution is, as yet, not known. There is usually a methoxy group at C-4, but maculosine (5-40), for example, has a 4-prenyloxy group instead. Oxygen containing substituents appear frequently at positions 6, 7, and 8, and very rarely at 5, such as in acronycidine (5-25). Frequently a prenyl moiety at C-5 is present either as an alkyl side-chain or in the form of a pyran or furan ring, e.g. tecleaverdoornine (5-34), medicosmine (6-1), and choisyine (6-5).

In the series of iso-alkaloids, glycarpine (**5**-18) shows again the rare 5-oxygenation pattern. Also of interest are acrophylline (**5**-24) and acrophyllidine (**5**-29) because of the presence of N-prenyl groups. The di- or tetra-hydrogenated benzene nucleus occurs in a few furoquinoline alkaloids obtained from *Haplophyllum perforatum*, e.g. perfamine (**5**-31) and haplophyllidine (**5**-33).

Another mode of cyclization of 3-prenylquinolones results in the pyranoquinolines, of which both the linear and angular anellated forms are found. The linear compounds have quinoline or quinolone structures, or they are quaternary quinolinium compounds, such as geibalansine (**7**-7), khaplofoline (**7**-2), and rutalinium (**7**-17). The known linear pyranoquinolines are all dihydropyrano derivatives, but the angular alkaloids are almost all pyrano derivatives like flindersine (**7**-1). Angular dihydropyrano compounds (**7**-11 and **7**-24) have been reported only recently, from *Euxylophora paraensis*[92] and *Zanthoxylum simulans*.[72]

Also recently the occurrence of a few dimeric pyranoquinolines have been reported. It seems that the dimerization occurs in three ways, resulting in the alkaloids of the pteledimerine (**7**-33), paraensidimerine D (**7**-31), and vepridimerine A (**7**-34) types.

3. *Acridone Derivatives* (Tables 8–10)

Acridone itself (**8**-1) occurs in *Thamnosma montana*[109] together with its N-methyl derivative, but generally the acridone alkaloids are oxygenated in positions 1 and 3, as in 1,3-dimethoxy-10-methylacridone (**8**-8). There are some alkaloids which have a fully substituted ring A, like melicopine (**8**-22) and atalaphylline (**8**-31). These substituents can be hydroxy, methoxy, methylenedioxy, or prenyl groups. Oxygenation in ring C is relatively rare, but there are some alkaloids with such a substitution pattern, e.g. 1-hydroxy-7-methoxyacridone (**8**-4) and atalaphylline (**8**-31).

A possible precursor of the acridones could be tecleanone (**8**-35) because, on the one hand, similar benzophenones are transformed *in vitro* into acridones and, on the other hand, tecleanone co-occurs in some rutaceous species together with acridones.

In the case of acridones pyrano, dihydrofuro, and furo derivatives are again known, all with an angular anellation. Acronycine (**10**-5) is a well known antitumour agent. Furacridone (**9**-1) and rutacridone (**9**-2) were the first representatives of their groups of acridone alkaloids, and from the roots of *Ruta graveolens* a series of rutacridone derivatives have been obtained. Among these there are some interesting compounds containing covalently bonded chlorine, e.g. gravacridonchlorine (**9**-9).

Two dimeric acridone alkaloids, atalanine (**9**-14) and ataline (**9**-15) have been reported from *Atalantia ceylanica*.[110] In these cases the dimerization occurs through ether linkages and not through carbon–carbon bonds as in the case of the dimeric quinoline alkaloids mentioned above.

4. *Quinazoline and Indolopyranoquinazoline Derivatives* (Tables 11 and 12)

Alkaloids are known within this group possessing a quinazoline structure such as in glycophymoline (**11**-4), a quinazolin-4-one structure, as in glycorine (**11**-1) and glycosminine (**11**-3), and also a quinazolin-2,4-dione structure, as in glycosmicine (**11**-2). Glycophymoline (**11**-4) and glycosminine (**11**-3) are derivatives of both anthranilic acid and phenylalanine.

The structurally more complicated indolo-pyrido-quinazolines represent another example of alkaloids which have two amino acids as precursors, in this case anthranilic acid and tryptophan. As well as the aromatic compounds like euxylophoricine B (**12**-13), dihydro and tetrahydro derivatives are known such as rutaecarpine (**12**-1) and evodiamine (**12**-6), and anhydronium bases such as euxylophorine B (**12**-16). 7-Carboxyevodiamine (**12**-15) is the only case in which the complete tryptophan molecule is conserved.

B. ALKALOIDS DERIVED FROM TRYPTOPHAN

1. *Simple Indole Derivatives* (Table 13)

The simplest representative, 3-formylindole (**13**-1), occurs in *Murraya paniculata*.[109] Tryptamine (**13**-2) and some simple tryptamine derivatives have also been reported. Very interesting is a series of dimeric 2-prenyltryptamine derivatives, which occurs in the leaves of *Flindersia fournieri*.[166,167,169,172] These compounds are derivatives of borreverine (**13**-6) and isoborreverine (**13**-7).

2. *Carbazoles* (Tables 14–16)

It is as yet only assumed that the carbazoles are also tryptophan derivatives and they may, in fact, be derived from 3-prenylquinolones. All carbazole alkaloids obtained from rutaceous species possess a 1-carbon unit in position 3. This can be a methyl, a formyl, or a carboxy group, such as in 3-methylcarbazole (**14**-1), lansine (**14**-6), or mukoeic acid (**14**-4), respectively. Hydroxy or methoxy substituents occur in all positions, except 4 and 5, and, naturally, position 3. Furthermore, some alkaloids contain C-prenyl or C-geranyl substituents, such as heptazoline (**14**-13) and mahanimbinol (**14**-15). The cyclization of the 1-prenyl- or 1-geranyl-2-hydroxycarbazoles leads to the pyranocarbazoles, such as girinimbine (**15**-1) and mahanine (**15**-12). Further cyclization in the alkaloids of the mahanine type results in the polycyclic carbazole alkaloids, as seen in cyclomahanimbine (**16**-1), bicyclomahanimbine (**16**-3), or mahanimbidine (**16**-5).

3. *β-Carbolines* (Table 17)

The β-carbolines isolated from species of the Rutales contain a 1C or 2C or 4C unit in position 1. The 1C unit can be a methyl (**17**-1, harmalan), hydroxymethylene (**17**-4), formyl (**17**-3, kumujian C), or carboxy (**17**-11); the 2C unit can be unsaturated (**17**-9), saturated such as in crenatidine (**17**-22), or oxidized to an acetyl group such as in 1-acetyl-β-carboline (**17**-5). Furthermore, there are known methoxylated derivatives with the substituent in positions 4, 7, and 8, and recently the occurrence of 9-methoxy-1-vinyl-β-carboline (**17**-10) in *Picrasma excelsa* has also been reported.[185,186] 3,4-Dihydro derivatives such as harmalan (**17**-1) and 1,2,3,4-tetrahydro derivatives such as borrerine (**17**-15) also occur in the Rutales.

4. *Canthin-6-ones* (Table 18)

The most common member of this class in species from the Rutaceae and Simaroubaceae is canthin-6-one (**18**-1) itself. Oxygen containing substituents appear in positions 1, 4, 5, and in all possible positions of the benzene nucleus. Some canthin-6-ones are dioxygenated, such as 4,5-dimethoxycanthin-6-one (**18**-20). Interesting is the isolation of two N-oxides (**18**-2 and **18**-15) in this series of alkaloids, because they are the only known N-oxides in the Rutales. The occurrence of some canthin-2,6-diones has also been reported, e.g. indacanthinone (**18**-19). Another very rare substituent, the S-methyl group, appears in 4-methylthiocanthin-6-one (**18**-14).

C. ALKALOIDS DERIVED FROM PHENYLALANINE AND/OR TYROSINE

1. *Simple Phenylethylamines and Oxazoles* (Tables 19 and 20)

Tyramine (**19**-1) occurs in *Citrus aurantium*[109] but the most common simple amines are hordenine (**19**-3) and the quaternary compound candicine (**19**-7). Some tyramine derivatives are N-benzoylated, such as tembamide (**19**-15), or they are acylated with cinnamic acid, such as marmeline (**19**-26). The occurrence of *N*-nicotinyl tyramines or styrylamines (e.g. **19**-16) is another example of alkaloids which belong to two alkaloid families at the same time, in this case to the families of phenylalanine and nicotinic acid.

The well known common prenylation of rutaceous constituents occurs in the case of these protoalkaloids too. The prenyl group appears as either *O*-prenyl or *C*-prenyl substituents; the two recently reported[35,160] dimeric alkaloids alfileramine (**19**-35) and culantraramine (**19**-36) are evidently in the latter category.

A very interesting compound is the recently described[73] 2,5-dibenzyl-1,4-dimethylpiperazine (**19**-34) which represents the first simple piperazine alkaloid found in the Order.

No work on the biosynthesis of the oxazole alkaloids has been reported, but the co-occurrence of balsamide (**19**-20) and balsoxin (**20**-3) in *Amyris balsamifera*[34] suggests a close relationship between the oxazoles and the aroyl tyramine derivatives. The other oxazole alkaloids known possess a 3-pyridyl substituent in position 2, e.g. halfordinol (**20**-1). Some corresponding *N*-nicotinyl-*β*-phenylethylamines have been obtained from *Amyris plumieri*,[33] but no reports on the co-occurrence of the last two alkaloid types are yet available.

2. *Isoquinolines, Aporphines, Protoberberines, and Protopines* (Tables 21–24)

The simple 2,2-dimethyl-6,7-methylenedioxy-1,2,3,4-tetrahydroisoquinolinium cation (**21**-1) has been reported from a *Zanthoxylum* species,[110] but more widespread are the benzylisoquinolinium salts, such as tembetarine (**21**-4). The phthalideisoquinoline narcotine (**21**-5) was reported from two *Citrus* species,[109] but in very low yield.

The oxoaporphine, liriodenine (**22**-1) and the normal aporphine corydine (**22**-4) have been found in *Zanthoxylum* species,[84,173] but the quaternary alkaloids of the magnoflorine (**22**-6) type are more frequent. Berberine (**23**-2), protoberberines such as *N*-methyl-*α*-canadine (**23**-8), and protopines such as *α*-allocryptopine (**24**-4) also occur in some *Zanthoxylum* species.

3. *Benzo*[c]*phenanthridines* (Tables 25 and 26)

The number of benzo[c]phenanthridines isolated from the Rutaceae is relatively large. They are quaternary metho salts, or 5,6-dihydro derivatives. Regarding the substitution pattern, all representatives are methylenedioxy substituted in positions 2 and 3, except fagaronine (**25**-14). Oxygenated substitution occurs also in positions 7 and 8 [the chelerythrine (**25**-10) type], or in 8 and 9 [the nitidine (**25**-11) type]. Some dihydro derivatives are additionally oxygenated in position 6, such as angoline (**25**-19) or oxynitidine (**25**-16) but these are probably artefacts. Furthermore, 6-alkylated derivatives are also known, e.g. 6-acetonyl-dihydrochelerythrine (**25**-23), and a recently reported[72] dimeric alkaloid (**25**-28).

The co-occurrence of some interesting amino-phenylnaphthalenes, which are evidently metabolites from the benzo[c]phenanthridines, have also been reported in recent years. One of these is arnottianamide (**26**-3).

D. ALKALOIDS DERIVED FROM OTHER AMINO ACIDS

1. *Histidine Derivatives* (Table 27)

Histidine derived alkaloids are confined to a few histamine derivatives found in *Casimiroa edulis* and to the *Pilocarpus* alkaloids. However, among the alkaloids isolated from *Casimiroa edulis*, casimiroedine (**27**-10) is a unique *N*-glycoside, and zapotidine (**27**-2) is the second sulphur containing alkaloid of this order. The *Pilocarpus* alkaloids differ among themselves mainly in the configuration of the asymmetric carbon atoms, e.g. pilocarpine (**27**-4) and isopilocarpine (**27**-5).

2. *Ornithine and Lysine Derivatives* (Tables 28–30)

The number of derivatives of the next two amino acids, ornithine and lysine, is also limited. Stachydrine (**28**-1) has been reported from several *Citrus* species.[109] More interesting are odorine (**28**-2) and odorinol (**28**-3), recently found in two *Aglaia* species[128,152] from the Meliaceae, and in which both nitrogens from ornithine are retained. Some putrescine derivatives are also known, an example being haplamide (**30**-3). The derivation of rohitukine (**29**-3) from lysine is plausible, but it could also be a pseudoalkaloid. This alkaloid too was recently reported to occur in a meliaceous species (*Amoora rohituka*).[77]

E. MISCELLANEOUS AMINES AND AMIDES

A few isobutylamine derivatives have been reported from rutaceous species (Tables 31 and 32). They are generally acylated with long-chain unsaturated fatty acids, such as γ-sanshoöl (**31**-5), but a cinnamate, fagaramide (**31**-3), is also known.

Tiglamide (**32**-3), benzamide (**32**-4), phenylacetamide (**32**-5), or its *N*-phenyl derivative glycomide (**32**-7) are further examples of simple amides found in the Rutales.

III. Distribution of Alkaloids in the Rutales

From the order Rutales more than 500 alkaloids were known in April 1982, these having been isolated from about 300 species belonging to 87 genera. The distribution of these alkaloids according to biogenetic types in the Rutales is presented in Fig. 1. The blocks in front represent the number of genera (within a family) from which the corresponding alkaloid types have been reported; the alkaloid types are indicated by columns in the background, together with number of known representatives.

As is seen, the picture is dominated by the alkaloids derived from anthranilic acid, tryptophan, and phenylalanine or tyrosine.

The alkaloids derived from anthranilic acid are the most widespread in the Rutaceae. They occur in 64 out of 74 genera from which alkaloids have been reported. Furthermore, they are distributed in 5 of the 7 subfamilies.

The tryptophan derived alkaloids show a more limited distribution; they occur in only 18 genera. Simple phenylalanine or tyrosine derivatives have been reported from 22 genera, mainly from the subfamilies Aurantioideae and Toddalioideae but more complex tyrosine derivatives are not found widely. The distribution of the nicotinic acid derived alkaloids is limited to four genera, and those derived from histidine to two, *Pilocarpus* and *Casimiroa*. Alkaloids derived from ornithine have been reported only from *Citrus* species, and lysine derivatives from the genus *Zanthoxylum*. The miscellaneous amines and amides occur in 11 genera, but they are reported only from the three major subfamilies of the Rutaceae, namely Rutoideae, Aurantioideae, and Toddalioideae.

From the family Simaroubaceae, alkaloids have been reported in 9 out of the 32 genera according to the Englerian system. All these genera belong to the major subfamily Simarouboideae. The occurrence of tryptophan derived alkaloids is common, but the occurrence of *N*-methylatanine, a 3-prenylated 2-quinolone alkaloid, in an *Ailanthus* species has also been reported[25].

IV. Tables

The 33 tables present knowledge of the distribution of alkaloids isolated from the three families of the Rutales, namely Rutaceae, Simaroubaceae, and

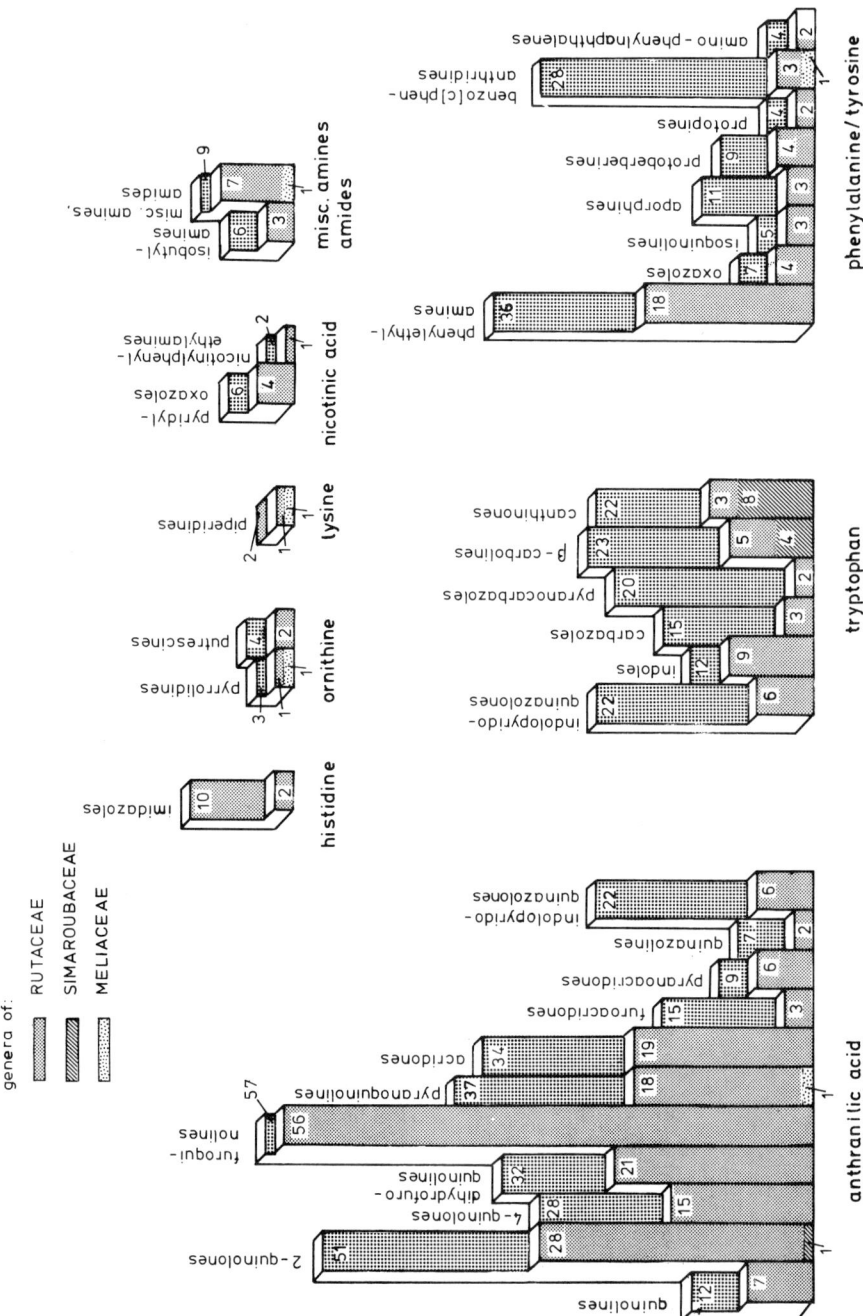

FIG. 1. Distribution of alkaloid types in the order Rutales.

Meliaceae. Tables 1–32 include the different types of alkaloids and in each table the alkaloids are included in the order of their increasing molecular mass, together with their distribution, showing the plant sources by codes. The corresponding plant names with codes are in Table 33, together with the alkaloids found in each species.

The following abbreviations are used to assist in indicating the formulas of the different types of alkaloids:

DH: dihydro

Ger^a: $CH_2CH=C(Me)CH_2CH_2CH(OH)C(OH)Me_2$

Ger^b: $CH(Me)C(Me)CH_2CH_2CH(OH)CMe_2$

$\underset{\displaystyle \text{O}}{\underbrace{}}$

Ger^c: $CH_2CH\overset{c}{=}C(Me)CH(OAc)CH_2CH=CMe_2$

Ger^d: $CH_2CH\overset{c}{=}C(Me)CH(OCOC_{15}H_{31}\text{-}n)CH_2CH\!-\!CMe_2$

$\underset{\displaystyle \text{O}}{\diagdown\!/}$

Ger^e: $CH_2CH=C(Me)CH_2CH_2CH=CMe_2$

HH: hexahydro

Ip^a: $CHMe_2$

Ip^b: $C(=CH_2)Me$

Ip^c: $C(OH)Me_2$

Ip^d: $C(OH)(CH_2OH)Me$

Ip^e: $C(OAc)Me_2$

Ip^f: $C(OH)(CH_2OAc)Me$

Ip^g: $C(Me)\!-\!CH_2$

$\underset{\displaystyle \text{O}}{\diagdown\!/}$

Ip^h: $C(CH_2OH)\!-\!CH_2$

$\underset{\displaystyle \text{O}}{\diagdown\!/}$

Ip^i: $C(OH)(CH_2OMe)Me$

Ip^j: $C(OH)(CH_2O\text{-glucose})Me$

Ip^k: $C(OH)(CH_2OH)_2$

Ip^l: $C(OH)(CH_2O\text{-glucose})CH_2OH$

Ip^m: $C(Cl)(CH_2OH)Me$

Ip^n: $C(OH)(CH_2Cl)Me$

Ip^o: $C(Cl)(CH_2OH)_2$

Ip^p: $C(=CH_2)CH_2OH$

Ipre: $CH(Me)C(=CH_2)Me$

MD: OCH_2O

Pre^a: $CH_2CH=CMe_2$

Pre^b: $CH_2CH\!-\!CMe_2$

$\underset{\displaystyle \text{O}}{\diagdown\!/}$

Pre^c: $CH_2CH(OH)C(OH)Me_2$

Pre^d: $CH_2CH(OH)C(=CH_2)Me$

Pre^e: $CH_2CH(OMe)C(=CH_2)Me$

Pre^f: $CH_2COCHMe_2$

Pre^g: $CH_2CH(OH)CHMe_2$

Pre^h: $CH_2COC(OH)Me_2$

Pre^i: $CH_2CH_2CH(Me)COOMe$

Pre^j: $CH_2CH_2C(OH)Me_2$

Pre^k: $CH_2CH(OH)C(OMe)Me_2$

Pre^l: $CH_2CH(OAc)C(OH)Me_2$

Pre^m: $CH_2CH=C(CH_2OH)Me$

TH: tetrahydro

TABLE 1

Quinolines

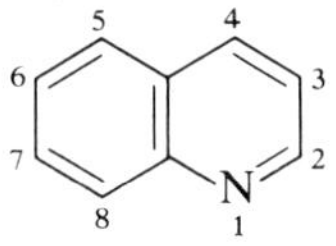

Name	Substituents	Occurrence
1. Quinoline (129)	—	Ci-1[109], Ga-1[109]
2. Quinaldine (143)	2-Me	Ga-1[109]
3. — (199)	2-C_4H_9-n	Ga-1[109]
4. — (229)	2-C_4H_9-n, 4-OMe	Ga-1[109]
5. (235)	2-Ph, 4-OMe	Lu-1[109]
6. Dubamine (249)	2-(3′,4′-MD-C_6H_3)	Di-2[109], Ha-8[109], Ha-12[110]
7. Graveolinine (279)	2-(3′,4′-MD-C_6H_3), 4-OMe	Lu-1[109], Ru-3[69], Ru-4[109]
8. Cusparine (307)	2-(3′,4′-MD-C_6H_3-CH_2CH_2), 4-OMe	Ga-1[109]
9. Cuspareine (311)	2-(3′,4′-MD-C_6H_3-CH_2CH_2), 1-Me, 1,2,3,4-TH	Ga-1[109]
10. Orixinone (317)	2,4-diOMe, 3-Pre^f, 7,8-MD	Or-5[109]
11. Galipine (323)	2-(3′,4′-diOMe-C_6H_3-CH_2CH_2), 4-OMe	Ga-1[109]
12. Orixine (335)	2,4-diOMe, 3-Pre^e, 7,8-MD	Or-5[109]

TABLE 2

2(1H)-Quinolinones

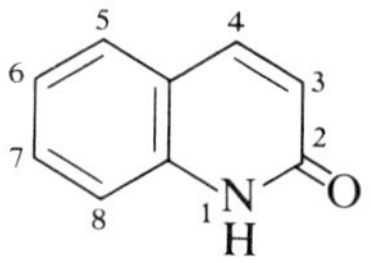

Name	Substituents	Occurrence
1. — (159)	1-Me	Ga-1[109]
2. — (175)	4-OMe	Ho-4[49]
3. — (189)	1-Me, 4-OMe	Fa-2[109], He-3[109], Ho-4[49], My-4[78,110], Za-17[84], Za-18[110]
4. Swietenidine B (205)	3,4-diOMe	Ch-1[23]
5. Edulitine (205)	4,8-diOMe	Ca-1[110], Ha-6[109], Ha-9[109], Ha-18[109]
6. Folifidine (205)	1-Me, 4-OMe, 8-OH	Ha-5[109], Ha-8[109], Ha-9[109]

TABLE 2—*contd.*

Name	Substituents	Occurrence
7. Folinine (219)	1-Me, 4,8-diOMe	Ha-9[109], Ha-15[131]
8. Casimiroine (233)	1-Me, 4-OMe, 7,8-MD	Ca-1[109]
9. Halfordamine (235)	4,6,8-triOMe	Ha-2[40]
10. Swietenidine A (235)	1-Me, 3,8-diOMe, 4-OH	Ch-1[23]
11. Ravenine (243)	1-Me, 4-OPre[a]	Ra-1[109]
12. Atanine (243)	3-Pre[a], 4-OMe	Af-1[133], Fa-26[109], Ra-1[109]
13. — (243)	1-Me, 3-Pre[a], 4-OH	Al-1[115]
14. Ravenoline (243)	1-Me, 3-Ipre, 4-OH	Ra-1[109]
15. — (249)	1-Me, 4,7,8-triOMe	Sp-1[110]
16. N-Methylatanine (257)	1-Me, 3-Pre[a], 4-OMe	Ai-3[25], Al-1[115], Me-4[62]
17. Glycosolone (273)	1-Me, 3-Pre[a], 4-OMe, 8-OH	Gl-6[53]
18. — (277)	1-Me, 4-OPre[c]	Eu-1[92]
19. — (287)	1-Me, 3-Pre[a], 4,8-diOMe	Gl-5[130]
20. — (287)	3-Pre[a], 4-OMe, 7,8-MD	Pt-3[109]
21. Edulinine (291)	1-Me, 3-Pre[c], 4-OMe	Ca-1[110], Ci-4[109], Du-1[18], Fa-12[175], Ru-4[109], Za-55[156]
22. — (297)	3-Pre[a], 4-OPre[a]	Ha-22[109]
23. Pteleprenine (301)	1-Me, 3-Pre[a], 4-OMe, 7,8-MD	Pt-3[110]
24. — (303)	3-Pre[f], 4-OMe, 7,8-MD	Pt-4[110]
25. Preskimmianine (303)	3-Pre[a], 4,7,8-triOMe	Di-1[109], Di-2[7]
26. Lunacridine (305)	1-Me, 3-Pre[g], 4,8-diOMe	Lu-1[109], Lu-2[109], Lu-3[109]
27. Foliosidine (307)	1-Me, 4-OMe, 8-OPre[c]	Ha-8[109], Ha-9[109] Ha-15[131]
28. — (311)	1-Me, 3,3-diPre[a], 1,2,3,4-TH, 4=O	Es-2[58]
29. — (317)	1-Me, 3-Pre[a], 4,6,8-triOMe	Pt-3[110]
30. Ptelefolidine (317)	1-Me, 3-Pre[d], 4-OMe, 7,8-MD	Pt-3[109], Pt-4[110]
31. Lunidonine (317)	1-Me, 3-Pre[f], 4-OMe, 7,8-MD	Lu-2[109], Pt-3[110], Pt-4[110]
32. — (317)	1-Me, 3-Pre[a], 4,7,8-triOMe	Ve-3[14]
33. Lunidine (319)	1-Me, 3-Pre[g], 4-OMe, 7,8-MD	Lu-2[109], Pt-3[109]
34. (−)Balfourolone (321)	1-Me, 3-Pre[c], 4,8-diOMe	Ba-1[109]
35. (+)Hydroxylunacridine (321)	1-Me, 3-Pre[c], 4,8-diOMe	Lu-1[109], Lu-2[109], Or-5[109]
36. Nororixine (321)	3-Pre[c], 4-OMe, 7,8-MD	Or-5[109]
37. Bucharaine (331)	4-OGer[a]	Ha-5[109]
38. Bucharidine (331)	3-Ger[b], 4-OH	Ha-5[109]
39. Ptelefolidine methyl ether (331)	1-Me, 3-Pre[c], 4-OMe, 7,8-MD	Pt-3[109]
40. Ptelecortine (331)	1-Me, 3-Pre[a], 4,8-diOMe, 6,7-MD	Pt-3[110]
41. Ptelefoline (333)	1-Me, 3-Pre[d], 4,6,8-triOMe	Pt-3[109]
42. Hydroxylunidonine (333)	1-Me, 3-Pre[h], 4-OMe, 7,8-MD	Pt-3[110]
43. Isoptelefoline (333)	1-Me, 3-Pre[d], 4,7,8-triOMe	Pt-3[109]
44. Hydroxylunidine (335)	1-Me, 3-Pre[c], 4-OMe, 7,8-MD	Lu-1[109], Lu-2[109], Pt-3[109]

TABLE 2—*contd.*

Name	Substituents	Occurrence
45. Ptelefoline methyl ether (347)	1-Me, 3-Pre[e], 4,6,8-triOMe	Pt-3[110]
46. Pteleoline (347)	1-Me, 3-Pre[i], 4-OMe, 7,8-MD	Pt-3[109]
47. Ptelefructine (347)	1-Me, 3-Pre[d], 4,8-diOMe, 6,7-MD	Pt-3[109]
48. — (347)	1-Me, 3-Pre[f], 4,6-diOMe, 7,8-MD	Pt-3[110]
49. — (349)	1-Me, 3-Pre[g], 4,6-diOMe, 7,8-MD	Pt-3[110]
50. Veprisilone (349)	1-Me, 3-Pre[h], 4,7,8-triOMe	Ve-3[10]
51. — (365)	1-Me, 3-Pre[e], 4,6-diOMe, 7,8-MD	Pt-3[110]

TABLE 3

4(1*H*)-Quinolinones

Name	Substituents	Occurrence
1. — (173)	1,2-diMe	Ac-2[109], Pl-1[109]
2. — (187)	2-C_3H_7-*n*	Bo-4[109]
3. — (235)	1-Me, 2-Ph	Ba-1[109], Fl-7[171], Ha-9[109], Ha-15[131], Lu-1[109]
4. Acutine (241)	2-$(CH_2)_2CH{=}CHEt$	Ha-4[109]
5. — (251)	1-Me, 2-Ph, 5-OH	Lu-4[109]
6. — (259)	1-CH_2OAc, 2-C_3H_7-*n*	Bo-4[109]
7. Eduline (265)	1-Me, 2-Ph, 6-OMe	Ca-1[109], Sk-3[109]
8. Eduleine (265)	1-Me, 2-Ph, 7-OMe	Ca-1[110], Lu-1[109], Lu-4[109]
9. — (267)	2-$(CH_2)_2CH{=}CHCH_2CH{=}CHEt$	Ve-1[109]
10. Pilokeanine (275)	1-Me, 3-Pre[g], 8-OMe	Pl-1[109]
11. Graveoline (279)	1-Me, 2-(3′,4′-MD-C_6H_3)	Ha-8[109], Ha-9[153], Ha-15[131], Ru-1[109], Ru-2[110], Ru-3[109], Ru-4[109]
12. Folimidine (281)	1-Me, 2-(3′-OMe-4′-OH-C_6H_3)	Ha-9[109]
13. — (285)	1-Me, 2-C_9H_{19}-*n*	Eu-8[93], Ru-4[75]
14. — (287)	2-$(CH_2)_8CH_2OH$	Ve-1[109]
15. Japonine (295)	1-Me, 2-Ph, 3,6-diOMe	Or-5[109,193]
16. — (299)	2-$C_{11}H_{23}$-*n*	Pt-3[110], Ru-4[110]
17. Lunasia I (309)	1-Me, 2-(3′,4′-MD-C_6H_3), 6-OMe	Lu-1[109]
18. Lunamarine (309)	1-Me, 2-(3′,4′-MD-C_6H_3), 7-OMe	Lu-1[109]

TABLE 3—*contd.*

Name	Substituents	Occurrence
19. Galipoline (309)	2-(3',4'-diOMe-C_6H_3-CH_2CH_2)	Ga-1[109]
20. — (313)	1-Me, 2-$C_{11}H_{23}$-n	Eu-8[110]
21. — (313)	2-$(CH_2)_9COCH_3$	Ve-1[109]
22. — (313)	2-$C_{12}H_{25}$-n	Ru-4[110]
23. Rutaverine (321)	2-(3',4'-MD-C_6H_3-$(CH_2)_4$)	Ru-4[109]
24. — (327)	2-$C_{13}H_{27}$-n	Ru-4[110]
25. Evocarpine (339)	1-Me, 2-$(CH_2)_7CH\!=\!CH(CH_2)_3Me$	Eu-8[109]
26. Dihydroevocarpine (341)	1-Me, 2-$C_{13}H_{27}$-n	Eu-8[110]
27. — (341)	2-$C_{14}H_{29}$-n	Ru-4[110]
28. — (369)	1-Me, 2-$C_{15}H_{31}$-n	Eu-8[110]

TABLE 4

Dihydrofuroquinolines

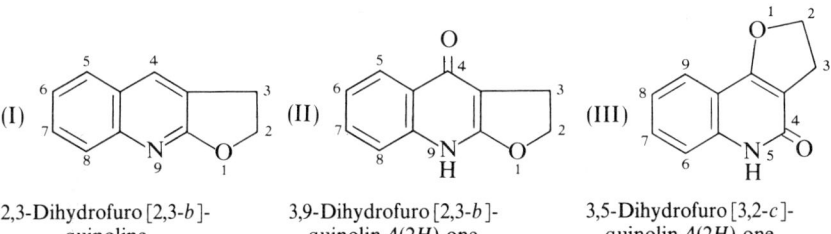

(I)

2,3-Dihydrofuro[2,3-*b*]-
quinoline

(II)

3,9-Dihydrofuro[2,3-*b*]-
quinolin-4(2*H*)-one

(III)

3,5-Dihydrofuro[3,2-*c*]-
quinolin-4(2*H*)-one

Name	Substituents	Occurrence
1. — (241)	III: 2-Ip[b], 5-Me	Al-1[155]
2. Ifflaiamine (243)	II: 2,3,3,9-tetraMe	Fl-8[109]
3. Lemobiline (243)	II: 2,2,3,9-tetraMe	Eu-1[92], Fl-8[109], Ra-1[109]
4. Platydesmine (259)	I: 2-Ip[c], 4-OMe	Ge-3[109], Ha-15[131], Me-7[110], Pl-1[109], Za-7[110], Za-40[109]
5. Isoplatydesmine (259)	II: 2-Ip[c], 9-Me	Ar-1[110], Ar-2[110], Pe-1[110]
6. Araliopsine (259)	III: 2-Ip[c], 5-Me	Ar-1[110]
7. Lunacrine (273)	II: 2-Ip[a], 8-OMe, 9-Me	Lu-1[109], Lu-2[109], Lu-3[109], Lu-4[109]
8. (+)(*R*)Platydesminium (cat. 274)	I: 2-Ip[c], 4-OMe, 9-Me	Ar-2[110], Fl-7[171], Ru-4[109,135], Sk-3[109]
9. (−)(*S*)Platydesminium (cat. 274)	I: 2-Ip[c], 4-OMe, 9-Me	Ch-4[135]
10. Dubinidine (275)	I: 2-Ip[d], 4-OMe	Di-2[109], Ha-8[109], Ha-9[109,153], Ha-15[135]

TABLE 4—*contd.*

Name	Substituents	Occurrence
11. Myrtopsine (275)	I: 2-Ip[c], 3-OH, 4-OMe	My-4[78,110]
12. Ribaline (275)	II: 2-Ip[c], 6-OH, 9-Me	Ba-1[109], Ru-4[109], Ru-5[181]
13. Folisine (275)	II: 2-Ip[d], 9-Me	Ha-9[109]
14. Ptelefolidone (285)	II: 2-Ip[p], 7,8-MD, 9-Me	Pt-3[110]
15. Lunine (287)	II: 2-Ip[a], 7,8-MD, 9-Me	Lu-1[109], Lu-4[109], Pt-3[109]
16. Lunasine (cat. 288)	I: 2-Ip[a], 4,8-diOMe, 9-Me	Lu-3[109], Lu-4[109]
17. (+)Balfourodine (289)	II: 2-Ip[c], 8-OMe, 9-Me	Ba-1[109], Pt-4[110]
18. (−)Hydroxylunacrine (289)	II: 2-Ip[c], 8-OMe, 9-Me	Lu-1[109]
19. Ribalinium (cat. 290)	I: 2-Ip[c], 4-OMe, 6-OH, 9-Me	Ba-1[109], Ru-4[109], Ru-5[181]
20. Pteleatinium (cat. 290)	I: 2-Ip[c], 4-OMe, 8-OH, 9-Me	Pt-3[109]
21. Platydesmine acetate (301)	I: 2-Ip[c], 4-OMe	Ge-3[109]
22. Ptelefolone (301)	II: 2-Ip[b], 6,8-diOMe, 9-Me	Pt-3[109]
23. (−)Luninium (cat. 302)	I: 2-Ip[a], 4-OMe, 7,8-MD, 9-Me	Pt-3[109]
24. Hydroxylunine (303)	II: 2-Ip[c], 7,8-MD, 9-Me	Lu-1[109], Or-5[109], Pt-3[109], Pt-4[110]
25. (+)(R)Balfourodinium (cat. 304)	I: 2-Ip[c], 4,8-diOMe, 9-Me	Ba-1[109], Or-5[109]
26. (−)(S)Balfourodinium (cat. 304)	I: 2-Ip[c], 4,8-diOMe, 9-Me	Ch-4[109]
27. (+)Ptelefolonium (cat. 316)	I: 2-Ip[b], 4,6,8-triOMe, 9-Me	Pt-3[110], Pt-5[135]
28. (+)Isoptelefolonium (cat. 316)	I: 2-Ip[b], 4,7,8-triOMe, 9-Me	Pt-5[135]
29. Dubinine (317)	I: 2-Ip[f], 4-OMe	Ha-7[109]
30. (−)(S)Hydroxyluninium (cat. 318)	I: 2-Ip[c], 4-OMe, 7,8-MD, 9-Me	Pt-3[110], Pt-5[135]
31. Bucharaminol (331)	II: 2,3-diMe, 3-CH$_2$-Pre[c]	Ha-4[20]
32. Veprisinium (cat. 334)	I: 2-Ip[c], 4,7,8-triOMe, 9-Me	Ve-3[15]

TABLE 5

Furoquinolines

(I)

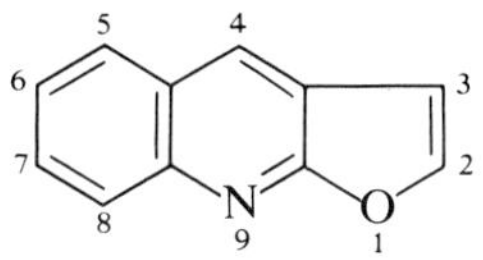

Furo[2,3-b]quinoline

(II)

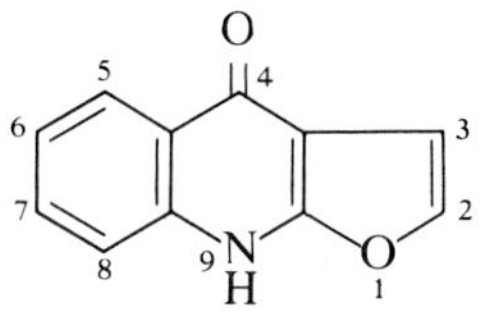

Furo[2,3-b]quinolin-4(9H)-one

Name	Substituents	Occurrence
1. Dictamnine (199)	I: 4-OMe	Ae-1[109], Af-1[110,133] Am-4[48], Ba-1[109], Bo-2[109], Bo-3[21], Ca-1[110], Ch-5[109], De-1[109], Di-1[109], Di-2[7,109], Di-3[109], Du-1[18], Du-2[18], Es-2[58], Es-4[58], Eu-3[109], Eu-6[109], Fa-12[110], Fl-1[109], Fl-7[170,171], Fl-10[109], Fl-11[110], Fl-12[109], Gl-2[31], Gl-5[130], Gl-6[109], Ha-1[40], Ha-5[109], Ha-6[109], Ha-15[131], Ha-17[109], Ha-18[20], Ha-20[109], He-1[109], Ho-1[109], Ho-4[49], Me-1[109], Me-4[62], Mo-1[110], My-1[78], My-2[78], My-3[78], My-4[78,110], Ph-1[109], Pi-12[109], Pt-3[109], Ru-1[109], Ru-3[69,109], Ru-4[109], Ru-6[109], Sk-2[109], Sk-3[109], Sk-5[109], To-1[110], To-2[150], Za-1[109], Za-2[109], Za-6[110], Za-8[110], Za-17[84], Za-18[110], Za-26[85], Za-40[109], Za-48[72]
2. Isodictamnine (199)	II: 9-Me	Di-1[109], Di-2[7], Di-3[109], He-1[109]

TABLE 5—contd.

Name	Substituents	Occurrence
3. Confusameline (215)	I: 4-OMe, 7-OH	Eu-4[109], Me-2[109], Me-5[165], My-4[110]
4. Robustine (215)	I: 4-OMe, 8-OH	Di-3[109], Ha-5[109], Ha-8[109], Ha-14[109], Ha-18[109], Th-1[110], To-1[110], Za-2[55], Za-6[110], Za-17[84]
5. Pteleine (229)	I: 4,6-diOMe	Du-1[18], Du-2[18], He-1[109], Me-1[109], Pl-1[109], Pt-3[109], Ru-4[109]
6. Evolitrine (229)	I: 4,7-diOMe	Ac-4[154], Al-1[115], Cu-1[109], Du-1[18], Du-2[18], Es-2[58], Es-4[58], Eu-3[109], Eu-6[109], Gl-4[145], Me-4[62], Me-5[165], Or-5[109], Ph-1[109], Pl-1[109]
7. γ-Fagarine (229)	I: 4,8-diOMe	Ae-1[109,110], Ca-1[110], Ch-1[109], Di-1[109], Di-2[7], Di-3[109], Er-6[88], Fa-6[109], Fa-12[110], Fl-7[170,171] Ge-3[109], Gl-1[109], Gl-6[109], Ha-5[109], Ha-7[17], Ha-11[110], Ha-14[109], Ha-18[109], Ha-19[110], Ha-21[110], Ha-22[8], Ha-23[110], Ho-1[109], My-2[78], My-3[78], My-4[78,110] Ph-1[109], Pi-12[109], Ra-1[109], Ru-3[109], Ru-4[109], Th-1[109], To-1[109,148], To-2[150], Ve-5[96], Za-2[109], Za-17[84], Za-39[55], Za-48[71], Za-52[110]
8. Isopteleine (229)	II: 6-OMe, 9-Me	Di-2[7], Di-3[109]
9. Maculine (243)	I: 4-OMe, 6,7-MD	Ar-1[110], Es-1[109], Es-4[58], Fl-1[109], Fl-3[109], Fl-6[109], Fl-10[109], Fl-13[109], Fl-14[109], He-1[109], Pt-1[109], Sa-2[33], Te-5[66], Te-6[110], Ve-2[109]

TABLE 5—contd.

Name	Substituents	Occurrence
10. Kokusagine (243)	I: 4-OMe, 7,8-MD	Lu-1[109], Or-5[109,193]
11. Taifine (243)	II: 7-OMe, 9-Et	Ru-3[61]
12. Delbine (245)	I: 4,7-diOMe, 6-OH	Mo-1[23]
13. Heliparvifoline (245)	I: 4,6-diOMe, 7-OH	He-2[110]
14. Haplopine (245)	I: 4,8-diOMe, 7-OH	Ae-1[110], Af-1[133], Ha-5[109], Ha-8[109], Ha-9[109], Ha-12[117,119] Ha-14[109], Ha-15[2,109] Ha-18[109], Me-5[165], Mo-1[114], Za-6[110], Za-17[84], Za-32[26]
15. Kokusaginine (259)	I: 4,6,7-triOMe	Ac-2[109], Ac-4[154], Ar-1[110], Ba-1[109], Ba-2[164], Ch-2[109], Ch-3[109], Ch-4[109], Du-1[18], Du-2[18], Es-4[58], Eu-2[109], Eu-3[109], Eu-6[109], Eu-10[109], Fl-5[109], Fl-10[109], Fl-12[109], Fl-13[109], Gl-2[30,31], Gl-6[109], Ha-1[40], Ha-20[109], He-1[109], He-2[109], Me-2[109], Me-5[165], Me-7[110], Or-2[94], Or-3[67], Or-5[109], Pe-1[110], Ph-1[109], Pl-1[109], Pt-1[109], Pt-3[109], Ru-3[69,109], Ru-4[109], Ru-6[109], Sa-2[56], Te-4[12], Te-7[110], Te-8[124], Ve-1[109], Ve-2[109], Ve-4[76], Za-42[109]
16. Maculosidine (259)	I: 4,6,8-triOMe	Ba-1[109], Er-1[109], Er-2[109], Er-3[109], Er-4[109], Er-5[109], Es-2[58], Es-3[109], Fl-10[109], Fl-12[109], Or-2[94], Pt-3[109]
17. Skimmianine (259)	I: 4,7,8-triOMe	Ac-2[109], Ae-1[109], Al-1[115], Am-4[48], Ar-1[110], Ar-2[110], Ba-1[109], Bo-4[109], Ca-1[110], Ch-1[109], Ch-2[109], Ch-3[109], Ch-4[109], De-1[109],

TABLE 5—contd.

Name	Substituents	Occurrence
		Di-1[109], Di-2[7,109], Di-3[109], Di-4[69], Di-7[110], Er-2[109], Er-3[109], Er-4[109], Er-5[109], Es-1[109], Es-2[58], Es-3[109], Eu-1[92], Eu-2[109], Eu-4[109], Fa-1[109], Fa-3[109], Fa-4[109], Fa-6[109], Fa-9[109], Fa-10[109], Fa-11[109], Fa-12[175], Fa-16[109], Fa-20[109], Fa-24[109], Fa-26[109], Fl-3[109], Fl-4[109], Fl-6[109], Fl-7[170], Fl-9[109], Fl-10[109], Fl-11[110], Fl-12[109], Ge-1[5], Ge-3[109], Ge-4[109] Gl-1[109], Gl-2[30], Gl-5[130], Gl-6[109], Ha-4[109], Ha-5[109], Ha-6[109], Ha-7[17], Ha-8[109], Ha-9[109], Ha-11[110], Ha-12[117,119] Ha-13[109], Ha-14[109], Ha-15[2,109], Ha-16[109], Ha-17[109], Ha-18[109], Ha-19[110], Ha-20[109], Ha-21[110], Ha-22[8,95], Ha-23[110], He-1[109], He-2[109], Ho-1[109], Ho-4[49], Lu-1[109], Me-2[109], Me-3[109], Me-5[165], Me-6[4], Me-7[110], Mo-1[109], Mu-2[109], Mu-3[63], My-1[78], My2[78], My-3[78], My-4[78,110], Or-1[94], Or-5[109], Ph-1[109], Pt-1[109], Pt-2[109], Pt-3[109], Ru-3[69,109], Ru-4[109], Ru-6[109], Sk-1[109], Sk-3[109], Sk-4[109], Te-6[66], Te-7[110], Te-8[65], Th-1[109,110], To-1[109], To-2[150],

TABLE 5—*contd.*

Name	Substituents	Occurrence
		Ve-2[109], Ve-5[96], Za-1[109], Za-2[109], Za-6[83], Za-8[110], Za-9[177], Za-11[134], Za-12[59], Za-16[160], Za-17[84], Za-18[110], Za-19[110], Za-23[59], Za-29[189], Za-30[59], Za-35[80], Za-38[138], Za-40[109], Za-42[109], Za-45[109], Za-47[109], Za-48[71], Za-49[176], Za-52[110], Za-55[156]
18. Glycarpine (259)	II: 5,7-diOMe, 9-Me	Gl-4[145]
19. Isomaculosidine (259)	II: 6,8-diOMe, 9-Me	Di-1[109], Di-3[109], Pt-3[110]
20. Flindersiamine (273)	I: 4,8-diOMe, 6,7-MD	Ar-1[110], Ba-1[109], Es-1[109], Es-2[58], Fl-3[109], Fl-4[109], Fl-5[109], Fl-6[109], Fl-10[109], Fl-12[109], Fl-14[109], He-1[109], He-2[110], Or-4[13], Te-4[12], Te-6[66,109], Te-8[124], Ve-2[109]
21. Isoflindersiamine (273)	II: 6,7-MD, 8-OMe, 9-Me	He-2[110]
22. Melineurine (283)	I: 4-OMe, 7-OPre[a]	Me-5[165]
23. Haplophydine (283)	I: 4-OMe, 8-OPre[a]	Ha-15[2,110]
24. Acrophylline (283)	II: 7-OMe, 9-Pre[a]	Ac-3[109]
25. Acronycidine (289)	I: 4,5,7,8-tetraOMe	Ac-2[109,158], Ba-2[50], Me-3[109]
26. Halfordinine (289)	I: 4,6,7,8-tetraOMe	Ar-2[110], Di-6[191], Ha-2[40], Me-7[110], Or-3[67], Te-8[65]
27. — (299)	I: 4-OMe, 7-OPre[d]	Eu-10[109]
28. Folifinine (301)	I: 4-OMe, 7-OPre[j], 8-OH	Ha-9[102]
29. Acrophyllidine (301)	II: 7-OMe, 9-Pre[j]	Ac-3[109]
30. — (313)	I: 4,8-diOMe, 7-OPre[a]	Ha-12[117,119] Ha-15[2,110], Pt-1[109]
31. Perfamine (313)	I: 4,7-diOMe, 7-Pre[c], 8 = O, 7,8-DH	Ha-15[110]
32. Evellerine (317)	I: 4-OMe, 7-OPre[c]	Eu-4[109]
33. Haplophyllidine (317)	I: 4,8-diOMe, 7-OH, 8-Pre[c], 5,6,7,8-TH	Ha-15[109]
34. Tecleaverdoornine (327)	I: 4-OMe, 5-Pre[a], 6,7-MD, 8-OH	Te-4[12], Te-8[124]
35. Tecleamine (327)	I: 4-OMe, 6,7-MD, 8-OPre[a]	Te-4[12]
36. Haplatine (329)	I: 4,8-diOMe, 7-OPre[m]	Ha-12[117]
37. Anhydroevoxine (329)	I: 4,8-diOMe, 7-OPre[b]	Eu-10[109]
38. Evodine (329)	I: 4,8-diOMe, 7-OPre[d]	Eu-10[109], Ha-15[2,6]
39. Evoxoidine (329)	I: 4,8-diOMe, 7-OPre[f]	Eu-10[109], Ha-15[6]
40. Maculosine (331)	I: 4-OPre[c], 6,7-MD	Fl-10[109]
41. Perforine (331)	I: 4,8-diOMe, 7-OH, 8-Pre[j], 5,6,7,8-TH	Ha-15[109]

TABLE 5—*contd.*

Name	Substituents	Occurrence
42. Evolatine (347)	I: 4,6-diOMe, 7-OPre[c]	Eu-2[109]
43. Montrifoline (347)	I: 4,7-diOMe, 6-OPre[c]	Mo-1[23], Te-4[12], Te-8[11]
44. Evoxine (347)	I: 4,8-diOMe, 7-OPre[c]	Ch-4[109], Di-2[109], Eu-2[109], Eu-10[109], Ha-8[109], Ha-10[109], Ha-12[110,118], Ha-13[109], Ha-15[109], Ha-16[109], Ha-17[109], Ha-20[109], Ha-22[8], Mo-1[114], Or-5[193], Te-1[180]
45. Methylevoxine (361)	I: 4,8-diOMe, 7-OPre[k]	Ha-15[110]
46. Evoxine acetate (389)	I: 4,8-diOMe, 7-OPre[l]	Ha-10[109]
47. Glycoperine (391)	I: 4,8-diOMe, 7-O-rhamnose	Ha-12[119], Ha-15[110]
48. Glycohaplopine (407)	I: 4,8-diOMe, 7-O-glucose	Ha-15[3]
49. Monoacetylglycoperine (433)	I: 4,8-diOMe, 7-O-rhamnose monoacetate	Ha-15[2]
50. Diacetylglycoperine (475)	I: 4,8-diOMe, 7-O-rhamnose diacetate	Ha-15[2]
51. Triacetylglycoperine (517)	I: 4,8-diOMe, 7-O-rhamnose triacetate	Ha-15[1,2]

TABLE 6

Difuroquinolines and pyranofuroquinolines

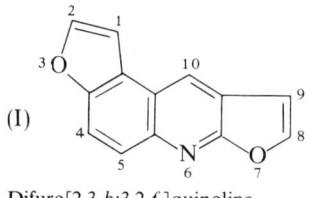

(I) Difuro[2,3-*b*:3,2-*f*]quinoline

(II) 3*H*-Furo[2,3-*b*]pyrano[3,2-*f*]quinoline

(III) 3*H*-Furo[2,3-*b*]pyrano[2,3-*h*]quinoline

(IV) 2*H*-Furo[2,3-*b*]pyrano[3,2-*h*]quinoline

Name	Substituents	Occurrence
1. Medicosmine (281)	II: 3,3-diMe 11-OMe	Me-1[109]
2. Dutadrupine (281)	III: 3,3-diMe, 7-OMe	Du-1[18]
3. Foliminine (283)	IV: 2,2-DiMe, 3,4-DH, 7-OMe	Ha-9[110]

TABLE 6—*contd.*

Name	Substituents	Occurrence
4. Acronydine (311)	II: 3,3-diMe, 5,11-diOMe	Ac-2[109]
5. Choisyine (313)	I: 1,2-DH, 2-Ip[a], 4,10-diOMe	Ch-2[109], Ch-3[109], Ch-4[109]
6. Anhydroperforine (317	III: 1,2,4a,5,6,11b-HH, 3,3-diMe, 7,11b-diOMe	Ha-15[19]

TABLE 7

Pyranoquinolines

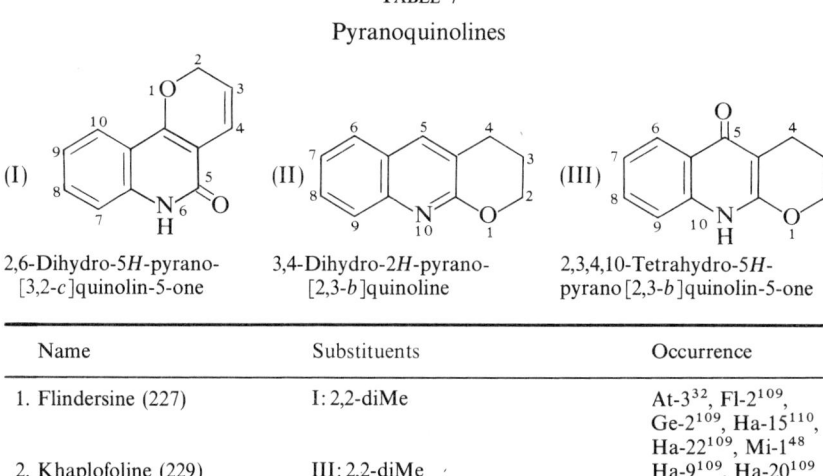

(I) 2,6-Dihydro-5*H*-pyrano-[3,2-*c*]quinolin-5-one

(II) 3,4-Dihydro-2*H*-pyrano-[2,3-*b*]quinoline

(III) 2,3,4,10-Tetrahydro-5*H*-pyrano[2,3-*b*]quinolin-5-one

Name	Substituents	Occurrence
1. Flindersine (227)	I: 2,2-diMe	At-3[32], Fl-2[109], Ge-2[109], Ha-15[110], Ha-22[109], Mi-1[48]
2. Khaplofoline (229)	III: 2,2-diMe	Ha-9[109], Ha-20[109]
3. *N*-Methylflindersine (241)	I: 2,2,6-triMe	Al-1[115], Eu-1[92], Fa-4[44,79] My-3[78], Pt-3[110], Sp-1[110], Xy-1[44]
4. — (243)	III: 2,2,10-triMe	Al-1[115]
5. Haplamine (257)	I: 2,2-diMe, 9-OMe	Ha-15[19,110]
6. — (257)	I: 2,2-diMe, 7-OMe	My-1[78]
7. Geibalansine (259)	II: 2,2-diMe, 3-OH, 5-OMe	Ge-1[5]
8. (+)Folifine (259)	III: 2,2,10-triMe, 3-OH	Ar-2[110], Ha-5[109], Ha-9[109]
9. (−)Ribalinine (259)	III: 2,2,10-triMe, 3-OH	Ar-1[110], Fa-12[175]
10. (±)Ribalinine (259)	III: 2,2,10-triMe, 3-OH	Ba-1[109]
11. − (259)	I: 2,2,6-triMe, 3,4-DH, 4-OH	Eu-1[92]
12. — (271)	I: 2,2,6-triMe, 8-OMe	Or-2[94]
13. Zanthobungeanine (271)	I: 2,2,6-triMe, 7-OMe	Za-11[134], Za-48[71]
14. (−)(*S*)Ribalinidine (275)	III: 2,2,10-triMe, 3,7-diOH	Ba-1[109], Ru-3[69], Ru-4[109], Ru-5[181]
15. (+)Isobalfourodine (289)	III: 2,2,10-triMe, 3-OH, 9-OMe	Ba-1[109]
16. (−)Lunacrinol (lunasia II) (289)	III: 2,2,10-triMe, 3-OH, 9-OMe	Lu-1[109]
17. Rutalinium (cat. 290)	II: 2,2,10-triMe, 3,7-diOH, 5-OMe	Ru-4[109], Ru-5[181]
18. Haplobucharine (297)	III: 2,2-diMe, 10-Pre[a]	Ha-5[110]
19. — (301)	II: 2,2-diMe, 3-OAc, 5-OMe	Ge-1[5]

TABLE 7—*contd.*

Name	Substituents	Occurrence
20. Oricine (301)	I: 2,2,6-triMe, 8,9-diOMe	Or-2[94], Or-3[109]
21. Veprisine (301)	I: 2,2,6-triMe, 7,8-diOMe	Or-2[94], Ve-3[14], Ve-5[96]
22. Pteleflorine (303)	II: 2,2-diMe, 3-OH, 5-OMe, 8,9-MD	Pt-3[110]
23. Neohydroxylunine (303)	III: 2,2,10-triMe, 3-OH, 8,9-MD	Pt-3[110]
24. — (305)	I:2,2,6-triMe, 7-OMe, 3,4-diOH, 3,4-DH	Za-48[72]
25. — (315)	I: 2,2-diMe, 6-CH$_2$OAc, 7-OH	Za-33[157]
26. — (325)	I: 2,2,6-triMe, 7-OPre[a]	Ve-5[96]
27. — (325)	I: 2,6-diMe, 2-CH$_2$-Pre[b]	Za-48[72]
28. Zanthophylline (329)	I: 2,2-diMe, 6-CH$_2$OAc, 7-OMe	Za-33[157]
29. — (355)	I: 2,2,6-triMe, 8-OMe, 7-OPre[a]	Ve-5[96]
30. — (371)	I: 2,2,6-triMe, 8-OMe, 7-OPre[b]	Ve-5[96]

31. Paraensidimerin D (482)   Eu-1[92]

32. Pteledimeridine (482)   Pt-3[113]

33. Pteledimerine (482)   Pt-3[113,132]

TABLE 7—*contd.*

Name	Substituents	Occurrence
34. Vepridimerine A (602)		Ve-3[120]
35. Vepridimerine B (602)		Or-2[120], Ve-3[120]
36. Vepridimerine C (602)		Or-2[120], Ve-3[120]

TABLE 7—*contd.*

Name	Substituents	Occurrence
37. Vepridimerine D (602)	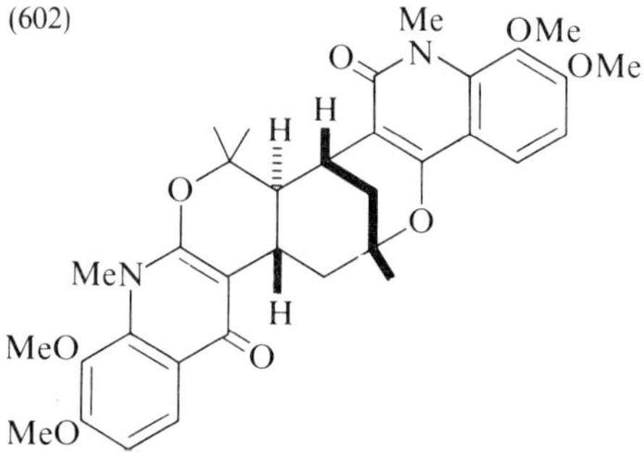	Or-2[120]

TABLE 8

Acridones

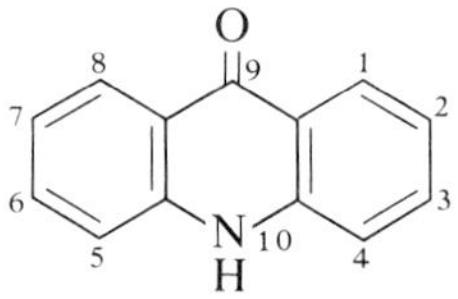

9(10*H*)-Acridinone

Name	Substituents	Occurrence
1. Acridone (195)	—	Th-1[110]
2. — (209)	10-Me	Th-1[109,110]
3. — (211)	1-OH	Bo-1[110,141]
4. — (225)	1-OH, 10-Me	Bo-1[110,141], Ru-4[109]
5. — (241)	1-OH, 7-OMe	Bo-1[110,141]
6. — (255)	1,3-diOMe	Ba-2[164]
7. — (255)	1-OH, 3-OMe, 10-Me	Bo-1[139], Es-4[58] Fa-9[109], Fa-20[109], Ru-4[110], Ve-4[76], Za-29[189]
8. — (269)	1,3-diOMe, 10-Me	Ac-2[109], Ba-2[164], Di-6[191], Eu-9[176], Or-3[67], Te-8[65], Ve-1[109], Ve-2[109], Ve-4[76]
9. Evoxanthidine (269)	1-OMe, 2,3-MD	Eu-10[109]
10. Norevoxanthine (269)	1-OH, 2,3-MD, 10-Me	Te-2[109]
11. Xanthoxoline (271)	1-OH, 2,3-diOMe	Eu-10[109]

TABLE 8—*contd.*

Name	Substituents	Occurrence
12. — (271)	1,3-diOH, 2-OMe, 10-Me	Ru-4[137]
13. Evoxanthine (283)	1-OMe, 2,3-MD, 10-Me	Ba-1[109], Di-7[110], Eu-2[109], Eu-9[181], Eu-10[109], Or-1[94], Or-2[94], Or-3[67], Te-1[110,180], Te-2[109], Te-3[109], Te-8[65], Ve-1[109], Ve-2[109]
14. Arborinine (285)	1-OH, 2,3-diOMe, 10-Me	Ac-3[109], Di-7[110], Eu-2[109], Eu-10[109], Fa-9[109], Fa-10[109], Fa-20[109], Gl-1[109], Gl-2[30], Gl-5[130], Mo-1[109,110], Or-2[94], Ra-1[109], Ru-3[69,109], Ru-4[109], Ru-6[109], Te-1[180], Te-3[109], Ve-2[109], Ve-4[76], Za-29[189]
15. — (299)	1,2,3-triOMe, 10-Me	Ba-2[50], Eu-2[109], Me-6[4], Ve-2[109]
16. — (299)	1,3,4-triOMe, 10-Me	Te-1[180]
17. — (299)	1,3,5-triOMe, 10-Me	Te-1[110]
18. Xanthevodine (299)	1,4-diOMe, 2,3-MD	Ac-2[109], Ba-2[164], Eu-10[109], Me-6[4]
19. Normelicopine (299)	1-OH, 2-OMe, 3,4-MD, 10-Me	Ac-2[158]
20. Normelicopidine (299)	1-OH, 4-OMe, 2,3-MD, 10-Me	Ac-2[158,159]
21. — (301)	1,5-diOH, 2,3-diOMe, 10-Me	Gl-2[30,31]
22. Melicopine (313)	1,2-diOMe, 3,4-MD, 10-Me	Ac-1[109], Ac-2[109], Ba-2[50,164], Me-3[109]
23. Melicopidine (313)	1,4-diOMe, 2,3-MD, 10-Me	Ac-2[109], Ba-2[110], Eu-2[109], Eu-10[109], Me-3[109], Me-6[4]
24. Tecleanthine (313)	1,5-diOMe, 2,3-MD, 10-Me	Di-6[191], Or-3[191], Te-1[110,180], Te-3[109], Te-8[65]
25. — (315)	1,2,3,4-tetraOMe	Ba-2[164]
26. Normelicopicine (315)	1-OH, 2,3,4-triOMe, 10-Me	Ac-2[158]
27. — (323)	1-OH, 2-Pre[a], 3-OMe, 10-Me	Gl-5[130]
28. Melicopicine (329)	1,2,3,4-tetraOMe, 10-Me	Ac-2[109], Ba-2[164], Me-3[109], Te-1[110]
29. Evoprenine (339)	1-OH, 2-OMe, 3-OPre[a], 10-Me	Eu-2[109]
30. — (343)	1,5,6-triOMe, 2,3-MD 10-Me	Te-1[110]
31. Atalaphylline (379)	1,3,5-triOH, 2,4-diPre[a]	At-2[109]
32. Atalaphyllidine A (379)	1,3,8-triOH, 2,7-diPre[a]	At-2[110]
33. — (393)	1,3,5-triOH, 2,4-diPre[a], 10-Me	At-2[109]
34. — (407)	1-OH, 2,4-diPre[a], 3,5-diOMe	At-2[101]

TABLE 8—*contd.*

Name	Substituents	Occurrence
35 Tecleanone (301)		Di-7[110], Or-2[94], Or-3[67], Te-2[110], Te-8[65,110]

TABLE 9

Furoacridones

(I)   (II)

Furo[2,3-*c*]acridin-6(11*H*)-one   1,11-Dihydrofuro[2,3-*c*]acridin-6(2*H*)-one

Name	Substituents	Occurrence
1. Furacridone (265)	I: 5-OH, 11-Me	Ru-4[110]
2. Rutacridone (307)	II: 2-Ip[b], 5-OH, 11-Me	Bo-1[141], Ru-3[69], Ru-4[109]
3. Rutacridone epoxide (323)	II: 2-Ip[g], 5-OH, 11-Me	Ru-4[116,140]
4. Gravacridonol (323)	II: 2-Ip[p], 5-OH, 11-Me	Ru-4[140]
5. Hydroxyrutacridone epoxide (339)	II: 2-Ip[h], 5-OH, 11-Me	Ru-4[192]
6. Gravacridondiol (341)	II: 2-Ip[d], 5-OH, 11-Me	Ru-4[109]
7. Gravacridondiol monomethyl ether (355)	II: 2-Ip[i], 5-OH, 11-Me	Ru-4[109]
8. Gravacridontriol (357)	II: 2-Ip[k], 5-OH, 11-Me	Ru-4[110]
9. Gravacridonchlorine (359/361)	II: 2-Ip[m], 5-OH, 11-Me	Ru-4[109]
10. Isogravacridonchlorine (359/361)	II: 2-Ip[n], 5-OH, 11-Me	Ru-4[110]
11. Gravacridonolchlorine (375/377)	II: 2-Ip[o], 5-OH, 11-Me	Ru-4[109]
12. Gravacridondiol glucoside (503)	II: 2-Ip[j], 5-OH, 11-Me	Ru-4[110]
13. Gravacridontriol glucoside (519)	II: 2-Ip[l], 5-OH, 11-Me	Ru-4[110]

TABLE 9—*contd.*

Name	Substituents	Occurrences
14. Atalanine (610)	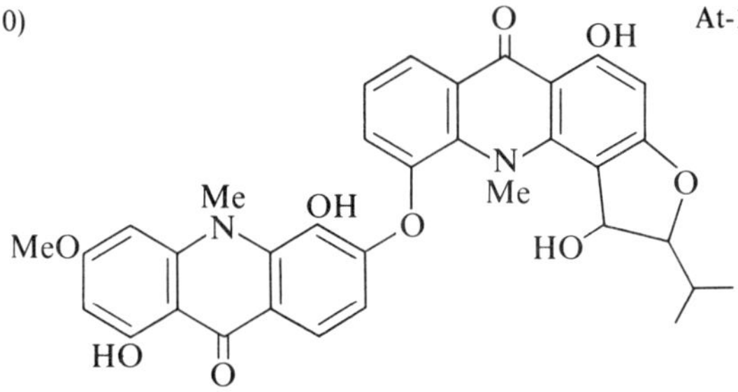	At-1[110]
15. Ataline (662)	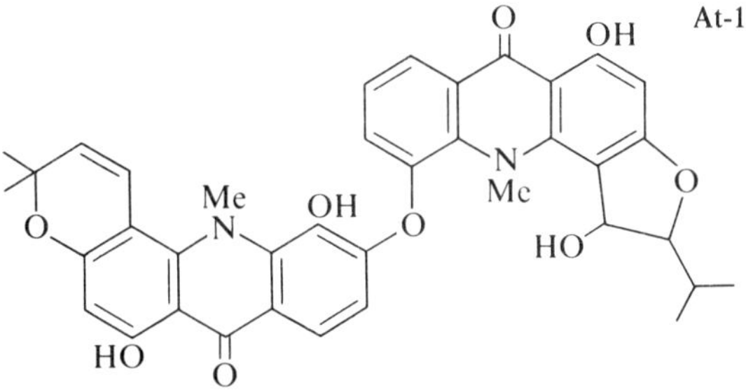	At-1[110]

TABLE 10

Pyranoacridones

(I)

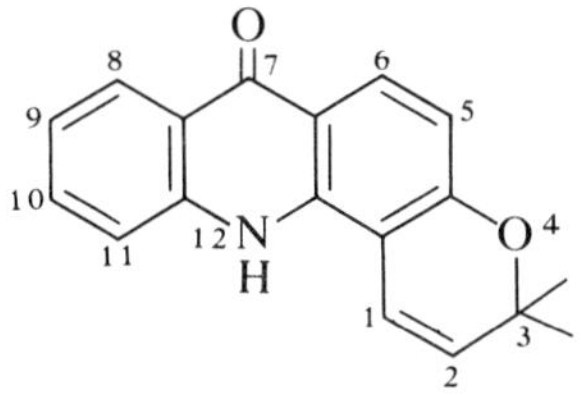

3,12-Dihydro-3,3-dimethyl-7*H*-pyrano-
[2,3-*c*]acridin-7-one

(II)

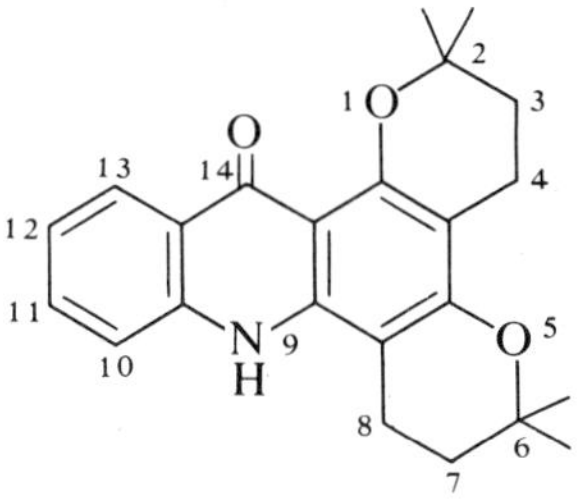

3,4,6,7,8,9-Hexahydro-2,2,6,6-tetramethyl-
2*H*,14*H*-dipyrano[2,3-*a*:2',3'-*c*-]-
acridin-14-one

Name	Substituents	Occurrence
1. — (293)	I: 6-OH	Mu-3[63]
2. — (307)	I: 6-OMe	Gl-6[109], Mu-3[63]
3. Noracronycine (307)	I: 6-OH, 12-Me	Bo-1[141], Gl-6[109], Mu-3[63]

TABLE 10—contd.

Name	Substituents	Occurrence
4. Atalaphyllidine (309)	I: 6,11-diOH	At-2[110]
5. Acronycine (321)	I: 6-OMe, 12-Me	Ac-2[109,159]
		Ba-2[50,110]
6. — (323)	I: 6,11-diOH, 12-Me	At-1[110]
7. Atalaphyllinine (377)	I: 5-Pre[a], 6,11-diOH	At-2[110]
8. — (391)	I: 5-Pre[a], 6,11-diOH, 12-Me	At-1[110]
9. — (393)	II: 9-Me, 10-OH	At-2[109]

TABLE 11

Quinazolines

(I)

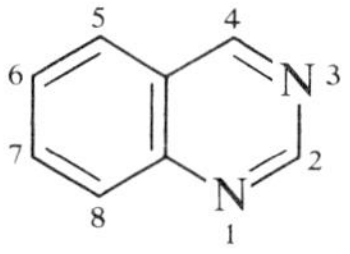

Quinazoline

(II)

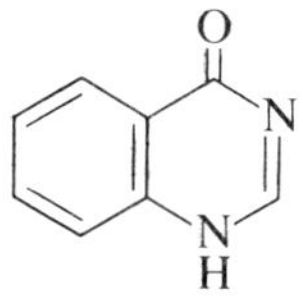

4(1H)-Quinazolinone

(III)

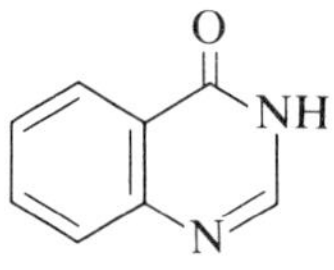

4(3H)-Quinazolinone

(IV)

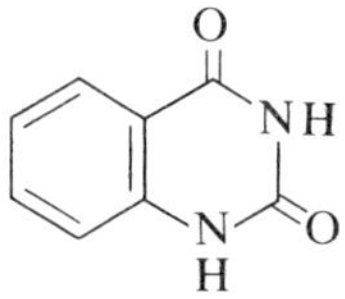

2,4(1H,3H)-Quinazolindione

Name	Substituents	Occurrence
1. Glycorine (160)	II: 1-Me	Gl-1[109], Gl-2[30]
2. Glycosmicine (176)	IV: 1-Me	Gl-1[109]
3. Glycosminine (236)	III: 2-CH$_2$Ph	Gl-1[109], Gl-6[109]
4. Glycophymoline (250)	I: 2-CH$_2$Ph, 4-OMe	Gl-6[144]
5. Arborine (250)	II: 1-Me, 2-CH$_2$Ph	Gl-1[109], Gl-2[30],
		Gl-6[109]
6. — (280)	IV: 1-Me, 3-CH$_2$CH$_2$Ph	Za-4[59]
7. — (310)	IV: 1-Me, 3-(4'-OMe-C$_6$H$_4$-CH$_2$CH$_2$)	Za-4[59]

TABLE 12

Indolopyridoquinazolines

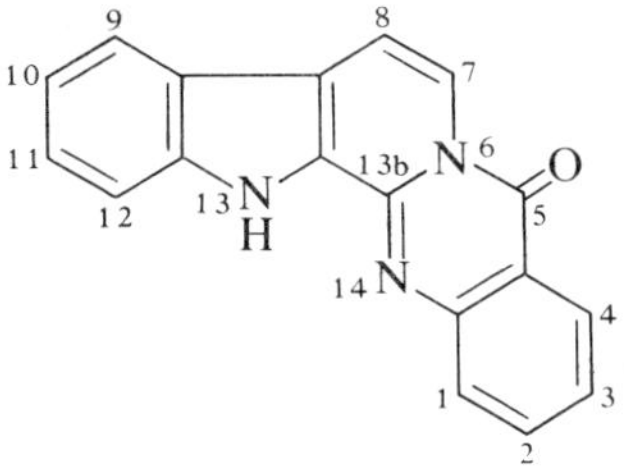

Indolo[2′,3′:3,4]pyrido[2,1-b]quinazolin-5(13H)-one

Name	Substituents	Occurrence
1. Rutaecarpine (287)	7,8-DH	Eu-8[109], Ho-1[109], Ho-2[110], Za-38[138], Za-42[109], Za-45[109]
2. Dihydrorutaecarpine (289)	7,8,13b,14-TH	Eu-8[93], Za-24[110]
3. — (301)	1-OH	Ve-3[10]
4. — (cat. 302)	7,8-DH, 14-Me	Eu-8[100]
5. — (303)	1-OH, 7,8-DH	Eu-1[110]
6. Evodiamine (303)	7,8,13b,14-TH, 14-Me	Ar-2[110], Eu-8[109], Za-45[109]
7. Hortiacine (317)	7,8-DH, 10-OMe	Ho-1[109], Ho-2[110], Ho-4[49]
8. — (317)	7,8,13b,14-TH, 14-CHO	Eu-8[93]
9. — (319)	7,8,13b,14-TH, 13b-OH, 14-Me	Za-45[109]
10. Euxylophoricine C (331)	2,3-MD, 7,8-DH	Eu-1[110]
11. Hortiamine (331)	7,8-DH, 10-OMe, 14-Me	Ho-1[109], Ho-3[109]
12. Euxylophoricine F (333)	2-OH, 3-OMe, 7,8-DH	Eu-1[110]
13. Euxylophoricine B (345)	2,3-diOMe	Eu-1[109]
14. Euxylophoricine A (347)	2,3-diOMe, 7,8-DH	Eu-1[109]
15. — (347)	7-COOH, 7,8,13b,14-TH, 14-Me	Eu-8[52]
16. Euxylophorine B (359)	2,3-diOMe, 14-Me	Eu-1[110]
17. Euxylophorine A (361)	2,3-diOMe, 7,8-DH, 14-Me	Eu-1[109]
18. Euxylophoricine E (375)	2,3,10-triOMe	Eu-1[110]
19. Euxylophoricine D (377)	2,3,10-triOMe, 7,8-DH	Eu-1[110]
20. Euxylophorine D (389)	2,3,10-triOMe, 14-Me	Eu-1[110]
21. Euxylophorine C (391)	2,3,10-triOMe, 7,8-DH, 14-Me	Eu-1[110]
22. Paraensine (399)	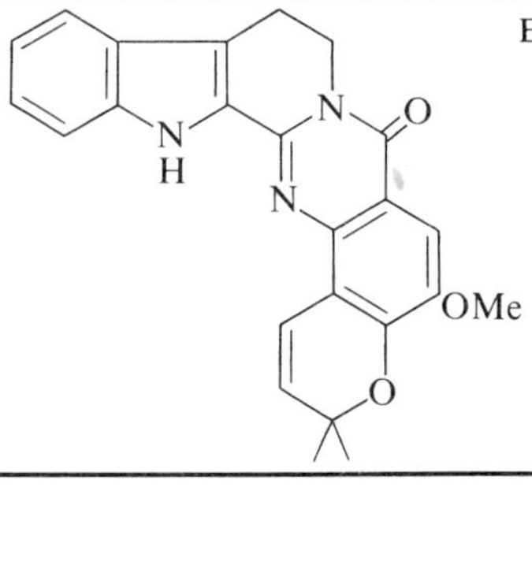	Eu-1[110]

TABLE 13

Simple indole derivatives and dimeric indole alkaloids

Name	Substituents	Occurrence
1. — (145)	3-CHO	Mu-3[109]
2. Tryptamine (160)	3-$CH_2CH_2NH_2$	Ci-1[109]
3. — (188)	3-$CH_2CH_2NMe_2$	Ve-1[109]
4. — (218)	3-$CH_2CH_2NMe_2$, 5-OMe	Di-5[109], Du-1[18], Du-2[18], Fu-8[162], Pi-8[16]
5. — (264)	3-$CH_2CH_2NHCOPh$	My-2[78]

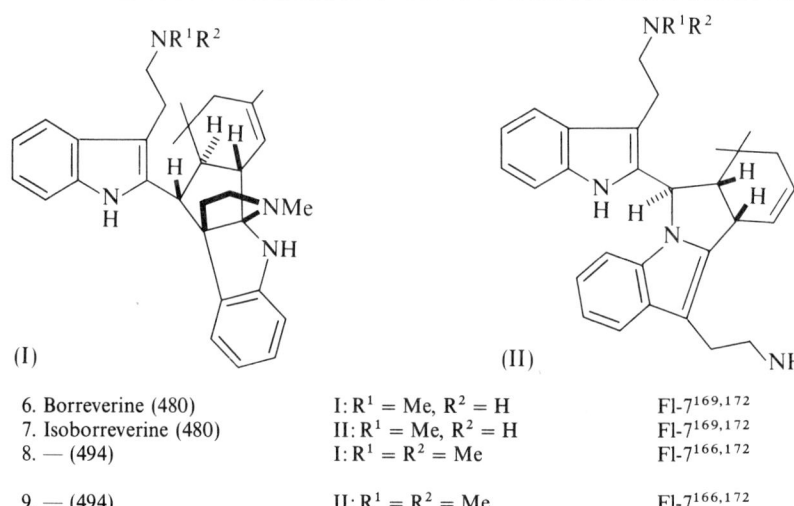

(I)   (II)

6. Borreverine (480)	I: R^1 = Me, R^2 = H	Fl-7[169,172]
7. Isoborreverine (480)	II: R^1 = Me, R^2 = H	Fl-7[169,172]
8. — (494)	I: R^1 = R^2 = Me	Fl-7[166,172]
9. — (494)	II: R^1 = R^2 = Me	Fl-7[166,172]

Table 13—*contd.*

Name	Substituents	Occurrence
10. — (498)		Fl-7[167,172]
11. — (498)		Fl-7[167,172]
12. — (508)		Fl-7[166,172]

TABLE 14

Carbazoles

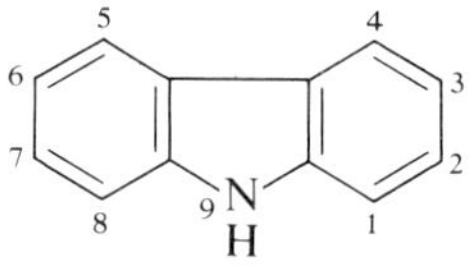

Name	Substituents	Occurrence
1. — (181)	3-Me	Cl-3[110], Cl-4[110]
2. Glycozoline (211)	3-Me, 6-OMe	Gl-5[130], Gl-6[109]
3. Murrayanine (225)	1-OMe, 3-CHO	Cl-3[110], Mu-1[109,129]
4. Mukoeic acid (241)	1-OMe, 3-COOH	Mu-1[109]
5. Mukonidine (241)	2-OH, 3-COOMe	Mu-1[39]
6. Lansine (241)	2-OH, 3-CHO, 6-OMe	Cl-5[127]
7. Glycozolidine (241)	2,6-diOMe, 3-Me	Gl-5[130], Gl-6[38,109]
8. Mukonine (255)	1-OMe, 3-COOMe	Mu-1[37]
9. Heptaphylline (279)	1-Pre[a], 2-OH, 3-CHO	Cl-3[109], Cl-5[127]
10. Atanisatine (279)	2-OH, 3-CHO, 7-Pre[a]	Cl-1[125]
11. Indizoline (293)	1-OMe, 2-Pre[a], 3-CHO	Cl-4[110]
12. Clausanitine (293)	2-OMe, 3-CHO, 8-Pre[a]	Cl-1[110]
13. Heptazoline (295)	1-Pre[a], 2,8-diOH, 3-CHO	Cl-3[109]
14. — (309)	1-Pre[a], 2-OH, 3-CHO, 6-OMe	Cl-4[110]
15. Mahanimbinol (333)	1-Ger[e], 2-OH, 3-Me	Mu-1[129]

TABLE 15

Pyranocarbazoles

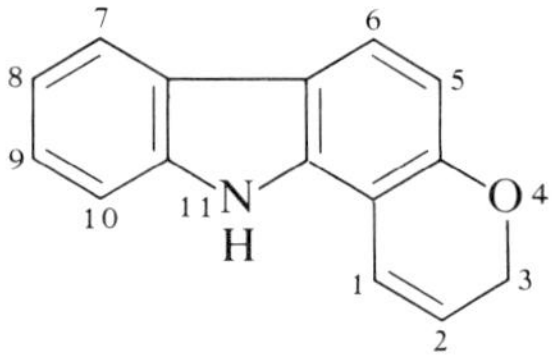

3,11-Dihydropyrano[3,2-a]carbazole

Name	Substituents	Occurrence
1. Girinimbine (263)	3,3,5-triMe	Cl-1[111], Cl-3[110], Mu-1[109]
2. Murrayacine (277)	3,3-diMe, 5-CHO	Cl-3[110], Mu-1[109]
3. Koenine (279)	3,3,5-triMe, 8-OH	Mu-1[109]
4. Koenimbine (293)	3,3,5-triMe, 8-OMe	Mu-1[109]
5. Mupamine (293)	3,3,5-triMe, 10-OMe	Cl-1[112]
6. Heptazolidine (293)	3,3,8-triMe, 5-OMe	Cl-3[110]

TABLE 15—*contd.*

Name	Substituents	Occurrence
7. Koenigine (309)	3,3,5-triMe, 8-OMe, 9-OH	Mi-2[29], Mu-1[109]
8. Koenigicine (koenidine, koenimbidine) (323)	3,3,5-triMe, 8,9-diOMe	Mu-1[109]
9. Mahanimbine (331)	3,5-diMe, 3-CH$_2$-Pre[a]	Mu-1[109], Mu-3[22]
10. Mahanimbicine (isomahanimbine) (331)	3,8-diMe, 3-CH$_2$-Pre[a]	Mu-1[109]
11. Murrayacinine (345)	3-Me, 3-CH$_2$-Pre[a], 5-CHO	Mu-1[110]
12. Mahanine (347)	3,5-diMe, 3-CH$_2$-Pre[a], 9-OH	Mu-1[109]
13. Mahanimboline (347)	3,5-diMe, 3-CH$_2$-Pre[d]	Mu-1[136]

TABLE 16

Penta- and hexa-cyclic carbazoles

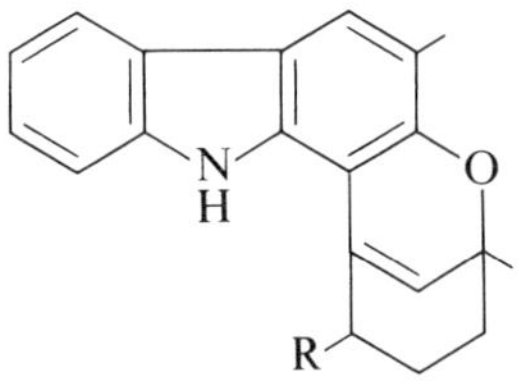

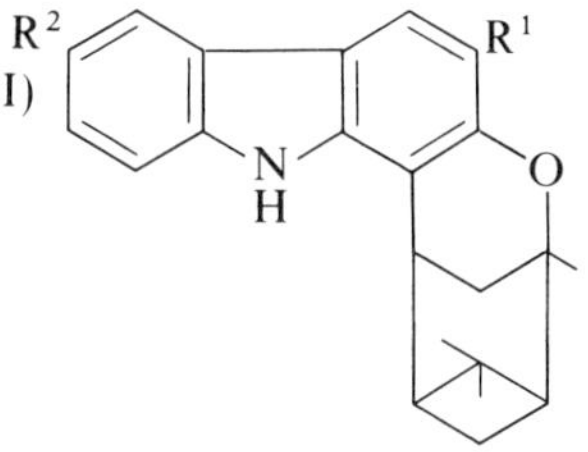

Name	Substituents	Occurrence
1. Cyclomahanimbine (329)	I: R = Ip[b]	Mu-1[109]
2. Murrayazolidine (curryanine) (331)	II: R = Ip[b]	Mu-1[109]
3. Bicyclomahanimbine (331)	III: R^1 = Me, R^2 = H	Mu-1[109]
4. Bicyclomahanimbicine (331)	III: R^1 = H, R^2 = Me	Mu-1[109]
5. Mahanimbidine (curryangine, murrayazoline) (331)	IV	Mu-1[109], Mu-3[22]
6. Exozoline (333)	II: R = Ip[a]	Mu-3[67a]
7. Murrayazolinine (349)	II: R = Ip[c]	Mu-1[110]

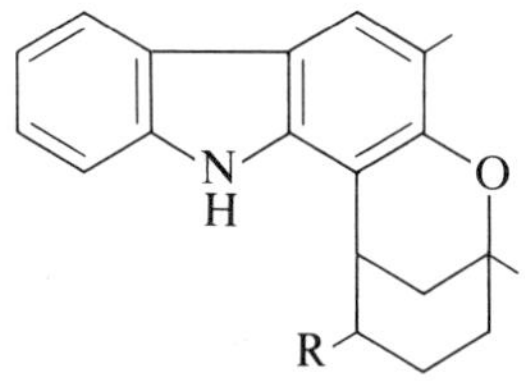

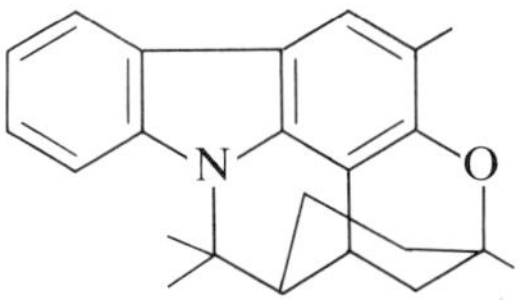

TABLE 17

β-Carbolines

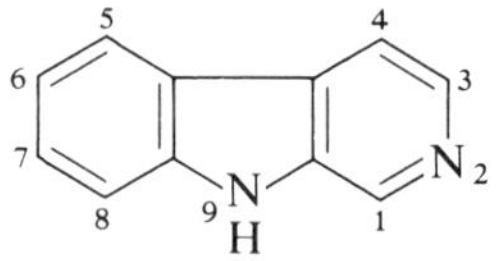

9H-Pyrido[3,4-b]indole

Name	Substituents	Occurrence
1. Harmalan (184)	1-Me, 3,4-DH	Fl-9[126]
2. Hannine (196)*	1-Et	Ha-3[16a]
3. Kumujian C (196)	1-CHO	Pi-4[194]
4. — (198)	1-CH_2OH	Pi-1[99]
5. — (210)	1-Ac	Ai-4[91]
6. — (211)	1-$CONH_2$	Ai-4[91]
7. — (212)	1-CHO, 4-OH	Pi-4[107]
8. — (216)	2-Me, 6-OMe, 1,2,3,4-TH	Du-2[18]
9. — (224)	1-CH=CH_2, 4-OMe	Ai-4[91], Pi-3[90]
10. — (224)	1-CH=CH_2, 9-OMe	Pi-2[185,186]
11. — (226)	1-COOMe	Ae-3[142], Ai-4[91], Pi-1[99], Pi-4[194]
12. Crenatine (226)	1-Et, 4-OMe	Ae-3[142], Ai-4[91]
13. Kumujian A (240)	1-COOEt	Pi-4[194]
14. — (240)	1-Ac, 4-OMe	Ai-4[91]
15. Borrerine (240)	1-CH=CMe_2, 2-Me, 1,2,3,4-TH	Fl-7[172]
16. Isoborrerine (240)	1-CH_2C(=CH_2)Me, 2-Me, 1,2,3,4-TH	Fl-7[168]
17. — (242)	1-CH_2CH_2OH, 4-OMe	Ai-1[122]
18. — (242)	1-COOMe, 4-OH	Pi-4[107]
19. — (254)	1-CH=CH_2, 4,7-diOMe	Pe-3[27]
20. Kumujian G (254)	1-CH=CH_2, 4,8-diOMe	Ai-4[91], Pi-4[194]
21. — (256)	1-COOMe, 4-OMe	Ai-1[182]
22. Crenatidine (256)	1-Et, 4,8-diOMe	Ae-3[142], Ai-4[91]
23. — (258)	1-CH(OH)CH_2OH, 4-OMe	Ai-1[122]
24. Rhetsinine (367)	1 = O, 2-(2′-NHMe-C_6H_4), 1,2,3,4-TH	Ar-2[110], Eu-8[109], Za-39[109], Za-45[109]

* Tentative identification.

TABLE 18

Canthin-6-ones and canthin-2,6-diones

(I)

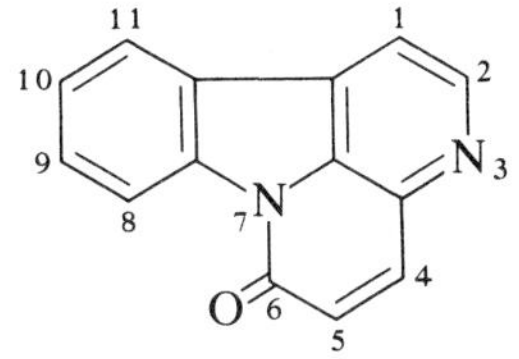

6H-Indolo[3,2,1-de][1,5]-
naphthyridin-6-one

(II)

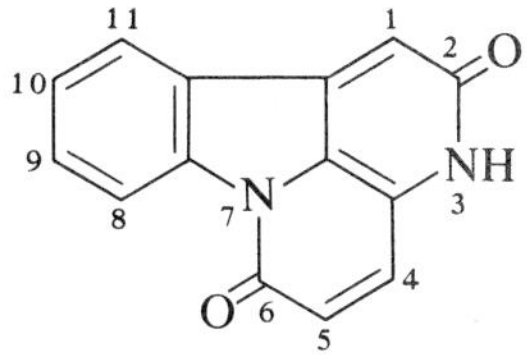

6H-Indolo[3,2,1-de][1,5]-
naphthyridin-2,6(3H)-dione

Name	Substituents	Occurrence
1. Canthin-6-one (220)	I: —	Ae-3[105], Ai-1[122,123,161,182], Ai-2[47], Fa-12[110], Fa-24[109], Fa-26[121], Pe-2[109], Pi-2[185,186], Pi-4[194], So-1[183], Za-8[110], Za-20[110], Za-21[109], Za-22[64,109], Za-24[110], Za-37[178], Za-38[110], Za-49[176], Za-50[109]
2. — (236)	I: 3→O	Ai-1[122,123]
3. — (236)	I: 1-OH	Ai-3[97]
4. — (236)	I: 5-OH	Si-3[103]
5. — (236)	I: 8-OH	Ai-2[47]
6. — (236)	I: 10-OH	Si-2[46]
7. Amarorine (236)	I: 11-OH	Am-1[45]
8. Canthin-2,6-dione (236)	II: —	Si-2[46]
9. — (250)	I: 1-OMe	Ai-1[122,123,182], Ai-2[47], So-1[183]
10. — (250)	I: 5-OMe	Ai-2[47], Am-1[45], Pe-2[109], Pi-2[186], Za-12[109], Za-22[109]
11. — (250)	I: 9-OMe	Si-1[68]
12. — (250)	I: 10-OMe	Si-2[46]
13. Amaroridine (250)	I: 11-OMe	Ai-1[122], Am-1[45]
14. — (266)	I: 4-SMe	Pe-2[109]
15. — (266)	I: 1-OMe, 3→O	Ai-1[122]
16. — (266)	I: 1-OMe, 11-OH	So-1[183]
17. Nigakinone (266)	I: 4-OMe, 5-OH	Pi-1[98,99], Pi-2[185,186], Pi-4[107,194]
18. — (266)	II: 3-OMe	Si-1[68], Si-2[46]
19. Indacanthinone (266)	II: 5-OMe	Sa-1[87]
20. — (280)	I: 4,5-diOMe	Pi-1[81,98,99], Pi-4[194]
21. — (282)	II: 3-OMe, 10-OH	Si-2[46]
22. — (398)	I: 11-O-glucose	So-2[46]

TABLE 19

2-Phenylethylamines and styrylamines

Name	Substituents	Occurrence
1. Tyramine (137)	I: 4'-OH	Ci-1[109]
2. — (151)	I: NMe, 4'-OH	Za-32[26]
3. Hordenine (165)	I: NMe$_2$, 4'-OH	Du-2[18], Ge-1[5], Za-15[160], Za-16[160], Za-32[26], Za-43[26]
4. (−)Synephrine (167)	I: NMe, 2,4'-diOH	Ci-6[109], Ci-7[109], Ci-8[109], Za-16[160], Za-23[156]
5. Noradrenaline (169)	I: 2,3',4'-triOH	Ci-1[109]
6. — (179)	I: NMe$_2$, 4'-OMe	Te-5[109]
7. Candicine (cat. 180)	I: NMe$_3$, 4'-OH	Fa-4[109], Fa-5[109], Fa-6[109], Fa-8[109], Fa-15[109], Fa-17[109], Fa-18[109], Fa-19[109], Fa-20[109], Gl-3[109], Ph-2[109], Te-6[66], Za-3[110], Za-7[110], Za-9[177], Za-13[110], Za-16[160], Za-22[64,109] Za-28[189], Za-29[189], Za-31[109], Za-46[189], Za-48[71]
8. — (195)	I: NMe$_2$, 3'-OMe, 4'-OH	Ge-1[5]
9. Coryneine (cat. 196)	I: NMe$_3$, 3',4'-diOH	Fa-8[109]
10. Rubesamide (235)	I: NCO-cyclopropyl, 3',4'-MD	Fa-20[50a]
11. — (241)	I: NCOPh, 4'-OH	Ca-1[109]
12. Alatamide (253)	II: NCOPh, 4'-OMe	Pl-2[42]
13. — (255)	I: NCOPh, 4'-OMe	My-2[78], Pl-2[42], Za-26[85]
14. Lansamide A (263)	II: NMe, NCOCH=CHPh	Cl-5[127]
15. Tembamide (271)	I: NCOPh, 2-OH, 4'-OMe	Ae-1[110], Cl-2[109], Fa-8[109], Za-14[109], Za-26[85], Za-36[109], Za-51[18a]
16. — (284)	II: N-nicotinyl, 2',4'-diOMe	Am-5[33]
17. — (285)	I: NCOPh, 2,4'-diOMe	Za-14[110]
18. Herclavine (295)	I: NMe, NCOCH=CHPh, 4'-OMe	Za-13[109]
19. Aegeline (297)	I: NCOCH=CHPh, 2-OH, 4'-OMe	Ae-1[41,109,146], Za-15[160], Za-36[109]
20. Balsamide (301)	I: NCOPh, 2-OH, 3',4'-diOMe	Am-3[34]
21. — (311)	I: NCOCH=CHPh, 2,4'-diOMe	Ae-1[106]
22. — (324)	I: N-nicotinyl, 3'-CH=CHCMe$_2$-O-4'	Am-5[33]

TABLE 19—*contd.*

Name	Substituents	Occurrence
23. — (325)	I: NCOCH=CHPh, 2-OEt, 4'-OMe	Ae-1[106]
24. — (325)	I: NCOCH=CH-(3″,4″-MD-C_6H_3), NMe, 4'-OH	Za-5[72]
25. — (339)	I: NCOCH=CH(3″,4″-MD-C_6H_3), NMe, 4'-OMe	Za-5[72], Za-48[72]
26. Marmeline (351)	I: NCOCH=CHPh, 2-OH, 4'-OPre[a]	Ae-1[146]
27. — (355)	I: NCOCH=CH-(3″,4″-diOMe-C_6H_3), NMe, 4'-OMe	Za-5[72]
28. — (365)	I: NCOCH=CHPh, 2-OMe, 4'-OPre[a]	Ae-1[106]
29. — (369)	I: NCOCH=CH-(3″,4″-MD-C_6H_3), NMe, 3',4'-diOMe	Za-5[72]
30. — (373)	I: NCOCH$_2$CH$_2$-(3″,4″-diOMe-C_6H_3), 3',4'-diOMe	Pl-2[101a]
31. — (393)	I: NCOCH=CH-(3″,4″-MD-C_6H_3), NMe, 4'-OPre[a]	Za-5[72]
32. — (435)	I: NCOPh, 4'-OGer[c]	Sw-1[109]
33. — (647)	I: NCOPh, 4'-OGer[d]	At-2[60], He-3[60], Se-1[109]

| 34. — (294) | 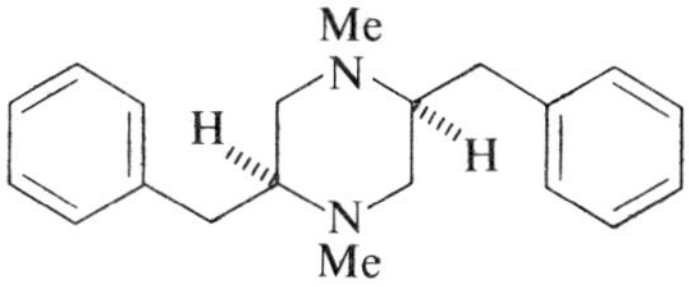 | Za-4[73] |

| 35. Alfileramine (462) | 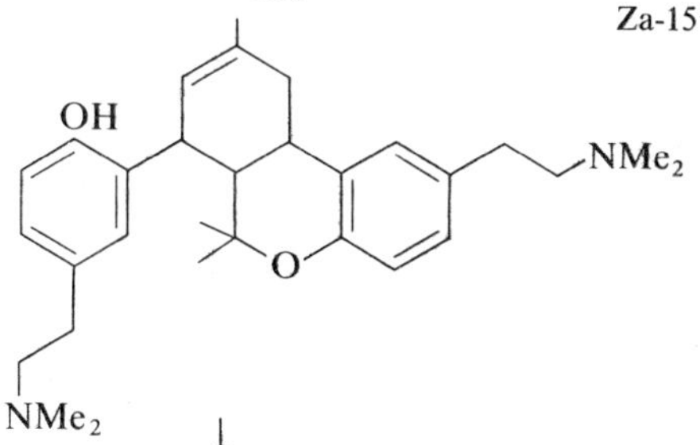 | Za-15[160], Za-44[35] |

| 36. Culantraramine (490) | 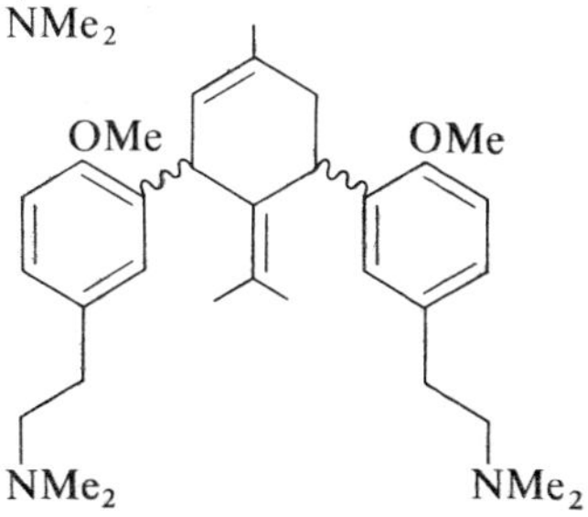 | Za-16[160] |

II: NCOCH=CHPh, NMe

TABLE 20

5-Phenyloxazoles

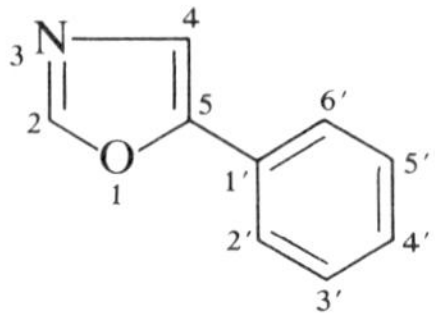

Name	Substituents	Occurrence
1. Halfordinol (aegelinine) (238)	2-(3-pyridyl), 4'-OH	Ae-1[109], Ha-2[109]
2. — (252)	2-(3-pyridyl), 4'-OMe	Ae-1[147], Mi-2[29]
3. Balsoxin (281)	2-Ph, 3',4'-diOMe	Am-3[34]
4. — (306)	2-(3-pyridyl), 4'-OPre[a]	Ae-1[106], Ae-2[109], Am-5[33]
5. Halfordinone (322)	2-(3-pyridyl), 4'-OPre[f]	Ha-1[109], Ha-2[109]
6. Halfordine (340)	2-(3-pyridyl), 4'-OPre[c]	Ha-1[109], Ha-2[109]
7. — (cat. 355)	2-[3-(1-methylpyridiniumyl)],4'-OPre[c]	Ha-1[109], Ha-2[109]

TABLE 21

Isoquinolines

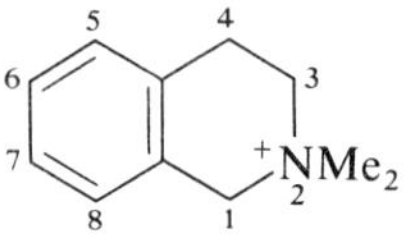

2,2-Dimethyl-1,2,3,4-tetra-
hydroisoquinoline

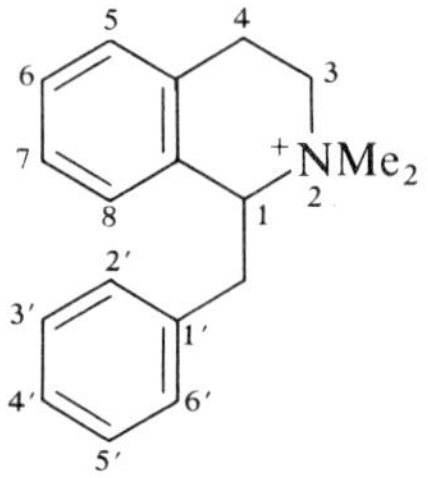

1-Benzyl-2,2-dimethyl-1,2,3,4-
tetrahydroisoquinoline

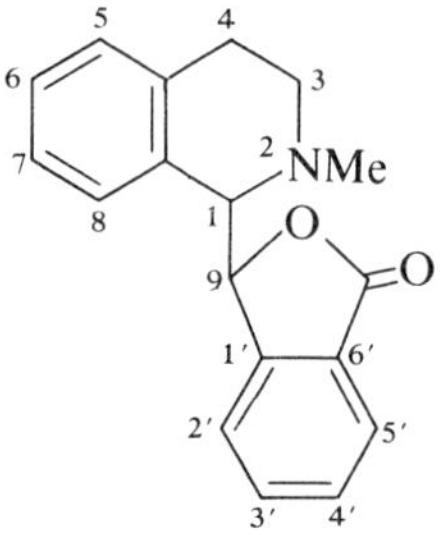

Phthalideisoquinoline

Name	Substituents	Occurrence
1. — (cat. 222)	I: 6,7-diOMe	Za-6[110]
2. (+)Armepavine metho salt (cat. 328)	II: 6,7-diOMe, 4'-OH	Fa-2[109]
3. Zanoxyline (cat. 342)	II: 4',6,7-triOMe	Za-39[174]

TABLE 21—*contd.*

Name	Substituents	Occurrence
4. Tembetarine (cat. 344)	II: 3′,6-diOMe, 4′,7-diOH	Fa-4[109], Fa-5[109], Fa-8[109], Fa-9[109], Fa-13[109], Fa-14[109], Fa-15[109], Fa-17[109], Fa-18[109], Fa-20[109], Za-3[110], Za-7[110], Za-9[177], Za-13[110], Za-16[160], Za-20[110], Za-22[64], Za-24[110], Za-28[189], Za-29[189], Za-31[109], Za-34[110], Za-46[189], Za-52[54]
5. Narcotine (413)	III: 6,7-MD, 4′,5′,8-triOMe	Ci-1[109], Ci-8[109]

TABLE 22

Aporphines

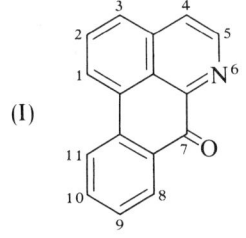

(I)

7*H*-Dibenzo[*de,g*]quinolin-7-one

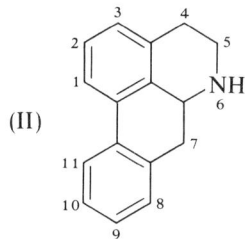

(II)

5,6,6a,7-Tetrahydro-4*H*-dibenzo[*de,g*]quinoline

Name	Substituents	Occurrence
1. Liriodenine (275)	I: 1,2-MD	Za-17[84]
2. — (307)	II: 1,2-MD, 6-Ac	Za-11[134], Za-48[71]
3. Zanthoxyphylline (cat. 340)	II: 1,2,11-triOMe, 6,6-diMe	Za-39[173]
4. Corydine (341)	II: 1-OH, 2,10,11-triOMe, 6-Me	Za-39[173]
5. Laurifoline (cat. 342)	II: 1,9-diOH, 2,10-diOMe, 6,6-diMe	Fa-5[109], Fa-8[109], Fa-17[109], Za-1[109], Za-3[110], Za-13[110], Za-22[109], Za-23[110], Za-36[109], Za-55[156]
6. Magnoflorine (cat. 342)	II: 1,11-diOH, 2,10-diOMe, 6,6-diMe	Fa-5[109], Fa-6[109], Fa-8[109], Fa-9[109], Fa-12[110], Fa-14[109],

TABLE 22—contd.

Name	Substituents	Occurrence
		Fa-17[109], Fa-18[109], Fa-19[109], Fa-20[109], Fa-24[109], Ph-2[109], Ph-4[110], Za-1[109], Za-2[109], Za-3[110], Za-7[110], Za-9[177], Za-13[110], Za-16[160], Za-22[64], Za-23[110], Za-28[189], Za-29[189], Za-32[26], Za-33[156], Za-34[110], Za-40[109], Za-41[109], Za-44[157], Za-46[189], Za-48[71], Za-55[156]
7. Thalphenine (cat. 352)	II: 2-OMe, 1-OCH$_2$-11, 9,10-MD, 6,6-diMe	Ph-4[110]
8. Xanthoplanine (cat. 356)	II: 1,2,10-triOMe, 6,6-diMe, 9-OH	Fa-8[109], Fa-15[109], Za-2[109]
9. — (cat. 356)	II: 1,9,10-triOMe, 6,6-diMe, 2-OH	Fa-18[109], Fa-23[109]
10. — (cat. 356)	II: 1-OH, 2,10,11-triOMe, 6,6-diMe	Fa-15[109]
11. — (cat. 356)	II: 1,2,10-triOMe, 6,6-diMe 11-OH	Fa-6[109], Fa-7[44], Fa-15[109], Fa-17[109], Fa-18[109], Fa-19[109], Za-10[109], Za-12[109], Za-15[160], Za-16[160], Za-28[189], Za-29[189], Za-44[157], Za-46[189], Za-53[109]

TABLE 23

Protoberberines

(I)

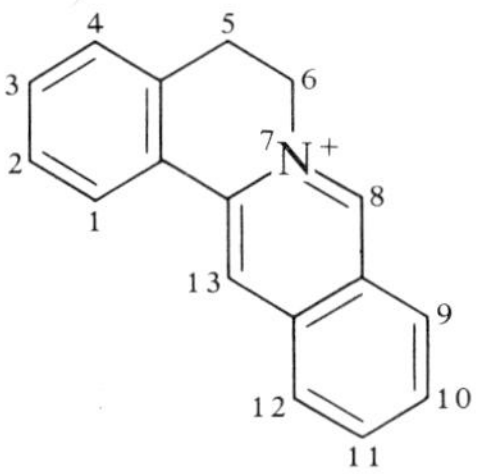

5,6-Dihydrodibenzo[a,g]quinolizinium

(II)

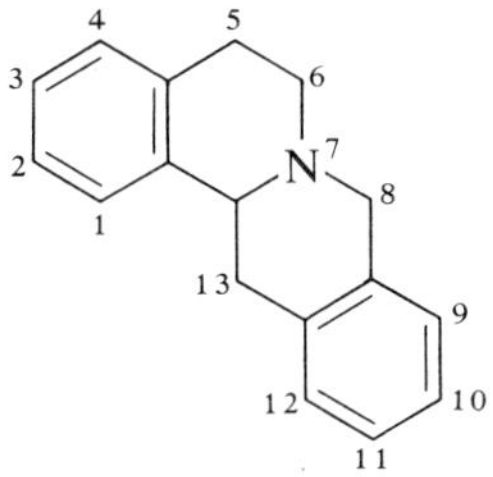

5,6,7,8,13,13a-Hexahydrodibenzo[a,g]-
quinolizine

Name	Substituents	Occurrence
1. Thalifendine (cat. 322)	I: 2,3-MD, 9-OMe, 10-OH	Ph-4[110]
2. Berberine (cat. 336)	I: 2,3-MD, 9,10-diOMe	Eu-7[109], Fa-6[109], Fa-26[121], Ph-2[109], Ph-3[109], Ph-4[110], Za-13[156,157]
3. Columbamine (cat. 338)	I: 2-OH, 3,9,10-triOMe	Ph-4[110]
4. Jatrorrhizine (cat. 338)	I: 3-OH, 2,9,10-triOMe	Ph-2[109], Ph-4[110]
5. α-Canadine (339)	II: 2,3-MD, 9,10-diOMe	Ph-2[109]
6. Phellodendrine (cat. 342)	II: 2,11-diOH, 3,10-diOMe, 7-Me	Ph-2[109], Ph-4[110]
7. Palmatine (cat. 352)	I: 2,3,9,10-tetraOMe	Fa-6[109], Ph-2[109], Ph-4[110]
8. (−)N-Methyl-α-canadine (cat. 354)	II: 2,3-MD, 9,10-diOMe, 7-Me	Fa-18[109], Za-10[109], Za-15[160], Za-31[109], Za-36[109], Za-40[109], Za-42[109], Za-53[109]
9. — (cat. 370)	II: 2,3,9,10-tetraOMe, 7-Me	Fa-3[109]

TABLE 24

Protopines

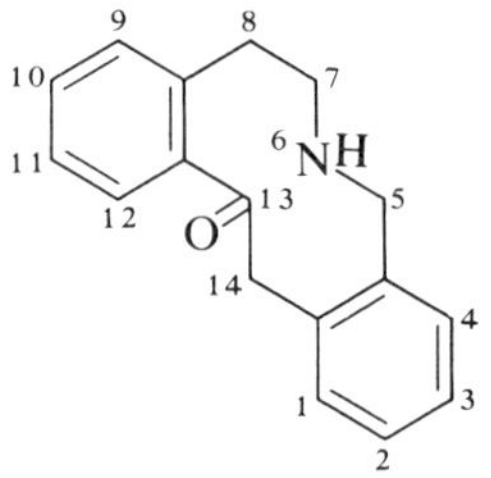

5,6,7,8,13,14-Hexahydrodibenz[c,g]azecin-13-one

Name	Substituents	Occurrence
1. — (353)	2,3:10,11-diMD, 6-Me	Fa-25[109], Za-14[109]
2. Protopine (353)	3,4:10,11-diMD, 6-Me	Za-42[109]
3. Fagarine II (369)	2,3-diOMe, 10,11-MD, 6-Me	Fa-6[109], Za-37[179], Za-49[176]
4. α-Allocryptopine (369)	3,4-diOMe, 10,11-MD, 6-Me	Fa-6[109], Fa-18[109], Za-10[109], Za-14[109], Za-35[80], Za-37[179], Za-49[176], Za-53[109]

TABLE 25

Benzo[c]phenanthridines

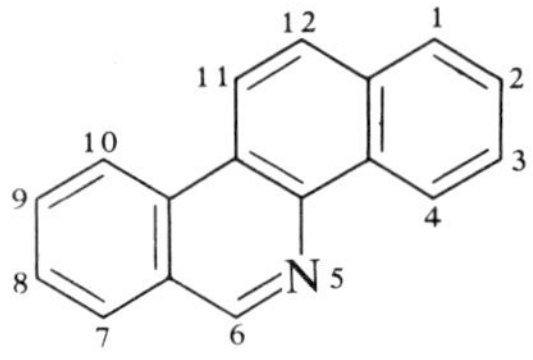

Name	Substituents	Occurrence
1. — (317)	2,3:8,9-diMD	Za-17[84]
2. Isodecarine (319)	2,3-MD, 7-OH, 8-OMe	Za-37[179], Za-49[176]
3. Sanguinarine (cat. 332)	2,3:7,8-diMD, 5-Me	Za-14[110]
4. Avicine (cat. 332)	2,3:8,9-diMD, 5-Me	Fa-2[109], Za-7[109]
5. — (333)	2,3-MD, 7,8-diOMe	To-1[109], Za-6[83], Za-11[134], Za-17[84], Za-35[187], Za-38[138]
6. — (333)	2,3-MD, 8,9-diOMe	Za-32[26]
7. Dihydroavicine (333)	2,3:8,9-diMD, 5-Me, 5,6-DH	To-1[148], Za-7[110]

TABLE 25—*contd.*

Name	Substituents	Occurrence
8. Decarine (cat. 334)	2,3-MD, 5-Me, 7-OMe, 8-OH	Za-6[83], Za-17[84], Za-18[110], Za-32[26], Za-54[110]
9. Fagaridine (cat. 334)	2,3-MD, 5-Me, 7-OH, 8-OMe	Fa-22[108], Fa-26[109,110]
10. Chelerythrine (cat. 348)	2,3-MD, 5-Me, 7,8-diOMe	Fa-2[109], Fa-3[24], Fa-4[44,109], Fa-5[109], Fa-6[109], Fa-8[109], Fa-9[109], Fa-10[109], Fa-12[110], Fa-13[109], Fa-14[109], Fa-17[109], Fa-19[109], Fa-20[109], Fa-21[109], Fa-22[108], Fa-24[109], Fa-25[109], Fa-26[109], To-1[109], Za-3[110], Za-6[83], Za-7[110], Za-9[177], Za-10[109], Za-13[110], Za-14[110], Za-15[160], Za-20[110], Za-22[64], Za-24[110], Za-28[110], Za-29[189], Za-31[109], Za-33[156], Za-36[109], Za-40[109], Za-42[109], Za-45[109], Za-46[189], Za-48[71], Za-52[110], Za-53[109], Za-55[156]
11. Nitidine (cat. 348)	2,3-MD, 5-Me, 8,9-diOMe	Fa-2[109], Fa-3[109], Fa-4[44,109], Fa-5[109], Fa-6[109], Fa-9[109], Fa-10[109], Fa-15[109], Fa-17[109], Fa-19[109], Fa-20[109], Fa-22[108], Fa-24[109], Fa-25[109], Za-1[109], Za-3[110], Za-7[110], Za-9[177], Za-13[110], Za-17[84], Za-19[110], Za-20[110], Za-24[109], Za-25[109], Za-26[85], Za-28[110], Za-29[189], Za-34[110], Za-35[109], Za-37[178], Za-46[189], Za-55[156]
12. Dihydrochelerythrine (349)	2,3-MD, 5-Me, 5,6-DH, 7,8-diOMe	Fa-4[44,79], Fa-7[44], Fa-10[109], Fa-21[109], Fa-26[109], To-1[109,149], Za-15[160], Za-22[64], Za-28[189], Za-29[189],

TABLE 25—*contd.*

Name	Substituents	Occurrence
		Za-46[189], Za-48[71], Za-52[110]
13. Dihydronitidine (349)	2,3-MD, 5-Me, 5,6-DH, 8,9-diOMe	To-1[148], Za-34[110], Za-35[80]
14. Fagaronine (cat. 350)	2-OH, 3,8,9-triOMe, 5-Me	Fa-26[109]
15. Oxychelerythrine (363)	2,3-MD, 5-Me, 5,6-DH, 6 = O, 7,8-diOMe	Za-6[83], Za-35[187], Za-48[71]
16. Oxynitidine (363)	2,3-MD, 5-Me, 5,6-DH, 6 = O, 8,9-diOMe	Za-17[84], Za-26[85], Za-35[109]
17. — (365)	2,3-MD, 5-Me, 5,6-DH, 6-OH, 7,8-diOMe	To-1[151]
18. — (365)	2,3-MD, 5-Me, 5,6-DH, 6-CH$_2$OH, 7-OH, 8-OMe	Za-32[26]
19. Angoline (379)	2,3-MD, 5-Me, 5,6-DH, 6,7,8-triOMe	Fa-1[109], Fa-4[109], Fa-9[109], Fa-10[109], Fa-24[109], Fa-26[109], To-1[148], Za-11[134], Za-35[80]
20. — (389)	2,3 : 7,8-diMD, 5-Me, 5,6-DH, 6-CH$_2$Ac	Fa-27[190]
21. — (391)	2,3-MD, 5-Me, 5,6-DH, 6-CH$_2$CHO, 7,8-diOMe	Za-48[72], Za-52[110]
22. — (cat. 392)	2,3-MD, 5-Me, 7,8-diOMe, 12-OEt	(?) Za-35[187]
23. — (405)	2,3-MD, 5-Me, 5,6-DH, 6-CH$_2$Ac, 7,8-diOMe	Fa-27[190], To-1[149], Xy-1[44], Za-14[110], Za-42[109], Za-52[110]
24. — (405)	2,3-MD, 5-Me, 5,6-DH, 6-CH$_2$Ac, 8,9-diOMe	Fa-27[190]
25. — (407)	2,3-MD, 5-Me, 5,6-DH, 6-CH$_2$COOH, 7,8-diOMe	Za-48[72]
26. — (419)	2,3-MD, 5-Me, 5,6-DH, 6-CH$_2$COEt, 7,8-diOMe	Fa-12[9]

| 27. Toddalidimerine (714) | R = H | To-1[149] |
| 28. — (728) | R = Me | Za-48[72] |

TABLE 26.

N-Formyl-N-methyl-1-amino-2-phenylnaphthalenes

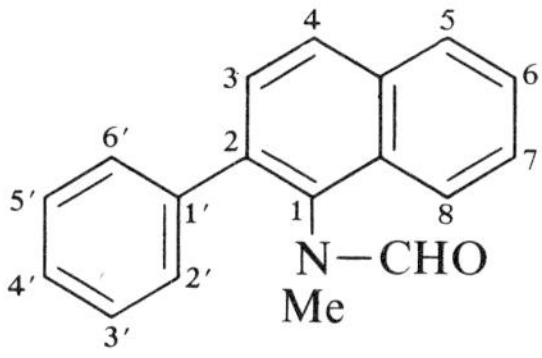

Name	Substituents	Occurrence
1. Integramide (365)	4',5': 6,7-diMD, 2'-OH	Za-27[82]
2. Iwamide (367)	6,7-MD, 2',4'-diOH, 3'-OMe	Za-6[83,86]
3. Arnottianamide (381)	6,7-MD, 2'-OH, 3',4'-diOMe	Fa-4[44], To-1[151], To-2[150], Za-6[83,110], Za-11[134], Za-17[84,110]
4. Isoarnottianamide (381)	6,7-MD, 2'-OH, 4',5'-diOMe	Za-17[84,110]

TABLE 27

Imidazoles

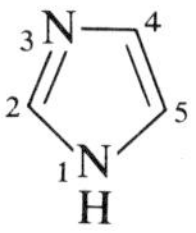

Name	Substituents	Occurrence
1. — (139)	4-CH$_2$CH$_2$NMe$_2$	Ca-1[109]
2. Zapotidine (167)	1-C(=S)N(Me)CH$_2$CH$_2$-5	Ca-1[109]
3. Pilocarpidine (194)		Pi-6[109]
4. Pilocarpine (208)	1-Me, 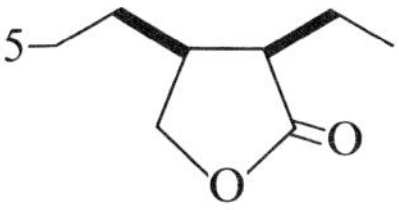 5-	Pi-5[109], Pi-6[109], Pi-7[109], Pi-9[109], Pi-10[109], Pi-11[109]
5. Isopilocarpine (208)	1-Me, 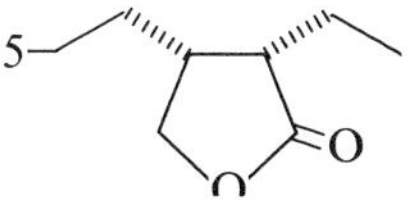 5-	Pi-6[109], Pi-7[109]

TABLE 27—*contd.*

Name	Substituents	Occurrence
6. (+)Pilosine (286)	1-Me, 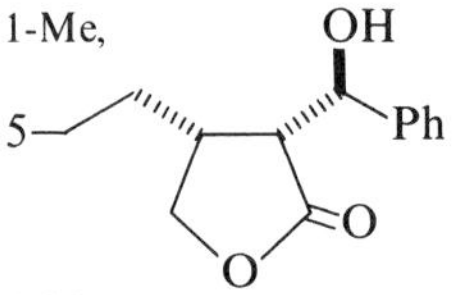	Pi-7[109]
7. (+)Isopilosine (286)	1-Me, 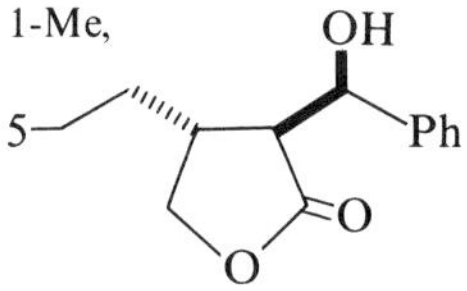	Pi-7[109]
8. (−)Epi-isopilosine (286)	1-Me, 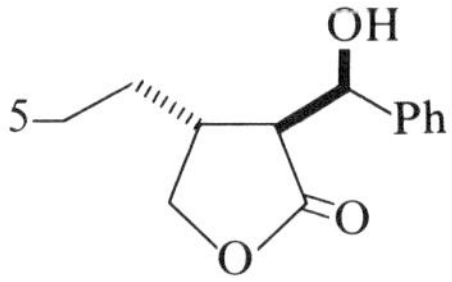	Pi-7[104]
9. Epi-isopiloturine (286)	1-Me, 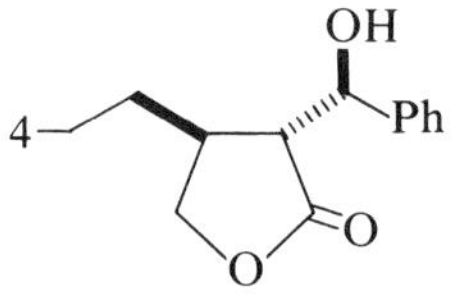	Pi-7[184]
10. Casimiroedine (417)	1-glucosyl, 4-CH$_2$CH$_2$N(Me)COCH=CHPh	Ca-1[109]

TABLE 28

Pyrrolidines

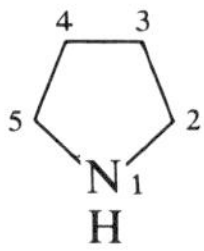

Name	Substituents	Occurrence
1. Stachydrine (143)	1,1-diMe, 2-COO$^-$	Ci-1[109], Ci-2[109], Ci-3[109], Ci-5[109]
2. Odorine (roxburghiline) (300)	1-COCH=CHPh, 2-NHCOCH(Me)CH$_2$Me	Ag-1[152], Ag-2[128]
3. Odorinol (316)	1-COCH=CHPh, 2-NHCOC(OH)(Me)CH$_2$Me	Ag-1[152], Ag-2[128]

TABLE 29

Pyridines and piperidines*

(I)

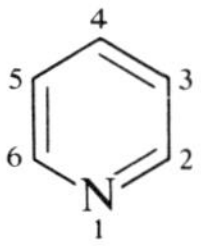

(II)

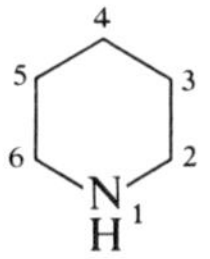

Name	Substituents	Occurrence
1. — (95)	I: 3-OH	En-1[43]
2. — (155)	II: 2,2,6,6-tetraMe, 4=O	Za-1[109]
3. Rohitukine (305)	II: 1-Me, 3-OH 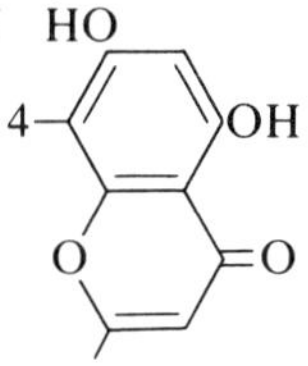	Am-2[77]

*See also **19**-16, **19**-22, **20**-1, **20**-2, **20**-4, **20**-5, **20**-6, and **20**-7.

TABLE 30

Putrescines

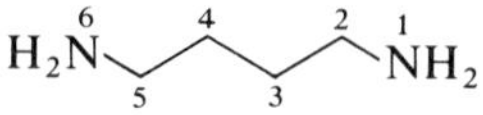

$$H_2\overset{6}{N}\underset{5}{\diagup}\overset{4}{\diagdown}\underset{3}{\diagup}\overset{2}{\diagdown}\overset{1}{NH_2}$$

Name	Substituents	Occurrence
1. Putrescine (88)	—	Ci-1[109], Ci-3[109]
2. — (264)	1-COCH=CH-(3'-OMe-4'-OH-C_6H_3)	Ci-7[109], Ci-8[109]
3. Haplamide (296)	1,6-diCOPh	Ha-12[117–119]
4. Haplamidine (322)	1-COPh, 6-COCH=CHPh	Ha-12[118,119]

TABLE 31

Acyl isobutylamines

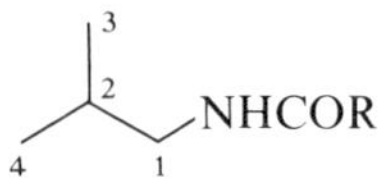

Name	Substituents	Occurrence
1. — (223)	R = $(CH=CH)_2(CH_2)_4Me$	Eu-5[110], Fa-26[28]
2. Neoherculine (247)	R = $CH=CH(CH_2)_2(CH=CH)_3Me$	Za-13[109], Za-41[109]
3. Fagaramide (247)	R = $CH=CH-(3',4'-MD-C_6H_3)$	Fa-4[109], Fa-10[70,109], Fa-22[108], Fa-26[109,163]
4. Sanshoamide (263)	R = $(CH=CH)_2(CH_2)_2(CH=CH)_2Me$, 2-OH	Za-41[109]
5. γ-Sanshoöl (273)	R = $(CH=CH)_2(CH_2)_2(CH=CH)_3Me$	Za-1[195]
6. Hydroxy γ- sanshoöl (289)	R = $(CH=CH)_2(CH_2)_2(CH=CH)_3Me$, 2-OH	Za-1[195]

TABLE 32

Miscellaneous simple amines and amides

Name	Formula	Occurrence
1. Acetamide (59)	$MeCONH_2$	To-2[150]
2. Guanidine (59)	$H_2NC(=NH)NH_2$	Ph-2[109]
3. Tiglamide (99)	$MeCH=C(Me)CONH_2$	Ag-3[89]
4. Benzamide (121)	$PhCONH_2$	Ha-5[109], My-1[78], My-2[78]
5. Phenylacetamide (135)	$PhCH_2CONH_2$	Ve-1[109]
6. — (150)	$2-MeNH-C_6H_4-CONH_2$	Eu-8[162]
7. Glycomide (211)	$PhCH_2CONHPh$	Gl-6[143]
8. Palmitamide (255)	$Me(CH_2)_{14}CONH_2$	Ca-1[109]

TABLE 33

Alkaloid bearing plants from the order Rutales and their contained alkaloids; plants belong to the Rutaceae except for those marked as belonging to the Simaroubaceae (*) or to the Meliaceae (†).

Code	Plant	Alkaloids
Ac-1	*Acronychia acidula*	**8**-22
Ac-2	*A. baueri*	**3**-1, **5**-15, **5**-17, **5**-25, **6**-4, **8**-8, **8**-18, **8**-19, **8**-20, **8**-22, **8**-23, **8**-26, **8**-28, **10**-5
Ac-3	*A. haplophylla*	**5**-24, **5**-29, **8**-14

TABLE 33—*contd.*

Code	Plant	Alkaloids
Ac-4	*A. pedunculata*	**5**-6, **5**-15
Ae-1	*Aegle marmelos*	**5**-1, **5**-7, **5**-14, **5**-17, **19**-15, **19**-19, **19**-21, **19**-23, **19**-26, **19**-28, **20**-1, **20**-2, **20**-4
Ae-2	*Aeglopsis chevalieri*	**20**-4
Ae-3	*Aeschrion crenata**	**17**-11, **17**-12, **17**-22, **18**-1
Af-1	*Afraegle paniculata*	**2**-12, **5**-1, **5**-14
Ag-1	*Aglaia odorata*†	**28**-2, **28**-3
Ag-2	*A. roxburghiana*†	**28**-2, **28**-3
Ag-3	*A. sp.*†	**32**-3
Ai-1	*Ailanthus altissima**	**17**-17, **17**-21, **17**-23, **18**-1, **18**-2, **18**-9, **18**-13, **18**-15
Ai-2	*A. excelsa**	**18**-1, **18**-5, **18**-9, **18**-10
Ai-3	*A. giraldii**	**2**-16, **18**-3
Ai-4	*A. malabarica**	**17**-5, **17**-6, **17**-9, **17**-12, **17**-14, **17**-20, **17**-22
Al-1	*Almeidea guyanensis*	**2**-13, **2**-16, **4**-1, **5**-6, **5**-17, **7**-3, **7**-4
Am-1	*Amaroria soulameoides**	**18**-7, **18**-10, **18**-13
Am-2	*Amoora rohituka*†	**29**-3
Am-3	*Amyris balsamifera*	**19**-20, **20**-3
Am-4	*A. pinnata*	**5**-1, **5**-17
Am-5	*A. plumieri*	**19**-16, **19**-22, **20**-4
Ar-1	*Araliopsis soyauxii*[a]	**4**-5, **4**-6, **5**-9, **5**-15, **5**-17, **5**-20, **7**-9
Ar-2	*A. tabouensis*[a]	**4**-5, **4**-8, **5**-17, **5**-26, **7**-8, **12**-6, **17**-24
At-1	*Atalantia ceylanica*	**9**-14, **9**-15, **10**-6, **10**-8
At-2	*A. monophylla*	**8**-31, **8**-32, **8**-33, **8**-34, **10**-4, **10**-7, **10**-9, **19**-33
At-3	*A. roxburghiana*	**7**-1
Ba-1	*Balfourodendron riedelianum*	**2**-34, **3**-3, **4**-12, **4**-17, **4**-19, **4**-25, **5**-1, **5**-15, **5**-16, **5**-17, **5**-20, **7**-10, **7**-14, **7**-15, **8**-13
Ba-2	*Bauerella simplicifolia* ssp. *neo-scotica*	**5**-15, **5**-25, **8**-6, **8**-8, **8**-15, **8**-18, **8**-22, **8**-23, **8**-25, **8**-28, **10**-5
Bo-1	*Boenninghausenia albiflora*	**8**-3, **8**-4, **8**-5, **8**-7, **9**-2, **10**-3
Bo-2	*B. albiflora* var. *japonica*	**5**-1
Bo-3	*Boronella* aff. *verticillata*	**5**-1
Bo-4	*Boronia ternata*	**3**-2, **3**-6, **5**-17
Ca-1	*Casimiroa edulis*	**2**-5, **2**-8, **2**-21, **3**-7, **3**-8, **5**-1, **5**-7, **5**-17, **19**-11, **27**-1, **27**-2, **27**-10, **32**-8

TABLE 33—*contd.*

Code	Plant	Alkaloids
Ch-1	*Chloroxylon swietenia*	**2**-4, **2**-10, **5**-7, **5**-17
Ch-2	*Choisya arizonica*	**5**-15, **5**-17, **6**-5
Ch-3	*C. mollis*	**5**-15, **5**-17, **6**-5
Ch-4	*C. ternata*	**4**-9, **4**-26, **5**-15, **5**-17, **5**-44, **6**-5
Ch-5	*Chorilaena quercifolia*	**5**-1
Ci-1	*Citrus aurantium*	**1**-1, **13**-2, **19**-1, **19**-5, **21**-5, **28**-1, **30**-1
Ci-2	*C. aurantium* ssp. *amara*	**28**-1
Ci-3	*C. grandis*	**28**-1, **30**-1
Ci-4	*C. macroptera*	**2**-21
Ci-5	*C. medica*	**28**-1
Ci-6	*C. nobilis* var *deliciosa*	**19**-4
Ci-7	*C. paradisi*	**19**-4, **30**-2
Ci-8	*C. sinensis*	**19**-4, **21**-5, **30**-2
Cl-1	*Clausena anisata*	**14**-10, **14**-12, **15**-1, **15**-5
Cl-2	*C. brevistyla*	**19**-15
Cl-3	*C. heptaphylla*	**14**-1, **14**-3, **14**-9, **14**-13, **15**-1, **15**-2, **15**-6
Cl-4	*C. indica*	**14**-1, **14**-11, **14**-14
Cl-5	*C. lansium*	**14**-6, **14**-9, **19**-14
Cu-1	*Cusparia macrocarpa*	**5**-6
De-1	*Decatropis bicolor*	**5**-1, **5**-17
Di-1	*Dictamnus albus*	**2**-25, **5**-1, **5**-2, **5**-7, **5**-17, **5**-19
Di-2	*D. angustifolius*	**1**-6, **2**-25, **4**-10, **5**-1, **5**-2, **5**-7, **5**-8, **5**-17, **5**-44
Di-3	*D. caucasicus*	**5**-1, **5**-2, **5**-4, **5**-7, **5**-8, **5**-17, **5**-19
Di-4	*D. hispanicus*	**5**-17
Di-5	*Dictyoloma incanescens*	**13**-4
Di-6	*Diphasia angolensis*	**5**-26, **8**-8, **8**-24
Di-7	*D. klaineana*	**5**-17, **8**-13, **8**-14, **8**-35
Du-1	*Dutaillyea drupacea*	**2**-21, **5**-1, **5**-5, **5**-6, **5**-15, **6**-2, **13**-4
Du-2	*D. oreophila*	**5**-1, **5**-5, **5**-6, **5**-15, **13**-4, **17**-8, **19**-3
En-1	*Entandrophragma cylindricum*†	**29**-1
Er-1	*Eriostemon brucei*	**5**-16
Er-2	*E. coccineus*	**5**-16, **5**-17
Er-3	*E. difformis*	**5**-16, **5**-17
Er-4	*E. thryptomenoides*	**5**-16, **5**-17
Er-5	*E. tomentellus*	**5**-16, **5**-17
Er-6	*Erythrochiton brasiliensis*	**5**-7

TABLE 33—*cont.*

Code	Plant	Alkaloids
Es-1	*Esenbeckia febrifuga*	5-9, 5-17, 5-20
Es-2	*E. flava*	2-28, 5-1, 5-6, 5-16, 5-17, 5-20
Es-3	*E. hartmanii*	5-16, 5-17
Es-4	*E. litoralis*	5-1, 5-6, 5-9, 5-15, 8-7
Eu-1	*Euxylophora paraensis*	2-18, 4-3, 5-17, 7-3, 7-11, 7-31, 12-5, 12-10, 12-12, 12-13, 12-14, 12-16, 12-17, 12-18, 12-19, 12-20, 12-21, 12-22
Eu-2	*Euodia alata*	5-15, 5-17, 5-42, 5-44, 8-13, 8-14, 8-15, 8-23, 8-29
Eu-3	*E. belahe*	5-1, 5-6, 5-15
Eu-4	*E. elleryana*	5-3, 5-17, 5-32
Eu-5	*E. hupehensis*	31-1
Eu-6	*E. litoralis*	5-1, 5-6, 5-15
Eu-7	*E. meliaefolia*	23-2
Eu-8	*E. rutaecarpa*	3-13, 3-20, 3-25, 3-26, 3-28, 12-1, 12-2, 12-4, 12-6, 12-8, 12-15, 13-4, 17-24, 32-6
Eu-9	*E. triphylla*	8-8, 8-13
Eu-10	*E. xanthoxyloides*	5-15, 5-27, 5-37, 5-38, 5-39, 5-44, 8-9, 8-11, 8-13, 8-14, 8-18, 8-23
Fa-1	*Fagara angolensis*[b]	5-17, 25-19
Fa-2	*F. beniensis*[b]	2-3, 21-2, 25-4, 25-10, 25-11
Fa-3	*F. capensis*[c]	5-17, 23-9, 25-10, 25-11
Fa-4	*F. chalybea*[c]	5-17, 7-3, 19-7, 21-4, 25-10, 25-11, 25-12, 25-19, 26-3, 31-3
Fa-5	*F. chiloperone* var. *angustifolia*[c]	19-7, 21-4, 22-5, 22-6, 25-10, 25-11
Fa-6	*F. coco*[c]	5-7, 5-17, 19-7, 22-6, 22-11, 23-2, 23-7, 24-3, 24-4, 25-10, 25-11
Fa-7	*F. holstii*[d]	22-11, 25-12
Fa-8	*F. hyemalis*[c]	19-7, 19-9, 19-15, 21-4, 22-5, 22-6, 22-8, 25-10
Fa-9	*F. leprieurii*[c]	5-17, 8-7, 8-14, 21-4, 22-6, 25-10, 25-11, 25-19
Fa-10	*F. macrophylla*[e]	5-17, 8-14, 25-10, 25-11, 25-12, 25-19, 31-3
Fa-11	*F. mantsurica*[c]	5-17
Fa-12	*F. mayu*[c]	2-21, 5-1, 5-7, 5-17, 7-9, 18-1, 22-6, 25-10, 25-26
Fa-13	*F. naranjillo*[c]	21-4, 25-10
Fa-14	*F. naranjillo* var. *paraguay-ensis*[c]	21-4, 22-6, 25-10
Fa-15	*F. nigrescens*[f]	19-7, 21-4, 22-8, 22-10, 22-11, 25-11
Fa-16	*F. okinawensis*[c]	5-17
Fa-17	*F. pterota*[g]	19-7, 21-4, 22-5, 22-6, 22-11, 25-10, 25-11

TABLE 33—*contd.*

Code	Plant	Alkaloids
Fa-18	*F. rhoifolia*[c]	**19**-7, **21**-4, **22**-6, **22**-9, **22**-11, **23**-8, **24**-4
Fa-19	*F. rhoifolia* var.	
	petiolulatum[c]	**19**-7, **22**-6, **22**-11, **25**-10, **25**-11
Fa-20	*F. rubescens*[c]	**5**-17, **8**-7, **8**-14, **19**-7, **19**-10, **21**-4, **22**-6, **25**-10, **25**-11
Fa-21	*F. semiarticulata*[c]	**25**-10, **25**-12
Fa-22	*F. tessmanii*[c]	**25**-9, **25**-10, **25**-11, **31**-3
Fa-23	*F. tingoassuiba*[c]	**22**-9
Fa-24	*F. viridis*[c]	**5**-17, **18**-1, **22**-6, **25**-10, **25**-11, **25**-19
Fa-25	*F. vitiensis*[c]	**24**-1, **25**-10, **25**-11
Fa-26	*F. xanthoxyloides*[c]	**2**-12, **5**-17, **18**-1, **23**-2, **25**-9, **25**-10, **25**-12, **25**-14, **25**-19, **31**-1, **31**-3
Fa-27	*Fagaropsis angolensis*	**25**-20, **25**-23, **25**-24
Fl-1	*Flindersia acuminata*	**5**-1, **5**-9
Fl-2	*F. australis*	**7**-1
Fl-3	*F. bennettiana*	**5**-9, **5**-17, **5**-20
Fl-4	*F. bourjotiana*	**5**-17, **5**-20
Fl-5	*F. collina*	**5**-15, **5**-20
Fl-6	*F. dissosperma*	**5**-9, **5**-17, **5**-20
Fl-7	*F. fournieri*	**3**-3, **4**-8, **5**-1, **5**-7, **5**-17, **13**-6, **13**-7, **13**-8, **13**-9, **13**-10, **13**-11, **13**-12, **17**-15, **17**-16
Fl-8	*F. ifflaiana*	**4**-2, **4**-3
Fl-9	*F. laevicarpa*	**5**-17, **17**-1
Fl-10	*F. maculosa*	**5**-1, **5**-9, **5**-15, **5**-16, **5**-17, **5**-20, **5**-40
Fl-11	*F. pimenteliana*	**5**-1, **5**-17
Fl-12	*F. pubescens*	**5**-1, **5**-15, **5**-16, **5**-17, **5**-20
Fl-13	*F. schottiana*	**5**-9, **5**-15
Fl-14	*F. xanthoxyla*	**5**-9, **5**-20
Ga-1	*Galipea officinalis*	**1**-1, **1**-2, **1**-3, **1**-4, **1**-8, **1**-9, **1**-11, **2**-1, **3**-19
Ge-1	*Geijera balansae*	**5**-17, **7**-7, **7**-19, **19**-3, **19**-8
Ge-2	*G. parviflora*	**7**-1
Ge-3	*G. salicifolia*	**4**-4, **4**-21, **5**-7, **5**-17
Ge-4	*Geleznowia verrucosa*	**5**-17
Gl-1	*Glycosmis arborea*	**5**-7, **5**-17, **8**-14, **11**-1, **11**-2, **11**-3, **11**-5
Gl-2	*G. bilocularis*	**5**-1, **5**-15, **5**-17, **8**-14, **8**-21, **11**-1, **11**-5
Gl-3	*G. cochinchinensis*	**19**-7
Gl-4	*G. cyanocarpa*	**5**-6, **5**-18
Gl-5	*G. mauritiana*	**2**-19, **5**-1, **5**-17, **8**-14, **8**-27, **14**-2, **14**-7
Gl-6	*G. pentaphylla*	**2**-17, **5**-1, **5**-7, **5**-15, **5**-17, **10**-2, **10**-3, **11**-3, **11**-4, **11**-5, **14**-2, **14**-7, **32**-7

TABLE 33—*contd.*

Code	Plant	Alkaloids
Ha-1	*Halfordia kendack*	**5**-1, **5**-15, **20**-5, **20**-6, **20**-7
Ha-2	*H. scleroxyla*	**2**-9, **5**-26, **20**-1, **20**-5, **20**-6, **20**-7
Ha-3	*Hannoa klaineana**	**17**-2
Ha-4	*Haplophyllum acutifolium*	**3**-4, **5**-17
Ha-5	*H. bucharicum*	**2**-6, **2**-37, **2**-38, **4**-31, **5**-1, **5**-4, **5**-7, **5**-14, **5**-17, **7**-8, **7**-18, **32**-4
Ha-6	*H. bungei*	**2**-5, **5**-1, **5**-17
Ha-7	*H. dauricum*	**5**-7, **5**-17
Ha-8	*H. dubium*	**1**-6, **2**-6, **2**-27, **3**-11, **4**-10, **4**-29, **5**-4, **5**-14, **5**-17, **5**-44
Ha-9	*H. foliosum*	**2**-5, **2**-6, **2**-7, **2**-27, **3**-3, **3**-11, **3**-12, **4**-10, **4**-13, **5**-14, **5**-17, **5**-28, **6**-3, **7**-2, **7**-8
Ha-10	*H. hispanicum*	**5**-44, **5**-46
Ha-11	*H. kowalenskyi*	**5**-7, **5**-17
Ha-12	*H. latifolium*	**1**-6, **5**-14, **5**-17, **5**-30, **5**-36, **5**-44, **5**-47, **30**-3, **30**-4
Ha-13	*H. obtusifolium*	**5**-17, **5**-44
Ha-14	*H. pedicellatum*	**5**-4, **5**-7, **5**-14, **5**-17
Ha-15	*H. perforatum*	**2**-7, **2**-27, **3**-3, **3**-11, **4**-4, **4**-10, **5**-1, **5**-14, **5**-17, **5**-23, **5**-30, **5**-31, **5**-33, **5**-38, **5**-39, **5**-41, **5**-44, **5**-45, **5**-47, **5**-48, **5**-49, **5**-50, **5**-51, **6**-6, **7**-1, **7**-5
Ha-16	*H. popovii*	**5**-17, **5**-44
Ha-17	*H. ramossisimum*	**5**-1, **5**-17, **5**-44
Ha-18	*H. robustum*	**2**-5, **5**-1, **5**-4, **5**-7, **5**-14, **5**-17
Ha-19	*H. schelkovnikovii*	**5**-7, **5**-17
Ha-20	*H. suaveolens*	**5**-1, **5**-15, **5**-17, **5**-44, **7**-2
Ha-21	*H. tenue*	**5**-7, **5**-17
Ha-22	*H. tuberculatum*	**2**-22, **5**-7, **5**-17, **5**-44, **7**-1
Ha-23	*H. villosum*	**5**-7, **5**-17
He-1	*Helietta longifoliata*	**5**-1, **5**-2, **5**-5, **5**-9, **5**-15, **5**-17, **5**-20
He-2	*H. parvifolia*	**5**-13, **5**-15, **5**-17, **5**-20, **5**-21
He-3	*Hesperethusa crenulata*	**2**-3, **19**-33
Ho-1	*Hortia arborea*	**5**-1, **5**-7, **5**-17, **12**-1, **12**-7, **12**-11
Ho-2	*H. badinii*	**12**-1, **12**-7
Ho-3	*H. braziliana*	**12**-11
Ho-4	*H. longifolia*	**2**-2, **2**-3, **5**-1, **5**-17, **12**-7
Lu-1	*Lunasia amara*	**1**-5, **1**-7, **2**-26, **2**-35, **2**-44, **3**-3, **3**-8, **3**-17, **3**-18, **4**-7, **4**-15, **4**-18, **4**-24, **5**-10, **5**-17, **7**-16
Lu-2	*L. amara* var. *repanda*	**2**-26, **2**-31, **2**-33, **2**-35, **2**-44, **4**-7
Lu-3	*L. costulata*	**2**-26, **4**-7, **4**-16

TABLE 33—*contd.*

Code	Plant	Alkaloids
Lu-4	*L. quercifolia*	3-5, 3-8, 4-7, 4-15, 4-16
Me-1	*Medicosma cunninghamii*	5-1, 5-5, 6-1
Me-2	*Melicope confusa*	5-3, 5-15, 5-17
Me-3	*M. fareana*	5-17, 5-25, 8-22, 8-23, 8-28
Me-4	*M. indica*	2-16, 5-1, 5-6
Me-5	*M. lasioneura*	5-3, 5-6, 5-14, 5-15, 5-17, 5-22, 5-30
Me-6	*M. leratii*	5-17, 8-15, 8-18, 8-23
Me-7	*M. perspicuinervia*	4-4, 5-15, 5-17, 5-26
Mi-1	*Micromelum minutum*	7-1
Mi-2	*M. zeylanicum*	15-7, 20-2
Mo-1	*Monnieria trifolia*	5-1, 5-12, 5-14, 5-17, 5-43, 5-44, 8-14
Mu-1	*Murraya koenigii*	14-3, 14-4, 14-5, 14-8, 14-15, 15-1, 15-2, 15-3, 15-4, 15-7, 15-8, 15-9, 15-10, 15-11, 15-12, 15-13, 16-1, 16-2, 16-3, 16-4, 16-5, 16-7
Mu-2	*M. omphalocarpa*	5-17
Mu-3	*M. paniculata*	5-17, 10-1, 10-2, 10-3, 13-1, 15-9, 16-5, 16-6
My-1	*Myrtopsis macrocarpa*	5-1, 5-17, 7-6, 32-4
My-2	*M. myrtoidea*	5-1, 5-7, 5-17, 13-5, 19-13, 32-4
My-3	*M. novae-caledoniae*	5-1, 5-7, 5-17, 7-3
My-4	*M. sellingii*	2-3, 4-11, 5-1, 5-3, 5-7, 5-17
Or-1	*Oricia gabonensis*	5-17, 8-13
Or-2	*O. renieri*	5-15, 5-16, 7-12, 7-20, 7-21, 7-35, 7-36, 7-37, 8-13, 8-14, 8-35
Or-3	*O. suaveolens*	5-15, 5-26, 7-20, 8-8, 8-13, 8-24, 8-35
Or-4	*Oriciopsis glaberrima*	5-20
Or-5	*Orixa japonica*	1-10, 1-12, 2-35, 2-36, 3-15, 4-24, 4-25, 5-6, 5-10, 5-15, 5-17, 5-44
Pe-1	*Pelea barbigera*	4-5, 5-15
Pe-2	*Pentaceras australis*	18-1, 18-10, 18-14
Pe-3	*Perriera madagascariensis**	17-19
Ph-1	*Phebalium nudum*	5-1, 5-6, 5-7, 5-15, 5-17
Ph-2	*Phellodendron amurense*	19-7, 22-6, 23-2, 23-4, 23-5, 23-6, 23-7, 32-2
Ph-3	*P. lavallei*	23-2
Ph-4	*P. wilsonii*	22-6, 22-7, 23-1, 23-2, 23-3, 23-4, 23-6, 23-7
Pi-1	*Picrasma ailanthoides**	17-4, 17-11, 18-17, 18-20

TABLE 33—*contd.*

Code	Plant	Alkaloids
Pi-2	*P. excelsa**	**17**-10, **18**-1, **18**-10, **18**-17
Pi-3	*P. javanica**	**17**-9
Pi-4	*P. quassioides**	**17**-3, **17**-7, **17**-11, **17**-13, **17**-18, **17**-20, **18**-1, **18**-17, **18**-20
Pi-5	*Pilocarpus heterophyllus*	**27**-4
Pi-6	*P. jaborandi*	**27**-3, **27**-4, **27**-5
Pi-7	*P. microphyllus*	**27**-4, **27**-5, **27**-6, **27**-7, **27**-8, **27**-9
Pi-8	*P. organensis*	**13**-4
Pi-9	*P. pennatifolius*	**27**-4
Pi-10	*P. racemosus*	**27**-4
Pi-11	*P. spicatus*	**27**-4
Pi-12	*Pitavia punctata*	**5**-1, **5**-7
Pl-1	*Platydesma campanulata*	**3**-1, **3**-10, **4**-4, **5**-5, **5**-6, **5**-15
Pl-2	*Pleiospermium alatum*	**19**-12, **19**-13, **19**-30
Pt-1	*Ptelea aptera*	**5**-9, **5**-15, **5**-17, **5**-30
Pt-2	*P. crenulata*	**5**-17
Pt-3	*P. trifoliata*	**2**-20, **2**-23, **2**-29, **2**-30, **2**-31, **2**-33, **2**-39, **2**-40, **2**-41, **2**-42, **2**-43, **2**-44, **2**-45, **2**-46, **2**-47, **2**-48, **2**-49, **2**-51, **3**-16, **4**-14, **4**-15, **4**-20, **4**-22, **4**-23, **4**-24, **4**-27, **4**-30, **5**-1, **5**-5, **5**-15, **5**-16, **5**-17, **5**-19, **7**-3, **7**-22, **7**-23, **7**-32, **7**-33
Pt-4	*P. trifoliata* ssp. *pallida* var. *confinis*	**2**-24, **2**-30, **2**-31, **4**-17, **4**-24
Pt-5	*P. trifoliata* ssp. *trifoliata* var. *trifoliata*	**4**-27, **4**-28, **4**-30
Ra-1	*Ravenia spectabilis*	**2**-11, **2**-12, **2**-14, **4**-3, **5**-7, **8**-14
Ru-1	*Ruta angustifolia*	**3**-11, **5**-1
Ru-2	*R. bracteosa*	**3**-11
Ru-3	*R. chalepensis*	**1**-7, **3**-11, **5**-1, **5**-7, **5**-11, **5**-15, **5**-17, **7**-14, **8**-14, **9**-2
Ru-4	*R. graveolens*	**1**-7, **2**-21, **3**-11, **3**-13, **3**-16, **3**-22, **3**-23, **3**-24, **3**-27, **4**-8, **4**-12, **4**-19, **5**-1, **5**-5, **5**-7, **5**-15, **5**-17, **7**-14, **7**-17, **8**-4, **8**-7, **8**-12, **8**-14, **9**-1, **9**-2, **9**-3, **9**-4, **9**-5, **9**-6, **9**-7, **9**-8, **9**-9, **9**-10, **9**-11, **9**-12, **9**-13
Ru-5	*R. graveolens* ssp. *hortensis*	**4**-12, **4**-19, **7**-14, **7**-17
Ru-6	*R. montana*	**5**-1, **5**-15, **5**-17, **8**-14
Sa-1	*Samadera indica**	**18**-19
Sa-2	*Sargentia greggii*	**5**-9, **5**-15

TABLE 33—*contd.*

Code	Plant	Alkaloids
Se-1	*Severinia buxifolia*	**19**-33
Si-1	*Simaba cuspidata**	**18**-11, **18**-18
Si-2	*S. multiflora**	**18**-6, **18**-8, **18**-12, **18**-18, **18**-21
Si-3	*Simarouba amara**	**18**-4
Sk-1	*Skimmia arisanensis*	**5**-17
Sk-2	*S. foremanii*	**5**-1
Sk-3	*S. japonica*	**3**-7, **4**-8, **5**-1, **5**-17
Sk-4	*S. laureola*	**5**-17
Sk-5	*S. repens*	**5**-1
So-1	*Soulamea pancheri**	**18**-1, **18**-9, **18**-16
So-2	*S. soulameoides**	**18**-22
Sp-1	*Spathelia sorbifolia*	**2**-15, **7**-3
Sw-1	*Swinglea glutinosa*	**19**-32
Te-1	*Teclea boiviniana*	**5**-44, **8**-13, **8**-14, **8**-16, **8**-17, **8**-24, **8**-28, **8**-30
Te-2	*T. grandifolia*	**8**-10, **8**-13, **8**-35
Te-3	*T. natalensis*	**8**-13, **8**-14, **8**-24
Te-4	*T. ouabanguiensis*	**5**-15, **5**-20, **5**-34, **5**-35, **5**-43
Te-5	*T. simplicifolia*	**19**-6
Te-6	*T. sudanica*	**5**-9, **5**-17, **5**-20, **19**-7
Te-7	*T. unifoliata*	**5**-9, **5**-15, **5**-17
Te-8	*T. verdoorniana*	**5**-15, **5**-17, **5**-20, **5**-26, **5**-34, **5**-43, **8**-8, **8**-13, **8**-24, **8**-35
Th-1	*Thamnosma montana*	**5**-4, **5**-7, **5**-17, **8**-1, **8**-2
To-1	*Toddalia aculeata*	**5**-1, **5**-4, **5**-7, **5**-17, **25**-5, **25**-7, **25**-10, **25**-12, **25**-13, **25**-17, **25**-19, **25**-23, **25**-27, **26**-3
To-2	*T. aculeata* var. *gracilus*	**5**-1, **5**-7, **5**-17, **26**-3, **32**-1
Ve-1	*Vepris ampody*	**3**-9, **3**-14, **3**-21, **5**-15, **8**-8, **8**-13, **13**-3, **32**-5
Ve-2	*V. bilocularis*	**5**-9, **5**-15, **5**-17, **5**-20, **8**-8, **8**-13, **8**-14, **8**-15
Ve-3	*V. louisii*	**2**-32, **2**-50, **4**-32, **7**-21, **7**-34, **7**-35, **7**-36, **12**-3
Ve-4	*V. pilosa*	**5**-15, **8**-7, **8**-8, **8**-14
Ve-5	*V. stolzii*	**5**-7, **5**-17, **7**-21, **7**-26, **7**-29, **7**-30
Xy-1	*Xylocarpus granatum†*	**7**-3, **25**-23
Za-1	*Zanthoxylum ailanthoides*	**5**-1, **5**-17, **22**-5, **22**-6, **25**-11, **29**-2, **31**-5, **31**-6
Za-2	*Z. alatum*	**5**-1, **5**-4, **5**-7, **5**-17, **22**-6, **22**-8
Za-3	*Z. americanum*	**19**-7, **21**-4, **22**-5, **22**-6, **25**-10, **25**-11
Za-4	*Z. arborescens*	**11**-6, **11**-7, **19**-34

TABLE 33—contd.

Code	Plant	Alkaloids
Za-5	Z. armatum	19-24, 19-25, 19-27, 19-29, 19-31
Za-6	Z. arnottianum	5-1, 5-4, 5-14, 5-17, 21-1, 25-5, 25-8, 25-10, 25-15, 26-2, 26-3
Za-7	Z. avicennae	19-7, 21-4, 22-6, 25-4, 25-7, 25-10, 25-11
Za-8	Z. belizense	4-4, 5-1, 5-17, 18-1
Za-9	Z. bouetense	5-17, 19-7, 21-4, 22-6, 25-10, 25-11
Za-10	Z. brachyacanthum[h]	22-11, 23-8, 24-4, 25-10
Za-11	Z. bungeanum	5-17, 7-13, 22-2, 25-5, 25-19, 26-3
Za-12	Z. caribaeum	5-17, 18-10, 22-11
Za-13	Z. clava-herculis	19-7, 19-18, 21-4, 22-5, 22-6, 23-2, 25-10, 25-11, 31-2
Za-14	Z. conspersipunctatum	19-15, 19-17, 24-1, 24-4, 25-3, 25-10, 25-23
Za-15	Z. coriaceum	19-3, 19-19, 19-35, 22-11, 23-8, 25-10, 25-12
Za-16	Z. culantrillo	5-17, 19-3, 19-4, 19-7, 19-36, 21-4, 22-6, 22-11
Za-17	Z. cuspidatum	2-3, 5-1, 5-4, 5-7, 5-14, 5-17, 22-1, 25-1, 25-5, 25-8, 25-11, 25-16, 26-3, 26-4
Za-18	Z. decaryi	2-3, 5-1, 5-17, 25-8
Za-19	Z. dinklagei	5-17, 25-11
Za-20	Z. dipetalum	18-1, 21-4, 25-10, 25-11
Za-21	Z. dominianum[i]	18-1
Za-22	Z. elephantiasis	18-1, 18-10, 19-7, 21-4, 22-5, 22-6, 25-10, 25-12
Za-23	Z. fagara	5-17, 19-4, 22-5, 22-6
Za-24	Z. flavum	12-2, 18-1, 21-4, 25-10, 25-11
Za-25	Z. hamiltonianum	25-11
Za-26	Z. inerme	5-1, 19-13, 19-15, 25-11, 25-16
Za-27	Z. integrifoliolum	26-1
Za-28	Z. lemairei	19-7, 21-4, 22-6, 22-11, 25-10, 25-11, 25-12
Za-29	Z. leprieurii[j]	5-17, 8-7, 8-14, 19-7, 21-4, 22-6, 22-11, 25-10, 25-11, 25-12
Za-30	Z. limoncillo	5-17
Za-31	Z. martinicense	19-7, 21-4, 23-8, 25-10
Za-32	Z. microcarpum	5-14, 19-2, 19-3, 22-6, 25-6, 25-8, 25-18
Za-33	Z. monophyllum	7-25, 7-28, 22-6, 25-10
Za-34	Z. myriacanthum	21-4, 22-6, 25-11, 25-13
Za-35	Z. nitidum	5-17, 24-4, 25-5, 25-11, 25-13, 24-15, 25-16, 25-19, 25-22
Za-36	Z. ocumarense	19-15, 19-19, 22-5, 23-8, 25-10
Za-37	Z. aff. oreophilum	18-1, 24-3, 24-4, 25-2, 25-11
Za-38	Z. ovalifolium[i]	5-17, 12-1, 18-1, 25-5
Za-39	Z. oxyphyllum	5-7, 17-24, 21-3, 22-3, 22-4
Za-40	Z. parviflorum	4-4, 5-1, 5-17, 22-6, 23-8, 25-10
Za-41	Z. piperitum	22-6, 31-2, 31-4

TABLE 33—*contd.*

Code	Plant	Alkaloids
Za-42	*Z. pluviatile*	**5**-15, **5**-17, **12**-1, **23**-8, **24**-2, **25**-10, **25**-23
Za-43	*Z. procerum*	**19**-3
Za-44	*Z. punctatum*	**19**-35, **22**-6, **22**-11
Za-45	*Z. rhetsa*	**5**-17, **12**-1, **12**-6, **12**-9, **17**-24, **25**-10
Za-46	*Z. rubescens*[k]	**19**-7, **21**-4, **22**-6, **22**-11, **25**-10, **25**-11, **25**-12
Za-47	*Z. schinifolium*	**5**-17
Za-48	*Z. simulans*	**5**-1, **5**-7, **5**-17, **7**-13, **7**-24, **7**-27, **19**-7, **19**-25, **22**-2, **22**-6, **25**-10, **25**-12, **25**-15, **25**-21, **25**-25, **25**-28
Za-49	*Z. sp.* Sévenet-Pusset 1345	**5**-17, **18**-1, **24**-3, **24**-4, **25**-2
Za-50	*Z. suberosum*[i]	**18**-1
Za-51	*Z. tingoassuiba*[l]	**19**-15
Za-52	*Z. tsihanimposa*	**5**-7, **5**-17, **21**-4, **25**-10, **25**-12, **25**-21, **25**-23
Za-53	*Z. veneficum*[h]	**22**-11, **23**-8, **24**-4, **25**-10
Za-54	*Z. viride*[m]	**25**-8
Za-55	*Z. williamsii*	**2**-21, **5**-17, **22**-5, **22**-6, **25**-10, **25**-11

[a] *Araliopsis soyauxii* and *A. tabouensis* are probably conspecific. [b] = *Fagara leprieurii*. [c] All *Fagara* species are often considered to be part of *Zanthoxylum* (check for reports under *Zanthoxylum* also). [d] This does not appear to be a published name. [e] *Zanthoxylum gilletii*. [f] = *Zanthoxylum friesii*. [g] = *Zanthoxylum fagara*. [h] *Zanthoxylum brachyacanthum* = *Z. veneficum*. [i] *Zanthoxylum ovalifolium* = *Z. dominianum* = *Z. suberosum*. [j] See also *Fagara leprieurii* and *F. angolensis*. [k] See also *Fagara rubescens*. [l] See also *Fagara tingoassuiba*. [m] See also *Fagara viridis*.

ACKNOWLEDGEMENTS

The author wishes to thank the following for supplying him with unpublished data or with manuscripts in the press: Dr. J. F. Ayafor, University of Yaounde, Cameroon; Prof. G. A. Cordell, University of Illinois, Chicago, U.S.A.; Dr. A. I. Gray, Trinity College, Dublin, Ireland; Prof. E. Stanislas, Université Toulouse III, France; Dr. F. Tillequin, Université René Descartes, Paris, France; Prof. J. Vaquette, Université de Franch-Comté, Besançon, France; and Dr. P. G. Waterman, University of Strathclyde, Glasgow, U.K.

REFERENCES

1. Abdullaeva, Kh. A., Bessonova, I. A. and Yunusov, S. Yu. (1977). *Khim. Prir. Soedin.* 425.
2. Abdullaeva, Kh. A., Bessonova, I. A. and Yunusov, S. Yu. (1978). *Khim. Prir. Soedin.* 219.

3. Abdullaeva, Kh. A., Bessonova, I. A. and Yunusov, S. Yu. (1979). *Khim. Prir. Soedin.* 873.
4. Ahond, A., Picot, F., Potier, P., Poupat, C. and Sévenet, T. (1978). *Phytochemistry* 17, 166.
5. Ahond, A., Poupat, C. and Pusset, J. (1979). *Phytochemistry* 18, 1415.
6. Akhmedzhanova, V. I., Bessonova, I. A. and Yunusov, S. Yu. (1977). *Khim. Prir. Soedin.* 289.
7. Akhmedzhanova, V. I., Bessonova, I. A. and Yunusov, S. Yu. (1978). *Khim. Prir. Soedin.* 476.
8. Al-Shamma, A., Al-Douri, N. A. and Phillipson, J. D. (1979). *Phytochemistry* 18, 1417.
9. Assem. E. M., Benages, I. A. and Albonico, S. M. (1979). *Phytochemistry* 18, 511.
10. Ayafor, J. F., Sondengam, B. L. and Ngadjui, B. T. (1982). *Phytochemistry* 21, 955.
11. Ayafor, J. F. and Okugun, J. I. (1982). *J. Nat. Prod.* 45, 182.
12. Ayafor, J. F., Sondengam, B. L., Bilon, A. N., Tsamo, E., Kimbu, S. F. and Okugun, J. I. (1982). *J. Nat. Prod.* 45, 714.
13. Ayafor, J. F., Sondengam, B. L., Kimbu, S. F., Tsamo, E. and Connolly, J. D. (1982). *Phytochemistry* 21, 2602.
14. Ayafor, J. F., Sondengam, B. L. and Ngadjui, B. (1980). *Tetrahedron Lett.* 3293.
15. Ayafor, J. F., Sondengam, B. L. and Ngadjui, B. T. (1981). *Tetrahedron Lett.* 2685.
16. Balsam, G. and Voigtländer, H. W. (1978). *Arch. Pharm.* 311, 1016.
16a. Bamgbose, S. O., Dramane, K. L. and Okogun, J. I. (1977). *Planta Med.* 31, 193.
17. Batsuren, D., Batirov, E. Kh. and Malikov, V. M. (1981). *Khim. Prir. Soedin.* 659.
18. Baudouin, G., Tillequin, F., Koch, M., Pusset, J. and Sévenet, T. (1981). *J. Nat. Prod.* 44, 546.
18a. Bernhard, H. O. and Thiele, K. (1978). *Helv. Chim. Acta* 61, 2269.
19. Bessonova, I. A., Abdullaeva, Kh. A. and Yunusov, S. Yu. (1974). *Khim. Prir. Soedin.* 682.
20. Bessonova, I. A. and Yunusov, S. Yu. (1977). *Khim. Prir. Soedin.* 303.
21. Bévalot, F., Vaquette, J. and Cabalion, P. (1980). *Plant. Med. Phytother.* 14, 218.
22. Bhattacharyya, P., Roy, S., Biswas, A., Bhattacharyya, L. and Chakraborty, D. P. (1978). *J. Indian Chem. Soc.* 55, 308.
23. Bhattacharyya, J. and Serur, L. M. (1981). *Heterocycles* 16, 371.
24. Bhide, K. S., Majumdar, R. B. and Rama Rao, A. V. (1977). *Indian J. Chem.* 15B, 440.
25. Bohlmann, F. and Bhaskar Rao, V. S. (1969). *Chem. Ber.* 102, 1774.
26. Boulware, R. T. and Stermitz, F. R. (1981). *J. Nat. Prod.* 44, 200.
27. Bourguignon-Zylber, N. and Polonsky, J. (1970). *Chim. Ther.* 5, 396.
28. Bowden, K. and Ross, W. J. (1963). *J. Chem. Soc.* 3503.
29. Bowen, I. H. and Christopher Perera, K. P. W. (1982). *Phytochemistry* 21, 433.
30. Bowen, I. H., Christopher Perera, K. P. W. and Lewis, J. R. (1978). *Phytochemistry* 17, 2125.
31. Bowen, I. H., Christopher Perera, K. P. W. and Lewis, J. R. (1980). *Phytochemistry* 19, 1566.
32. Bowen, I. H. and Lewis, J. R. (1978). *Lloydia* 41, 184.
33. Burke, B. A. and Parkins, H. (1978). *Tetrahedron Lett.* 2723.
34. Burke, B., Parkins, H. and Talbot, A. M. (1979). *Heterocycles* 12, 349.
35. Caolo, M. A. and Stermitz, F. R. (1979). *Tetrahedron* 35, 1487.
36. Chakraborty, D. P. (1977). *Fortschr. Chem.* 34, 299.

37. Chakraborty, D. P., Bhattacharyya, P., Roy, S., Bhattacharyya, S. P. and Biswas, A. K. (1978). *Phytochemistry* **17**, 834.
38. Chakraborty, D. P., Das, B. P. and Basak, S. P. (1974). *Plant. Biochem. J.* **1**, 73.
39. Chakraborty, D. P., Roy, S. and Guha, R. (1978). *J. Indian Chem. Soc.* **55**, 1114.
40. Chakravarti, D. and Chakravarti, R. N. (1967). *J. Proc. Inst. Chem.* (India) **39**, 131.
41. Chatterjee, A., Bose, S. and Srimany, S. K. (1959). *J. Org. Chem.* **24**, 687.
42. Chatterjee, A., Chakrabarty, M. and Kundu, A. B. (1975). *Aust. J. Chem.* **28**, 457.
43. Chavarria, R. G. (1971). *An Asoc. Quim. Argentina* **59**, 371.
44. Chou, F. Y., Hostettmann, K., Kubo, I., Nakanishi, K. and Taniguchi, M. (1977). *Heterocycles* **7**, 969.
45. Clarke, P. J., Jewers, K. and Jones, H. F. (1980). *J. Chem. Soc. Perkin Trans.* **1**, 1614.
46. Cordell, G. A. and Farnsworth, N. R. (1981), unpublished results.
47. Cordell, G. A., Ogura, M. and Farnsworth, N. R. (1978). *Lloydia* **41**, 166.
48. Cordell, G. A. and Tantivatana, P. (1981), unpublished results.
49. Correa, D. de B., Gottlieb, O. R., de Padua, A. P. and Da Rocha, A. I. (1976). *Rev. Latinoam. Quim.* **7**, 43; cf. *Chem. Abs.* **84**, 161790g (1976).
50. Cougé, B., Tillequin, F., Koch, M. and Sévenet, T. (1980). *Plant. Med. Phytother.* **14**, 208.
50a. Dadson, B. A. and Minta, A. (1976). *J. Chem. Soc. Perkin Trans.* **1**, 146.
51. Dahlgren, R. (1977). *Plant Syst. Evol.* Suppl. 1, 253.
52. Danieli, B., Lesma, G. and Palmisano, G. (1979). *Experientia* **35**, 156.
53. Das, B. P. and Chowdhury, D. N. (1978). *Chem. Ind.* (London) 272.
54. Decaudain, N., Kunesch, N. and Poisson, J. (1977). *Ann. Pharm. Fr.* **35**, 521.
55. Deshpande, V. H. and Shastri, R. K. (1977). *Indian J. Chem.* **15B**, 95.
56. Dominguez, X. A., Butruille, D., Rudy, A. and Sergio, G. G. (1977). *Rev. Latinoam. Quim.* **8**, 47; cf. *Chem. Abs.* **86**, 117636d (1977).
57. Dreyer, D. L. (1977). *Rev. Latinoam. Quim.* **8**, 11.
58. Dreyer, D. L. (1980). *Phytochemistry* **19**, 941.
59. Dreyer, D. L. and Brenner, R. C. (1980). *Phytochemistry* **19**, 935.
60. Dreyer, D. L., Rigod, J. F., Basa, S. C., Mahanty, P. and Das, D. P. (1980). *Tetrahedron* **36**, 827.
61. El-Tawil, B. A. H., El-Beih, F. K. A., Budzikiewicz, H. and Mohr, N. (1981). *Z. Naturforsch.* **36b**, 1169.
62. Fauvel, M. T., Gleye, J., Moulis, C., Blasco, F. and Stanislas, E. (1981). *Phytochemistry* **20**, 2059.
63. Fauvel, M. T., Gleye, J., Moulis, C. and Fouraste, I. (1978). *Plant. Med. Phytother.* **12**, 207.
64. Fish, F., Gray, A. I. and Waterman, P. G. (1976). *J. Pharm. Pharmacol.* **28**, Suppl. 69P.
65. Fish, F., Meshal, I. A. and Waterman, P. G. (1976). *J. Pharm. Pharmacol.* **28**, Suppl. 72P.
66. Fish, F., Meshal, I. A. and Waterman, P. G. (1977). *Fitoterapia* (Milano) **48**, 170.
67. Fish, F., Meshal, I. A. and Waterman, P. G. (1978). *Planta Med.* **33**, 228.
67a. Ganguly, S. N. and Sarkar, A. (1978). *Phytochemistry* **17**, 1816.
68. Giesbrecht, A. M., Gottlieb, H. E., Gottlieb, O. R., Goulart, M. O. F., De Lima, R. A. and Santana, A. E. G. (1980). *Phytochemistry* **19**, 313.
69. Gonzales Gonzales, A., Diaz Chico E., Lopez Dorta, H., Luis, J. R. and Rodriguez

Luis, F. (1977). *An. Quim.* **73**, 430.
70. Goodson, J. A. (1921). *Biochem. J.* **15**, 123.
71. Gray, A. I. and O'Sullivan, J. J. (1980). *Planta Med.* **39**, 209.
72. Gray, A. I. and O'Sullivan, J. J. (1982), unpublished results.
73. Grina, J. A. and Stermitz, F. R. (1981). *Tetrahedron Lett.* 5257.
74. Grundon, M. F. (1978). In "The Alkaloids" (Manske, R. H. F. and Rodrigo, R. G. A. eds.), Vol. 17, p. 105. Academic Press, New York.
75. Grundon, M. F. and Okely, H. M. (1979). *Phytochemistry* **18**, 1768.
76. Hänsel, R. and Cybulski, E. M. (1978). *Arch. Pharm.* **311**, 135.
77. Harmon, A. D., Weiss, U. and Silverton, J. V. (1979). *Tetrahedron Lett.* 721.
78. Hifnawy, M. S., Vaquette, J., Sévenet, T., Pousset, J. L. and Cavé, A. (1977). *Phytochemistry* **16**, 1035.
79. Hostettmann, K., Pettei, M. J., Kubo, I. and Nakanishi, K. (1977). *Helv. Chim. Acta* **60**, 670.
80. Huang, Z. X. and Li, Z. H. (1980). *Hua Hsueh Hsueh Pao* **38**, 535; cf. *Chem. Abs.* **94**, 99773e (1981).
81. Inamoto, N., Masuda, S., Shimamura, O. and Tsuyuki, T. (1961). *Bull. Chem. Soc. Japan* **34**, 888.
82. Ishii, H., Chen, I. S., Ishikawa, T., Ishikawa, M. and Lu, S. T. (1979). *Heterocycles* **12**, 1037.
83. Ishii, H., Ishikawa, T. and Haginiwa, J. (1977). *Yakugaku Zasshi* **97**, 890; cf. *Chem. Abs.* **87**, 197250g (1977).
84. Ishii, H., Ishikawa, T., Lu, S. T. and Chen, I. S. (1976). *Yakugaku Zasshi* **96**, 1458; cf. *Chem. Abs.* **86**, 136297k (1977).
85. Ishii, H., Murakami, K., Takeishi, K., Ishikawa, T. and Haginiwa, J. (1981). *Yakugaku Zasshi* **101**, 504; cf. *Chem. Abs.* **95**, 111726x (1981).
86. Ishikawa, T. and Ishii, H. (1976). *Heterocycles* **5**, 275.
87. Iyer, V. S. and Rangaswami, S. (1972). *Curr. Sci.* (Calcutta) **41**, 140.
88. Johne, S. and Härtling, S. (1977). *Pharmazie* **32**, 415.
89. Johns, S. R. and Lamberton, J. A. (1969). *Aust. J. Chem.* **22**, 1315.
90. Johns, S. R., Lamberton, J. A. and Sioumis, A. A. (1970). *Aust. J. Chem.* **23**, 629.
91. Joshi, B. S., Kamat, V. N. and Gawad, D. H. (1977). *Heterocycles* **7**, 193.
92. Jurd, L. and Wong, R. Y. (1981). *Aust. J. Chem.* **34**, 1625.
93. Kamikado, T., Murakoshi, S. and Tamura, S. (1978). *Agric. Biol. Chem.* **42**, 1515.
94. Khalid, S. A. and Waterman, P. G. (1981). *Phytochemistry* **20**, 2761.
95. Khalid, S. A. and Waterman, P. G. (1981). *Planta Med.* **43**, 148.
96. Khalid, S. A. and Waterman, P. G. (1982). *J. Nat. Prod.* **45**, 343.
97. Khan, S. A. and Shamsuddin, K. M. (1981). *Phytochemistry* **20**, 2062.
98. Kimura, Y., Takido, M. and Koizumi, S. (1967). *Yakugaku Zasshi* **87**, 1371; cf. *Chem. Abs.* **68**, 87441v (1968).
99. Kondo, Y. and Takemoto, T. (1973). *Chem. Pharm. Bull.* **21**, 837.
100. Kong, Y. C. and King, C. L. (1979). *Recent Adv. Nat. Prod. Res., Proc. Int. Symp.* 104; cf. *Chem. Abs.* **95**, 18341 h (1981).
101. Kulkarni, G. H. and Sabata, B. K. (1981). *Phytochemistry* **20**, 867.
101a. Kundu, A. B. and Chakrabarty, M. (1975). *Chem. Ind.* (London) 433.
102. Kurbanov, D., Bessonova, I. A. and Yunusov, S. Yu. (1968). *Khim. Prir. Soedin.* 373.
103. Lassak, E. V., Polonsky, J. and Jaquemin, H. (1977). *Phytochemistry* **16**, 1126.
104. Löwe, W. and Pook, K. H. (1973). *Liebigs Ann. Chem.* 1476.

105. MacPhillamy, H. B. (1965), personal communication cited by Taylor, W. I., in "The Alkaloids" (ed. R. H. F. Manske), Vol. 8, p. 250. Academic Press, New York.
106. Manandar, M. D., Shoeb, A., Kapil, R. S. and Popli, S. P. (1978). *Phytochemistry* **17**, 1814.
107. Matsumura, S., Enomoto, H., Aoyagi, Y., Nomiyama, Y., Kono, T., Matsuda, M. and Tanaka, H. (1980). *Ger. Offen.* 2,941,449; cf. *Chem. Abs.* **93**, 114495r (1980).
108. Mensah, I. A. and Sofowora, E. A. (1979). *Planta Med.* **35**, 94.
109. Mester, I. (1973). *Fitoterapia* (Milano) **44**, 123.
110. Mester, I. (1977). *Fitoterapia* (Milano) **48**, 268.
111. Mester, I. (1982), unpublished results.
112. Mester, I. and Reisch, J. (1977). *Liebigs Ann. Chem.* 1725.
113. Mester, I., Reisch, J., Szendrei, K. and Körösi, J. (1979). *Liebigs Ann. Chem.* 1785.
114. Moulis, C., Gleye, J., Fourasté, I. and Stansislas, E. (1981). *Planta Med.* **42**, 400.
115. Moulis, C., Wirasutisna, K., Gleye, J., Moretti, C. and Stanislas, E. (1979), poster communication, Colloque International CNRS-ORSTOM: Substances Naturelles d'Intérêt Biologique du Pacifique, Nouméa.
116. Nahrstedt, A., Eilert, U., Wolters, B. and Wray, V. (1981). *Z. Naturforsch.* **36c**, 200.
117. Nesmelova, E. F., Bessonova, I. A. and Yunusov, S. Yu. (1977). *Khim. Prir. Soedin.* 289.
118. Nesmelova, E. F., Bessonova, I. A. and Yunusov, S. Yu. (1977). *Khim. Prir. Soedin.* 427.
119. Nesmelova, E. F., Bessonova, I. A. and Yunusov, S. Yu. (1978). *Khim. Prir. Soedin.* 758.
120. Ngadjui, T. B., Ayafor, J. F., Sondengam, B. L., Connolly, J. D., Rycroft, D. S., Khalid, S. A., Waterman, P. G., Brown, N. M. D., Grundon, M. F. and Ramachandran, V. N. (1982). *Tetrahedron Lett.* 2041.
121. Odebiyi, O. O. and Sofowora, E. A. (1979). *Planta Med.* **36**, 204.
122. Ohmoto, T., Koike, K. and Sakamoto, Y. (1981). *Chem. Pharm. Bull.* **29**, 390.
123. Ohmoto, T., Tanaka, R. and Nikaido, T. (1976). *Chem. Pharm. Bull.* **24**, 1532.
124. Okogun, J. I. and Ayafor, J. F. (1977). *J. Chem. Soc. Chem. Commun.* 652.
125. Okorie, D. A. (1975). *Nigerian J. Sci.* **9**, 201.
126. Picker, K., Ritchie, E. and Taylor, W. C. (1976). *Aust. J. Chem.* **29**, 2023.
127. Prakash, D., Raj, K., Kapil, R. S. and Popli, S. P. (1980). *Indian J. Chem.* **19B**, 1075.
128. Purushothaman, K. K., Sarada, A., Connolly, J. D. and Akinniyi, J. A. (1979). *J. Chem. Soc., Perkin Trans.* 1, 3171.
129. Rama Rao, A. V., Bhide, K. S. and Mujumdar, R. B. (1980). *Chem. Ind.* (London) 697.
130. Rastogi, K., Kapil, R. S. and Popli, S. P. (1980). *Phytochemistry* **19**, 945.
131. Razakova, D. M., Bessonova, I. A. and Yunusov, S. Yu. (1976). *Khim. Prir. Soedin.* 682.
132. Reisch, J., Mester, I., Körösi, J. and Szendrei, K. (1978). *Tetrahedron Lett.* 3681.
133. Reisch, J., Müller, M. and Mester, I. (1981). *Planta Med.* **43**, 285.
134. Ren, L. and Xie, F. (1981). *Yaoxue Xuebao* **16**, 672; cf. *Chem. Abs.* **96**, 48974e (1982).
135. Rideau, M., Verchère, C., Hibon, P., Chénieux, J. C., Maupas, P. and Viel, C. (1979). *Phytochemistry* **18**, 155.
136. Roy, S., Ghosh, S. and Chakraborty, D. P. (1979). *Chem. Ind.* (*London*) 669.
137. Rózsa, Zs. (1977). Dissertation, Szeged.

138. Rózsa, Zs., Hohmann, J. and Reisch, J. (1979). *Planta Med.* **36**, 260.
139. Rózsa, Zs., Reisch, J., Mester, I. and Szendrei, K. (1981). *Fitoterapia* (Milano) **52**, 37.
140. Rózsa, Zs., Reisch, J., Szendrei, K. and Minker, E. (1981). *Fitoterapia* (Milano) **52**, 93.
141. Rózsa, Zs., Szendrei, K. Kovács, Z., Novák, I., Minker, E. and Reisch, J. (1978). *Phytochemistry* **17**, 169.
142. Sanchez, E. and Comin, J. (1971). *Phytochemistry* **10**, 2155.
143. Sarkar, M. and Chakraborty, D. P. (1977). *Phytochemistry* **16**, 2007.
144. Sarkar, M. and Chakraborty, D. P. (1979). *Phytochemistry* **18**, 694.
145. Sarkar, M., Kundu, S. and Chakraborty, D. P. (1978). *Phytochemistry* **17**, 2145.
146. Sharma, B. R., Rattan, R. K. and Sharma, P. (1981). *Phytochemistry* **20**, 2606.
147. Sharma, B. R. and Sharma, P. (1981). *Planta Med.* **43**, 102.
148. Sharma, P. N., Shoeb, A., Kapil, R. S. and Popli, S. P. (1979). *Indian J. Chem.* **17B**, 299.
149. Sharma, P. N., Shoeb, A., Kapil, R. S. and Popli, S. P. (1981). *Phytochemistry* **20**, 2781.
150. Sharma, P. N., Shoeb, A., Kapil, R. S. and Popli, S. P. (1981). *Indian J. Chem.* **20B**, 936.
151. Sharma, P. N., Shoeb, A., Kapil, R. S. and Popli, S. P. (1982). *Phytochemistry* **21**, 252.
152. Shiengthong, D., Ungphakorn, A., Lewis, D. E. and Massy Westropp, R. A. (1979). *Tetrahedron Lett.* 2247.
153. Sidyakin, G. P., Eskairov, M. and Yunusov, S. Yu. (1960). *Zh. Obshch. Khim.* **30**, 338; cf. *Chem. Abs.* **54**, 22697h (1960).
154. Silva, L. B. de, Silva, U. L. L. de, Mahendran, M. and Jennings, R. (1979). *Phytochemistry* **18**, 1255.
155. Stanislas, E. (1982), personal communication
156. Stermitz, F. R., Caolo, M. A. and Swinehart, J. A. (1980). *Phytochemistry* **19**, 1469.
157. Stermitz, F. R. and Sharifi, I. A. (1977). *Phytochemistry* **16**, 2003.
158. Svoboda, G. H. (1966). *Lloydia* **29**, 206.
159. Svoboda, G. H., Poore, G. A., Simpson, P. J. and Boder, G. B. (1966). *J. Pharm. Sci.* **55**, 758.
160. Swinehart, J. A. and Stermitz, F. R. (1980). *Phytochemistry* **19**, 1219.
161. Szendrei, K., Korbély, T., Krenzien, H., Reisch, J. and Novák, I. (1977). *Herba Hung.* **16**, (No. 3), 15.
162. Takagi, S., Akiyama, T., Konoshita, T., Sankawa, U. and Shibata, S. (1979). *Shoyakugaku Zasshi* **33**, 30; cf. *Chem. Abs.* **92**, 55040t (1980).
163. Thoms, H. and Thümen, F. (1911). *Ber. Dtsch. Chem. Ges.* **44**, 3717.
164. Tillequin, F., Baudouin, G., Koch, M. and Sévenet, T. (1980). *J. Nat. Prod.* **43**, 498.
165. Tillequin, F., Baudouin, G., Ternoir, M., Koch, M., Pusset, J. and Sévenet, T. (1982). *J. Nat. Prod.* **45**, 486.
166. Tillequin, F. and Koch, M. (1979). *Phytochemistry* **18**, 1559.
167. Tillequin, F. and Koch, M. (1979). *Phytochemistry* **18**, 2066.
168. Tillequin, F. and Koch, M. (1980). *Phytochemistry* **19**, 1282.
169. Tillequin, F., Koch, M., Bert, M. and Sévenet, T. (1979). *J. Nat. Prod.* **42**, 92.
170. Tillequin, F., Koch, M. and Sévenet, T. (1980). *Plant. Med. Phytother.* **14**, 4.
171. Tillequin, F., Koch, M. and Sévenet, T. (1980). *Planta Med.* **38**, 383.
172. Tillequin, F., Rousselet, R., Koch, M., Bert, M. and Sévenet, T. (1979). *Ann.*

Pharm. Fr. **37**, 543.
173. Tiwari, K. P. and Masood, M. (1978). *Phytochemistry* **17**, 1068.
174. Tiwari, K. P. and Masood, M. (1979). *Phytochemistry* **18**, 517.
175. Torres, R. and Cassels, B. K. (1978). *Phytochemistry* **17**, 838.
176. Vaquette, J. (1981), personal communication.
177. Vaquette, J., Cavé, A. and Waterman, P. G. (1978). *Plant. Med. Phytother.* **12**, 235.
178. Vaquette, J., Cavé, A. and Waterman, P. G. (1979). *Planta Med.* **35**, 42.
179. Vaquette, J., Cavé, A., Waterman, P. G., Sévenet, T. and Pusset, J. (1979), poster communication, Colloque International CNRS-ORSTOM: Substances Naturelles d'Intérêt Biologique du Pacifique, Nouméa.
180. Vaquette, J., Hocquemiller, R., Pousset, J. L. and Cavé, A. (1978). *Planta Med.* **33**, 78.
181. Varga, E., Kuzovkina, I. N., Rózsa, Zs. and Szendrei, K. (1978). *Acta Pharm. Hung.* **48**, 193.
182. Varga, E., Szendrei, K., Reisch, J. and Maróti, G. (1980). *Planta Med.* **40**, 337.
183. Viala, B. (1971). Thesis, Paris (cited in ref. 103).
184. Voigtländer, H. W., Balsam, G., Engelhardt, M. and Pohl, L. (1978). *Arch. Pharm.* **311**, 927.
185. Wagner, H., Nestler, T. and Neszmélyi, A. (1978). *Tetrahedron Lett.* 2777.
186. Wagner, H., Nestler, T. and Neszmélyi, A. (1979). *Planta Med.* **36**, 113.
187. Wang, M. H. (1981). *Yao Hsueh T'ung Pao* **16**, 48; cf. *Chem. Abs.* **95**, 192260r (1981).
188. Waterman, P. G. (1975). *Biochem. Syst. Ecol.* **3**, 149.
189. Waterman, P. G., Gray, A. I. and Crichton, E. G. (1976). *Biochem. Syst. Ecol.* **4**, 259.
190. Waterman, P. G. and Khalid, S. A. (1981). *Biochem. Syst. Ecol.* **9**, 45.
191. Waterman, P. G., Meshal, I. A., Hall, J. B. and Swaine, M. D. (1978). *Biochem. Syst. Ecol.* **6**, 239.
192. Wolters, B. and Eilert, U. (1981). *Planta Med.* **43**, 166.
193. Yajima, T., Kato, N. and Munakata, K. (1977). *Agric. Biol. Chem.* **41**, 1263.
194. Yang, J. S., Luo, S. R., Shen, X. L. and Li, Y. X. (1979). *Yao Hsueh Hsueh Pao* **14**, 167; cf. *Chem. Abs.* **92**, 72679a (1980).
195. Yasuda, I., Takeya, K. and Itokawa, H. (1981). *Chem. Pharm. Bull.* **29**, 1791.

CHAPTER 4

Structural Diversity and Distribution of Coumarins and Chromones in the Rutales

A. I. GRAY

School of Pharmacy, Trinity College, Dublin, Eire

I. INTRODUCTION

Some five years ago the distribution and taxonomic significance of coumarins in the Rutaceae was reviewed.[70] At that time coumarins had been encountered in only one other family of the Rutales, the Ptaeroxylaceae (*Ptaeroxylon* and *Cedrelopsis* spp.). Coumarins have now been found in all of the families that are characteristic of the order (Rutaceae, Simaroubaceae, Meliaceae, Cneoraceae, Ptaeroxylaceae, and Burseraceae). Chromones, specifically the 2-methyl-4-chromones, have now been isolated from the

97

Ptaeroxylaceae, Cneoraceae, *Spathelia* spp. (Rutaceae), and *Harrisonia abyssinica* (Simaroubaceae). Although not closely related to the coumarins in a biogenetic sense, these chromones share many structural features in common with the coumarins. In this chapter the known distribution of both classes of compounds in the order is reviewed.

II. STRUCTURAL DIVERSITY OF COUMARINS

Tables 1–29 list the coumarins which have been isolated from the Rutales together with their sources and include pre-1978 research on coumarins listed in two reviews.[70,100] Table 35 is an alphabetical listing of the coumarin bearing plants of the order and their contained coumarins. Thus some 200 species from about 60 genera have so far been found to produce almost 300 different coumarins. These compounds vary in complexity from "simple" oxy, alkoxy, alkyl, and acyl substituted types to dimeric forms and variously substituted furo- and pyrano-coumarins.

A. SIMPLE COUMARINS

The "simple" coumarins, based on the umbelliferone (**2-1**) nucleus, are listed in Tables 2–14, and are classified principally on the basis of the number and orientation of attached oxygen atoms and alkyl substituents. The alkyl and alkoxy substituents frequently take the form of variously oxidized/modified prenyl groupings. Some of the types encountered may be artefacts of the isolation procedure. For example, the author has found that the prenyl epoxide **2-4**, on prolonged solution in chloroform, is converted cleanly to the corresponding glycol (**2-5**). However, not all glycols of this type can be considered as artefacts. The co-occurrence of the glycol obtusinin (**8-8**) with its 3-β-glucoside obtusoside (**8-9**) in *Haplophyllum obtusifolium*[89,91] suggests that **8-8** is not an artefact but has been involved *in vivo* in the biosynthesis of **8-9**.

Prenyl epoxides may also undergo solvolysis during extraction and work-up with, for example, methanol and ethanol, to give ether artefacts of the types known to occur in some substituted cinnamides.[87] In the coumarin series compounds such as the ethanolate **3-17** and the methanolate **21-8** may also be artefacts formed from the corresponding epoxides, suberosin epoxide (**3-10**) and oxypeucedanin (**21-6**), by addition of ethanol and methanol, respectively.

A rather unusual "simple" coumarin, crenulatin (**3-1**), has been "isolated" from several rutaceous genera (Table 3). The most recent report of **3-1** was from *Micromelum zeylanicum*[21] where it was found to co-occur with micromelin (3-22). According to the isolation procedure, micromelin (3-22) was obtained by crystallization directly from a light petrol extract of the plant material whereas crenulatin (3-1) was obtained after Al_2O_3 chromatography of the supernatant extract. Mock *et al.*[94] have shown that an "epoxy alcohol", 6-(2',3'-epoxy-1-hydroxy-3'-methylbutyl)-7-methoxycoumarin, on chromatography over Al_2O_3 gives 3-1 in high yield (Scheme 1). Crenulatin (angelical) is also formed when the umbelliferous coumarin angelol is heated.[102] Crenulatin (3-1) may therefore be considered as an artefact formed by the removal of a four-carbon moiety from the side-chain of a 6-prenylated 7-methoxycoumarin rather than a direct introduction of a formyl group into position-6.

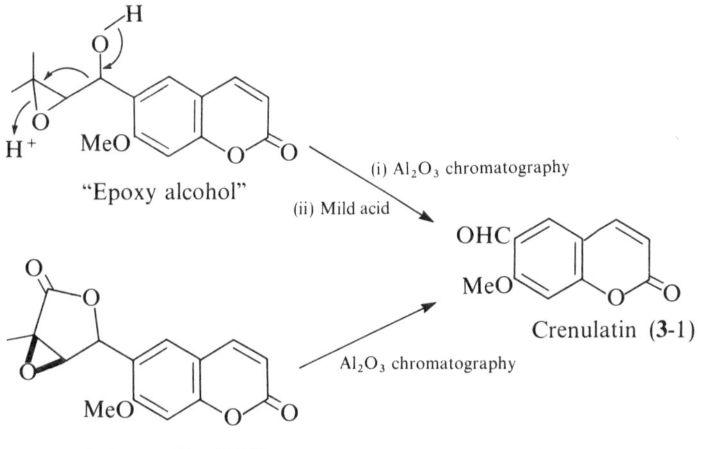

"Epoxy alcohol"

(i) Al_2O_3 chromatography
(ii) Mild acid

OHC ... MeO

Crenulatin (3-1)

Al_2O_3 chromatography

Micromelin (3-22)

SCHEME 1. Some alternative routes[21,94] to crenulatin (3-1).

A coumarin which retains a formyl group as part of a five-carbon side-chain (Pent^a type) is citrusal (4-8), from *Citrus paradisi.*[70,100] The total synthesis of 4-8 has been reported,[29] the key intermediate being the epoxide meranzin (4-4), which rearranges in the presence of Lewis acids to give citrusal. It is noteworthy that meranzin has also been found in *C. paradisii* and that citrusal could be an artefact.

Among the multi-oxygenated "simple" coumarins some, such as the 5,7-dioxygenated 6-substituted types of *Toddalia* spp. (Table 6), are restricted in their distribution whereas others, such as the 6,7-dioxygenated types (Table 8), have widespread distribution within the Rutales. Some of the latter exhibit an unusual ethylenedioxy grouping, cf. obliquin (8-18) and obliquol (8-19). The ethylenedioxy group appears to have arisen from the cyclization of a 7-prenyloxy substituent with an available 6-hydroxy group. In this regard prenyletin (**8**-3) is an obvious candidate as precursor to **8**-18 while obtusinol (**8**-7) may be regarded as a blocked (by 6-O-methylation) precursor to **8**-19. This hypothesis receives support from the co-occurrence of obtusinol (**8**-7) and two ethylenedioxy substituted coumarins, obtusifol (**13**-8) and obtusin (**13**-9), in *Haplophyllum obtusifolium*.[2,90]

B. DIMERIC COUMARINS

Three types of dimeric coumarin are recognized, all based on "simple" coumarin monomers, viz. the ether dimers (Table 15), the bicoumarins (Table 16), and the "Diels–Alder" dimers (Table 17). Formation of the latter probably arises through a Diels–Alder type reaction involving the prenyl sidechains of two monomers (see Chapter 2).

Both **7**-2 and **17**-4 have been isolated from *Murraya paniculata*[28,70,100] but, surprisingly, no 5,7-dioxy-8-prenylated coumarins (Table 7) have yet been reported in *Toddalia* spp. Mock *et al.*[94] have considered the possibility of thamnosin (cyclobisuberodiene) (**17**-2) being an artefact. They attempted to synthesize it from suberosin (**3**-8) and some of its oxygenated derivatives without success and therefore concluded that **17**-2 was a natural product. It remains to be seen whether the same may be said for the other "Diels–Alder" dimers.

C. FUROCOUMARINS

The furocoumarins (Tables 18–25) have so far been found only in the Rutaceae except for a single report for Meliaceae. The most common types are linear furocoumarins among which bergapten (**21**-4) and xanthotoxin (**22**-7) are of widespread occurrence. Some are more restricted in distribution. For example, the 8-prenylated (Table 20) and 5-oxygenated 8-prenylated compounds (Table 23) have been found only in *Chloroxylon swietenia*[13,70,107] and the 5-geranyloxy types (**21**-11 to **21**-14) are typical constituents of *Citrus* spp.[70]

D. PYRANOCOUMARINS

The pyranocoumarins (Tables 26 and 27) have also been found only in the Rutaceae, the majority being of the 2,2-dimethylpyran type. An interesting exception to this is the angular hydroxydihydropyranocoumarin lomatin isovalerate (**26**-7) isolated from *Haplophyllum* spp.[5] which might be classed as a coumarin typical of the family Umbelliferae.[70,102]

Two major types are recognizable, the 3-substituted coumarins (Table 28) and the 3,4-disubstituted coumarins (Table 29). Among the former group all of the major coumarin classes are represented, the most common 3-substituent being the 1,1-dimethylallyl moiety.[70] The 3,4-disubstituted coumarins are all 3,4-dioxyfurocoumarins that have been found so far only in *Halfordia* spp. (Rutaceae).

III. DISTRIBUTION AND CHEMOTAXONOMIC SIGNIFICANCE OF COUMARINS IN THE RUTALES

The Rutaceae continues to be the source of new and interesting coumarins. In the past 5 years the majority of almost 80 "first reports" of coumarins from the Rutaceae were of novel structures. Even so, the overall distribution of coumarin types within the family has changed little since the review by Gray and Waterman.[70] Thus "simple" coumarins account for about 55 % of the total, with furocoumarins (*ca.* 28 %) and pyranocoumarins (*ca.* 11 %) making up the bulk of the remainder.

In comparison to the Rutaceae, the other families of the Rutales have yielded very few coumarins. In the Simaroubaceae a single 6,7-dioxygenated coumarin, scopoletin (**8**-4), has been isolated from *Ailanthus* and *Picrasma* species. In the Burseraceae, a family better known for the utilization of cinnamate in the production of lignans, resins, and volatile oils (see Chapter 10), scopoletin and the interesting coumarin propacin have been recorded. Propacin (**13**-10) appears to have been formed by the coupling of a 6,7,8-trioxygenated coumarin with an isoeugenol moiety.

Within the Meliaceae a few cinnamate derived "simple" coumarins and two unusual coumarins have been encountered. Thus *Melia azedarach* is reported[81] to contain the 6,7-dioxygenated coumarins aesculetin (**8**-1), isoscopoletin (**8**-2), and scopoletin (**8**-4). The "abnormal" coumarins ekersenin (**1**-2) and siderin (**1**-3) have been isolated from *Ekebergia senegalensis*[11,103,134] and *Cedrela toona*,[101] respectively. Both **1**-2 and **1**-3 are examples of 4-methoxy-5-methylcoumarins that may be derived from acetate/pentaketide intermediates[103] rather than from cinnamate.

The Ptaeroxylaceae and Cneoraceae show striking similarity in the production of several 6,7-dioxygenated (Table 8) and 6,7-dioxygenated 8-prenylated coumarins (Table 9), particularly with ethylenedioxy substituents. For example, obliquin (8-18) has been isolated from *Ptaeroxylon obliquum* (Ptaeroxylaceae)[100] and from *Cneorum tricoccon* (Cneoraceae)[60] while cedrelopsin (9-3) has been detected in *Cedrelopsis grevei* ((Ptaeroxylaceae)[100] and *Neochamaelea pulverulenta* (Cneoraceae).[100] Both of these coumarin types have also been detected in rutaceous genera. For instance, prenyletin methyl ether (8-6) has been isolated from *Agathosma, Diosma* (tribe Diosmae), and *Haplophyllum* spp. (tribe Rutoideae) (Rutaceae) as well as *Ptaeroxylon* (Table 8), and brayleanin (9-4) has been detected in *Flindersia brayleana* (Rutaceae).[70,100]

Outwith the Rutales the prenyletin and obliquin type coumarins have been isolated from members of the family Compositae. Recent work by Bohlmann and coworkers has shown the presence of prenyletin (8-3) and obliquin (8-18) in *Nidorella* spp. (tribe Asterae)[16,17] and of a dihydrohydroxy obliquin derivative (1) in *Acritopappus* spp.[19] and *Conocliniopsis* spp.[18] A similarly substituted coumarin, obtusifol (13-8) has been detected in *Haplophyllum obtusifolium* (Rutaceae).

(1) Dihydrohydroxyobliquin

Several variously prenylated 6,7-dioxy, 5,6,7-trioxy, and 6,7,8-trioxy substituted "simple" coumarins have been recently isolated from *Pterocaulon* spp. (Compositae)[85] while very similarly substituted compounds have been found in the rutaceous genera *Agathosma, Diosma,* and *Phyllosma* (tribe Diosmae) and in *Haplophyllum* (Tables 8, 11, and 13). The above observations appear to suggest links between some members of the Rutales and some Compositae.

Gray and Waterman[70] have noted the high degree of similarity between the coumarins of the Rutaceae and Umbelliferae[102] but drew attention to the fact that differences existed between the types of prenyl-unit modifications encountered in the families. These differences have been reduced somewhat by the discovery of lomatin isovalerate (26-7) in three *Haplophyllum* species,[5] paniculatin (4-23) in *Murraya paniculata*,[70,100] and a linear pyranochromone (31-8), bearing a senecioyl side-chain (Pre^x), in *Spathelia sorbifolia*.[128]

However, the biogenetic origin of these side-chains (either mevalonate or amino acid[70,102]) in both families is uncertain and it would be imprudent to place too great an emphasis on these observations at this time.

IV. Structural Diversity, Distribution, and Chemotaxonomic Significance of Chromones in the Rutales

The chromones are acetate derived compounds[74] and the basic nucleus can be regarded as 5,7-dihydroxy-2-methyl-4-chromone (Table 30). Tables 30–33 list the 2-methyl-4-chromones which have so far been reported from the Rutales together with their sources. Table 34 lists the unusual 2-phenethylchromones encountered in *Flindersia laevicarpa* (Rutaceae). Table 36 gives an alphabetical listing of the chromone bearing plants of the order and their contained chromones.

So far typical chromones are known from only nine species belonging to seven genera but these have yielded over 40 different compounds. In general terms the chromones found can be subdivided into "simple" chromones, pyranochromones, 2-hydroxymethylchromones, and oxepinochromones.

A. SIMPLE CHROMONES

The majority of "simple" chromones are prenylated (Pre[a]) in positions 6 and/or 8 (Table 30), the only exception being rohitukine (**30-1**) isolated from *Amoora rohituka* (Meliaceae)[73] which bears an unusual 8-(3-hydroxy-1-methylpiperid-4-yl) grouping. The other "simple" chromones are restricted in distribution to *Cneorum tricoccon* and *Neochamaelea pulverulenta* (Cneoraceae) and to *Cedrelopsis grevei* and *Ptaeroxylon obliquum* (Ptaeroxylaceae) (Table 30).

B. PYRANOCHROMONES

The pyranochromones, listed in Table 31, show remarkable similarities to their pyranocoumarin counterparts (cf. Tables 26 and 27). They occur widely in *Spathelia* spp. (Rutaceae), Cneoraceae, and Ptaeroxylaceae, and one, methylspatheliachromene (**31-3**), has been detected in *Harrisonia abyssinica* (Simaroubaceae).[137] Most of the pyranochromones bear the 2,2-dimethyl-chromene ring system but greveiglycol (**31-17**) and Cneorum chromone Q (**31-18**) have been oxidized to 3′,4′-dihydroxydihydropyrans. One dimeric type,

Cneorum chromone R (**31**-19), isolated from *Cneorum tricoccon*,[43] appears to have arisen by dimerisation of spatheliabischromene (**31**-16).

A number of pyranochromones, which have undergone oxidation at the 2-methyl position, have been included in Table 32.

C. 2-HYDROXYMETHYLCHROMONES

This class of chromone (Table 32), comprised mainly of pyrano derivatives, includes the only example of a furochromone yet encountered in the Rutales. Thus umtatin (**32**-1) has been isolated from the heartwood of *Ptaeroxylon obliquum*.[33] The distribution of 2-hydroxymethylchromones is, so far, restricted to Ptaeroxylaceae and Cneoraceae. The 2-hydroxymethyl oxepino derivative, karenin (**33**-8), has been included in Table 33.

D. OXEPINOCHROMONES

The oxepinochromones are listed in Table 33. This ring system appears to result from oxidative cyclization involving a prenyl terminal olefinic or prenyl dienyl group at C-6 and a 7-hydroxy substituent[71] (cf. ptaeroxylin, **33**-2). The major structural modifications arise through oxidation of the carbons of the oxepine ring and attached 3'-methyl groups in ptaeroxylinol (**33**-3) and ptaeroglycol (**33**-5); ptaeroxylone (**33**-1), isolated from *Ptaeroxylon obliquum*,[34] may have been derived by oxidative cleavage of the 1,2-glycol ptaeroglycol (**33**-5). Further modification of the chromone nucleus, except for oxidation of the 2-methyl group in one instance (**33**-8), is absent in this chromone type. The oxepinochromones have been found only in Ptaeroxylaceae and Cneoraceae. The acetylated types **33**-4 and **33**-6 are restricted to *Neochamaelea pulverulenta* (Cneoraceae).[43]

It is interesting to note that these chromones, and all the other types encountered so far in the Rutales, have not undergone oxidation of the benzene ring carbons (positions 6 and 8, partial structure Table 30). This seems to be a major difference between the chromones and their coumarin analogues.

E. CHROMONES OF *Flindersia laevicarpa*

These unusual 2-phenethylchromones (Table 34) co-occur in *F. laevicarpa*[105] with 1,5-diphenylpentane1,3-diol (**2**) and other 1,5-diphenylpentanes from which the chromones may be derived.

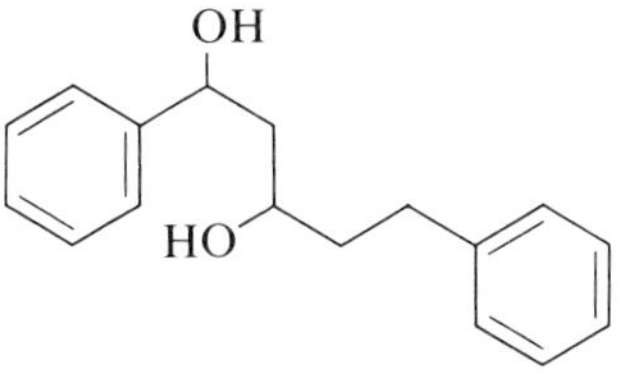

(2) 1,5-Diphenylpentane-1,3-diol

Outwith the Rutales chromones have been encountered frequently in the families Umbelliferae and Ranunculaceae, and very recently[15] the simple chromone 5,7-dihydroxy-2-methyl-4-chromone was isolated from the aerial parts of *Mikania alvimii* (tribe Eupatorieae, family Compositae).

The Umbelliferae is particularly well known for the production of a series of substituted furochromones,[74] e.g. visnagin (3), khellin (4), khellol (5), and ammiol (6) in *Ammi* spp.

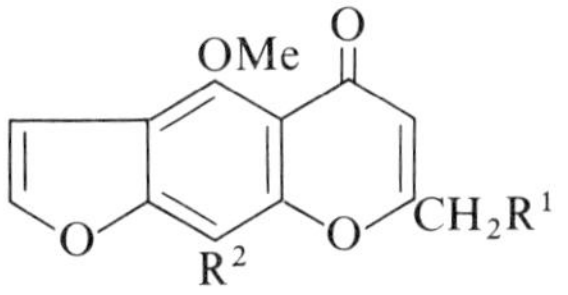

(3) Visnagin ($R^1 = R^2 = H$)

(4) Khellin ($R^1 = H$, $R^2 = OMe$)

(5) Khellol ($R^1 = OH$, $R^2 = H$)

(6) Ammiol ($R^1 = OH$, $R^2 = OMe$)

The Ranunculaceae, especially *Cimicifuga* and *Eranthis* spp.,[74] produce a similar range of chromone types including cimifugin (7) (cf. umtatin, 32-1, from *Ptaeroxylon* sp.). Recently, angular annelated oxepino types such as eranthin (8) have been detected in *Eranthis hiemalis*.[79]

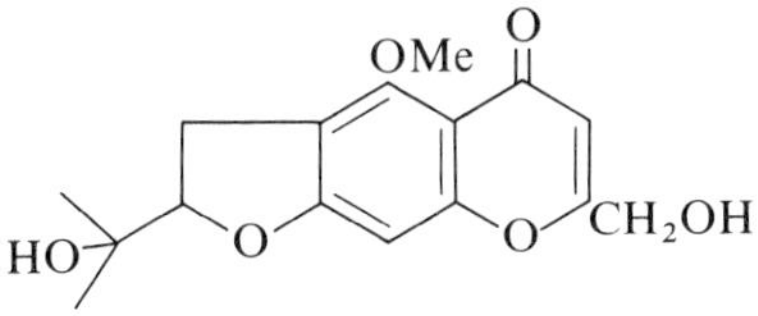

(7) Cimifugin

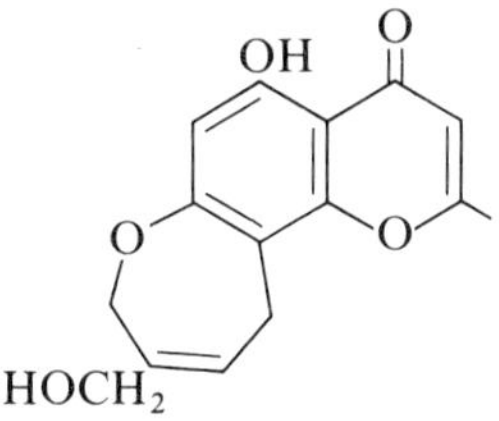

(8) Eranthin

The chromone data therefore support the coumarin data which indicate probable chemosystematic links between Rutaceae and Umbelliferae. They also point to possible ties between some genera of Ranunculaceae and those of Ptaeroxylaceae and Cneoraceae.

V. Tables

The coumarins and chromones of the Rutales, and their distribution, are listed in a series of 36 tables. The following abbreviations are employed:

Ac: $COCH_3$
Bu^a: $COCH_2CH_2CH_3$
Bu^b: $CH{=}CHCOCH_3$
DMA: $C(CH_3)_2CH{=}CH_2$
Et: CH_2CH_3
Frn: $CH_2CH{=}C(CH_3)CH_2CH_2CH{=}C(CH_3)CH_2CH_2CH{=}C(CH_3)_2$
Ger^a: $CH_2CH{=}C(CH_3)CH_2CH_2CH{=}C(CH_3)_2$
Ger^b: $CH_2CH{=}C(CH_3)CH_2CH_2CH{-}C(CH_3)_2$
$\qquad\qquad\qquad\qquad\qquad\qquad \underset{O}{\diagdown\diagup}$
Ger^c: $CH_2CH{=}C(CH_3)CH_2CH_2CH(OH)C(OH)(CH_3)_2$
Ger^d: $CH_2CH{=}C(CH_3)CH_2CH_2CH{-}C(CH_3)_2$
$\qquad\qquad\qquad\qquad\qquad\qquad O\quad O$
$\qquad\qquad\qquad\qquad\qquad\qquad \diagdown C(CH_3)_2$
Ger^e: $CH_2CH{=}C(CH_3)CH_2CH_2COC(OH)(CH_3)_2$
Ger^f: $CH_2CH{=}C(CH_3)CH_2CH_2CH(OH)C({=}CH_2)CH_3$
Ger^g: $CH_2CH_2C({=}CH_2)CH_2CH_2CH{-}C(CH_3)_2$
$\qquad\qquad\qquad\qquad\qquad\qquad \underset{O}{\diagdown\diagup}$
Ger^h: $CH_2CH_2C({=}CH_2)CH_2CH_2CH(OH)C(OH)(CH_3)_2$
Ger^i: $CH_2CH_2CH(CH_3)C{=}CHCOC(CH_3)_2$
$\qquad\qquad\qquad\qquad\quad \lfloor\!\!-\!O\!-\!\!\rfloor$
Ger^j: $CH_2CH{=}C(CH_3)C{=}CHCOC(CH_3)_2$
$\qquad\qquad\qquad\qquad\quad \lfloor\!\!-\!O\!-\!\!\rfloor$
Ger^k: $CH_2CH{=}C(CH_3)CH_2CHCH{=}C(CH_3)CO$
$\qquad\qquad\qquad\qquad\quad \lfloor\!\!-\!O\!-\!\!\rfloor$
Ger^l: $CH_2CH(OH)C(OH)(CH_3)CH_2CHCH{=}C(CH_3)CO$
$\qquad\qquad\qquad\qquad\qquad\qquad \lfloor\!\!-\!O\!-\!\!\rfloor$
Ger^m: $CH_2CH{=}C(CH_3)CH_2CH_2CH_2C(CHO)(CH_3)_2$

Glu: β-D-glucose
Ip: $CH(CH_3)_2$
Ip^a: $C(OH)(CH_3)_2$
Ip^b: $C({=}CH_2)CH_3$
Ip^c: $C(OAc)(CH_3)_2$
Ip^d: $C(OGlu)(CH_3)_2$
Ip^e: $C({=}CH_2)CH_2OH$
Ip^f: ${=}C(COOCH_3)CH_3$
Ip^g: ${=}C(COOCH_3)CH_2OH$

MD: OCH_2O
Me: CH_3
$Pent^a$: $CH_2C(CHO)(CH_3)_2$
$Pent^b$: $C(CH_3){=}C(CH_3)CHO$
$Pent^c$: $COC(CH_3)_3$
$Pent^d$: $CH{-}C(CH_3)_2$
$\qquad\qquad \diagdown CH_2\diagup$
PhAc: $COCH_2C_6H_5$

Pre^a: $CH_2CH{=}C(CH_3)_2$
Pre^b: $CH_2CH{-}C(CH_3)_2$ (epoxide, O bridge)

Pre^c: $CH_2CH(OH)C(OH)(CH_3)_2$
Pre^d: $CH_2CH(OH)C(OMe)(CH_3)_2$
Pre^e: $CH_2CH(OMe)C(OH)(CH_3)_2$
Pre^f: $CH_2CH(OH)C(OEt)(CH_3)_2$
Pre^g: $CH_2CH(OH)C(OGlu)(CH_3)_2$
Pre^h: $CH_2CH(OH)C(Cl)(CH_3)_2$
Pre^i: $CH_2CH{=}C(CH_2OH)CH_3$
Pre^j: $CH_2CH_2CH(COOH)CH_3$
Pre^k: $CH_2CH{=}C(COOH)CH_3$
Pre^l: $CH_2CH{=}C(COOMe)CH_3$
Pre^m: $CH_2CH_2C(OH)(CH_3)_2$
Pre^n: $CH{-}CHC(OH)(CH_3)_2$ (epoxide)
Pre^o: $CH{=}CHC(OH)(CH_3)_2$
Pre^p: $CH{=}CHC(OMe)(CH_3)_2$
Pre^q: $CH_2CH(OH)CH(CH_3)_2$
Pre^r: $CH_2CH(OH)C({=}CH_2)CH_3$
Pre^s: $CH_2COCH(CH_3)_2$
Pre^t: $CH_2COC({=}CH_2)CH_3$

Pre^u: $CH{=}CHC({=}CH_2)CH_3$
Pre^v: $COCH_2CH(CH_3)_2$
Pre^w: $COCH_2C({=}CH_2)CH_3$
Pre^x: $COCH{=}C(CH_3)_2$
Pre^y: $CH{-}CHCH(CH_3)_2$ (epoxide)
Pre^z: $CH{-}CHC({=}CH_2)CH_3$ (epoxide)
Pre^{aa}: $COCH{-}C(CH_3)_2$ (epoxide)
Pre^{bb}: $CH(OPre^v)COCH(CH_3)_2$
Pre^{cc}: $CH(OH)CH(OH)CH(OH)(CH_3)_2$
Pre^{dd}: $CH(OH)CH(OH)C({=}CH_2)CH_3$
Pre^{ee}: $CH(OMe)CH(OH)C({=}CH_2)CH_3$
Pre^{ff}:

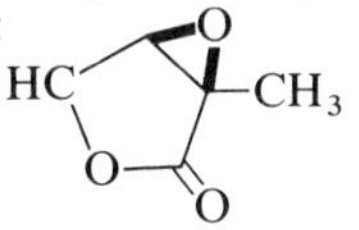

Rhm: L-rhamnose

TABLE 1

Abnormal coumarins

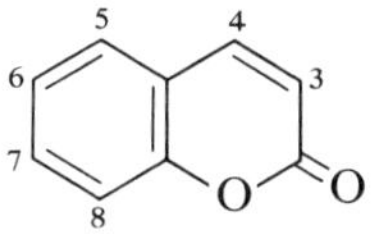

Name	Substituents	Occurrence*
1. Coumarin	—	Ru-3[70], Ru-7[70], Ru-8[70]
2. Ekersenin	4-OMe, 5-Me	Ek-1[11,103,134]
3. Siderin†	4,7-diOMe, 5-Me	Ce-1[101]
4. —	3-OMe, 7,8-diOH	Ha-14[1]
5. —	4-DMA, 7-OMe, 8-OH	Ru-9[56,70]

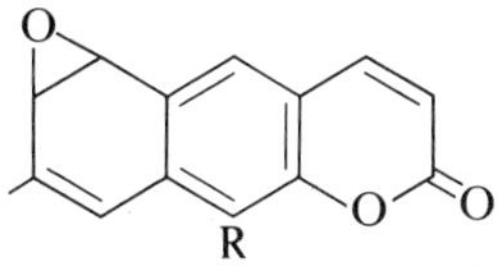

6. —	R = H	Za-3[70]
7. —	R = OMe	Za-3[70]

* Two reviews[70,100] have been used in compiling Tables 1–29 to locate pre-1978 research on coumarins.
† See also Chapter 10.

TABLE 2

7-Oxygenated "simple" coumarins

Name	Substituent	Occurrence
1. Umbelliferone	7-OH	Ae-1[70], Bo-1[70], Ci-3[70], Ci-5[70], Ci-9[82], Ci-17[70], Er-1[70], Eu-1[70], Ge-2[70], Ha-7[70], Ru-3[70,133], Ru-7[70], Ru-8[70], Ru-9[58], Sk-3[70], Sk-4[70,122], Tr-1[37]
2. Herniarin	7-OMe	Ci-5[70], Ru-3[70], Ru-6[70], Ru-7[70], Ru-8[70]
3. —	7-OPre[a]	Co-1[68], Co-2[69], Di-5[25], Eu-4[70]
4. —	7-OPre[b]	Co-1[68], Co-2[69],
5. —	7-OPre[c]	Co-1[68], Co-2[69]
6. —	7-OPre[j]	Eu-4[70,100]
7. —	7-OPre[k]	Eu-4[70,100]
8. —	7-OPre[l]	Co-2[69]
9. Aurapten	7-OGer[a]	Ae-1[70], Af-1[70], At-2[70], At-4[70], Ci-3[70,100], Ci-17[70], Di-1[70], Fe-1[70], Ge-1[70], Ha-7[70], Po-1[70], Pt-4[70], Pt-5[70], Za-16[70]
10. —	7-OGer[b]	Ci-17[127]
11. Marmin	7-OGer[c]	Ae-1[70,100], Ci-17[70], Ge-1[70]
12. —	7-OGer[e]	Ge-1[70,100]
13. —	7-OGer[k]	Mi-4[36]
14. 2′,3′-Dihydro-geiparvarin	7-OGer[i]	Ge-1[70,100]
15. Geiparvarin	7-OGer[j]	Ge-1[70,100], Ge-2[70]
16. Umbelliprenin	7-OFrn	Ci-9[82], Th-1[70]
17. Skimmin	7-OGlu	Ae-1[70], Sk-2[70], Sk-3[70,100], Sk-4[70,122]
18. —	7-OCOCH$_2$Ph	He-2[30]

TABLE 3

7-Oxygenated 6-substituted "simple" coumarins

Name	Substituents	Occurrence
1. Crenulatin	6-CHO, 7-OMe	Bo-1[70,100], Ci-16[108], Gl-1[70], He-2[70,100] Mi-4[21], Ru-9[58], Za-16[70,100]
2. Tenuidin	6-Bu^a, 7-OMe	Ha-8[5], Ha-15[5], Ha-17[5]
3. Suberenone	6-Bu^b, 7-OMe	Bo-1[70], Ru-3[70,100]
4. Demethylsuberosin	6-Pre^a, 7-OH	Ch-1[70,100], Za-3[70]
5. Swietenol	6-Pre^m, 7-OH	Ch-1[70]
6. —	6-Pre^n, 7-OH	Bo-1[70]
7. —	6-Pre^o, 7-OH	Bo-1[70]
8. Suberosin	6-Pre^a, 7-OMe	Ch-1[13], Ci-16[108], Cl-5[106], He-2[70], Za-3[70], Za-10[70], Za-16[70,100], Za-22[70]
9. Dihydrosuberenol	6-Pre^m, 7-OMe	Li-1[50]
10. Suberosin epoxide	6-Pre^b, 7-OMe	Co-1[68], Co-2[69], He-2[70,100]
11. Suberenol	6-Pre^n, 7-OMe	Am-3[24], Ci-16[108], He-2[70], Za-16[70,100]
12. Geijerin	6-Pre^v, 7-OMe	Ge-2[70,100]
13. Thamnosmin	6-Pre^z, 7-OMe	Th-1[70,100]
14. Dehydrogeijerin	6-Pre^x, 7-OMe	Am-3[24], Am-4[70], Ge-1[70,100], Ge-2[70]
15. (−)-Peucedanol	6-Pre^c, 7-OH	Eu-2[70]
16. Ulopterol	6-Pre^c, 7-OMe	Co-1[68,70,100], Er-1[70], Mi-3[70,100], Ru-8[70] Za-3[70,100]
17. —	6-Pre^f, 7-OMe	Ru-8[66]
18. Hopeyhopine	6-Pre^{aa}, 7-OMe	Am-3[24], Am-4[70]
19. —	6-Pre^s, 7-OMe	Ru-8[66]
20. Tamarin	6-Pre^r, 7-OMe	Am-2[23], Ru-8[66]
21. Thamnosmonin	6-Pre^{dd}, 7-OMe	Th-1[70,100]
22. Micromelin	6-Pre^{ff}, 7-OMe	Mi-1[27], Mi-2[70,100], Mi-3[70,100], Mi-4[21]
23. Ostruthin	6-Ger^a, 7-OH	At-1[70], Er-5[70], Lu-1[70]

TABLE 3—*contd.*

Name	Substituents	Occurrence
24. Naphthoherniarin	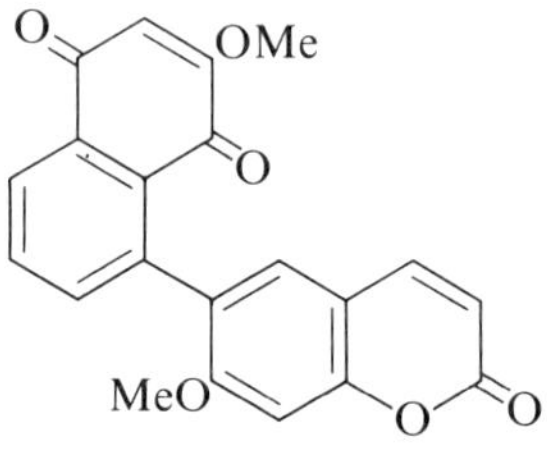	Ru-3[112]

TABLE 4

7-Oxygenated 8-substituted "simple" coumarins

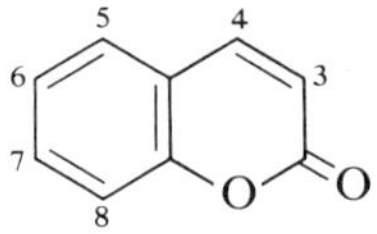

Name	Substituents	Occurrence
1. Osthenol	7-OH, 8-Pre[a]	Za-3[70]
2. Osthol	7-OMe, 8-Pre[a]	Cl-1[70], Cn-1[70], Fl-1[70], Fl-9[70], Fl-10[70], Ha-4[49], Mi-2[70], Mu-4[70], My-1[70], My-3[70]
3. Ramosin	7-OPre[a], 8-Pre[a]	Ch-2[70,100], Ch-3[70,100], Ha-13[47]
4. Meranzin	7-OMe, 8-Pre[b]	Ci-17[127], Mu-4[70], Sk-3[70]
5. Poncimarin	7-OPre[b], 8-Pre[b]	Po-1[70,100]
6. Auraptenol	7-OMe, 8-Pre[r]	Ci-3[70,100], Ci-4[70], Ci-5[70], Ci-19[70]
7. Arnottinin	7-OMe, 8-Pre[i]	Za-3[70,100]
8. Citrusal	7-OMe, 8-Pent[a]	Ci-17[70,100]
9. Isomeranzin	7-OMe, 8-Pre[s]	Sk-3[70,100], Tr-1[37]
10. Isoponcimarin	7-OPre[b], 8-Pre[s]	Po-1[70,100]
11. Triphasiol	7-OPre[c], 8-Pre[s]	Tr-1[37]
12. Meranzin hydrate	7-OMe, 8-Pre[c]	Ci-17[127], Mu-4[70,86], Sk-3[70]
13. —	7-OMe, 8-Pre[e]	Mu-4[86]
14. Myrselline	7-OPre[b], 8-Pre[a]	My-4[70]
15. Myrsellinol	7-OPre[c], 8-Pre[a]	My-4[70]
16. *trans*-Dehydroosthol	7-OMe, 8-Pre[u]	Ch-4[100]

TABLE 4—*contd.*

Name	Substituents	Occurrence
17. Phebalosin	7-OMe, 8-Pre[z]	Mu-4[70], Ph-2[70], Ph-3[70], Ph-6[70,100]
18. Murralongin	7-OMe, 8-Pent[b]	Mu-1[70,100], Mu-4[70]
19. Micropubescin	7-OMe, 8-Pre[w]	Bo-1[12], Mi-3[70,100]
20. Murrayone	7-OMe, 8-Pre[t]	Mu-4[70,100]
21. Murrangatin	7-OMe, 8-Pre[dd]	Mu-1[70,100], Mu-4[70,100]
22. —	7-OMe, 8-Pre[ee]	Mu-4[86]
23. Paniculatin	7-OMe, 8-Pre[bb]	Mu-4[70,100]
24. Versicolin	7-OMe, 8-Ger[a]	Ha-16[48]

TABLE 5

5,7-Dioxygenated "simple" coumarins

Name	Substituents	Occurrence
1. —	5-OH, 7-OMe	Ha-4[49]
2. Limettin (citropten)	5,7-diOMe	Ci-7[70,100], Ci-9[52,82], Ci-12[70], Ci-17[70], Er-4[70], Gl-1[70], Ru-7[70], Ru-8[70], Ru-9[58]
3. —	5-OPre[a], 7-OMe	Ci-9[70,100]
4. —	5-OGer[a], 7-OMe	Ci-2[70,100], Ci-6[70], Ci-7[70], Ci-9[58]
5. —	5-OGer[f], 7-OMe	Ci-2[100]

TABLE 6

5,7-Dioxygenated 6-substituted "simple" coumarins

Name	Substituents	Occurrence
1. Toddaculin	5,7-diOMe, 6-Pre[a]	To-1[70,100]
2. Aculeatin	5,7-diOMe, 6-Pre[b]	To-1[70,100]
3. Toddalolactone	5,7-diOMe, 6-Pre[c]	To-1[70,100], To-2[116,118]
4. Toddanol	5,7-diOMe, 6-Pre[r]	To-1[119]
5. Toddanone	5,7-diOMe, 6-Pre[s]	To-1[119], To-2[118]
6. —	5,7-diOMe, 6-Pre[h]	To-1[119], To-2[118]

TABLE 7

5,7-Dioxygenated 8-substituted "simple" coumarins

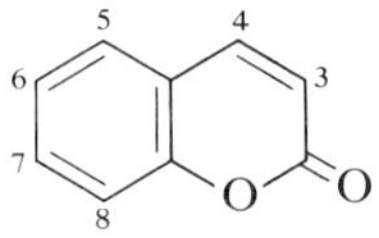

Name	Substituents	Occurrence
1. Pinnarin	5,7-diOMe, 8-DMA	Ru-8[70,100]
2. Coumurrayin	5,7-diOMe, 8-Pre[a]	Cl-1[70], Ho-2[70], Ho-3[70], Mu-3[138], Mu-4[70,100]
3. —	5,7-diOMe, 8-Pre[v]	Mu-3[138], Se-1[70,100]
4. Omphamurin	5,7-diOMe, 8-Pre[r]	Mu-3[138]
5. Mexoticin	5,7-diOMe, 8-Pre[c]	Mu-3[138], Mu-4[70,100], Se-1[131]

TABLE 8

6,7-Dioxygenated "simple" coumarins

Name	Substituents	Occurrence
1. Aesculetin	6,7-diOH	Me-1[81], Ru-7[70], Ru-8[70], Za-1[70]
2. Isoscopoletin	6-OH, 7-OMe	Af-1[109], Me-1[81]
3. Prenyletin	6-OH, 7-OPre[a]	Pt-1[100]
4. Scopoletin	6-OMe, 7-OH	Ae-1[70], Ai-1[126], Am-5[6], Bu-1[139], Bu-2[139], Ca-1[70], Ch-5[28a], Ci-9[82], Ha-7[70], Ha-8[5], Ha-11[70], Ha-12[140], Ha-15[5], Ha-17[5], Me-1[8], Mi-1[27], Mu-2[72], Mu-4[70], Pe-1[70], Pi-1[135], Pt-1[35], Pt-5[70], Ru-3[70], Ru-8[70], Sk-3[70], Sk-4[70,122], Sw-1[8], Za-21[40], Za-23[70]
5. Scoparon	6,7-diOMe	Ae-1[70], Ae-2[38,70], Af-1[70], Ag-1[25], Ag-2[25], Ag-3[25], Ag-4[25], Ag-7[25], Ch-1[70], Ch-5[28a], Ci-15[79], Ci-18[84], Di-5[25], Ho-3[70], Za-1[70], Za-6[70], Za-11[70], Za-17[70], Za-23[70], Za-25[70,100], Za-26[124]

TABLE 8—*contd.*

Name	Substituents	Occurrence
6. —	6-OMe, 7-OPre[a]	Ag-3[25], Ag-5[25], Di-5[25], Ha-4[49], Ha-10[90], Ha-13[47], Pt-1[100]
7. Obtusinol	6-OMe, 7-OPre[i]	Ha-10[90]
8. Obtusinin	6-OMe, 7-OPre[e]	Ha-10[89]
9. Obtusoside	6-OMe, 7-OPre[g]	Ha-10[91]
10. Scopolin	6-OMe, 7-OGlu	Ha-12[140], Mu-4[70], Pt-5[70], Ru-3[70], Sk-4[122]
11. Haploperoside A	6-OMe, 7-OGlu-2←1-Rhm	Ha-12[140]
12. Acetylhaploperoside A	6-OMe, 7-OGlu-2←1-Rhm(Ac)	Ha-12[140]
13. —	6-OMe, 7-OGer[a]	Fe-1[70], Ha-7[70], Ha-11[70,100], Po-1[70],
14. Pedicellone	6-OMe, 7-OGer[e]	Ha-11[70,100]
15. 6-Methoxymarmin	6-OMe, 7-OGer[e]	Ha-11[70,100]
16. 6-Methoxymarmin acetonide	6-OMe, 7-OGer[d]	Ha-11[48]
17. —	6-OGer[a], 7-OMe	Ha-11[4]

18. Obliquin	R = CH$_3$	Cn-2[60], Pt-1[100]
19. Obliquol	R = CH$_2$OH	Pt-1[100]

20. Bethancorin	R = CH$_3$	Cn-2[60]
21. Bethancorol	R = CH$_2$OH	Cn-2[100]

TABLE 9

6,7-Dioxygenated 8-substituted "simple" coumarins

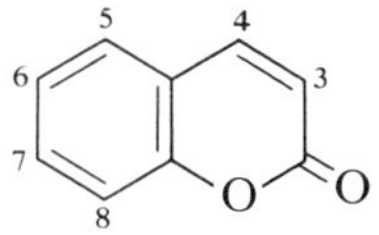

Name	Substituents	Occurrence
1. Obliquetol	6,7-diOH, 8-DMA	Pt-1[100]
2. Obliquetin	6-OMe, 7-OH, 8-DMA	Pt-1[100]
3. Cedrelopsin	6-OMe, 7-OH, 8-Pre[a]	Ce-2[100], Ne-1[100]
4. Brayleanin	6-OMe, 7-OPre[a], 8-Pre[a]	Fl-2[70,100]
5. Cneorum coumarin B	6-OMe, 7-OH, 8-Pre[r]	Ne-1[96,100]

TABLE 10.

7,8-Dioxygenated "simple" coumarins.

Name	Substituents	Occurrence
1. Daphnetin 8-methyl ether	7-OH, 8-OMe	Bo-1[70], Fl-4[14]
2. —	7,8-MD	Di-5[25]
3. —	7-OMe, 8-OPre[a]	Co-2[69]
4. —	7-OMe, 8-OPre[b]	Co-1[68], Co-2[69]
5. —	7-OMe, 8-OPre[c]	Co-2[69]
6. Villosin	7-OMe, 8-Oger[g]	Ha-8[5], Ha-15[5], Ha-17[5]
7. Tenudiol	7-OMe, 8-OGer[h]	Ha-8[5], Ha-15[5], Ha-17[5]
8. —	7-OPre[a], 8-OMe	Di-5[25]
9. Collinin	7-OGer[a], 8-OMe	Fl-3[70,100], Fl-4[70], Fl-7[70], Ha-3[130]
10. Brosiparin (arnottianol)	6-Pre[a], 7-OMe, 8-OH	Za-3[70,100]

TABLE 11

5,6,7-Trioxygenated "simple" coumarins

Name	Substituents	Occurrence
1. Isofraxetin	5,6-diOH, 7-OMe	Ha-10[89]
2. —	5,6-diOMe, 7-OH	Ha-7[54]
3. —	5,6,7-triOMe	Di-6[25]
4. —	5,6-diOMe, 7-OGer[a]	Ha-7[70,100]

TABLE 12

5,7,8-Trioxygenated "simple" coumarins

Name	Substituents	Occurrence
1. —	5,7-diOMe, 8-OH	Ru-9[70,100]
2. —	5,7,8-triOMe	Ru-9[70,100], To-1[70,100]
3. Sabandinin	5-OMe, 7,8-MD	Ru-7[70], Ru-8[70,100]
4. Sabandinone	5-OPre[s], 7,8-MD	Ru-8[70,100]
5. Sabandinol	5-OPre[e], 7,8-MD	Ru-8[70,100]

TABLE 13

6,7,8-Trioxygenated "simple" coumarins

Name	Substituents	Occurrence
1. —	6,7-MD, 7-OMe	Ag-3[25]
2. —	6-OMe, 7,8-MD	Ag-3[25]
3. —	6,7,8-triOMe	Ru-7[70], Ru-9[56], Za-1[70,76], Za-11[70,100], Za-12[70], Za-20[20]
4. Capensin	6-OMe, 7-OPre[a], 8-OH	Ha-10[9], Ph-7[26]
5. Obtusitsin	6-OMe, 7-OPre[i], 8-OH	Ha-10[9]
6. Puberulin	6,8-diOMe, 7-OPre[a]	Ag-6[70]
7. Haptusinol	6-OMe, 7-OPre[q], 8-OH	Ha-10[3]
8. Obtusifol	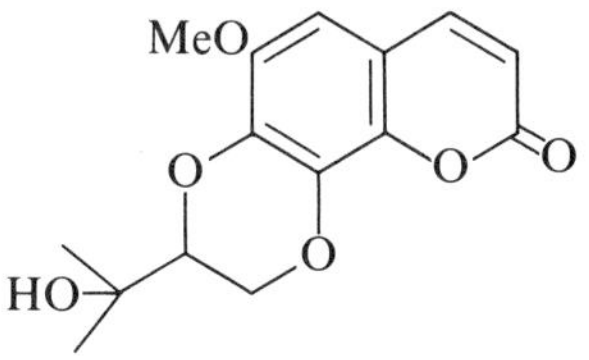	Ha-10[2,70]
9. Obtusin	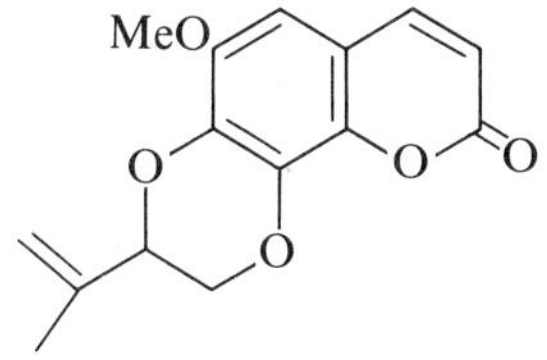	Ha-10[2]

TABLE 13—*contd.*

Name	Substituents	Occurrence
10. Propacin	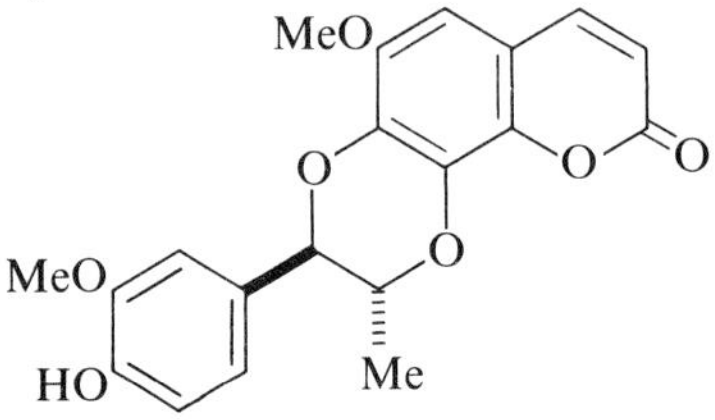 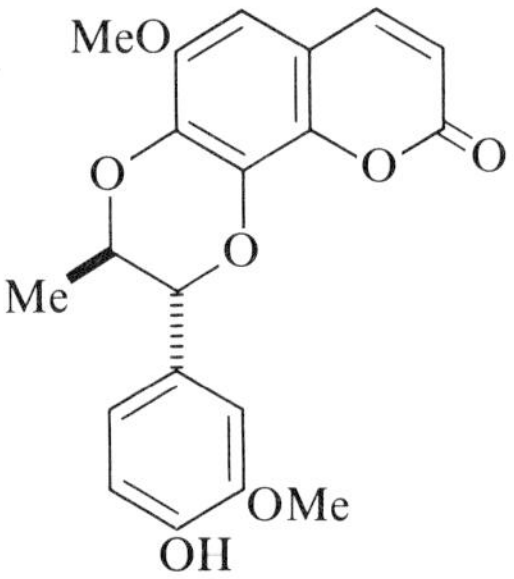	Pr-1[141]

TABLE 14

5,6,7,8-Tetraoxygenated "simple" coumarins

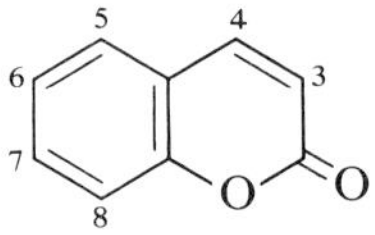

Name	Substituents	Occurrence
1. Sabandin	5,8-diOMe, 6,7-MD	Ru-8[110], Ru-9[65]

TABLE 15

Coumarin ether dimers

Name	Substituents	Occurrence
	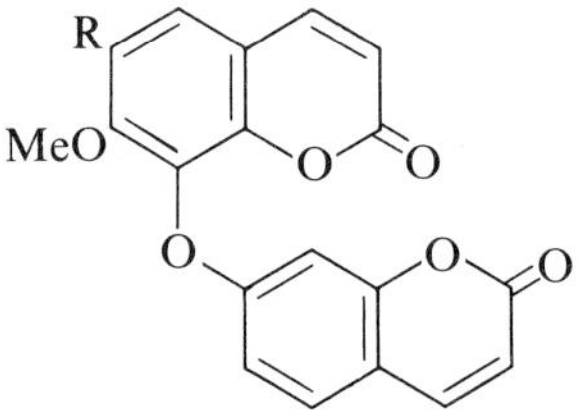	
1. Fatagenin	R = H	Ru-7[70]
2. Oreojasmin	R = OMe	Ru-7[70]

Table 15—*contd.*

Name	Substituents	Occurrence

Name	Substituents	Occurrence
3. Daphnoretin	R = OH	Bo-1[70], Ru-3[70]
4. Daphnoretin methyl ether	R = OMe	Ru-3[70]
5. Daphnorin	R = OGlu	Ru-3[70]

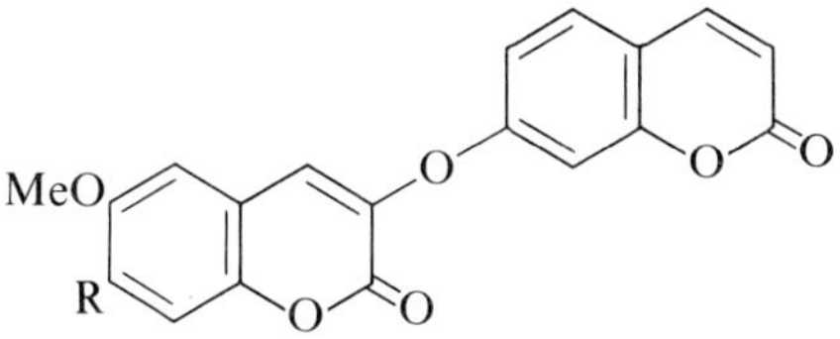

Table 16

Bicoumarins*

Name	Structure	Occurrence

Name	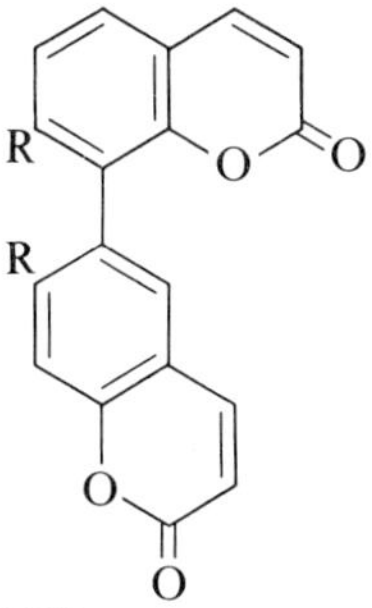 Structure	Occurrence
1. Bicoumol	R = OH	Ru-9[58]
2. Matsukazelactone	R = OMe	Bo-1[70]

* See also castanguyone (**14** in Chapter 16).[121]

118 A. I. GRAY

TABLE 17

"Diels–Alder" dimeric coumarins

Name	Structure	Occurrence
1. Phebalin	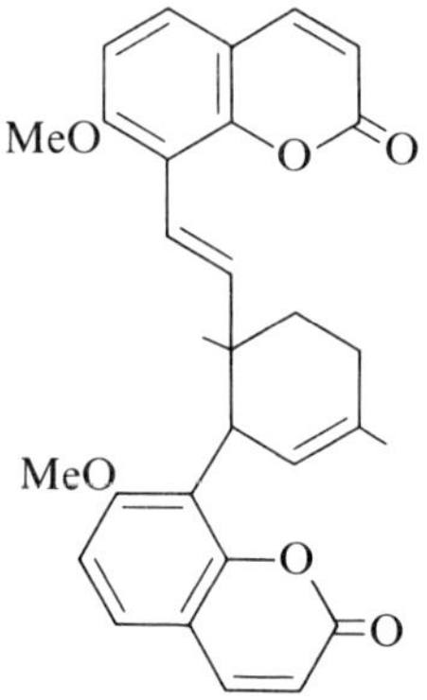	Ph-5[70]
2. Thamnosin (cyclobisuberodiene)	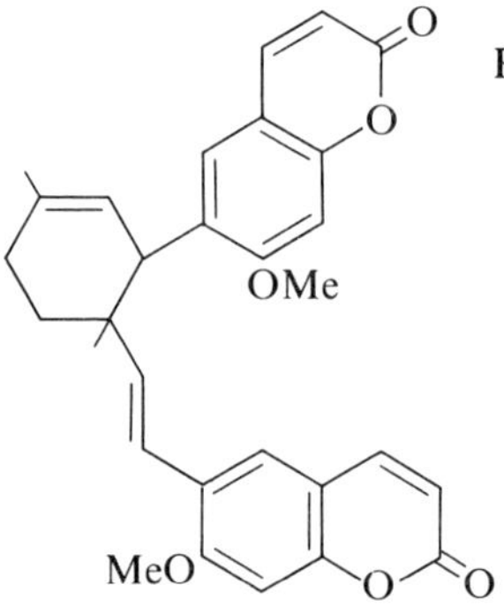	Ru-9[58], Th-1[70], Za-16[70]
3. Isothamnosin A	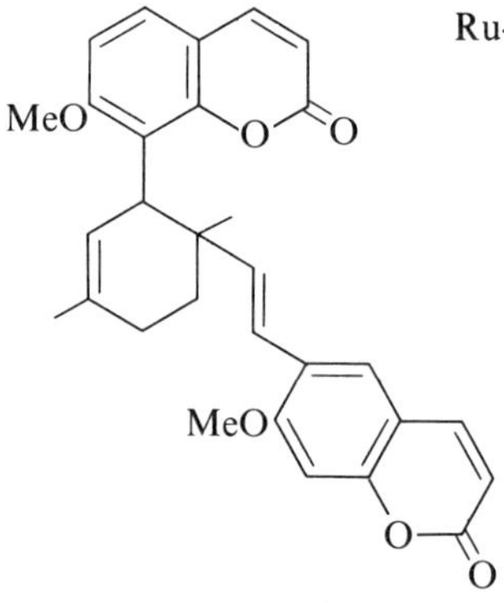	Ru-9[53]

TABLE 17—*contd.*

Name	Structure	Occurrence
4. Toddasin (mexolide)	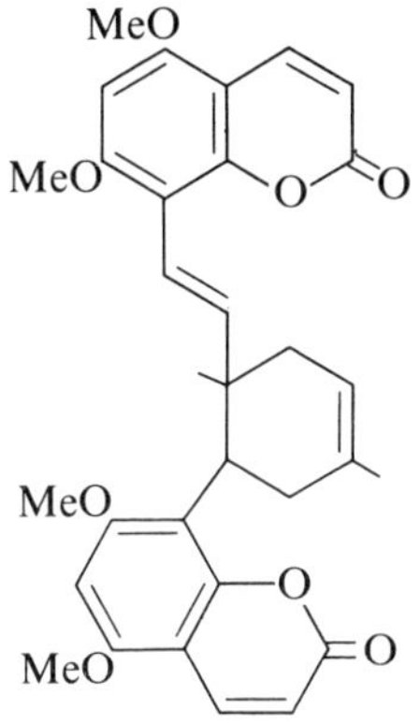	Mu-4[28], To-1[117], To-2[118]

TABLE 18

Angular furocoumarins

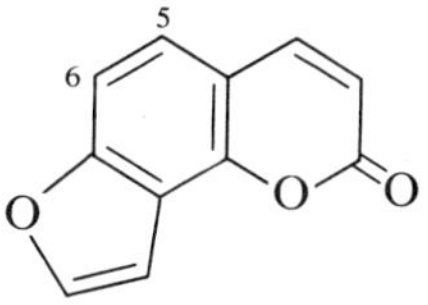

Name	Substituents	Occurrence
1. (−)-Columbianetin	I: 2′-Ip[a]	Ru-9[58], Za-3[70,100], Za-14[70]
2. Angenomalin	I: 2′-Ip[b]	Bo-1[70], Bo-2[10]
3. Angelicin	II	At-2[83]
4. Isobergapten	II: 5-OMe	Ru-8[70]
5. Sphondin	II: 6-OMe	Ru-8[70]
6. Nieshoutin (cycloobliquetin)	I: 6-OMe, 2′,3′,3′-triMe	Pt-1[100]
7. Nieshoutol	I: 5-OMe, 6-OH, 2′,3′,3′-triMe	Pt-1[100]
8. Pimpinellin	II: 5,6-diOMe	To-1[70], To-2[116,136]

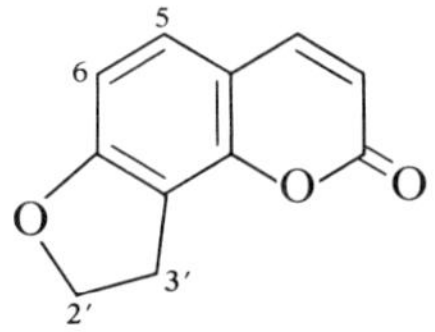

TABLE 19

Linear furocoumarins

Name	Substituents	Occurrence
1. (+)-Marmesin (−)-Nodakenetin (±)-Prangeferol	I: 2′-Ip[a]	Ae-1[70,100], Af-1[70], Am-3[24], At-2[70], Ch-1[70], Cn-1[70], Eu-2[70], Fe-1[70], He-2[70], Pe-1[70], Po-1[70], Pt-4[70], Pt-5[70], Pt-6[70], Ru-3[70,133], Ru-8[70], Ru-9[70], Za-3[70], Za-5[70]
2. Isoangenomilin	I: 2′-Ip[b]	Ho-1[95]
3. (−)-Nodakenetin acetate (marmesin acetate)	I: 2′-Ip[c]	Am-2[24], Bo-1[70,100]
4. (−)-Marmesinin (+)-Nodakenin	I: 2′-Ip[d]	Ae-1[80,113], Pt-6[70], Ru-3[70]
5. Xanthoarnol	I: 2′-Ip[a], 3′-OH	Za-3[70,100]
6. Psoralen	II	Ae-1[70], At-2[83], Di-1[70], Di-2[70], Di-3[70], Di-4[70], Pe-1[70], Ph-1[70], Pt-2[70], Pt-4[70], Ru-1[70], Ru-2[70], Ru-3[70], Ru-6[70], Ru-7[70], Ru-8[70], Th-1[70], Za-3[70], Za-10[70], Za-13[70]

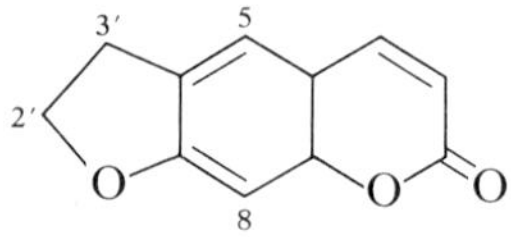

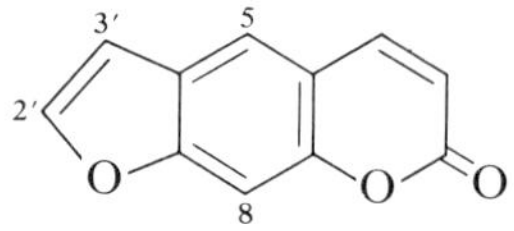

TABLE 20

8-Prenylated furocoumarins

Name	Substituents	Occurrence
1. —	I: 2′-Ip[a], 8-Pre[a]	Ch-1[13]
2. Swietenocoumarin A	II: 8-Pre[a]	Ch-1[13]
3. Swietenocoumarin C	II: 8-Pre[b]	Ch-1[13]
4. Swietenocoumarin E	II: 8-Pre[c]	Ch-1[13]
5. Swietenone	II: 8-Pent[c]	Ch-1[13,70]
6. Swietenocoumarin H	II: 8-Pre[n]	Ch-1[107]

TABLE 21

5-Oxygenated furocoumarins

Name	Substituents	Occurrence
1. Hortinone	I: 2′-Ip[b], 5-OMe	Ho-1[95]
2. —	II: 2′-Ip[a], 5-OMe	Pe-1[70]
3. Bergaptol	II: 5-OH	Ci-2[70], Ci-3[70], Ci-6[70,100] Ci-17[70]
4. Bergapten	II: 5-OMe	Bo-1[70], Ca-1[70], Ce-1[28b], Ch-1[70], Ci-1[70], Ci-2[70], Ci-3[70], Ci-6[70,100], Ci-9[52,70], Ci-17[70], Cn-1[70], Di-1[70], Di-4[70], Er-4[70], Er-6[70], Es-1[39], Es-2[39], Eu-3[70], Fe-1[70], My-2[70], My-3[70], My-4[70], Or-1[70], Ph-1[70], Po-1[70], Pt-4[70], Pt-5[70], Ru-1[70], Ru-2[70], Ru-3[70], Ru-4[104], Ru-5[70], Ru-6[70], Ru-7[70], Ru-8[70], Ru-9[70], Se-1[70], Sk-3[70], Sk-4[70,121], Th-1[70], To-1[70], Za-13[70], Za-23[70], Za-27[129]
5. Isoimperatorin	II: 5-OPre[a]	Ci-2[70], Ci-9[70], Cl-2[67], Cn-1[70], Eu-3[70], Ru-2[42], Ru-3[70], Ru-8[70], Ru-9[70], Sk-1[70], Sk-3[70], Sk-4[122]
6. Oxypeucedanin	II: 5-OPre[b]	Ci-2[123], Ru-8[70], Sk-1[70], Sk-3[70]
7. Prangol (oxypeucedanin hydrate)	II: 5-OPre[c]	Ci-2[70], Ci-9[70], Ru-8[70], Sk-3[70]
8. Oxypeucedanin methanolate	II: 5-OPre[d]	Sk-3[70,100]
9. Pangelin	II: 5-OPre[r]	Ru-3[70], Ru-8[70]
10. Isooxypeucedanin	II: 5-OPre[s]	Ru-8[70]
11. Bergamottin	II: 5-OGer[a]	Ci-2[70], Ci-6[70,100], Ci-7[70], Ci-9[70], Ci-10[70], Ci-11[51], Ci-17[70]
12. —	II: 5-OGer[b]	Ci-17[127]
13. —	II: 5-OGer[c]	Ci-10[70,100], Ci-17[127]
14. —	II: 5-OGer[m]	Ci-17[70,100]

TABLE 22

8-Oxygenated furocoumarins

Name	Substituents	Occurrence
1. Rutaretin	I: 2′-Ip[a], 8-OH	Ru-3[70,100], Ru-9[58]
2. Rutaretin methyl ether	I: 2′-Ip[a], 8-OMe	Za-3[77]
3. Isorutarin	I: 2′-Ip[d], 8-OH	Ru-3[70,100,133]
4. Rutarin	I: 2′-Ip[a], 8-OGlu	Ru-3[70,100,133]
5. Arnocoumarin	I: 2′-Ip[b], 8-OMe	Za-3[75]
6. Xanthotoxol	II: 8-OH	Ae-1[114], Cl-5[106], Eu-3[70], Po-1[70]
7. Xanthotoxin	II: 8-OMe	Ae-1[70], Af-1[70], At-1[7], At-3[78], Bo-1[70], Cn-1[70], Di-1[70], Di-2[70], Di-3[70], Di-4[70], Eu-3[70], Lu-2[70], Or-1[70], Ph-1[70], Ph-4[70], Ru-1[70], Ru-2[70], Ru-3[70], Ru-4[104], Ru-6[70], Ru-7[70], Ru-8[70], Sk-4[122], Th-1[70], Za-1[70], Za-13[70], Za-24[100], Za-27[70]
8. Imperatorin	II: 8-OPre[a]	Ae-1[70], Ae-2[38,70], Af-1[70], Am-5[6], Ci-2[70], Ci-9[70], Ci-14[70], Cl-1[70] Cl-2[70], Cl-5[70], Cn-1[70], Es-2[39], Mu-4[46], Or-1[70], Po-1[70], Pt-4[70], Pt-5[70], Ru-7[70], Ru-8[70], Se-1[70], Za-10[70], Za-18[70]
9. Heraclenin (imperatorin oxide, prangenin)	II: 8-OPre[b]	Af-1[70], Ph-3[70,100], Po-1[70], Pt-3[70], Pt-4[70], Ru-8[70], Sk-4[122]
10. Heraclenol (prangenin hydrate)	II: 8-OPre[c]	Am-5[6], Po-1[70], Ru-8[70], Ru-9[58]
11. —	II: 8-OPre[f]	Ru-9[58]
12. Isoheraclenin	II: 8-OPre[s]	Ru-9[58]
13. 8-Geranyloxypsoralen	II: 8-OGer[a]	Ci-2[70], Ci-9[52,70,100]
14. Indicolactonediol	II: 8-OGer[l]	Cl-5[106]

TABLE 23

5-Oxygenated 8-substituted furocoumarins

Name	Substituents	Occurrence
1. Hortiolone	I: 2'-Ip[b], 5-OH, 8-DMA	Ho-1[70,100]
2. Furopinnarin	II: 5-OMe, 8-DMA	Ru-8[70,100], Ru-9[70]
3. Swietenocoumarin B	II: 5-OMe, 8-Pre[a]	Ch-1[13]
4. Swietenocoumarin D	II: 5-OMe, 8-Pre[b]	Ch-1[13]
5. Swietenocoumarin F	II: 5-OMe, 8-Pre[c]	Ch-1[13]
6. Swietenocoumarin G	II: 5-OMe, 8-Pre[n]	Ch-1[107]

TABLE 24

8-Oxygenated 5-substituted furocoumarins

Name	Substituents	Occurrence
1. Alloimperatorin	II: 5-Pre[a], 8-OH	Ae-1[70,100], Es-2[39], Po-1[70]
2. Benahorin	II: 5-DMA, 8-OMe	Ru-7[70], Ru-8[70,100], Ru-9[70]
3. Alloimperatorin methyl ether	II: 5-Pre[a], 8-OMe	Ae-1[115], Th-1[70,100]
4. —	II: 5-Pre[b], 8-OMe	Th-1[70,100]
5. —	II: 5-Pre[c], 8-OMe	Th-1[70,100]
6. Thamontanin	II: 5-Pre[cc], 8-OMe	Th-1[70,100]

TABLE 25

5,8-Dioxygenated furocoumarins

Name	Substituents	Occurrence
1. —	II- 5-OMe, 8-OH	Ca-1[70,100], Cl-5[106], Es-2[39], Ph-5[70], To-2[118]
2. Isopimpinellin	II: 5,8-diOMe	Ae-2[38,70], At-1[70], Bo-1[70], Ca-1[70], Ch-1[70], Ci-1[70], Ci-2[70], Ci-7[70], Ci-9[70], Cn-1[70], Es-2[39], Eu-3[70], Fe-1[70], Fl-1[70], Lu-2[70], Me-3[70], Po-1[70], Pt-5[70], Ru-2[70], Ru-3[70], Ru-7[70], Ru-8[70], Ru-9[58], Se-1[70], Sk-4[70,122], Th-1[70], To-1[70],

TABLE 25—*contd.*

Name	Substituents	Occurrence
		To-2[118,136], Tr-1[41], Za-1[70], Za-5[70], Za-13[70], Za-16[70]
3. Phellopterin	II: 5-OMe, 8-OPre[a]	Am-5[6], Ca-1[70], Ci-2[70], Ci-9[70], Cl-5[70], Es-2[39], My-3[70], My-4[70], Ph-5[70], Pt-2[70], Pt-5[70], Pt-6[70], Th-1[70]
4. Byakangelicol	II: 5-OMe, 8-OPre[b]	Ci-9[52]
5. Byakangelicin	II: 5-OMe, 8-OPre[c]	Ci-7[70], Cl-5[106], Pt-5[70], Pt-6[70], Ru-3[70], Ru-5[70], Ru-8[70], Ru-9[70], Ru-10[88], Th-1[70]
6. —	II: 5-OMe, 8-OPre[r]	Pt-6[70]
7. —	II: 5-OMe, 8-OGer[a]	Ca-1[70]
8. Tederin	II: 5-OIp, 8-OMe	Ru-8[70]
9. —	II: 5-OGer[a], 8-OMe	Ci-2[70]

TABLE 26

Angular pyranocoumarins

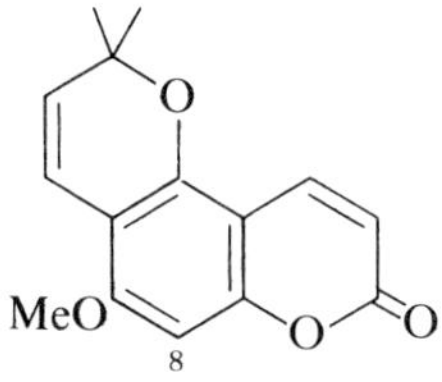

Name	Substituents	Occurrence
1. Alloxanthoxyletin	I	Ch-1[70], Za-2[70,100]
2. Dipetaline	I: 8-Pre[a]	Za-2[70], Za-7[70,100]
3. Avicennol	I: 8-Pre[n]	Za-4[70,100], Za-7[70], Za-8[70]
4. Cis-avicennol	I: 8-Pre[o]	Za-8[70,100]
5. Avicennin	I: 8-Pre[u]	Er-3[70], Za-4[70,100], Za-8[70]

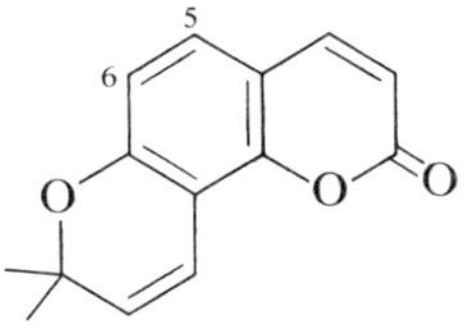

TABLE 26—contd.

Name	Substituents	Occurrence
6. Clausenidin		Cl-3[70], Cl-4[70,100], Cl-6[70]
7. Lomatin isovalerate		Ha-8[5], Ha-15[5], Ha-17[5]
8. Seselin	II	Am-5[6], Ci-3[70], Ci-4[70], Ci-7[70], Ci-17[70], Fl-1[70], Fl-5[70], Fl-8[70], Ha-5[130], Ha-6[70], Ha-9[70], My-1[70], My-2[70], My-3[70], Ph-1[70], Pi-2[93], Po-1[70], Ru-7[70], Ru-8[70], Sk-3[70], Sk-5[70,100]
9. Norbraylin	II: 6-OH	To-1[70,100]
10. Braylin	II: 6-OMe	Fl-2[70,100], Pi-2[70]
11. Dipetalolactone (hortiline)		Ho-1[70,100], Za-7[70,100], Za-8[70]

TABLE 27

Linear pyranocoumarins

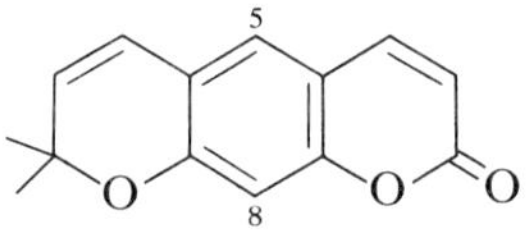

Name	Substituents	Occurrence
1. (−)-Decursinol (+)-Aegelinol	I: 3′-OH	Ae-1[31,70]
2. Xanthyletin	II	Af-1[70], At-1[7], At-2[70], At-3[78], Bo-1[70], Ch-1[70], Ch-2[70], Ch-3[70], Ci-1[70], Ci-3[70], Ci-8[70], Ci-9[70], Ci-12[70], Ci-13[70], Ci-16[70,108], Ci-17[70], Ci-19[70], Ci-20[70], Fl-8[70], Gl-1[70], Ha-6[70], Ha-9[70], Ho-1[70], Lu-1[70], Lu-2[70], Ru-2[70], Ru-3[70], Ru-5[70], Ru-7[70], Ru-8[70], Ru-9[58], Za-1[70], Za-2[70,100], Za-3[70], Za-9[99], Za-19[70]
3. Clausenin	I: 5-OH, 4′,4′-O	Cl-3[70], Cl-4[70,100]
4. Xanthoxyletin	II: 5-OMe	Af-1[70], Ch-1[70], Cl-1[70], Er-4[70], Er-6[70], Ha-2[70], Me-2[70], Me-4[70], Ru-10[88], Za-2[70], Za-7[70], Za-8[70], Za-15[99]
5. Arnottianin	I: 8-OMe, 3′-OH	Za-3[70,100]
6. —	II: 8-OH	Ch-1[13]
7. Luvangetin	II: 8-OMe	Ch-1[70], He-2[70], Lu-2[70,100], Ru-5[70], Ru-7[70], Ru-8[70], Ru-9[88], Ru-10[88], To-1[70]
8. Nordentatin	II: 5-OH, 8-DMA	Cl-2[70,100]
9. Trachyphyllin	II: 5-OH, 8-Pre[a]	Er-6[70,100]
10. Poncitrin (dentatin)	II: 5-OMe, 8-DMA	Ci-5[70], Cl-2[70,100], Cl-4[70], Cl-6[70], Po-1[70,100]
11. Racemosin	II: 5,8-diOMe	At-3[78]

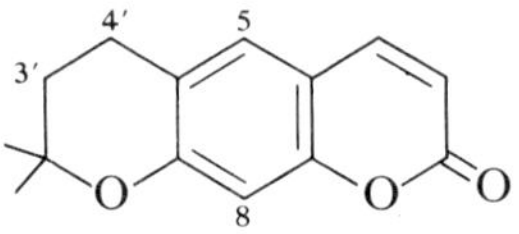

TABLE 27—*contd.*

Name	Substituents	Occurrence

	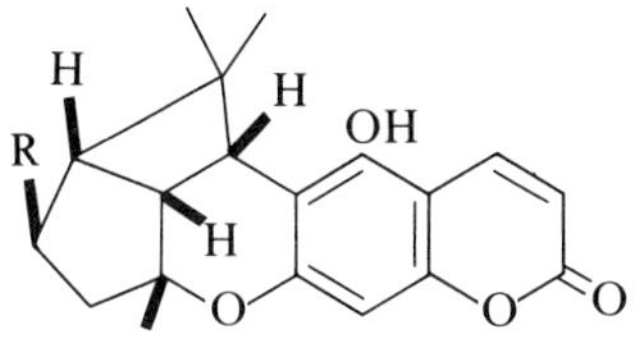	
12. Eriobrucinol	R = H	Er-2[70,100]
13. Hydroxyeriobrucinol	R = OH	Er-2[70,100]

	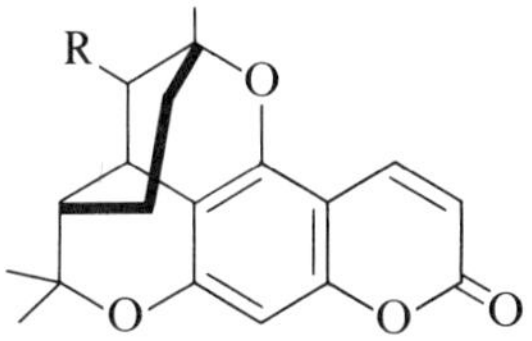	
14. Deoxybruceol	R = H	Er-2[70,100]
15. Bruceol	R = OH	Er-2[70,100]

TABLE 28

3-Substituted coumarins

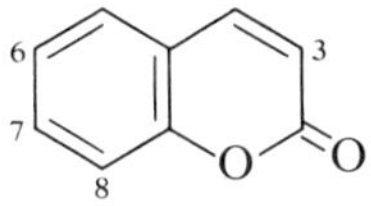

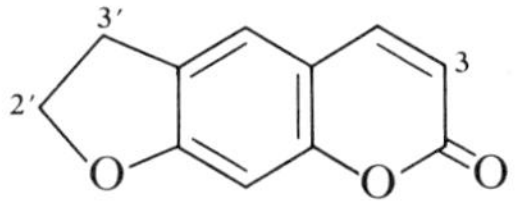

Name	Substituents	Occurrence
1. Balsamiferone	I: 3,6-diPre[a], 7-OH	Am-2[23]
2. —	II: 3-Pre[a], 2'-Ip[a]	Am-3[24]
3. —	IV: 3-Pre[a]	Am-6[32], Ho-1[95]
4. Rutolide (clausindine)	III: 3-Pent[d]	Cl-5[70,100], Ho-1[95], Ru-6[70,100]
5. —	I: 3-DMA, 7-OMe	Ru-3[70,100]
6. Pinnaterin	I: 3-DMA, 6-CH$_2$OH, 7-OMe	Ru-8[70,100]

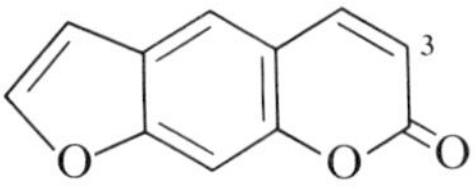

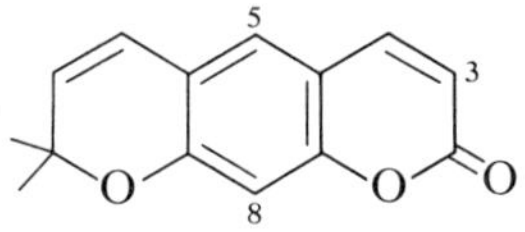

TABLE 28—*contd.*

Name	Substituents	Occurrence
7. —	I: 3-DMA, 6-CHO, 7-OMe	Ru-9[57,58]
8. —	I: 3-DMA, 6-Ip, 7-OMe	Ru-9[57,58]
9. Gravelliferone	I: 3-DMA, 6-Pre[a], 7-OH	Ru-3[70,100]
10. Gravelliferone methyl ether	I: 3-DMA, 6-Pre[a], 7-OMe	Ru-3[70,100], Ru-7[59], Ru-8[45], Ru-9[58]
11. Swietenocoumarin I	I: 3-DMA, 6-Pre[c], 7-OMe	Ch-1[107]
12. Ramosinin	I: 3-DMA, 7-OMe, 8-Pre[a]	Ha-13[47]
13. —	I: 3-DMA, 6-OMe, 7-OH	Ru-3[70,100]
14. Rutacultin	I: 3-DMA, 6,7-diOMe	Ru-3[70,100]
15. —	I: 3-DMA, 7-OMe, 8-OH	Ru-9[70,100]
16. —	I: 3-DMA, 7,8-diOMe	Ru-3[70,100]
17. 8-Methoxygravelliferone	I: 3-DMA, 6-Pre[a], 7-OH, 8-OMe	Ru-3[70,100]
18. Chalepin (heliettin, rutamarin alcohol)	II: 3-DMA, 2′-Ip[a]	Ch-1[70], Cl-1[70], Cl-5[70], He-1[70,100], Ho-1[95], Ru-2[70,100], Ru-3[70,100], Ru-6[70]
19. Chalepin acetate (rutamarin)	II: 3-DMA, 2′Ip[c]	Bo-1[70,111], Ch-1[13], Ru-2[70,100], Ru-3[70,100]
20. Chalepensin (xylotenin)	III: 3-DMA	Bo-1[70], Ch-1[70,100], Cl-5[70], Ho-1[70], Ru-2[70,100], Ru-3[70], Ru-6[70]
21. —	IV: 3-DMA	Bo-1[70,100], Cl-1[70], Cl-7[125]
22. —	IV: 3-DMA, 8-Pre[a]	Ru-3[70,100]
23. Clausarin	IV: 3-DMA, 5-OH, 8-DMA	Cl-6[70]
24. —		Cl-6[120]

TABLE 29

3,4-Disubstituted coumarins

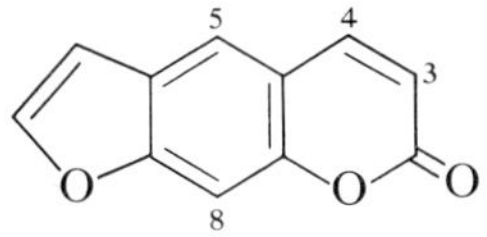

Name	Substituents	Occurrence
1. Halkendin	3,4-diOMe	Ha-1[70,100]
2. Halfordin	3,4,5-triOMe	Ha-1[70], Ha-2[70,100]
3. Halfordinin	3,5-diOMe, 4-ODMA	Ha-1[70,100]
4. Isohalfordin	3,4,8-triOMe	Ha-1[70], Ha-2[70,100]

TABLE 30

"Simple" chromones

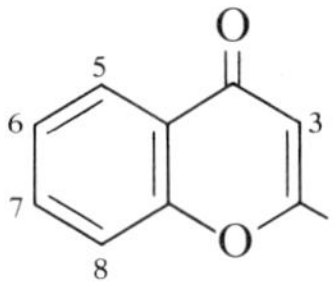

Name	Substituents	Occurrence
1. Rohitukine	5,7-diOH, 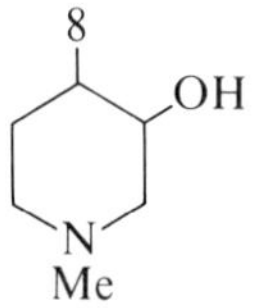	Am-1[73]
2. Peucenin	5,7-diOH, 6-Pre[a]	Ce-2[35], Pt-1[33,92]
3. Peucenin 7-methyl ether	5-OH, 6-Pre[a], 7-OMe	Pt-1[35]
4. Heteropeucenin	5,7-diOH, 8-Pre[a]	Ce-2[35], Pt-1[33]
5. Heteropeucenin 7-methyl ether (cneorum chromone O)	5-OH, 7-OMe, 8-Pre[a]	Ne-1[43], Pt-1[33]
6. Heteropeucenin 5,7-dimethyl ether	5,7-diOMe, 8-Pre[a]	Pt-1[33]
7. Pulverin (cneorum chromone H)	5,7-diOH, 6,8-diPre[a]	Ne-1[62,96]
8. Pulverin 7-methyl ether	5-OH, 7-OMe, 6,8-diPre[a]	Ne-1[43,63]

TABLE 31

Pyranochromones

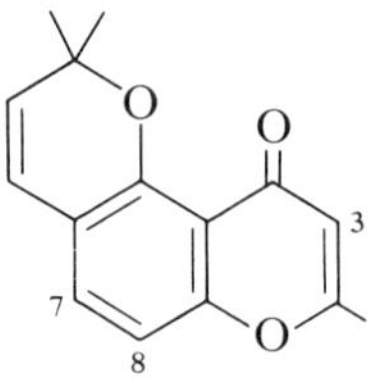

(I) 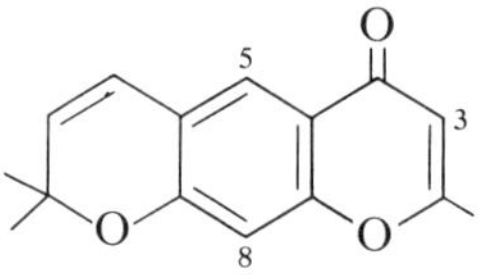 (II) 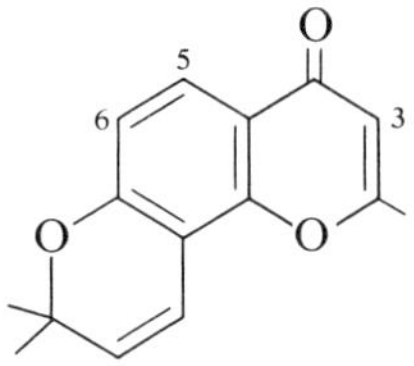 (III)

Name	Substituents	Occurrence
1. Methyliso-spatheliachromene	I: 7-OMe	Sp-2[128]
2. Spatheliachromene (cneorum chromone M)	II: 5-OH	Ne-1[43,97], Sp-2[128]
3. Methylspathelia-chromene	II: 5-OMe	Ha-18[137]
4. 8-(3,3-Dimethylallyl)-spatheliachromene (cneorum chromone B)	II: 5-OH, 8-Pre[a]	Cn-2[61,64], Ne-1[55,98], Sp-2[128]
5. Cneorum chromone K	II: 5-OH, 8-Pre[n]	Ne-1[132]
6. Cneorum chromone K 3″-methyl ether	II: 5-OH, 8-Pre[p]	Ne-1[43]
7. Cneorum chromone K 5-methyl ether	II: 5-OMe, 8-Pre[n]	Cn-2[43]
8. —	II: 5-OH, 8-Pre[x]	Sp-2[128]
9. Alloptaeroxylin (cneorum chromone D)	III: 5-OH	Ce-2[35], Ne-1[55,96], Sp-1[22], Sp-2[128]
10. Alloptaeroxylin methyl ether	III: 5-OMe	Ce-2[35], Cn-2[43,61], Sp-2[128]
11. —	III: 5-OH, 6-Pre[a]	Sp-1[22], Sp-2[128]
12. Sorbifolin (cneorum chromone I)	III: 5-OH, 6-Pre[n]	Cn-2[132], Ne-1[132], Sp-1[22], Sp-2[128]
13. Sorbifolin 5-methyl ether	III: 5-OMe, 6-Pre[n]	Sp-2[128]
14. Sorbifolin 3'-methyl ether (cneorum chromone I 3'-methyl ether)	III: 5-OH, 6-Pre[p]	Cn-2[43], Ne-1[43]
15. Anhydrosorbifolin (cneorum chromone G)	III: 5-OH, 6-Pre[u]	Cn-2[132], Ne-1[55,96], Sp-2[128]

TABLE 31—*contd.*

Name	Substituents	Occurrence
16. Spatheliabischromene (cneorum chromone A)		Cn-2[61,64], Ne-1[62,98], Sp-1[22], Sp-2[128], Sp-3[128]
17. Greveiglycol		Ce-2[35]
18. Cneorum chromone Q		Cn-2[43]
19. Cneorum chromone R		Cn-2[43]

TABLE 32

2-Hydroxymethylchromones

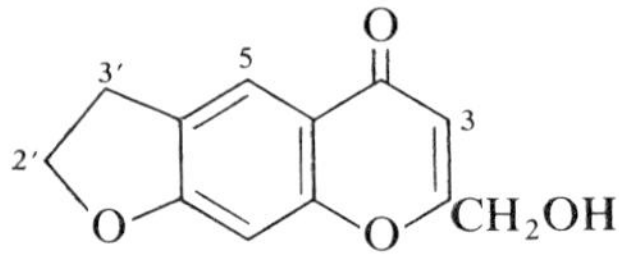

(I) 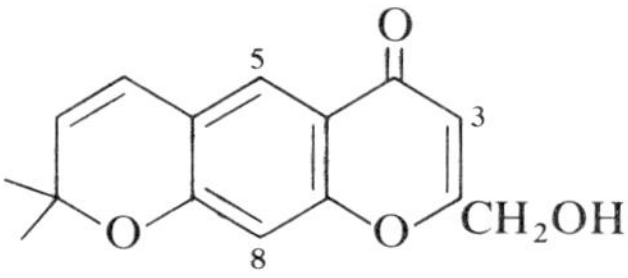   (II)

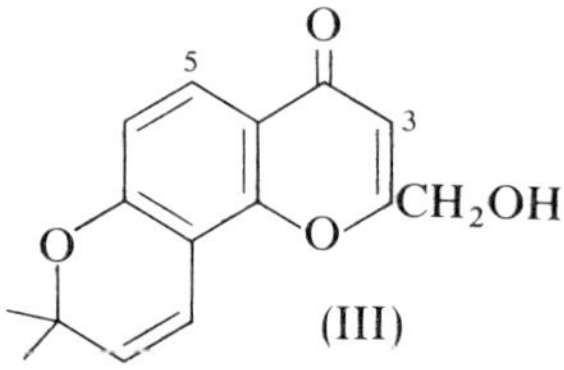

(III)

Name	Substituents	Occurrence
1. Umtatin	I: 5-OH, 2'-Ip[b]	Pt-1[33]
2. Greveichromenol	II: 5-OH	Ce-2[35]
3. Cneorum chromone C	II: 5-OH, 8-Pre[a]	Ne-1[55,96]
4. Cneorum chromone E	II: 5-OH, 8-Pre[e]	Ne-1[96]
5. Ptaerochromenol	III: 5-OH	Ne-1[63], Pt-1[33]
6. Ptaerochromenol 5-methyl ether	III: 5-OMe	Cn-2[61], Ne-1[63]
7. Cneorum chromone P	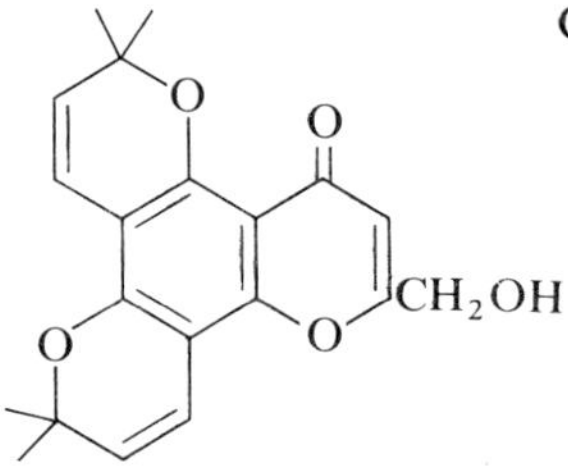	Cn-2[43]

TABLE 33

Oxepinochromones

Name	Structure	Occurrence
1. Ptaeroxylone	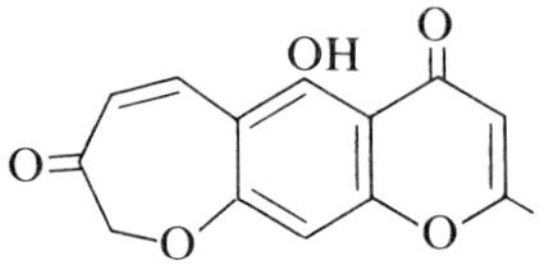	Pt-1[34]

TABLE 33—*contd.*

Name	Structure	Occurrence
2. Ptaeroxylin (desoxykarenin)		Ce-2[44], Pt-1[34,92]
3. Ptaeroxylinol R = H 4. Cneorum chromone N R = Ac (ptaeroxylinol acetate)		Pt-1[34] Ne-1[43]
5. Ptaeroglycol R = H (cneorum chromone F) 6. Cneorum chromone F primary acetate (ptaeroglycol primary acetate) R = Ac		Ce-2[35], Ne-1[63,96] Pt-1[34] Ne-1[43]
7. Ptaerocyclin		Pt-1[33]
8. Karenin		Pt-1[92]
9. Dehydroptaeroxylin		Pt-1[34]

TABLE 34

Chromones of *Flindersia laevicarpa*[105]

Name	Structure
	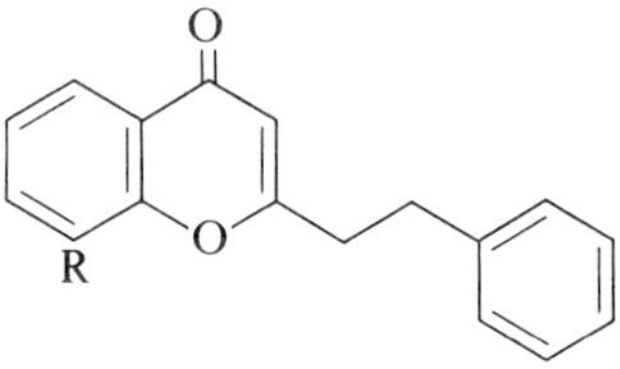
1. Flindersiachromone	R = H
2. 8-Methoxyflindersia- chromone	R = OMe
3. Dihydroflindersia- chromone	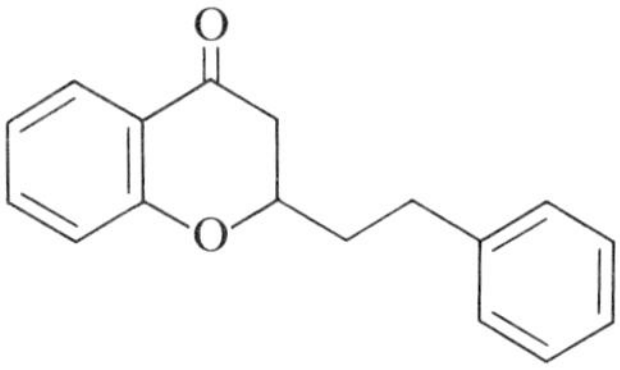

TABLE 35

Coumarin bearing plants of the order Rutales and their contained coumarins; plants belong to the Rutaceae except for those marked as belonging to the Simaroubaceae (*), Meliaceae (†), Cneoraceae (‡), Ptaeroxylaceae (§), or Burseraceae (¶)

Code	Plant	Coumarins
Ae-1	*Aegle marmelos*	**2**-1, **2**-9, **2**-11, **2**-17, **8**-4, **8**-5, **19**-1, **19**-4, **19**-6, **22**-6 **22**-7, **22**-8, **24**-1, **24**-3, **27**-1
Ae-2	*Aegleopsis chevalieri*	**8**-5, **22**-8, **25**-2
Af-1	*Afraegle paniculata*	**2**-9, **8**-2, **8**-5, **19**-1, **22**-7, **22**-8, **22**-9, **27**-2, **27**-4
Ag-1	*Agathosma collina*	**8**-5
Ag-2	*A. eriantha*	**8**-5
Ag-3	*A. imbricata*	**8**-5, **8**-6, **13**-1, **13**-2
Ag-4	*A. lanceolata*	**8**-5
Ag-5	*A. mundtii*	**8**-6
Ag-6	*A. puberula*	**13**-6
Ag-7	*A. scaperula*	**8**-5

TABLE 35—*contd.*

Code	Plant	Coumarins
Ai-1	*Ailanthus altissima**	**8**-4
Am-2	*Amyris balsamifera*	**19**-3, **28**-1
Am-3	*A. elemifera*	**3**-11, **3**-14, **3**-18, **3**-20, **19**-1, **28**-2
Am-4	*A. madrensis*	**3**-14, **3**-18
Am-5	*A. pinnata*	**8**-4, **22**-8, **22**-10, **25**-3, **26**-8
Am-6	*A. simplicifolia*	**28**-3
At-1	*Atalantia missiones*	**3**-23, **22**-7, **25**-2, **27**-2
At-2	*A. monophylla*	**2**-9, **18**-3, **19**-1, **19**-6, **27**-2
At-3	*A. racemosa*	**22**-7, **27**-2, **27**-11
At-4	*A. wightii*	**2**-9
Bo-1	*Boenninghausenia albiflora*	**2**-1, **3**-1, **3**-3, **3**-6, **3**-7, **4**-19, **10**-1, **15**-3, **16**-2, **18**-2, **19**-3, **21**-4, **22**-7, **25**-2, **27**-2, **28**-19, **28**-20, **28**-21
Bo-2	*Boronella aff. verticillata*	**18**-2
Bu-1	*Bursera aptera* ¶	**8**-4
Bu-2	*B. morelense*¶	**8**-4
Ca-1	*Casimiroa edulis*	**8**-4, **21**-4, **25**-1, **25**-2, **25**-3, **25**-7
Ce-1	*Cedrela toona*†	**1**-3, **21**-4
Ce-2	*Cedrelopsis grevei*§	**9**-3
Ch-1	*Chloroxylon swietenia*	**3**-4, **3**-5, **3**-8, **8**-5, **19**-1, **20**-1, **20**-2, **20**-3, **20**-4, **20**-5, **20**-6, **21**-4, **23**-3, **23**-4, **23**-5, **23**-6, **25**-2, **26**-1, **27**-2, **27**-4, **27**-6, **27**-7, **28**-11, **28**-18, **28**-19, **28**-20
Ch-2	*Choisya arizonica*	**4**-3, **27**-2
Ch-3	*C. mollis*	**4**-3, **27**-2
Ch-4	*C. ternata*	**4**-16
Ch-5	*Chukrasia tabularis*†	**8**-4, **8**-5
Ci-1	*Citrus acida*	**5**-2, **21**-4, **25**-2, **27**-2
Ci-2	*C. aurantifolia*	**5**-2, **5**-4, **5**-5, **21**-3, **21**-4, **21**-5, **21**-6, **21**-7, **21**-11, **22**-8, **22**-13, **25**-2, **25**-3, **25**-9,
Ci-3	*C. aurantium*	**2**-1, **2**-9, **4**-6, **21**-3, **21**-4, **26**-8, **27**-2
Ci-4	*C. aurantium* var. *amara*	**4**-6, **26**-8

TABLE 35—*contd.*

Code	Plant	Coumarins
Ci-5	*C. aurantium* var. *natsudaidai*	**2**-1, **2**-2, **4**-6, **27**-10
Ci-6	*C. bergamia*	**5**-2, **5**-4, **21**-3, **21**-4, **21**-11
Ci-7	*C. limetta*	**5**-2, **5**-4, **21**-11, **25**-2, **25**-5, **26**-8
Ci-8	*C. limettoides*	**27**-2
Ci-9	*C. limon*	**2**-1, **2**-16, **5**-2, **5**-3, **5**-4, **8**-4, **21**-4, **21**-5, **21**-7, **21**-11, **22**-8, **22**-13, **25**-2, **25**-3, **25**-4, **27**-2
Ci-10	*C. macroptera*	**21**-11, **21**-13
Ci-11	*C. medica*	**21**-11
Ci-12	*C. medica* var. *sarcodactylis*	**5**-2, **27**-2
Ci-13	*C. medica* var. *vulgaris*	**27**-2
Ci-14	*C. meyeri*	**22**-8
Ci-15	*C. mitis*	**8**-5
Ci-16	*C. nobilis*	**3**-1, **3**-8, **3**-11, **27**-2
Ci-17	*C. paradisi*	**2**-1, **2**-9, **2**-10, **2**-11, **4**-4, **4**-8, **4**-12, **5**-2, **21**-3, **21**-4, **21**-11, **21**-12, **21**-13, **21**-14, **26**-8, **27**-2
Ci-18	*C. reticulata*	**8**-5
Ci-19	*C. sinensis*	**4**-6, **27**-2
Ci-20	*C. tankan*	**27**-2
Cl-1	*Clausena anisata*	**4**-2, **7**-2, **22**-8, **27**-4, **28**-18, **28**-21
Cl-2	*C. dentata*	**21**-5, **22**-8, **27**-8, **27**-10
Cl-3	*C. excavata*	**26**-6, **27**-3
Cl-4	*C. heptaphylla*	**26**-6, **27**-3, **27**-10
Cl-5	*C. indica*	**3**-8, **22**-6, **22**-8, **22**-14, **25**-1, **25**-3, **25**-5, **28**-4, **28**-18, **28**-20
Cl-6	*C. pentaphylla*	**26**-6, **27**-10, **28**-23, **28**-24
Cl-7	*C. wildenovii*	**28**-21
Cn-1	*Cneoridium dumosum*	**4**-2, **19**-1, **21**-4, **21**-5, **22**-7, **22**-8, **25**-2
Cn-2	*Cneorum tricoccon*‡	**8**-18, **8**-20, **8**-21
Co-1	*Coleonema album*	**2**-3, **2**-4, **2**-5, **3**-10, **3**-16, **10**-4
Co-2	*C. calycinum*	**2**-3, **2**-4, **2**-5, **2**-8, **3**-10, **10**-3, **10**-4, **10**-5
Di-1	*Dictamnus albus*	**2**-9, **19**-6, **21**-4, **22**-7
Di-2	*D. dasycarpus*	**19**-6, **22**-7

TABLE 35—*contd.*

Code	Plant	Coumarins
Di-3	*D. gymnostylis*	**19**-6, **22**-7
Di-4	*D. hispanicus*	**19**-6, **21**-4, **22**-7
Di-5	*Diosma acmaeophylla*	**2**-3, **8**-5, **8**-6, **10**-2, **10**-8
Di-6	*D. pilosa*	**11**-3
Ek-1	*Ekebergia senegalensis†*	**1**-2
Er-1	*Eremocitrus glauca*	**2**-1, **3**-16
Er-2	*Eriostemon brucei*	**27**-12, **27**-13, **27**-14, **27**-15
Er-3	*E. coccineus*	**26**-5
Er-4	*E. obovalis*	**5**-2, **21**-4, **27**-4
Er-5	*E. tomentellus*	**3**-23
Er-6	*E. trachyphyllus*	**21**-4, **27**-4, **27**-9
Es-1	*Esenbeckia berlandieri*	**21**-4
Es-2	*E. litoralis*	**21**-4, **22**-8, **24**-1, **25**-1, **25**-2, **25**-3
Eu-1	*Euodia alata*	**2**-1
Eu-2	*E. belahe*	**3**-15, **19**-1
Eu-3	*E. hupehensis*	**21**-4, **21**-5, **22**-6, **22**-7, **25**-2
Eu-4	*E. viteflora*	**2**-3, **2**-6, **2**-7
Fe-1	*Feronia elephantum*	**2**-9, **8**-13, **19**-1, **21**-4, **25**-2
Fl-1	*Flindersia bennettiana*	**4**-2, **25**-2, **26**-8
Fl-2	*F. brayleana*	**9**-4, **26**-10
Fl-3	*F. collina*	**10**-9
Fl-4	*F. dissosperma*	**10**-1, **10**-9
Fl-5	*F. ifflaiana*	**26**-8
Fl-6	*F. laevicarpa*	no coumarins found todate
Fl-7	*F. maculosa*	**10**-9
Fl-8	*F. pimenteliana*	**26**-8, **27**-2
Fl-9	*F. pubescens*	**4**-2
Fl-10	*F. schottiana*	**4**-2
Ge-1	*Geijera parviflora*	**2**-1, **2**-9, **2**-11, **2**-12, **2**-14, **2**-15, **3**-14
Ge-2	*G. salicifolia*	**2**-15, **3**-12, **3**-14
Gl-1	*Glycosmis cyanocarpa*	**3**-1, **5**-2, **27**-2
Ha-1	*Halfordia kendack*	**29**-1, **29**-2, **29**-3, **29**-4
Ha-2	*H. scleroxyla*	**27**-4, **29**-2, **29**-4
Ha-3	*Haplophyllum alberty-regelli*	**10**-9
Ha-4	*H. bungei*	**4**-2, **5**-1, **8**-6
Ha-5	*H. dubium*	**26**-8
Ha-6	*H. dzhungaricum*	**26**-8, **27**-2

TABLE 35—*contd.*

Code	Plant	Coumarins
Ha-7	*H. hispanicum*	**2**-1, **2**-9, **8**-4, **8**-13, **11**-2, **11**-4
Ha-8	*H. kowalenskyi*	**3**-2, **8**-4, **10**-6, **10**-7, **26**-7
Ha-9	*H. multicaule*	**26**-8, **27**-2
Ha-10	*H. obtusifolium*	**8**-6, **8**-7, **8**-8, **8**-9, **11**-1, **13**-4, **13**-5, **13**-7, **13**-8, **13**-9
Ha-11	*H. pedicellatum*	**8**-4, **8**-13, **8**-14, **8**-15, **8**-16, **8**-17
Ha-12	*H. perforatum*	**8**-4, **8**-10, **8**-11, **8**-12
Ha-13	*H. ramosissimum*	**4**-3, **8**-6, **28**-12
Ha-14	*H. schelkovnikovii*	**1**-4
Ha-15	*H. tenue*	**3**-2, **8**-4, **10**-6, **10**-7, **26**-7
Ha-16	*H. versicolor*	**4**-24
Ha-17	*H. villosum*	**3**-2, **8**-4, **10**-6, **10**-7, **26**-7
He-1	*Helietta longifoliata*	**28**-18
He-2	*Hesperethusa crenulata*	**2**-18, **3**-1, **3**-8, **3**-10, **3**-11, **19**-1, **27**-7
Ho-1	*Hortia arborea*	**19**-2, **21**-1, **23**-1, **26**-11, **27**-2, **28**-3, **28**-4, **28**-18, **28**-20
Ho-2	*H. badinii*	**7**-2
Ho-3	*H. longifolia*	**7**-2, **8**-5
Li-1	*Limonia acidissima*	**3**-9
Lu-1	*Luvunga eleutherandra*	**3**-23, **27**-2
Lu-2	*L. scandens*	**22**-7, **25**-2, **27**-2, **27**-7
Me-1	*Melia azedarach*†	**8**-1, **8**-2, **8**-4
Me-2	*Melicope mantellii*	**27**-4
Me-3	*M. melanophloia*	**25**-2
Me-4	*M. ternata*	**27**-4
Mi-1	*Micromelum integerrimum*	**3**-22, **8**-4
Mi-2	*M. minutum*	**3**-22, **4**-2
Mi-3	*M. pubescens*	**3**-16, **3**-22, **4**-19
Mi-4	*M. zeylanicum*	**2**-13, **3**-1, **3**-22
Mu-1	*Murraya elongata*	**4**-18, **4**-21
Mu-2	*M. koenigii*	**8**-4
Mu-3	*M. omphalocarpa*	**7**-2, **7**-3, **7**-4, **7**-5
Mu-4	*M. paniculata*	**4**-2, **4**-4, **4**-12, **4**-13, **4**-17, **4**-18, **4**-20, **4**-21, **4**-22, **4**-23, **7**-2, **7**-5, **8**-4, **8**-10, **17**-4, **22**-8
My-1	*Myrtopsis macrocarpa*	**4**-2, **26**-8
My-2	*M. myrtoidea*	**21**-4, **26**-8

TABLE 35—*contd.*

Code	Plant	Coumarins
My-3	*M. novae-caledonica*	**4**-2, **21**-4, **25**-3, **26**-8
My-4	*M. sellingii*	**4**-14, **4**-15, **21**-4, **25**-3
Ne-1	*Neochamaelea pulverulenta*‡	**9**-3, **9**-5
Or-1	*Orixa japonica*	**21**-4, **22**-7, **22**-8,
Pe-1	*Pelea barbigera*	**8**-4, **19**-1, **19**-6, **21**-2
Ph-1	*Phebalium argenteum*	**19**-6, **21**-4, **22**-7, **26**-8
Ph-2	*P. dentatum*	**4**-17
Ph-3	*P. drummondii*	**4**-17, **22**-9
Ph-4	*P. filiforme*	**22**-7
Ph-5	*P. nudum*	**17**-1, **25**-1, **25**-3
Ph-6	*P. tuberculatum*	**4**-17
Ph-7	*Phyllosma capensis*	**13**-4
Pi-1	*Picrasma excelsa**	**8**-4
Pi-2	*Pitavia punctata*	**26**-8, **26**-10
Po-1	*Poncirus trifoliata*	**2**-9, **4**-5, **4**-10, **8**-13, **19**-1, **21**-4, **22**-6, **22**-8, **22**-9, **22**-10, **24**-1, **25**-2, **26**-8, **27**-10
Pr-1	*Protium opacum* ¶	**13**-10
Pt-1	*Ptaeroxylon obliquum*§	**8**-3, **8**-4, **8**-6, **8**-18, **8**-19, **9**-1, **9**-2, **18**-6, **18**-7
Pt-2	*Ptelea aptera*	**19**-6, **25**-3
Pt-3	*P. baldwinii*	**22**-9
Pt-4	*P. crenulata*	**2**-9, **19**-1, **19**-6, **21**-4, **22**-8, **22**-9
Pt-5	*P. trifoliata*	**2**-9, **8**-4, **8**-10, **19**-1, **21**-4, **22**-8, **25**-2, **25**-3, **25**-5
Pt-6	*P. trifoliata* ssp. *pallida* var. *confinis*	**19**-1, **19**-4, **25**-3, **25**-5, **25**-6
Ru-1	*Ruta bracteosa*	**19**-6, **21**-4, **22**-7
Ru-2	*R. chalepensis*	**19**-6, **21**-4, **21**-5, **22**-7, **25**-2, **27**-2, **28**-18, **28**-19, **28**-20
Ru-3	*R. graveolens*	**1**-1, **2**-1, **2**-2, **3**-3, **3**-24, **8**-4, **8**-10, **15**-3, **15**-4, **15**-5, **19**-1, **19**-4, **19**-6, **21**-4, **21**-5, **21**-9, **22**-1, **22**-3, **22**-4, **22**-7, **25**-2, **27**-2, **28**-5, **28**-9, **28**-10, **28**-13, **28**-14, **28**-16, **28**-17, **28**-18, **28**-19, **28**-20, **28**-21
Ru-4	*R. macrophylla*	**21**-4, **22**-7

TABLE 35—*contd.*

Code	Plant	Coumarins
Ru-5	*R. microcarpa*	**21**-4, **25**-5, **27**-2, **27**-7
Ru-6	*R. montana*	2-2, **19**-6, **21**-4, **22**-7, **28**-4, **28**-18, **28**-20,
Ru-7	*R. oreojasme*	1-1, 2-1, 2-2, 5-2, 8-1, **12**-3, **13**-3, **15**-1, **15**-2, **19**-6, **21**-4, **22**-7, **22**-8, **24**-2, **25**-2, **26**-8, **27**-2, **27**-7, **28**-10
Ru-8	*R. pinnata*	1-1, 2-1, 2-2, 3-16, 3-17, 3-19, 3-20, 5-2, 7-1, 8-1, 8-4, **12**-3, **12**-4, **12**-5, **14**-1, **18**-4, **18**-5, **19**-1, **19**-6, **21**-4, **21**-5, **21**-6, **21**-7, **21**-9, **21**-10, **22**-7, **22**-8, **22**-9, **22**-10, **23**-2, **24**-2, **25**-2, **25**-5, **25**-8, **26**-8, **27**-2, **27**-7, **28**-6, **28**-10
Ru-9	*R.* sp. *Tene 29662*	1-5, 2-1, 3-1, 5-2, **12**-1, **12**-2, **13**-3, **14**-1, **16**-1, **17**-2, **17**-3, **18**-1, **19**-1, **21**-4, **21**-5, **22**-1, **22**-10, **22**-11, **22**-12, **23**-2, **24**-2, **25**-2, **25**-5, **27**-2, **27**-7, **28**-7, **28**-8, **28**-10, **28**-15,
Ru-10	*R.* sp. *46782*	**25**-5, **27**-4, **27**-7
Se-1	*Severinia buxifolia*	**7**-3, **7**-5, **21**-4, **22**-8, **25**-2
Sk-1	*Skimmia foremanii*	**21**-5, **21**-6
Sk-2	*S. fortunii*	2-17
Sk-3	*S. japonica*	2-1, 2-17, 4-4, 4-9, 4-12, 8-4, **21**-4, **21**-5, **21**-6, **21**-7, **21**-8, **26**-8
Sk-4	*S. laureola*	2-1, 2-17, 8-4, 8-10, **21**-4, **21**-5, **22**-7, **22**-9, **25**-2
Sk-5	*S. repens*	**26**-8
Sw-1	*Swietenia mahogoni†*	8-4
Th-1	*Thamnosma montana*	2-16, 3-13, 3-21, **17**-2, **19**-6, **21**-4, **22**-7, **24**-3, **24**-4, **24**-5, **24**-6, **25**-2, **25**-3, **25**-5
To-1	*Toddalia aculeata*	6-1, 6-2, 6-3, 6-4, 6-5, 6-6, **12**-2, **17**-4, **18**-8, **21**-4, **25**-2, **26**-9, **27**-7
To-2	*T. asiatica*	6-3, 6-5, 6-6, **17**-4, **18**-8, **25**-1, **25**-2
Tr-1	*Triphasia trifolia*	2-1, 4-9, 4-11, **25**-2
Za-1	*Zanthoxylum ailanthoides*	8-1, 8-5, **13**-3, **22**-7, **25**-2, **27**-2

TABLE 35—*contd.*

Code	Plant	Coumarins
Za-2	*Z. americanum*	**26**-1, **26**-2, **27**-2, **27**-4
Za-3	*Z. arnottianum*	1-6, **1**-7, **3**-4, **3**-8, 3-16, **4**-1, 4-7, **10**-10, **18**-1, **19**-1, **19**-5, 19-6, **22**-2, **22**-5, 27-2, 27-5
Za-4	*Z. avicennae*	26-3, **26**-5
Za-5	*Z. belizense*	**19**-1, **25**-2
Za-6	*Z. decaryi*	**8**-5
Za-7	*Z. dipetalum*	**26**-2, **26**-3, **26**-11, **27**-4
Za-8	*Z. elephantiasis*	**26**-3, **26**-4, **26**-5, **26**-11, **27**-4,
Za-9	*Z. faurei*	**27**-2
Za-10	*Z. flavum*	3-8, **19**-6, **22**-8
Za-11	*Z. gillettii*	**8**-5, **13**-3
Za-12	*Z. leprieurii*	**13**-3
Za-13	*Z. mayu*	**19**-6, **21**-4, **22**-7, **25**-2
Za-14	*Z. monophyllum*	**18**-1
Za-15	*Z. okinawense*	**27**-4
Za-16	*Z. ovalifolium*	2-9, **3**-1, 3-8, 3-11, **17**-2, **25**-2
Za-17	*Z. piperitum*	**8**-5
Za-18	*Z. piperitum* var. *inerme*	**22**-8
Za-19	*Z. pluviatile*	**27**-2
Za-20	*Z. procerum*	**13**-3
Za-21	*Z. pterota*	**8**-4
Za-22	*Z. rhetsa*	3-8
Za-23	*Z. schinifolium*	**8**-4, **8**-5, **21**-4
Za-24	*Z. senegalense*	**22**-7
Za-25	*Z. setosum*	**8**-5
Za-26	*Z. williamsii*	**8**-5
Za-27	*Z. xanthoxyloides*	**21**-4, **22**-7

Coumarin (**1**-1) has been reported from *Bursera balsamifera, B. leptophloeos, Protium* sp. "ubirasiqua" and *P. haplophyllum* (all Burseraceae), siderin (**1**-3) from *Balsamodendron commiphora* (Burseraceae), and scopeletin (**8**-4) from *Cedrela sinensis* (Meliaceae); see Murray, R. D. H., Mendez, J. and Brown, S. A. (1982). "The Natural Coumarins". Wiley-Interscience, New York.

TABLE 36

Chromone bearing plants of the Rutales and their contained chromones; plants not belonging to the Rutaceae are indicated by symbols as in Table 35.

Code	Plant	Chromones
Am-1	*Amoora rohituka*†	**30**-1
Ce-2	*Cedrelopsis grevei*§	**30**-2, **30**-4, **31**-9, **31**-10, **31**-17, **32**-2, **33**-2, **33**-5

TABLE 36—contd.

Code	Plant	Chromones
Cn-2	Cneorum tricoccon‡	31-4, 31-7, 31-10, 31-12, 31-14, 31-15, 31-16, 31-18, 31-19, 32-6, 32-7
Fl-6	Flindersia laevicarpa	34-1, 34-2, 34 3
Ha-18	Harrisonia abyssinica*	31-3
Ne-1	Neochamaelea pulverulenta‡	30-5, 30-7, 30-8, 31-2, 31-4, 31-5, 31-6, 31-9, 31-12, 31-14, 31-15, 31-16, 32-3, 32-4, 32-5, 32-6, 33-4, 33-5, 33-6
Pt-1	Ptaeroxylon obliquum§	30-2, 30-3, 30-4, 30-5, 30-6, 32-1, 32-5, 33-1, 33-2, 33-3, 33-5, 33-7, 33-8
Sp-1	Spathelia glabrescens	31-9, 31-11, 31-12, 31-16
Sp-2	S. sorbifolia	31-1, 31-2, 31-4, 31-8, 31-9, 31-10, 31-11, 31-12, 31-13, 31-15, 31-16
Sp-3	S. sp. 88/2	31-16

REFERENCES

1. Abyshev, A. Z., Denisenko, P. P., Isaev, N. Ya. and Kerimov, Yu. B. (1978). Khim. Prir. Soedin. 654.
2. Abyshev, A. Z. and Gashimov, N. F. (1979). Khim. Prir. Soedin. 401, 403.
3. Abyshev, A. Z. and Gashimov, N. F. (1979). Khim. Prir. Soedin. 845.
4. Abyshev, A. Z. and Gashimov, N. F. (1979). Khim. Prir. Soedin. 846.
5. Abyshev, A. Z., Isaev, N. Ya, and Kerimov, Yu. B. (1980). Khim. Prir. Soedin. 800.
6. Badawi, M. M., Seida, A. A., Kinghorn, A. D., Cordell, G. A. and Farnsworth, N. R. (1981). J. Nat. Prod. 44, 331.
7. Barua, A. K., Ghosh, S., Chakrabarti, P. and Bose, P. K. (1980). Planta Med. 38, 188.
8. Basak, S. P. and Chakraborty, D. P. (1970). J. Indian Chem. Soc. 47, 722.
9. Batirov, E. Kh., Matkarimov, A. D., Malikov, V. M., Yagudaev, M. R. and Seitmuratov, E. (1980). Khim. Prir. Soedin. 785.
10. Bevalot, F., Vaquette, J. and Cabalion, P. (1980). Plant. Med. Phytother. 14, 218.
11. Bevan, C. W. L. and Ekong, D. E. U. (1965). Chem. Ind. (London) 383.
12. Bhan, M. K., Raj, S., Nayar, M. N. S. and Handa, K. L. (1973). Phytochemistry 12, 3010.
13. Bhide, K. S., Majumdar, R. D. and Rao, A. V. R. (1977) Indian J. Chem. 15B, 440.
14. Binns, S. V., Halpern, B., Hughes, G. K. and Ritchie, E. (1957). Aust. J. Chem. 10, 480.
15. Bohlmann, F., Adler, A., King, R. M. and Robinson, H. (1982). Phytochemistry 21,

173.

16. Bohlmann, F. and Wagner, P. (1982). *Phytochemistry* **21**, 1175.
17. Bohlmann, F., Wagner, P. and Jakupovic, J. (1982). *Phytochemistry* **21**, 1109.
18. Bohlmann, F., Zdero, C., King, R. M. and Robinson, H. (1980). *Phytochemistry* **19**, 1547.
19. Bohlmann, F., Zdero, C., King, R. M. and Robinson, H. (1982). *Phytochemistry* **21**, 147.
20. Boulware, R. T. and Stermitz, F. R. (1981). *J. Nat. Prod.* **44**, 200.
21. Bowen, I. H. and Perera, K. P. W. C. (1982). *Phytochemistry* **21**, 433.
22. Box, V. G. and Taylor, D. R. (1973). *Phytochemistry* **12**, 956.
23. Burke, B. A. and Parkins, H. (1979). *Phytochemistry* **18**, 1073.
24. Burke, B. A. and Philip, S. (1981). *Heterocycles* **16**, 897.
25. Campbell, W. E., unpublished results.
26. Campbell, W. E. and Cragg, G. M. L. (1979). *Phytochemistry* **18**, 688.
27. Cassady, J. M., Ojima, N., Chang, C.-J. and McLaughlin, J. L. (1979). *J. Nat. Prod.* **42**, 274.
28. Chakraborty, D. P., Roy, S., Chakraborty, A., Mandal, A. K. and Chaudhury, B. K. (1980). *Tetrahedron* **36**, 3563.
28a. Chatterjee, A., Banerjee, B., Ganguly, S. N. and Sircar, S. M. (1974). *Phytochemistry* **13**, 2012.
28b. Chatterjee, A., Chakraborty, T. and Chandrasekharan, S. (1971). *Phytochemistry* **10**, 2533.
29. Chatterjee, A., Roy, R. N., Banerji, A. and Banerji, J. (1976). *Chem. Ind.* (London) 410.
30. Chatterjee, A., Sarkar, S. and Shoolery, J. N. S. (1980). *Phytochemistry* **19**, 2219.
31. Chatterjee, A., Sen, R. and Ganguly, D. (1978). *Phytochemistry* **17**, 328.
32. Cordova, H. E. and Garelli, L. E. (1974). *Phytochemistry* **13**, 758.
33. Dean, F. M., Parton, B., Price, A. W., Somvichien, N. and Taylor, D. A. H. (1967). *Tetrahedron Lett.* 2737.
34. Dean, F. M., Parton, B., Somvichien, N. and Taylor, D. A. H. (1967). *Tetrahedron Lett.* 3459.
35. Dean, F. M. and Robinson, M. L. (1971). *Phytochemistry* **10**, 3221.
36. De Silva, L. B., De Silva, U. L. L., Mahendran, M. and Jennings, R. C. (1980). *Indian J. Chem.* **19B**, 820.
37. De Silva, L. B., Herath, W. H. M. W., Jennings, R. C., Mahendran, M. and Wannigama, G. P. (1981). *Phytochemistry* **20**, 2776.
38. Dreyer, D. L. (1968). *J. Org. Chem.* **33**, 3658.
39. Dreyer, D. L. (1980). *Phytochemistry* **19**, 941.
40. Dreyer, D. L. and Brenner, R. C. (1980). *Phytochemistry* **19**, 935.
41. Dreyer, D. L. and Lee, A. (1972). *Phytochemistry* **11**, 763.
42. El-Tawil, B. A. H., Baghlef, A. O. and Barbood, S. O. (1980). *Pharmazie* **35**, 503; *Chem. Abs.* (1981). **94**, 200989m.
43. Epe, B., Oelbermann, U. and Mondon, A. (1981). *Chem. Ber.* **114**, 757.
44. Eshiett, I. T. and Taylor, D. A. H. (1968). *J. Chem. Soc.* (C), 481.
45. Estevez, R. R. and Gonzalez, A. G. (1971). *An. Quim.* **67**, 207.
46. Ganguly, S. N., Ghosh, S. and Basak, A. (1977). *Trans. Bose. Res. Inst.* **40**, 123; *Chem. Abs.* (1979). **90**, 19071d.
47. Gashimov, N. F., Abyshev, A. Z., Kagramanov, A. A. and Rozhkova, L. I. (1979). *Khim. Prir. Soedin.* 15.

48. Gashimov, N. F., Abyshev, A. Z., Kagramanov, A. A. and Rozhkova, L. I. (1979). *Khim. Prir. Soedin.* 87.
49. Gashimov, N. F. and Orazmikhamedova, N. O. (1978). *Khim. Prir. Soedin.*, 653.
50. Ghosh, P., Sil, P., Majumdar, S. G. and Thakur, S. (1982). *Phytochemistry* **21**, 240.
51. Glandian, R., Corneteau, H., Drouet, S. and Rouzet, M. (1978). *Plant. Med. Phytother.* **12**, 112; *Chem. Abs.* (1979). **90**, 76390f.
52. Glandian, R., Corneteau, H., Drouet, S. and Rouzet, M. (1978). *Labo-Pharma-Probl. Tech.* **26**, 503; *Chem. Abs.* (1978). **89**, 117524c.
53. Gonzalez, A. G., Cardona, R. J., Diaz-Chico, E., Lopez-Dorta, H. and Rodriguez-Luis, F. (1977). *An. Quim.* **73**, 1510.
54. Gonzalez, A. G., Cardona, R. J., Moreno, O. R. and Rodriguez-Luis, F. (1973). *An. Quim.* **69**, 781.
55. Gonzalez, A. G., Castaneda, J. P. and Fraga, B. M. (1972). *An. Quim.* **68**, 447.
56. Gonzalez, A. G., Diaz-Chico, E., Lopez-Dorta, H., Medina, J. M. and Rodriguez-Luis, F. (1976). *An. Quim.* **72**, 191.
57. Gonzalez, A. G., Diaz-Chico, E., Lopez-Dorta, H., Luis, J. R. and Rodriguez-Luis, F. (1977). *An. Quim.* **73**, 607.
58. Gonzalez, A. G., Diaz-Chico, E., Lopez-Dorta, H., Medina, J. M. and Rodriguez-Luis, F. (1977). *An. Quim.* **73**, 1015.
59. Gonzalez, A. G., Estevez, R. R. and Jaraiz, I., (1972). *An. Quim.* **68**, 415.
60. Gonzalez, A. G., Fraga, B. M., Hernandez, M. G., Pino, O. and Ravelo, A. G. (1978). *Rev. Latinoamer. Quim.* **9**, 205.
61. Gonzalez, A. G., Fraga, B. M. and Pino, O. (1974). *Phytochemistry* **13**, 2305.
62. Gonzalez, A. G., Fraga, B. M. and Pino, O. (1975). *Phytochemistry* **14**, 1656.
63. Gonzalez, A. G., Fraga, B. M. and Pino, O. (1977). *An. Quim.* **73**, 557.
64. Gonzalez, A. G., Fraga, B. M. and Torres, R. (1974). *An. Quim.* **70**, 91.
65. Gonzalez, A. G., Lopez-Dorta, H., Martinez-Iniguez, M. A., Melian, R. M. and Rodriguez-Luis, F. (1972). *An. Quim.* **68**, 1139.
66. Gonzalez, A. G., Reyes, R. E. and Espino, M. R. (1977). *Phytochemistry* **16**, 2033.
67. Govindachari, T. R., Pai, B. R., Subramaniam, P. S. and Muthukumaraswamy, N. (1968). *Tetrahedron* **24**, 753.
68. Gray, A. I. (1981). *Phytochemistry* **20**, 1711.
69. Gray, A. I., Meegan, C. J. and O'Callaghan, N. B., unpublished results
70. Gray, A. I. and Waterman, P. G. (1978). *Phytochemistry* **17**, 845.
71. Grundon, M. F. (1978). *Tetrahedron* **34**, 143; Baird, K. J. and Grundon, M. F. (1980). *J. Chem. Soc. Perkin Trans. 1*, 1820.
72. Gupta, G. L. and Nigam, S. S. (1971). *Planta Med.* **19**, 83.
73. Harmon, A. D., Weiss, U. and Silverton, J. V. (1979). *Tetrahedron Lett.* 721.
74. Hegnauer, R. (1973). "Chemotaxonomie der Pflanzen", Vol. 6, Birkhauser, Basel.
75. Ishii, H. and Ishikawa, T. (1978). *Chem. Pharm. Bull.* **26**, 2598.
76. Ishii, H., Murakame, K., Takeishi, K., Ishikawa, T. and Haginawa, T. (1981). *Yakugaku Zasshi* **101**, 504.
77. Ishii, H., Sekiguchi, F. and Ishikawa, T. (1981). *Tetrahedron* **37**, 285.
78. Joshi, B. S., Gawad, D. H. and Ravindranath, K. R. (1978). *Proc. Indian Acad. Sci.* **87A**, 173; *Chem. Abs.* (1978). **89**, 215258p.
79. Junior, P. (1979). *Phytochemistry* **18**, 2053.
80. Khalegue, A., Ismail, K. M., Rahman, A. K. M. M., Amin, M. S., Fritz, H. and Besch, E. (1980). *Bangladesh J. Sci. Ind. Res.* **15**, 1224; *Chem. Abs.* (1981). **95**, 200604y.

81. Khalil, A. M., Ashy, M. A., Tawfik, N. I. and El-Tawil, B. A. H. (1979). *Pharmazie* **34**, 106; *Chem. Abs.* (1979). **91**, 181258.
82. Khalil, A. M., El-Tawil, B. A. H., Ashy, M. A. and Elbeih, F. K. A. (1981). *Pharmazie* **36**, 569; *Chem. Abs.* (1981). **95**, 200551d.
83. Kulkarni, G. H., Haribal, M. M. and Sabata, B. K. (1980). *Indian J. Chem.* **19B**, 424; *Chem. Abs.* (1980). **93**, 66153d.
84. Lu, S. T., Wu, T. S. and Chang, S. I. (1977). *Taiwan Yao Ikuah Tsu Chih* **29**, 1; *Chem. Abs.* (1979). **90**, 148494t.
85. Magalhaes, A. F., Magalhaes, E. G., Leitao Filho, H. F., Frighetto, R. T. S. and Barros, S. M. G. (1981). *Phytochemistry*, **20**, 1369.
86. Manandhar, M. D. (1980). *Indian J. Chem.* **19B**, 1006; *Chem. Abs.* (1981). **94**, 136115g.
87. Manandhar, M. D., Shoeb, A., Kapil, R. S. and Popli, S. P. (1978). *Phytochemistry* **17**, 1814.
88. Martinez, E. A., Reyes, R, E., Gonzalez, A. G. and Rodriguez-Luis, F. (1967). *An. Acad. Soc. Espan. Fis. Quim.*, Ser. B. **63**, 197; *Chem. Abs.* (1967). **67**, 47086f.
89. Matkarimov, A. D., Batirov, E. Kh., Malikov, V. M. and Seitmuratov, E. (1980). *Khim. Prir. Soedin.*, 328; *Chem. Abs.* (1981). **94**, 164316d.
90. Matkarimov, A. D., Batirov, E. Kh., Malikov, V. M. and Seitmuratov, E. (1980). *Khim. Prir. Soedin.* 565; *Chem. Abs.* (1981). **94**, 200982d.
91. Matkarimov, A. D., Batirov, E. Kh., Malikov, V. M. and Seitmuratov, E. (1980). *Khim. Prir. Soedin.* 831; *Chem. Abs.* (1981). **94**, 171049t.
92. McCabe, P. H., McCrindle, R. and Murray, R. D. H. (1967). *J. Chem. Soc.* (*C*), 145.
93. Millan, H. N. and Silva, O. M. (1969). *Rev. Real. Acad. Ci-Exact., Fis. Nat. Madrid* **63**, 635; *Chem. Abs.* (1970). **72**, 107836.
94. Mock, J. R., Senior, R. G. and Taylor, W. C. (1980). *Aust. J. Chem.* **33**, 395.
95. Monache, F. D., Valera, G. C., Marini-Bettolo, G. B., DeMelle, J. F. and De-Lima, O. G. (1977). *Gazz. Chim. Ital.* **107**, 399.
96. Mondon, A. and Callsen, H. (1975). *Chem. Ber.* **108**, 2005.
97. Mondon, A., Callsen, H. and Hartmann, P. (1975). *Chem. Ber.* **108**, 1989.
98. Mondon, A. and Schwarzmaier, U. (1975). *Chem. Ber.* **108**, 925.
99. Morita, N., Arisawa, M. and Takezaki, T. (1967). *Yakugaku Zasshi* **87**, 1017.
100. Murray, R. D. H. (1978). *Fortschr. Chem.* **35**, 199.
101. Nagasampagi, B. A., Sriraman, M. C., Yankov, L. and Dev, S. (1975). *Phytochemistry* **14**, 1673.
102. Nielsen, B. E. (1970). *Dansk. Tidsskr. Farm.* **44**, 111.
103. Okogun, J. I., Enyenihi, V. U. and Ekong, D. E. U. (1978). *Tetrahedron* **34**, 1221.
104. Petrushenko, N. A. and Bandyukova, V. A. (1981). *Khim. Prir. Soedin.* 393; *Chem. Abs.* (1981). **95**, 147133r.
105. Picker, K., Ritchie, E. and Taylor, W. C. (1976). *Aust. J. Chem.* **29**, 2023.
106. Prakash, D., Roy, K., Kapil, R. J. and Popli, S. P. (1978). *Phytochemistry* **17**, 1194.
107. Rao, A. V. R., Bhide, K. S. and Majumdar, R. B. (1980). *Indian J. Chem.*, **19B**, 1046; *Chem. Abs.* (1981). **94**, 117796w.
108. Reisch, J., Mester, I. and Sofowara, E. A. (1980). *Planta Med.*, Suppl., 56–59.
109. Reisch, J., Mueller, M. and Mester, I. (1981). *Planta Med.* **45**, 285.
110. Reyes, R. E. and Gonzalez, A. G. (1970). *Phytochemistry* **9**, 833.
111. Rizvi, S. H., Shoeb, A., Kapil, R. S. and Popli, S. P. (1979). *Indian J. Pharm. Sci.* **41**, 205; *Chem. Abs.* (1980). **92**, 177441.
112. Rosza, Zs., Mester, I., Reisch, J. and Szendrei, K. (1980). *Planta Med.* **39**, 219.

113. Sharma, B. R., Rattan, R. K. and Sharma, P. (1980). *Indian J. Chem.* **19B**, 162.
114. Sharma, B. R., Rattan, R. K. and Sharma, P. (1981). *Phytochemistry* **20**, 2606.
115. Sharma, B. R. and Sharma, P. (1981). *Planta Med.* **43**, 102.
116. Sharma, P. N., Shoeb, A., Kapil, R. S. and Popli, S. P. (1979). *Indian J. Chem.*, **17B**, 299.
117. Sharma, P. N., Shoeb, A., Kapil, R. S. and Popli, S. P. (1980). *Phytochemistry* **19**, 1258.
118. Sharma, P. N., Shoeb, A., Kapil, R. S. and Popli, S. P. (1981). *Indian J. Chem.* **20B**, 936; *Chem. Abs.* (1982). **96**, 65684c.
119. Sharma, P. N., Shoeb, A., Kapil, R. S. and Popli, S. P. (1981). *Phytochemistry* **20**, 335.
120. Shoeb, A., Kapil, R. S., Popli, S. P., Patnaik, G. K. and Dhaman, B. (1977). *Indian Patent* No. 145322; *Chem. Abs.* (1980). **92**, 153133v.
121. Snyder, J., Nakanishi, K., Chaverria, G., Leal, Y., Ocha, C. C. and Dominguez, X. A. (1981). *Tetrahedron Lett.* 5015.
122. Sood, S., Gupta, B. D. and Banerjee, S. K. (1978). *Planta Med.* **34**, 338.
123. Stanley, W. L. and Vannier, S. H. (1967). *Phytochemistry* **6**, 585.
124. Stermitz, F. R., Caola, M. A. and Swinehart, J. A. (1980). *Phytochemistry* **19**, 1469.
125. Subba Rao, G. S. R., Raj, K. and Kumar, V. P. S. (1981). *Indian J. Chem.* **20B**, 88; *Chem. Abs.* (1981). **94**, 153468z.
126. Szendrei, K., Korbely, T., Krenzien, H., Reisch, J. and Novak, I. (1977). *Herba Hung.* **16**, 15; *Chem. Abs.* (1978). **88**, 166719h.
127. Tatum, J. H. and Berry, R. E. (1979). *Phytochemistry* **18**, 500.
128a. Taylor, D. R., Warner, J. M. and Wright, J. A. (1977). *J. Chem. Soc., Perkin Trans.* 1, 397. (b) Taylor, D. R. and Wright, J. A. (1971). *Rev. Latinoam. Quim.* **2**, 84.
129. Thoms, H. (1911). *Ber.* **44**, 3325.
130. Tikhomirova, L. I., Kuznetsova, G. A. and Pimenov, M. G. (1977). *Khim. Prir. Soedin.* 859.
131. Tin-wa, M., Bonomo, S. and Scora, R. W. (1979). *Planta Med.* **37**, 379.
132. Trautmann, D., Epe, B., Oelbermann, U. and Mondon, A. (1976). *Chem. Ber.* **109**, 2963.
133. Varga, E., Kuzovkina, I. N., Rozsa, Z. and Szendrei, K. (1978). *Acta. Pharm. Hung.* **48**, 193.
134. Venturella, P., Bellino, A. and Piozzi, F. (1974). *Heterocycles* **2**, 345; *Chem. Abs.* (1974). **81**, 120385y.
135. Wagner, H., Nestler, T. and Neszmelyi, A. (1979). *Planta Med.* **36**, 113.
136. Wang, C.-T. and Chang, H.-J. (1981). *Shan-hsi Hsiu I Yao* **10**, 51; *Chem. Abs.* (1981). **95**, 175631v.
137. Waterman, P. G. (1982), personal communication.
138. Wu, T. S., Tien, H.-J., Arisawa, M., Shimizu, M. and Morita, N. (1980). *Phytochemistry* **19**, 2227; Wu, T. S. (1981). *Phytochemistry* **20**, 178.
139. Young, D. A. (1976). *Syst. Bot.* **1**, 149.
140. Yuldashev, M. P., Batirov, E. Kh. and Malikov, V. M. (1980). *Khim. Prir. Soedin.* 168, 412.
141. Zoghbi, M. D. G. B., Roque, N. F. and Gottlieb, O. R. (1981). *Phytochemistry* **20**, 180.

CHAPTER 5

The Flavonoids of the Rutales

JEFFREY B. HARBORNE

Plant Science Laboratories, The University, Reading, U.K.

I. INTRODUCTION

The genus *Ruta*, the type genus of the Rutaceae, has special significance to flavonoid chemists since the most common of all flavonoids, rutin (**1**), was named after it. Rutin (quercetin 3-rutinoside) was first isolated in crystalline condition from the Southern European herb, *Ruta graveolens* L. or rue, in 1842 by Weiss.[62] The use of rue as a herbal tea in the past[22] may be connected in part with the considerable rutin content, although other secondary substances, especially the volatiles, must contribute to its bitter, acrid, nauseous taste. Rutin is regarded by some as being beneficial to man and it is used medicinally in some European countries in treating capillary fragility. A more familiar herb plant with rutin is the buckwheat *Fagopyrum esculentum* (Polygonaceae), the leaves

147

of which contain 3 % dry weight of this flavonoid. Nevertheless the occurrence of rutin in *Ruta graveolens* is highly characteristic of the Rutales, and subsequent investigation has indicated that rutin is probably the most common single leaf flavonoid known in the order.

Thus rutin is known to be a regular constituent in the Rutaceae, the only rutalean family to have been widely investigated for flavonoids. Rutin has also been recorded in Cneoraceae (*Cneorum, Neochamaelea*), Meliaceae (*Melia, Soymida*), and Simaroubaceae (*Suriana*) (see Section IIB). However, the fact that rutin is a typical constituent of the order has little taxonomic relevance, since rutin is generally found in most if not all angiosperm orders.

Two much more distinctive flavonoids characteristic of the Rutaceae and rarely found elsewhere are the methylated flavone exoticin (**2**) and the flavanone glycoside naringin (**3**). Extensive *O*-methylation of highly oxygenated flavones and flavonols is a very special biosynthetic trait in these plants. The production of exoticin (3,5,6,7,8,3′,4′,5′-octamethoxyflavone), the most highly substituted flavonoid known, in the leaves of *Murraya exotica* is typical. In particular, a unique range of such methylated flavones, including nobiletin (**4**) and tangeretin (**5**), have been characterized in the genus *Citrus* (see Section IIIB).

The taxonomic significance of this biosynthetic feat of the Rutaceae is enhanced by the recent recognition that separate *O*-methylating enzymes are probably needed for the methylation of each of the eight hydroxyl substituents in the molecule of exoticin (**2**). At least, fractionation of enzymic activities from

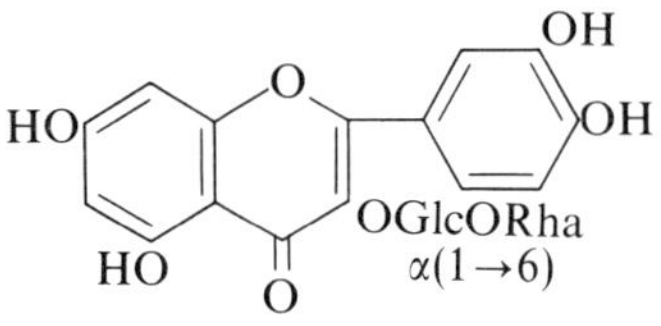

(**1**)

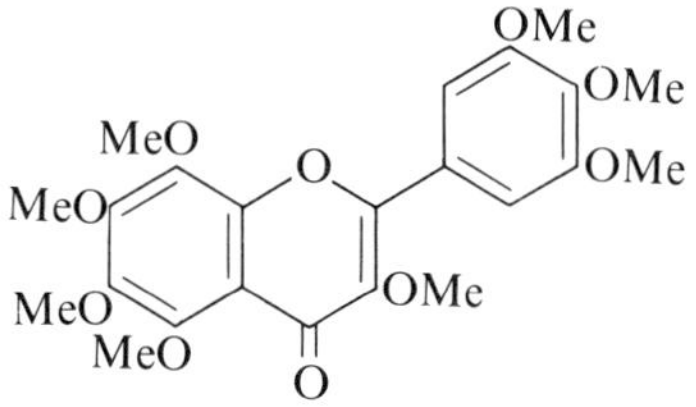

(**2**) Exoticin

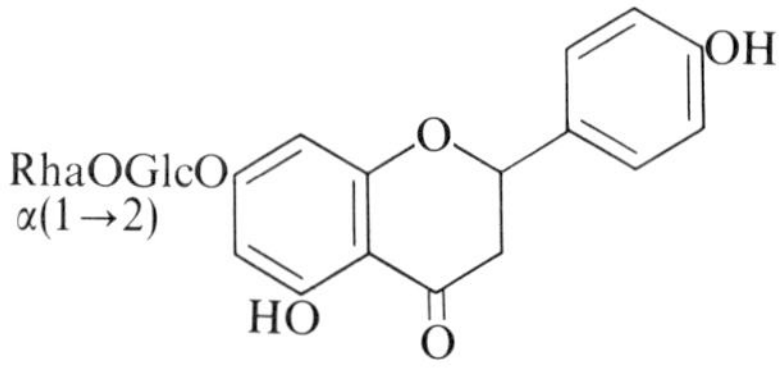

(**3**) Naringin

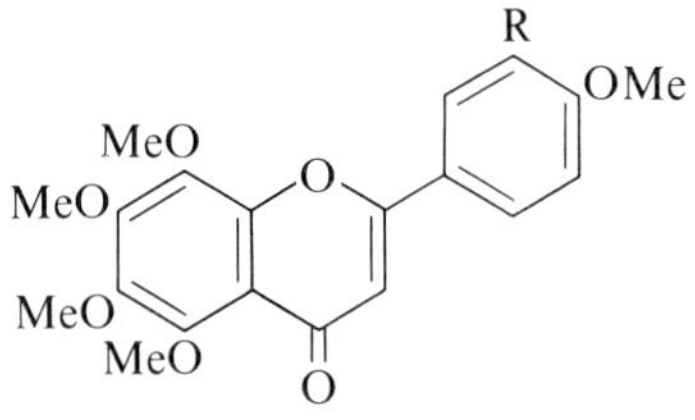

(**4**) Nobiletin (R = OMe)
(**5**) Tangeretin (R = H)

Citrus extracts has indicated that four *O*-methyltransferases for introducing methylation at the 3,7,3' and 4' positions of quercetin can all be separately distinguished.[10,17]

The purpose of this highly unusual *O*-methylation in the flavone series that occurs in the Rutaceae is as yet far from clear. However, such compounds are biologically active and are probably more toxic to animals than the common leaf flavonoids. It is likely because of their high lipid solubility that their distribution in the leaf is confined either to special resin ducts or to the leaf surface, where they may be present in the wax or cutin. Here they may be important as insect antifeedants or as pre-infectional antifungal agents.

Some evidence in favour of the latter possibility has been obtained by Piattelli and Impellizzeri,[47] who found that leaves of *Citrus* cultivars resistant to Mal-secco disease contain larger quantities of the two methylated flavones nobiletin and tangeretin than the leaves of susceptible forms. Furthermore, these two flavones significantly inhibit the growth *in vitro* of the Mal-secco organism *Deuterophoma tracheiphila* and of other pathogenic fungi.[49] The correlation between methylated flavone content and disease response is, however, not complete (Table 1) so that other factors, besides the simple presence of these antifungal agents on the leaf surface, are probably concerned in the resistance or susceptibility of *Citrus* plants to this fungal disease. Indeed, in the case of infection of *Citrus* plants with *Phytophthora citrophthora*, a phytoalexin response has been recorded and the post-infectional production of xanthoxylin (2-hydroxy-4,6-dimethoxyacetophenone) among other phenols has been described.[29]

TABLE 1

Relationships between resistance to Mal-secco disease and methylated flavone content of citrus leaves

Cultivar	Content (µg/g dry wt) of nobiletin	tangeretin	Disease response
Mandarin	1110	420 ⎫	
Sweet orange	115	30 ⎬	Resistant
Volkamerian lemon	220	90 ⎭	
Sour orange	88	56 ⎫	
Myrtle leaf orange	90	52 ⎪	
Citron	0	12 ⎬	Susceptible
Lime	0	0 ⎪	
Trifoliate orange	0	20 ⎭	

Two incompletely methylated flavones also accumulate with nobiletin and tangeretin in certain of the resistant cultivars. Data modified from ref. 47.

The other characteristic rutaceous flavonoid mentioned earlier, namely the flavanone glycoside naringin, may have a role in plant–animal interactions in *Citrus* in that it has an intensely bitter taste. Naringin is the major water-soluble bitter compound of citrus fruits and its control and reduction in amount have important practical implications in the manufacture of citrus juices (see Chapter 12). The relationship between structure and taste in the flavanone series has been extensively investigated (Table 2) and it is known that the flavanone nucleus and the specific attachment of the disaccharide neohesperidose (present in naringin) are both essential for maximal bitterness.[32] Replacement of neohesperidose (rhamnosyl-α-1 $\rightarrow$ 2-glucose) by rutinose (rhamnosyl-α-1 $\rightarrow$ 6-glucose), the disaccharide of rutin, in naringin to give naringenin 7-rutinoside produces a compound lacking all bitterness. Interestingly, related dihydrochalcones (which are not naturally occurring in *Citrus*) are very sweet (Table 2) and have been developed as sweetening agents for use in the food industry.

TABLE 2

Relation between taste and structure among citrus flavanones and dihydrochalcones

No taste	Bitter	Sweet
Naringenin	Neohesperidin (0.02)	Naringin
Hesperidin	Phloridzin (0.1)	dihydrochalcone (1)
Naringenin	Poncirin (0.2)	Neohesperidin
7-rutinoside	Naringin (0.2)	dihydrochalcone (20)
	Quinine (1)	Saccharin (1)

Values in parentheses refer to comparative bitterness or sweetness on a molar basis, relative to standards. Data summarized from ref. 32.

Flavanone glycosides, such as the bitter naringin and the related non-bitter hesperidin, in addition to occurring in citrus fruits are also found in other plant parts, including the leaves. Such flavanones are also widespread throughout the Rutaceae, and their presence in these plants, like that of the methylated flavones, constitutes a significant taxonomic marker for this family.

In the present review, attention will be given to the distribution of both common and rare flavonoids within these plants. It is proposed to consider first the basic patterns within the Rutales and then enumerate some of the special structural features attached to the flavonoid nucleus which distinguish the flavonoids of the Rutales from those of other plant orders. Finally, the possible chemotaxonomic significance of the flavonoids recorded in the Rutales will be considered.

II. Basic Patterns

A. ANTHOCYANIN PIGMENTS

Anthocyanin pigmentation is presumably present throughout the Rutales, as in almost all angiosperms, providing colour in flower and fruit. Rarely, however, have the pigments present been characterized. Thus, there are complete identifications of the fruit anthocyanins of three species[12,33,34] and the partial identifications of floral anthocyanins of three further species[38] (Table 3). It would appear from these very limited results that the pattern is a relatively simple one, with unmethylated anthocyanidins occurring usually in association with a simple glycosidic pattern.

TABLE 3

Anthocyanins of the Rutaceae

Plant source	Pigments identified*	Ref.
	Fruit pigments	
Citrus sinensis, "blood" orange	Cy and Dp 3-glucosides	12
Skimmia japonica var. *repens*	Pg 3-glucoside	33
	Pg 3-rhamnosylglucoside	
Zanthoxylum piperitum	Cy 3-rhamnosylglucoside	34
	Flower pigments	
Boronia elatior	Mv 3,5-dimonoside	38
Correa speciosa var. *ventricosa*	Cy 3-rhamnoglucoside(?)	38
Dictamnus albus var. *caucasicus*	Dp 3-rhamnoglucoside(?)	38

* Pg = pelargonidin; Cy = cyanidin; Dp = delphinidin; Mv = malvidin

Typical, perhaps, is the finding of Chandler[12] that coloration of the juice and skin of the so-called blood orange is due to a mixture of cyanidin and delphinidin 3-glucosides. It should be mentioned that this type of coloration in citrus fruit is unusual in that most juice and peel coloration is due to carotenoid, and citrus flowers also have yellow carotenoids.[21] Skin coloration in other rutaceous genera may, however, be regularly due to anthocyanin, as in *Skimmia* and *Zanthoxylum* (Table 3).

Nothing appears to be known about anthocyanins in families other than the Rutaceae. Certainly, more work is needed on anthocyanin production in the Rutales before any firm conclusions can be drawn about pigment patterns in these plants.

B. LEAF FLAVONOIDS

From a very limited survey of flavonoids in acid-hydrolysed leaf tissue of 37 species (Table 4), it is apparent that the aglycone pattern in the Rutales is that recognized as being characteristic of the more primitive woody angiosperm families.[4,44] Thus flavonols are commonly present, as are condensed tannins based on procyanidin and prodelphinidin.[4] In addition, the flavonol with a trihydroxy substituted B-ring, namely myricetin, which is a specific marker for woody families,[4] is present in a minority (*ca.* 30%) of the species sampled. In addition, hydrolysable tannins based on hexahydroxydiphenic acid (giving rise to ellagic acid in acid-treated leaf extracts) are found specifically in one of the five families, in the Simaroubaceae.[44]

TABLE 4

Flavonoid patterns in leaves of the Rutales

| Family | Frequency of occurrence in species surveyed of | | | |
	myricetin	quercetin and/or kaempferol	procyanidin and/or prodelphinidin	ellagitannin
Burseraceae	—	2/2	2/2	—
Cneoraceae	—	1/1	—	—
Meliaceae	—	4/4	2/4	—
Rutaceae	5/14	10/14	5/14	—
Simaroubaceae	{ —	3/4	—	3/4
	5/12	7/12	5/12	6/12

— = not recorded. Data from Bate-Smith,[4] except that second analysis of Simaroubaceae of 12 spp. is that of Nooteboom.[44]

The predominance of flavonols in the leaves of these plants apparent from the above surveys is supported by detailed investigations of a variety of species which have revealed a range of common flavonol glycosides (Table 5). As already mentioned, quercetin 3-rutinoside (rutin) is relatively widespread. Other quercetin glycosides identified in the Rutaceae include the 3-rhamnoside, the 3-α-arabinofuranoside (avicularin), the 3-galactoside, and the 3,7-diglucoside. The pattern in the Meliaceae is not very different (Table 5), although one may note the presence of the rare myricetin 3'-arabinoside in *Azadirachta indica* and of xylose based glycosides in two of the five species examined.

The methods of leaf aglycone surveys employed by Bate-Smith[4] do not specifically reveal glycoflavones or flavones, so that data on their general

TABLE 5

Common flavonol glycosides identified in the Rutales

Plant	Compounds identified*	Ref.
Meliaceae		
Azadirachta indica petal	Qu 3-galactoside,	
	Km 3-galactoside,	
	My 3'-arabinoside	56
Cedrelopsis grevei leaf	Qu-3-xyloside,	
	Qu 3-rhamnoside,	46
	Qu 3-glucoside,	
	Qu 3-rutinoside	
Melia azedarach leaf	Qu 3-rhamnoside,	
	Qu 3-rutinoside	41
Neobeguea mahafalensis leaf	Qu 3-glucoside,	
	Km 3-xylosylglucoside,	
	Qu 3-xylosylglucoside	46
Soymida febrifuga leaf	Qu 3-rhamnoside,	
	Qu 3-rutinoside	41
Rutaceae		
Barosma betulina	Qu 3,7-diglucoside,	
	Qu 3-rutinoside	50
Boenninghausenia albiflora leaf	Qu 3-rutinoside	30
Pitavia punctata stem/leaf	Qu 3-arabinoside	
	(avicularin),	55
	Qu 3-rhamnosylarabinoside	
Ruta graveolens L. leaf (and other *Ruta* spp.)	Qu 3-rutinoside	62
Zanthoxylum piperitum var. *inerme* leaf	Qu 3-rhamnoside,	
	Qu 3-galactoside,	
	Km 3-rhamnoside	36
Simaroubaceae		
Ailanthus altissima leaf	Qu 3-glucoside	42
Suriana maritima whole plant	Qu 3-rutinoside,	
	rhamnetin 3-rutinoside	40

* Qu = quercetin; Km = kaempferol; My = myricetin; Qu 3-rutinoside = rutin; rhamnetin = quercetin 7-methyl ether

distribution is still lacking. However, in a recent survey of 35 species belonging to the subfamily Aurantioideae of the Rutaceae, Grieve and Scora[23] reported that, in addition to the flavonols widely present, all taxa contained glycoflavones; in addition, a few species had flavones in *O*-glycosidic combination as well. In more detailed studies of mainly rutaceous species, a variety of glycoflavones have been characterized (Table 6). The compounds present are mainly based on apigenin or its methyl ethers; in addition, the rare occurrence of C-glucosides of diosmetin in *Citrus* is to be noted.[14] Flavone O-glycosides that have been characterized in *Citrus* include apigenin 7-glucoside, apigenin 7-rutinoside, and luteolin 7-rutinoside.[30] Acacetin 7-rutinoside has been obtained from petals of *Fortunella* species.[43] The characteristic occurrence of diosmetin 7-rutinoside (diosmin) in the Rutaceae will be discussed later (see Section IIIE) in relation to its co-occurrence with the flavanone glycoside hesperidin.

TABLE 6

Glycoflavones identified in the Rutales

Plant	Compounds identified*
Rutaceae	
Almeidea guyanensis bark	6,8-diarabinosylgenkwanin,
	8-arabinosylgenkwanin 2″-glucoside,
	6-arabinosylgenkwanin 2″-glucoside
Citrus fruit	vitexin, vitexin 2″-xyloside,
	6-glucosyldiosmetin, 8-glucosyldiosmetin,
	6,8-diglucosyldiosmetin
Fagara nitida root bark	vitexin
Fortunella margarita	6-glucosylacacetin 2″-rhamnoside,
	8-glucosylacacetin 2″-rhamnoside
Vepris sudanica leaf	vitexin, isovitexin, orientin, isoorientin
Simaroubaceae	
Ailanthus excelsa leaf	vitexin

* genkwanin = apigenin 7-methyl ether; diosmetin = luteolin 4′-methyl ether; acacetin = apigenin 4′-methyl ether; vitexin = 8-glucosylapigenin; orientin = 8-glucosyl-luteolin. For references, see ref. 14.

In summary then, the leaves (and other tissues) of rutalean plants contain a pattern of common flavonoid constituents which is similar to that present in the majority of other woody angiosperm groups. The pattern is dominated by the presence of flavonols and of proanthocyanidins occurring in major amount. Glycoflavones are also probably regular constituents, although this has not yet been fully established. By contrast, flavone O-glycosides are relatively uncommon and when present tend to be minor components. A small shift away from the predominantly "woody" pattern to a more specialized flavone-based pattern may occur in some groups. It is apparent, at least in the Aurantioideae, that the more primitive subtribes (e.g. the Clauseninae) lack flavones and have myricetin while the more specialized subtribes (e.g. the Citrinae) have flavones and lack myricetin.[23]

Superimposed on this pattern of common flavonoids are a number of rarer, more highly substituted flavonoids more characteristically confined to plants of the rutalean alliance. The chemistry and distribution of these more specialized flavonoids are detailed in the next section.

III. Special Structural Features

A. Yellow Flavonols

Introduction of a 6- or 8-hydroxyl group into the common flavonols, e.g. kaempferol or quercetin, causes a shift in the spectrum from a pale cream to a primrose yellow colour, and such 6- and 8-hydroxyflavonols are responsible for yellow flower colour in a variety of angiosperm plants (e.g. *Primula vulgaris* Primulaceae, *Chrysanthemum segetum* Compositae, etc.). Such compounds also occur more widely in the leaves of the same or related plants, where their biological function is more obscure. They have a limited natural distribution, having been recorded so far in some 20–30 plant families, so that they have some potential as taxonomic markers.[25] They have been found in several Rutaceae, although their value as taxonomic indicators has yet to be exploited.

One of the first gossypetin (8-hydroxyquercetin) derivatives to be reported in a rutaceous plant was tambuletin, a yellow pigment extracted from the seed of *Zanthoxylum acanthopodium*. This was originally incorrectly formulated as herbacetin 8-methyl ether[3] but later, more critical studies showed that it was in fact a glucoside, not an aglycone, and that it was the 8-glucoside (6) of gossypetin 7,4'-dimethyl ether.[28] This structure was subsequently confirmed by synthesis and by re-isolation from the original source.[13] The structure of a co-occurring pigment in the seed of *Z. acanthopodium* called tambulin as the 7,8,4'-trimethyl ether of herbacetin (8-hydroxykaempferol) has been confirmed by later

(6) Tambuletin

investigation.[13] A second herbacetin methyl ether has been reported in the family, in *Flindersia maculosa*, and this is the 3,7,8,4'-tetramethyl ether, flindulatin.[9]

Another source of gossypetin derivatives in the Rutaceae is the genus *Citrus*, from which the 8,3'-dimethyl and the 3,7,3',4'-tetramethyl ether have been obtained.[16,31] The former was isolated from the peel of the lemon, *Citrus limon*, while the latter occurred in the peel of several species. The isomeric 3,7,8,3'-tetramethyl ether, interestingly enough, has been reported in *Melicope*[8] (Table 7).

TABLE 7

Occurrence of 6- and 8-hydroxyflavonols in the Rutaceae

Flavonoid	Plant source	Ref.
Herbacetin (8-hydroxykaempferol)		
7,8,4'-trimethyl ether (tambulin)	*Zanthoxylum acanthopodium* fruit	13
3,7,8,4'-tetramethyl ether (flindulatin)	*Flindersia maculosa* leaf	9
6-Hydroxykaempferol		
6-methyl ether	as 3-glucoside, in *Haplophyllum tuberculatum* aerial parts	35
Gossypetin (8-hydroxyquercetin)		
3'-methyl ether	as 7-glucoside, in *Haplophyllum perforatum* leaf in glycosidic form in flowers of *Haplophyllum* spp.	5,6 *
7-methyl ether	in glycosidic form in flowers of *Ruta* spp.	*
7,4'-dimethyl ether	as 8-glucoside in *Zanthoxylum acanthopodium* fruit	13,28
8,3'-dimethyl ether	in bound form in lemon peel. *Citrus limon*	31
3,7,3',4'-tetramethyl ether	free in *Citrus* peel	16
3,7,8,3'-tetramethyl ether (ternatin)	in *Melicope mantellii* and *M. simplex* bark	8

* Harborne, J. B. and Boardley, M. (1983). *Z. Naturforsch.* **38C**, 148.

More recently, another gossypetin derivative, the 3'-methyl ether has been characterized in the leaves of *Haplophyllum perforatum*, where it is reported to occur as the 7-glucoside[6] and the 7-(6''-acetylglucoside).[5] A derivative of 6-hydroxykaempferol, namely the 6-methyl ether 3-glucoside, has also been reported in *Haplophyllum*, in the aerial parts of *H. tuberculatum*.[35] This is the only report so far of a 6-hydroxyflavonol as opposed to the more common 8-hydroxyflavonols in the Rutaceae.

It appears from our own preliminary investigations of some four *Haplophyllum* species, namely *H. linifolium, H. buxbaumii, H. suaveolens*, and *H. tuberculatum*, that yellow flavonols are important as flower pigments in this genus. The dried yellow coloured corollas of these four species have yielded a yellow flavonol glycoside, which on hydrolysis gave gossypetin 3'-methyl ether. Common flavonols, kaempferol, quercetin, and isorhamnetin, were also present in glycosidic form. Further work is in progress to identify the glycosidic components. Another source of gossypetin based colours in the Rutaceae appears to be the genus *Ruta*, since here again yellow flowers of *R. graveolens, R. chalepensis, R. angustifolia*, and *R. montana* gave a yellow flavonol glycoside. This time the aglycone detected was the 7-methyl ether of gossypetin. These identifications are being confirmed in more detailed studies.

The characterization of these yellow flavonols is hampered to some extent by their lability in solution, and in general, because of the ease of oxidation, they are best handled in solution under nitrogen. Location of the substituents in the partial methyl ethers can also present problems. However, a number of different mono- and di-methyl ethers of gossypetin are now known, and it is possible to distinguish different isomers in both series by means of paper and thin-layer chromatography (Table 8).[27] Identifications can be confirmed on a micro scale by ultraviolet spectroscopy and mass spectrometry (Table 9).[27] Where larger samples are available, ^{13}C NMR spectroscopy is valuable for locating the positions of methyl substitutions.[39]

TABLE 8

Chromatographic properties of gossypetin and its methyl ethers

| | R_f values ($\times 100$) | | | | | | |
| | | Paper | | | Si gel | Polyamide | Colour in |
	Forestal	50% HOAc	BAW	CAW	BPF	BEM	UV light
Gossypetin	26	16	22	03	06	09	brown-black
3'-methyl ether	34	22	50	19	22	28	brown-black
4'-methyl ether	43	31	50	23	18	31	brown-black
7-methyl ether	43	30	27	15	20	35	brown-black
8-methyl ether	46	34	54	39	26	26	brown-yellow
8,3'-dimethyl ether	61	46	74	62	48	60	brown-yellow

*

TABLE 9

Spectral and mass spectrometric properties of gossypetin and various methyl ethers

	λ_{max} (nm) in MeOH	MeOH–AlCl$_3$	M	(M − 15)	MS (m/z) (M − 43)	A ring	B ring
Gossypetin	264, 277, 338, 386	370, 449	318 (100)		289 (14)	169 (25)	137 (28)
3′-methyl ether	260, 272, 338, 385	377, 484	332 (100)	317 (3.5)	289 (1.4)	169 (6.4)	151 (10.6)
4′-methyl ether	261, 272 337, 384	374, 498	332 (100)	317 (20.3)	289 (5.2)	169 (7.3)	151 (8.5)
7-methyl ether	260, 275 342, 392	388, 470	332 (100)	317 (20.4)	289 (13.4)		137 (19.2)
8-methyl ether	260, 271, 340, 381	367, 444	332 (48)	317 (100)	289 (13)		137 (12)
8,3′-dimethyl ether	258, 272 340, 380	370, 438	346 (30)	331 (100)	307 (20)		151 (12)

No systematic investigation of yellow flower colour in the Rutaceae has yet been attempted. While yellow colour may be due to carotenoid (and such pigments have been characterized in *Citrus* flowers), it now appears that it may also be based on yellow flavonols; at least this appears to be true in *Haplophyllum* and *Ruta*. A wider investigation is clearly called for and may well reveal a range of further gossypetin derivatives in these plants.

B. METHYLATED FLAVONES

Most plant flavonoids are polyphenolic, have one or more sugars attached to them, are water-soluble, and are located in the cell vacuole. Methylation of the phenolic hydroxyl groups, if present, is usually restricted to one or two such groups and no more. By contrast, a more limited number of natural flavonoids are highly methylated, usually lack glycosylation, are lipophilic, and are located in secretory structures or are deposited externally on leaf surfaces. Such highly methylated flavones are only found regularly in some 10–12 angiosperm families, although they do appear occasionally in some 30 others.[64]

The regular occurrence of highly methylated flavones in a given family is thus a distinctive chemical feature of those plants, and this character marks off the Rutaceae from nearly all other neighbouring taxa. The ability to *O*-methylate both flavones and flavonols is most pronounced in the genus *Citrus*, where these compounds are often secreted in the oil glands of the fruit peel. A unique range of 11 fully methylated and 22 highly methylated flavones have been characterized to date from species and cultivars of this genus (Table 10).

TABLE 10

Fully and partly methylated flavones that have been characterized in *Citrus*

Fully methylated flavones

5,7,4'-OMe (apigenin trimethyl ether); 5,6,7,4'-OMe; 5,7,8,4'-OMe;
5,6,7,8,4'-OMe (tangeretin); 5,6,7,3',4'-OMe (sinensetin); 5,7,8,3',4'
OMe (isosinensetin); 3,6,7,8,4'-OMe (auranetin); 5,6,7,8,3',4'-OMe (nobiletin);
3,5,6,7,3',4'-OMe; 3,5,7,8,3',4'-OMe; 3,5,6,7,8,3',4'-OMe

Partly methylated flavones

5,7,4'-OH-6-OMe (hispidulin); 5,4'-OH-7,8-OMe; 5-OH-7,8,4'-OMe;
5,7,4'-OH-6,8-OMe; 5.4'-OH-6,7,8-OMe (xanthomicrol); 5,7,4'-OH-6,3'-OMe
(jaceosidin);
5-OH-6,7,8,4'-OMe (gardenin B); 5-OH-6,7,3',4'-OMe; 5-OH-7,8,3',4'-OMe;
5,6,7,8-OH-3',4'-OMe; 3,5,7,4'-OH-8,3'-OMe (limocitrin);
5,7,4'-OH-6,8,3'-OMe (sudachitin); 5,4'-OH-6,7,8,3'-OMe; 5,8-OH-3,7,3',4'-OMe;
5-OH-6,7,8,3',4'-OMe (5-desmethylnobiletin); 4'-OH-5,6,7,8,3'-OMe;
5-OH-3,6,7,8,4'-OMe; 3,5,7,4'-OH-6,8,3'-OMe (limocitrol);
3,5,7,3'-OH-6,8,4'-OMe (isolimocitrol); 5-OH-3,6,7,8,3',4'-OMe;
3-OH-5,6,7,8,3',4'-OMe (natsudaidain)

Individual references are given in ref. 64.

Most of the *Citrus* compounds have extra oxygen substitution in both the 6- and 8-positions, the hydroxyl groups thus inserted usually carrying methylation as well. As already mentioned earlier, enzymic evidence,[10,17] although incomplete, indicates that separate *O*-methyltransferases are required for these methylations. There is no one single enzyme which is able to catalyse the transfer of methyl groups to all the hydroxyls of a presumed precursor, such as 5,6,7,8,3',4'-hexahydroxyflavone. The fact that a wide range of related incompletely methylated flavones accompany the fully methylated derivatives in *Citrus* peels would support the view that methylation takes place sequentially, one hydroxyl being methylated at a time in a given order, related perhaps to the relative reactivities of the various hydroxyl functions.

The 5-hydroxyl of flavones in particular is partly inactivated by hydrogen bonding with the adjacent 4-carbonyl group and is chemically difficult to substitute with methyl. Even enzymically the process seems to present some difficulties in that the most frequently encountered partly methylated flavones in *Citrus* are those in which all but the 5-hydroxyl are methylated. Eight such compounds have been described, such as 5-desmethylnobiletin, 5-desmethylsinensetin, etc. A variety of other incompletely methylated analogues are also known (Table 10) and one suspects that a missing member of any given series not yet characterized could be found eventually by continued analyses of *Citrus* peel extracts.

At least two other genera, i.e. *Murraya* and *Merrillia*, in the same subfamily as *Citrus* (the Aurantioideae), have yielded *O*-methylated flavones. In *Murraya*, six such compounds have been isolated from three species (Table 11).[64] These substances (e.g. exoticin) differ from those in *Citrus* in that the flavonoid B-ring is trisubstituted in *Murraya* whereas it is mostly disubstituted in *Citrus*. In the fruit of *Merrillia caloxylon*, a monotypic species from Malaysia, a number of methylated flavones and their related chalcones have been obtained.[18] Four flavones have been identified in the fruit skin: luteolin 5,7,3′,4′-tetramethyl ether, 5,7,3′,4′,5′-pentamethoxyflavone, and the two related 6-methoxy-5-desmethyl analogues, i.e. 5-hydroxy-6,7,3′,4′-tetramethoxyflavone (**7**) and 5-hydroxy-6,7,3′,4′,5′-pentamethoxyflavone (**8**). Three related chalcones are present, including compounds **9** and **10**. In addition, the flavone eupatorin (5,3′-dihydroxy-6,7,4′-trimethoxyflavone) has been characterized in the fruit oil of the same plant.[1] The discovery of chalcones in *Merrillia* represents the first indisputable record of this class of flavonoid in the Rutaceae.

(**7**) R = H
(**8**) R = OMe

(**9**) R = H
(**10**) R = OMe

TABLE 11

Methylated flavones isolated from *Murraya*

Flavone substitution	Source
5,4′-Dihydroxy-3,6,7,3′,5′-pentamethoxy	*M. omphalocarpa* fruit
4′-Hydroxy-3,5,6,7,3′,5′-hexamethoxy	*M. paniculata* leaf
3,5,6,7,3′,4′,5′-Heptamethoxy	*M. paniculata* leaf,
	M. exotica leaf,
	M. omphalocarpa fruit
6-Hydroxy-3,5,7,8,3′,4′,5′-heptamethoxy	*M. exotica* leaf
8-Hydroxy-3,5,6,7,3′,4′,5′-heptamethoxy	*M. exotica* leaf
3,5,6,7,8,3′,4′,5′-Octamethoxy (exoticin)	*M. exotica* leaf

For references see ref. 64.

The ability to extensively methylate flavonoid phenolic groups is also expressed in the Australian genus *Eriostemon* (subfamily Rutoideae). Analysis of two species, *E. hispidulus* and *E. buxifolius* subsp. *buxifolius*, has yielded from the leaves of both a new flavone, eriostemin, which is the 5,6,7,4'-tetramethyl ether of 6,8-dihydroxykaempferol.[37] Extra methylation of common flavonols is expressed in *Euodia glabra*, the bark of which contains quercetin 3,7,3'-trimethyl ether, and in *Melicope perspicuinerva*, the leaves of which have yielded quercetin 3,4'-dimethyl ether.[64]

Mention may be made here of methylenedioxy substitution in rutalean flavonoids, since this process is clearly related to *O*-methylation in that an *ortho*-hydroxymethoxy precursor is probably required for methylenedioxy bridge formation. Furthermore, almost all the compounds with methylenedioxy bridges are also highly methylated elsewhere in the molecule. Such substances have so far only been recorded in the Rutales in the Australian genus *Melicope* (subfamily Rutoideae). Typical of the flavonoids of these plants are melinervin (**11**) from *M. perspicuinerva* leaf with methylenedioxy substitution in the B-ring and meliternatin (**12**) from *M. ternata* bark with methylenedioxy substitution in both A- and B-rings. Other methylenedioxy flavonoids of *Melicope* are listed in the review of Wollenweber and Dietz.[64] Since methylenedioxy bridges are

(**11**) Melinervin   (**12**) Meliternatin

present in the alkaloids of a considerable number of rutaceous genera,[60] it is evident that the enzyme or enzymes catalysing this synthesis are widespread in the family. If the enzymes are normally specific for alkaloid substrates, then it is conceivable that in *Melicope*, the only known source of methylenedioxy flavonoids so far, the substrate specificity may have broken down. An alternative possibility is that the enzyme system is relatively non-specific, in which case future research should reveal methylenedioxyflavones in many more rutaceous plants.

None of the characteristic methylated or methylenedioxy substituted flavones of the Rutaceae has yet been characterized anywhere else in the Rutales. However, the reports of two relatively uncommon flavonol methyl ethers, tabularin and syringetin, in the Meliaceae may be taken to indicate that enzymes for extra O-methylation may be present in plants of this family too. Tabularin (5,7-dihydroxy-6,2',4',5'-tetramethoxyflavone) has been obtained from leaves of *Chukrasia tabularis*[51] while syringetin (myricetin 3',5'-dimethyl ether) is recorded in root and heartwood of *Soymida febrifuga*.[45]

C. UNUSUAL OXYGENATION PATTERNS

As a result of many tracer and enzymic experiments, the pathway of flavonoid biosynthesis is now well defined and the route followed appears to be universally identical, at least within all higher plant groups.[26] In this pathway, oxygen functions are introduced into the 5, 7, and 4' positions of the molecule, so that the vast majority of naturally occurring pigments inevitably have hydroxyl (or otherwise O-substituted) groups at these positions. Flavonoids lacking such oxygenation are generally rare and are presumably either formed by a different pathway of else suffer deoxygenation at these positions subsequent to their formation. 5-Deoxyflavonoids are not too uncommon and are known to accumulate occasionally, for example, in species of the Anacardiaceae and Leguminosae. 7-Deoxyflavonoids, on the other hand, are quite rare and their production in two species of the Rutaceae, namely *Casimiroa edulis* and *Sargentia greggii*, is very unusual. As will be seen from Table 12, seven compounds of this type have been identified from one or other of these plants. Typical structures are zapotin (**13**) from *Casimiroa* and cerrosillin (**14**) from *Sargentia*. The unusualness of the oxygenation in these compounds is also

TABLE 12

7-Deoxyflavones of *Casimiroa* and *Sargentia*

Flavone	Source
5,6-Dimethoxy	*C. edulis* bark;
	S. greggii fruit, stem, leaf
5,6,2'-Trimethoxy	*C. edulis* fruit peel
5,6,3'-Trimethoxy	*C. edulis* seed
5-Hydroxy-6,2',6'-trimethoxy	*C. edulis* fruit, bark
5,6,2',6'-Tetramethoxy (zapotin)	*C. edulis* fruit, bark
5,6,3',5'-Tetramethoxy (cerrosillin)	*C. edulis* seed;
	S. greggii fruit, stem, leaf
5,6,3',4',5'-Pentamethoxy	*S. greggii* leaf

For references, see ref. 64.

(13) Zapotin   (14) Cerrosillin

apparent from the fact that, in six of the seven, oxygen is additionally absent from the 4'-position. Another rare structural feature exhibited in these compounds is the presence of 2'-oxygenation in three of them. Such 2'-substitution is relatively infrequent and is confined to a limited number of plant families, notably the Anacardiaceae, Amarantaceae, Compositae, Datis-cataceae, Leguminosae, and Moraceae. It is interesting that such 2'-oxygenation is present in the flavone tabularin of *Chukrasia tabularis* (Meliaceae), which has already been mentioned in the previous section.

One final point may be made about the 7-deoxyflavones of *Casimiroa* and *Sargentia*. A high degree of O-methylation also characterizes these substances. Thus both *Casimiroa* and *Sargentia*, which are closely related genera in the subfamily Toddalioideae, can be added to the list of rutaceous genera (see Section IIIB) which have highly methylated flavones.

D. ISOPRENOID FLAVONOIDS

Isoprenylation of aromatic nuclei is a common structural feature among the coumarins and alkaloids of the Rutaceae. It is not surprising, therefore, that isoprenylated flavonoids are occasionally encountered. Isoprenylation, like methylation, introduces lipophilic properties into compounds which are otherwise highly polar in character. So far, isopentenyl flavonoids have not been found in many species, but no deliberate surveys have been initiated to determine their frequency of occurrence. Isoprenylated flavonoids so far encountered are of two types, those with nuclear isoprenylation and those which are isoprenyl ethers. It may be of significance that so far no compounds of this type have been found in which the isopentenyl side-chain has further condensed with an adjacent hydroxyl function to produce a benzofuran or benzopyran derivative. Such compounds are however known among the isoprenoid substituted coumarins and alkaloids of Rutaceae.

Isoprenylation of the flavonoid nucleus in the 8-position has given rise to a notable series of 8-isoprenyl dihydroflavonols and flavonols which characterize the genus *Phellodendron* (Toddalioideae). Four such structures based on dihydrokaempferol and kaempferol have been characterized: the flavanonols phellamurin (**15**) and the related exocyclic glucoside dihydrophelloside (**16**) together with the related flavonols amurensin and phelloside. The C-6 substituted isomers of phellamurin and amurensin have also been characterized.[7] These substances are variously present in leaves and other parts of *P. amurense*, *P. japonicum*, *P. lavalei*, and *P. sachalinense*.

The possibility of a more widespread occurrence of 8-isopentenyl flavonoids in the Rutaceae has been opened up by the discovery of 8-(3-methylbut-2-enyl)quercetin 3,7,3'-trimethyl ether in the leaf of *Phebalium dentatum* (subfamily Rutoideae).[48] In this compound, the isopentenyl side-chain differs from that in the *Phellodendron* compounds in carrying a double bond, rather than being saturated and having a hydroxyl attachment. Furthermore, a partly characterized isopentenylflavanone has been reported in the leaves of another Rutoideae species, *Euodia rutaecarpa*.[24] It is either the 6- or 8-(3-methylbut-2-enyl) derivative of naringenin 7,4'-diglucoside.

Isoprenyl ethers are represented in the Rutaceae by two flavanones with C_5 and C_{10} substitution at the 4'-hydroxyl group. These compounds (**17** and **18**) occur in the leaves of *Melicope sarcococca*.[11,20] These are rare compounds, since the only other occurrence of flavonoid O-prenyl ethers appears to be in members of the Moraceae.

(**15**) Phellamurin (R = H)
(**16**) Dihydrophelloside (R = Glc)

(**17**) R =
(**18**) R =

E. FLAVANONES

The characteristic occurrence of flavanones, some of which like naringin taste bitter, in the Rutaceae has already been alluded to (Section I). The greatest concentration of such compounds occurs in *Citrus*. Here a variety of flavanone glycosides derived from naringenin (5,7,4'-trihydroxyflavanone), iso-sakuranetin (naringenin 4'-methyl ether), eriodictyol (5,7,3',4'-tetrahydroxy-flavanone), and hesperitin (5,7,3'-trihydroxy-4'-methoxyflavanone) have been characterized. Their distribution in peel, fruit, and leaves of many species and cultivars has been exhaustively tabulated.[7] Hesperidin, the tasteless 7-rutinoside of hesperitin, and naringin, the bitter 7-neohesperidoside of naringenin, are the two most widespread flavanones in *Citrus*.

Of the latter two compounds, the only one which is widely reported outside *Citrus* is hesperidin (**19**). Flavanones such as hesperidin may be *in vivo* precursors of the corresponding flavones; additionally, the *in vitro* oxidation of flavanone to flavone occurs rather readily. For both these reasons, it is not surprising to find that hesperidin, when it has been detected in a given rutaceous plant, is sometimes accompanied by the related flavone 7-rutinoside, which is known as diosmin (**20**). It is conceivable that the two compounds occur in plants in an oxidative–reductive equilibrium, with one or other predominating at different times in growth, development, and maturity.

(19) Hesperidin   (20) Diosmin

From the taxonomic viewpoint, it is probably better therefore to consider a record of either compound in a plant as being representative that both may actually be produced *in vivo*. As can be seen from Table 13, one or other or both have already been recorded in representative genera of the three largest rutaceous subfamilies, i.e. the Rutoideae, Aurantioideae, and Toddalioideae. To this list of hesperidin–diosmin containing genera should be added *Melicope* and *Phellodendron* (with isopentenylflavanones) to produce a complete account of all flavanone synthesizing genera. The ability to produce flavanones is very pronounced, therefore, in the family as a whole. In contrast, there are only two records of flavanones in other members of the rutalean alliance. Naringenin has been isolated from *Soymida febrifuga* (Meliaceae),[53] while the same flavanone and its 7-(6"-*p*-coumaroylglucoside) have been characterized from nut shells of *Anacardium occidentale* (Anacardiaceae),[52] this being a family which may or may not be included with Rutales (see Section IVc).

TABLE 13

Rutaceous genera in which hesperidin and/or diosmin have been recorded

Aurantioideae	Rutoideae	Toddalioideae
Citrus*†	Barosma†	Ptelea†
Clymenia*	Dictamnus†	Skimmia*†
Eremocitrus*	Euodia*†	Toddalia†
Fortunella*	Fagara*†	Vepris†
Microcitrus*	Zanthoxylum*†	
Pleiospermum*		
Poncirus*		
	Flindersioideae	
	Flindersia*	

*Hesperidin recorded. † Diosmin recorded.
For references, see Hegnauer.[30] Flindersioideae is sometimes now referred to as a separate family from Rutaceae.

IV. CHEMOTAXONOMY

A. PATTERNS WITHIN RUTACEAE

Flavonoids have only been exhaustively investigated in a single rutaceous genus, namely *Citrus*. The concentration of chemical effort on this genus is largely because of the economic value of these plants, of the possible importance of methylated flavones in the disease resistance of this crop, and of the property of some flavanone glycosides to impart bitterness to citrus products. Our view of flavonoid synthesis in the Rutaceae as a whole is therefore highly distorted since no other genus has been analysed to anything like the same extent. We do not know how widely the characteristic *Citrus* flavonoids occur elsewhere in the family. Nor do we know whether specific flavonoid structures such as the methylenedioxyflavonols of *Melicope* are distinctive of these plants or whether they have a more general distribution. Additionally, the general screening techniques used up to now for flavonoid surveys are generally incapable of revealing the presence of most of the more distinctive flavonoid components of these plants.

It is therefore premature to draw any conclusions from the flavonoid data about subfamily relationships. Nevertheless, it is worth noting that flavanones, as useful markers for the family, have been detected in the four major subfamilies (Table 13) and that methylated flavones extend to at least two (Aurantioideae and Rutoideae). A search for flavanones in the Spathelioideae and for methylated flavones in the Toddalioideae would clearly be of interest.

Below subfamily level, some correlations between flavonoid distribution and taxonomy have emerged in the Aurantioideae. In a survey of fresh leaf tissue in some 35 taxa, representing five subtribes in the tribes Clauseneae and Citreae, Grieve and Scora[23] noted a progressive changeover in flavonoid pattern with advancement in morphology (Table 14). In particular, flavones (as O-glycosides) replace flavonols, except in the Balsamocitrinae where flavonols predominate. The flavonoid data additionally are in general accord with Swingle's treatment[57] of the subtribes Citrinae and Balsamocitrinae. Interestingly, myricetin was recorded in only one taxon, *Clausena excavata*. Flavanones were almost entirely restricted to the *Citrus* group, apart from a single occurrence in *Pleiospermum sumatranum*. As will be seen (Table 14), glycoflavones occur throughout and are useless as taxonomic markers within these tribes.

TABLE 14

Flavonoid trends in the Aurantioideae

Flavonoid type	Clauseninae	Merrilliinae	Triphasiinae	Citrinae	Balsamo-citrinae
Flavonol*					
Km/Qu	+	–	+	+	+
My	+	–	–	–	–
Glycoflavone	+	+	+	+	+
Flavanone	–	–	–	+	–
Flavone	–	–	+	+	+

*Km = kaempferol; Qu = quercetin; My = myricetin. Data from Grieve and Scora.[23]

A similar change in pattern observed within the genera of the Aurantioideae, the replacement of flavonol by flavone, has also been observed within the genus *Acmadenia* at the species level.[61] *Acmadenia* is included in the tribe Diosmeae of the Rutoideae and consists of some 33 species of shrub, which grow in the Cape province of South Africa. An analysis of hydrolysed leaf extracts of 15 of these species showed a satisfying correlation between sectional classification and the presence of flavonols or of flavones (Table 15). As will be seen, the flavonols quercetin and kaempferol present in the sections *Peltatiglandula* and *Patentipetalum* are almost completely replaced by the flavones luteolin and apigenin in the sections *Glandula* and *Acmadenia*. Again (see above), it will be noted that myricetin has only a limited presence in these plants, being confined to 3 of the 8 flavonol containing species. These data have some practical taxonomic value in that they corroborate a recent revision of the sectional classification, developed by Williams[63] from biological criteria.

J. B. HARBORNE

TABLE 15

Presence or absence of flavonol and flavone aglycones in leaf hydrolysates of *Acmadenia* species

Section and species	myricetin	Flavonols		Flavones	
		quercetin	kaempferol	luteolin	apigenin
Peltatiglandula					
A. bodkinii	+	+	−	−	−
A. rourkeana	+	+	−	−	−
Patentipetalum					
A. wittebergensis	−	+	+	−	−
A. tenax	−	+	+	−	−
*A. teretifolia**	+	+	+	−	−
*A. matroosbergensis**	−	+	+	−	−
A. argillophila	−	+	+	−	−
A. patentifolia	−	+	+	−	+
Glandula					
A. latifolia	−	−	−	+	+
A. kiwanensis	−	−	−	+	+
A. obtusata	−	+	−	+	+
Acmadenia					
A. heterophylla	−	+	−	+	+
A. rupicola	−	−	−	+	+
A. niveni	−	+	−	+	+
A. sheilae	−	+	−	+	+

* More than one specimen of these taxa were examined, with identical results.

The only other rutaceous genus where flavonoids have been demonstrated to have taxonomic potential is *Citrus*, where most, if not all, species have recognizably different patterns of methylated flavones in the peel and the leaves. The possibility of using methylated flavone profiles for investigating the parental origin of *Citrus* hybrids has been pointed out by Tatum and Berry.[58] These authors found, for example, that the concentrations of eight methoxyflavones in the peel of the *Citrus* cultivar calamondin were intermediate between the values observed in the two most probable parents, a Valencia orange and a mandarin type. Little practical use, however, has been made of these varying flavonoid patterns in *Citrus*. Other biochemical characters, such as isozyme patterns, are at present preferred for the chemotaxonomic analysis of hybridization in this complex genus (see Chapter 13).

B. PATTERNS WITHIN RUTALES

The Rutaceae stands out from other rutalean families in respect of its extremely versatile synthetic capacity to produce unusual, highly substituted, and in some cases unique, flavonoid constituents. The rutaceous flavonoids include substances which have a high degree of O-methylation, extra substitution in the 6,8, and 2' positions, the presence of methylenedioxy groups, and the attachment of isoprenoid sidechains. While many of these structural modifications occur on their own in other families, the combination of them in the Rutaceae is highly distinctive and marks off these plants from all other angiosperm groups.

By contrast, the flavonoids as far as they are known in the Burseraceae, Cneoraceae, Meliaceae, and Simaroubaceae show few, if any, of the above special features and there is little at present in their flavonoid profiles which links these four families to the Rutaceae. Clearly more attention deserves to be given to the flavonoids of these four families but there are as yet few hints to indicate that the richness of structural variation encountered in the Rutaceae extends to these other families.

Exceptionally, the Burseraceae contain a class of flavonoid not found elsewhere in the Rutales, namely a biflavonoid. Thus the apigenin dimer amentoflavone has recently been reported in leaves of *Garuga pinnata*.[2] Biflavonoids, although well known as gymnosperm constituents, are rare in the angiosperms and have so far been found in only 11 such families.[19] The nearest other source of biflavonoids is the Anacardiaceae (*Rhus*, *Semecarpus*, *Toxicodendron*), although some of the biflavonoids reported here are structurally distinct from the amentoflavone of *Garuga*. It is perhaps premature to assess the significance of this finding to the taxonomy of the Rutales. However, if biflavonoids are found to be more widely present in Burseraceae, the character would tend to separate the family from the order, since these constituents are not so far known in Rutaceae, Meliaceae, Cneoraceae, or Simaroubaceae.

C. COMPARISON BETWEEN RUTALES AND RELATED ORDERS

While most taxonomists (e.g. Dahlgren[15]) consider the Rutales to consist of a discrete cohort of five families, there have been suggestions that the families of the Sapindales should be associated with the Rutales, and even the suggestion of a close link with the Leguminosae (Fabaceae) has been promoted.[59] The flavonoids of these two orders have been compared by Young,[65] who has suggested that the presence of 7-deoxyflavonoids in both the Rutaceae and Anacardiaceae supports the merging of the Rutales and Sapindales.

However, a more detailed comparison of the flavonoid profiles of Rutaceae and Anacardiaceae (Table 16) indicates a number of discrepancies as well as similarities. The presence of 5-deoxyflavonoids in Anacardiaceae and their clear absence so far from the Rutaceae must be noted. Also, aurones are found in Anacardiaceae but not in Rutaceae. By contrast, few of the special flavonoids of Rutaceae have actually been recorded in Anacardiaceae. Another sapindaceous family, the Zygophyllaceae, has been investigated for flavonoids[54] and again few obvious links with the Rutaceae are apparent, apart from the presence of herbacetin derivatives in both groups.

TABLE 16

Flavonoid profiles of the Anacardiaceae and Rutaceae

Character	Anacardiaceae	Rutaceae
Biflavonoids	+ +	−
7-Deoxyflavonoids	+	+
5-Deoxyflavonoids	+ + +	?
Aurones	+ +	−
Flavanones	(+)	+ +
Methylenedioxyflavonols	−	+

In summary then, the present available data are equivocal regarding any direct merging of the Rutales *s.s.* and Sapindales *s.s.* and other chemical characters must be used, if this association is to be pressed (cf. Chapter 16).

V. Conclusion

It is clear from a summary view (Table 17) of the flavonoids reported in the Rutales that the chemistry of these constituents is now well defined. The present picture is very incomplete, however, in relation to the large number of plants not yet studied at all. Further unusual structures will almost certainly be uncovered in future researches, particularly in members of the Rutaceae.

By contrast with the chemistry, the chemotaxonomy of the rutalean flavonoids has hardly been developed at all so far, and this would seem to be an important area for future activity. By comparison with limonoids, coumarins, and alkaloids, the flavonoids have been rather neglected, and when more

TABLE 17

Genera of the Rutales and the different flavonoid classes which have been characterized in them

---

Rutaceae

Barosma (flavonol glyc., flavanone); *Boenninghausenia* (flavanol glyc.);
Boronia (anthocyanin); *Casimiroa* (meth. flavone, 2'-OH flavone, 7-deoxy flavone);
Citrus (anthocyanin, glycoflavone, meth. flavone, flavanone, 7-OH flavonol);
Correa (anthocyanin); *Dictamnus* (anthocyanin, flavanone);
Eriostemon (meth. flavonol); *Euodia* (meth. flavonol, flavanone, C_5 flavanone);
Fagara (glycoflavone, flavanone); *Flindersia* (8-OH flavonol, flavanone);
Haplophyllum (6-OH flavonol, 8-OH flavonol);
Melicope (8-OH flavonol, meth. flavonol, C_5 flavanone, MD flavonol);
Merrillia (meth. flavone, chalcone); *Murraya* (meth. flavone);
Phebalium (C_5 flavonol); *Phellodendron* (C_5 flavonol, C_5 flavanone);
Ptelea (flavanone); *Pitavia* (flavonol glyc.); *Ruta* (8-OH flavonol, flavonol glyc.);
Sargentia (meth. flavone, 2'-OH flavone, 7-deoxy flavone);
Skimmia (anthocyanin, flavanone); *Toddalia* (flavanone);
Zanthoxylum (anthocyanin, flavonol glyc., flavanone, 8-OH flavonol).

Meliaceae

Azadirachta (flavonol glyc.); *Cedrelopsis* (flavonol glyc.); *Chukrasia* (meth. flavone);
Melia (flavonol glyc.); *Neobeguea* (flavonol glyc.);
Soymida (flavonol glyc., meth. flavonol, flavanone).

Simaroubaceae

Ailanthus (flavonol glyc., glycoflavone); *Suriana* (flavonol glyc.).

Burseraceae

Garuga (biflavonoid)

This table excludes plants which have only been generally surveyed for their flavonoid pattern. meth. = methylated; glyc. = glycoside; MD = methylenedioxy; C_5 = isopentenyl. Flavanone implies presence as such or in glyc. form.

distributional data become available it is likely that useful systematic correlations will emerge. Finally, the biological activities, particularly in plant–animal interactions, of the unusual and distinctive rutaceous flavonoids deserve more attention than they have received so far.

REFERENCES

1. Adams, J. H. and Lewis, J. R. (1978). *Malaysian J. Sci.* **5B**, 173.
2. Ansari, F. R., Ansari, W. H. and Rahman, W. (1978). *Indian J. Chem.* **16B**, 846.
3. Balakrishna, K. J. and Seshadri, T. R. (1947). *Proc. Indian Acad. Sci.* **26A**, 234.
4. Bate-Smith, E. C. (1962). *J. Linn. Soc. Bot.* **58**, 95.

5. Batirov, E. K. and Malikov, V. M. (1980). *Khim. Prir. Soedin.* 330.
6. Batirov, E. K., Malikov, V. M. and Mirzamatov, R. T. (1980). *Khim. Prir. Soedin.* 836.
7. Bohm, B. A. (1975). *In* "The Flavonoids" (Harborne, J. B., Mabry, T. J. and Mabry, H., eds.), pp. 560–631. Chapman and Hall, London.
8. Briggs, L. H. and Locker, R. H. (1969). *J. Chem. Soc.* 2157.
9. Brown, R. F. C., Gilham, P. T., Hughes, G. K. and Ritchie, E. (1954). *Aust. J. Chem.* **7**, 181.
10. Brunet, G. and Ibrahim, R. K. (1980). *Phytochemistry* **19**, 741.
11. Brune, W. and Geissman, T. A. (1965). *Aust. J. Chem.* **18**, 1649.
12. Chandler, B. V. (1958). *Nature* (Lond.) **182**, 933.
13. Chatterjee, A., Malaker, D. and Ganguly, D. (1976). *Indian J. Chem.* **14B**, 233.
14. Chopin, J., Bouillant, M. L. and Besson, E. (1982). *In* "The Flavonoids: Recent Advances" (Harborne, J. B. and Mabry, T. J., eds.), pp. 449–503. Chapman and Hall, London.
15. Dahlgren, R. M. T. (1975) *Bot. Notiser* **128**, 119.
16. D'Amore, G. and Calapaj, R. (1965). *Rass. Chim.* (Rome) **17**, 264.
17. de Luca, V., Brunet, G., Khouri, H., Ibrahim, R. and Hrazdina, G. (1982). *Z. Naturforsch.* **37C**, 134.
18. Fraser, A. W. and Lewis, J. R. (1974). *Phytochemistry* **13**, 1561.
19. Geiger, H. and Quinn, C. (1982). *In* "The Flavonoids: Recent Advances" (Harborne, J. B. and Mabry, T. J., eds.), pp. 505–534, Chapman and Hall, London.
20. Geissman, T. A. (1958). *Aust. J. Chem.* **11**, 376.
21. Goodwin, T. W. (1981). "The Biochemistry of the Carotenoids", Volume 1, Plants, 2nd edn., Chapman and Hall, London.
22. Greive, M. (1931). "A Modern Herbal". Jonathan Cape, London.
23. Grieve, C. M. and Scora, R. W. (1980). *Syst. Bot.* **5**, 39.
24. Grimshaw, J. and Lamer-Zarawska, E. (1975). *Phytochemistry* **14**, 838.
25. Harborne, J. B. (1975). *In* "The Flavonoids" (Harborne, J. B., Mabry, T. J. and Mabry, H., eds.), pp. 1056–1095. Chapman and Hall, London.
26. Harborne, J. B. (1980). *In* "Encyclopedia of Plant Physiology", new series, Vol. 8, pp. 329–402. Springer Verlag, Berlin.
27. Harborne, J. B. (1981). *Phytochemistry* **20**, 1117.
28. Harborne, J. B., Lebreton, P., Combier, H., Mabry, T. J. and Hammam, Z. (1971). *Phytochemistry* **10**, 883.
29. Hartmann, G. and Nienhaus, F. (1974). *Phytopath. Z.* **81**, 971
30. Hegnauer, R. (1973). "Chemotaxonomie der Pflanzen", Vol. 6, pp. 174–239. Birkhauser Verlag, Basel.
31. Horowitz, R. M. (1957). *J. Amer. Chem. Soc.* **79**, 6561.
32. Horowitz, R. M. (1964). *In* "Biochemistry of Phenolic Compounds" (J. B. Harborne, ed.), pp. 545–572. Academic Press, London.
33. Ishikura, N. (1971). *Experientia* **27**, 1006.
34. Ishikura, N. (1975). *Bot. Mag. Tokyo* **88**, 41.
35. Khalid, S. A. and Waterman, P. G. (1981). *Planta Med.* **43**, 148.
36. Komasaki, H., Kusumoto, S., Ohsuka, A. and Kotake, M. (1968). *Nippon Kagaku Zasshi* **89**, 717.
37. Lassak, E. V. and Southwell, I. A. (1972). *Aust. J. Chem.* **25**, 2517.
38. Lawrence, W. J. C., Price, J. R., Robinson, G. M. and Robinson, R. (1939). *Phil. Trans. Roy. Soc.* **230**, 149.
39. Markham, K. R. and Chari, V. M. (1982). *In* "The Flavonoids: Recent Advances", pp. 19–51. Chapman and Hall, London.

40. Mitchell, R. E. and Geissman, T. A. (1971). *Phytochemistry* **10**, 1559.
41. Nair, A. G. R. and Subramanian, S. S. (1975). *Indian J. Chem.* **13**, 527.
42. Nakoaki, T. and Morita, N. (1958). *J. Pharm. Soc. Japan* **78**, 558.
43. Natsuno, T. (1958). *J. Pharm. Soc. Japan* **78**, 1311.
44. Nooteboom, H. P. (1966). *Blumea* **14**, 309.
45. Pardhasaradhi, M. and Sidhu, G. S. (1972). *Phytochemistry* **11**, 1520.
46. Paris, R. R. and Debray, M. (1972). *Plant Med. Phytother.* **6**, 311.
47. Piattelli, M. and Impellizzeri, G. (1971). *Phytochemistry* **10**, 2657.
48. Pinhey, J. T. and Southwell, I. A. (1973). *Aust. J. Chem.* **26**, 409.
49. Pinkas, J., Lavie, D. and Chorin, M. (1968). *Phytochemistry* **7**, 169.
50. Pinkas, M., Bezanger-Beauquesne, L. and Lallemont, N. (1968). *Compt. Rend.* **267D**, 1656.
51. Puroskothaman, K. K., Sarada, A., Saraswathi, G. and Connolly, J. D. (1977). *Phytochemistry* **16**, 398.
52. Rahman, W., Ishratullah, K., Wagner, H., Seligmann, O., Chari, V. H. and Osterdahl, B. G. (1978). *Phytochemistry* **17**, 1064.
53. Rao, M. M., Gupta, P. S., Krishna, E. M. and Singh, P. P. (1979). *Indian J. Chem.* **17B**, 178.
54. Saleh, N. A. M. and El-Hadidi, M. N. (1977). *Biochem. Syst, Ecol.* **5**, 121.
55. Silva, M., Cruz, M. A., and Sammes, P. G. (1971). *Phytochemistry* **10**, 3255.
56. Subramanian, S. S. and Nair, A. G. R. (1972). *Indian J. Chem.* **10**, 452.
57. Swingle, W. T. (1943). "The Botany of Citrus and its Wild Relatives of the Orange Subfamily". Univ. Calif. Press, Berkeley.
58. Tatum, J. H. and Berry, R. E. (1978). *Phytochemistry* **17**, 447.
59. Thorne, R. F. (1981). *In* "Phytochemistry and Angiosperm Phylogeny" (Young, D. A. and Seigler, D. S., eds.), pp. 233–294. Praeger, New York.
60. Waterman, P. G. (1975). *Biochem. Syst. Ecol.* **3**, 149.
61. Waterman, P. G. and Hussain, R. A. (1983). *Bot. J. Linn. Soc.*, in the press.
62. Weiss, A. (1842). *Chem. Zentr.* 305.
63. Williams, I. (1982). *J.S.Afr. Bot.* **48**, in press.
64. Wollenweber, E. and Dietz, V. H. (1981). *Phytochemistry* **20**, 869.
65. Young, D. A. (1981). *In* "Phytochemistry and Angiosperm Phylogeny" (Young, D. A. and Seigler, D. S., eds.), pp. 205–232. Praeger, New York.

CHAPTER 6

Chemistry of the Limonoids of the Meliaceae and Cneoraceae

JOSEPH D. CONNOLLY

Department of Chemistry, University of Glasgow, Glasgow, U.K.

I. INTRODUCTION

The publication of the structure of limonin (**1**)[7] marked the beginning of the era of limonoid chemistry and acted as a stimulus for extensive investigations of the Meliaceae and Rutaceae. These resulted in the isolation of many limonoids.[28,32] Despite the virtual absence of biosynthetic results it is possible to draw an impressive biogenetic picture of the relationships within this group. In recent years a new dimension has been added by the appearance of the complex C_{25} compounds (pentanortriterpenoids) of the Cneoraceae. A brief overall picture of the current state of limonoid chemistry as it concerns the Meliaceae and Cneoraceae is given, and then some interesting new compounds from *Carapa* species which we have been investigating in collaboration with our colleagues in the University of Yaoundé are described.

175

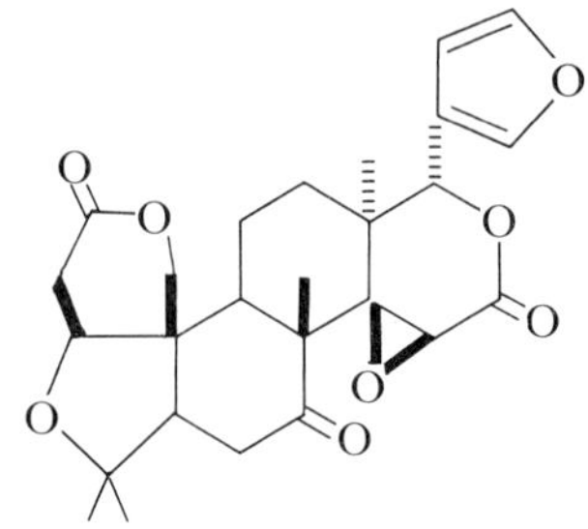

(1) Limonin

An inspection of the structure of limonin (1) suggests a derivation from a tetracyclic triterpenoid whose side-chain has been converted to a furan with the loss of four carbon atoms, hence the name tetranortriterpenoid. The generally accepted biogenesis of limonoids, originally proposed by Barton and his colleagues,[7] is shown in Scheme 1. A tirucallol (20αH) [or euphol (20βH)] precursor (2) undergoes rearrangement to the corresponding apotirucallol (3) with introduction of an oxygen substituent at C-7. Loss of four carbon atoms

cleavage of rings A,B,C,D

pentanortriterpenoids

quassinoids

(2)

(3)

(4)

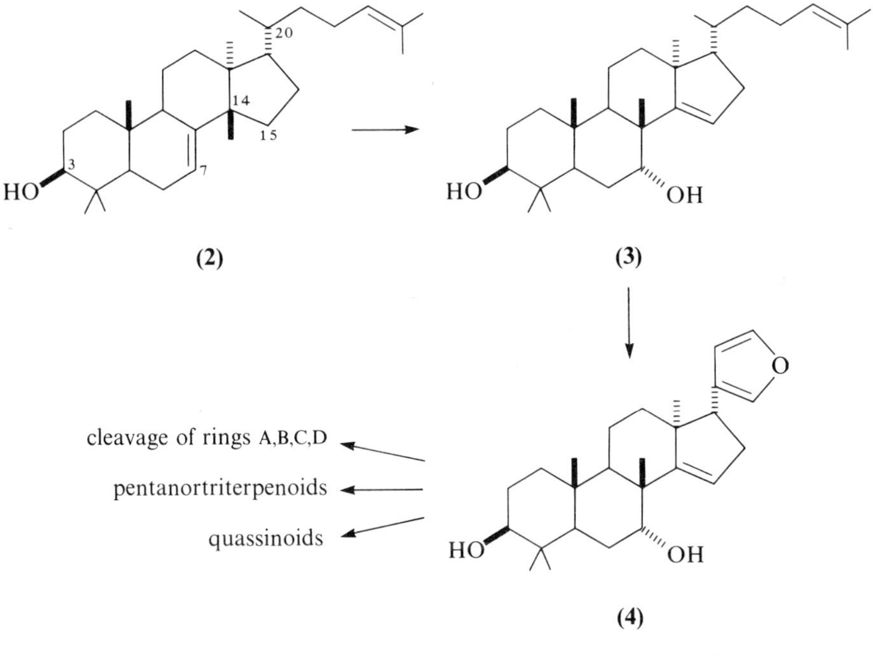

SCHEME 1. Biogenesis of limonoids, as proposed by Barton et al.[7]

and modification of the side-chain to a furan ring leads to the simplest tetranortriterpenoid (4). Subsequent oxidations, especially Baeyer–Villiger ring cleavages, and rearrangements afford the wide range of structural types which has been described. This simple scheme can be extended by further degradation to give the pentanortriterpenoids of the Cneoraceae and the quassinoids of the Simaroubaceae. Overwhelming support for this biogenetic picture comes from the isolation of compounds representing the various stages. A biogenetic approach will be adopted as a basis for my discussion of the tetranortriterpenoids.

II. C_{30} Group: Derivatives of Tirucallane or Euphane (Table 1)

Most of the compounds with an intact triterpenoid skeleton are derivatives of tirucallol. Turreanthin (5)[10] from *Turreanthus africanus* has a pattern of oxygenation in the side-chain which suggests a potential furan ring. Halsall and his colleagues successfully converted[13] this side-chain into a furan by treatment with $NaIO_4$/perchloric acid followed by TsOH in benzene. The triol 6 from *Entandrophragma* species[86] has a different side-chain arrangement and on treatment with BF_3 afforded sapelin A (7), isolated from *E. utile* and *E. cylindricum*, with a tetrahydropyran side-chain. This conversion was first achieved by Taylor and his colleagues in Jamaica using thermolytic conditions.[64] The product of alternative opening of the epoxide has not been obtained *in vitro* but is represented in nature by sapelin B (8) from *E. cylindricum*.[16] Several authentic euphane derivatives exemplified by sendanolactone (9) (6-oxokulactone) have been reported from *Melia azedarach*.[81]

(5) Turreanthin   (6)

(7) Sapelin A   (8) Sapelin B   (9) Sendanolactone

III. C$_{30}$ GROUP: DERIVATIVES OF APOTIRUCALLANE (Table 2)

The biogenetic scheme clearly suggests the intermediacy of apotirucallol derivatives in the formation of tetranortriterpenoids. The first example of this type was grandifoliolenone (10), from *Khaya grandifoliola*,[26] with the same side-chain as sapelin A (7). Further examples of naturally occurring apotirucallol derivatives have since been reported. The apo rearrangement is probably triggered by opening of a 7α,8α-epoxide with simultaneous migration of the methyl group from C-14 to C-8. In this context it is appropriate to point out that all tetranortriterpenoids have oxygenation on C-7. Both Halsall[13] and Lavie[59] successfully mimicked this process *in vitro* by Lewis acid catalysed opening of a 7α,8α-epoxide. A possible variation in this rearrangement is represented by glabretal (11) and its congeners from *Guarea glabra*.[40] The C-14 cation is captured by the C-13 methyl group with formation of a cyclopropane ring

(10) Grandifoliolenone   (11) Glabretal

instead of simple proton loss to yield a 14,15 double bond. It is interesting to note that the pentanortriterpenoids of the Cneoraceae, which we shall discuss later, also have this cyclopropane structural feature.

IV. C$_{26}$ GROUP: INTACT CARBON SKELETON (Table 3)

The next step in the process is loss of four carbon atoms from the side-chain and formation of the furan ring, and there are many examples of this type of tetranortriterpenoid with an intact carbon skeleton. Azadirone (12), azadiradione (13), and epoxyazadiradione (14)[60] have increasing oxygenation of ring D which leads eventually, as we shall see, to ring cleavage and lactone formation. Recently 17α- and 17β-hydroxy compounds (15[99] and 16[50]) have been reported. It is interesting to note that two compounds, the hemiacetal 17 and the γ-lactone 18, which appear to represent stages between the intact side-chain and the furan ring, have been isolated from *Chisocheton paniculatus*.[22] In

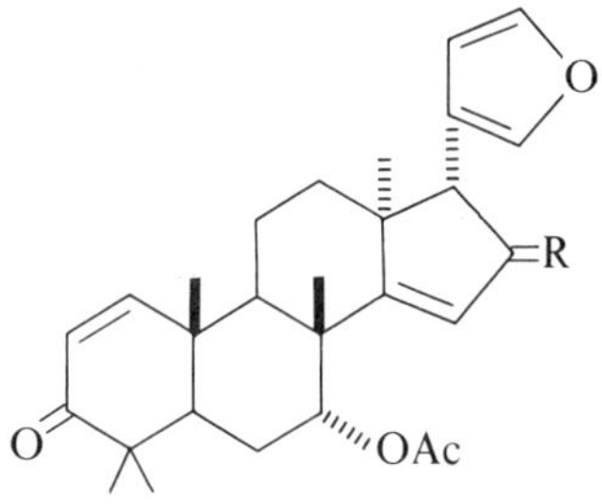

(12) Azadirone (R = H$_2$)
(13) Azadiradione (R = O)

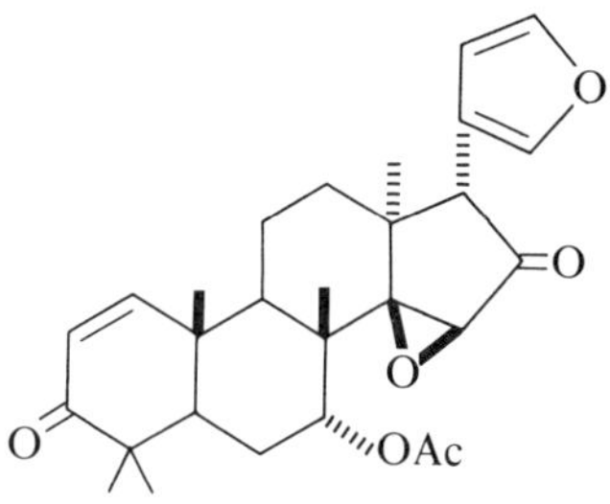

(14) Epoxyazadiradione

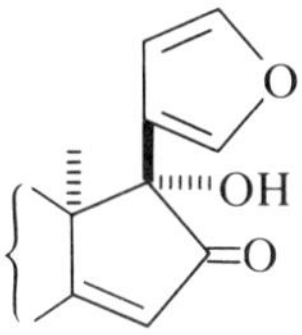

(15)

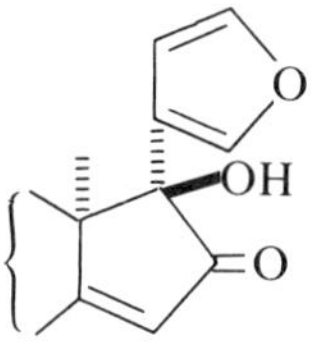

(16)

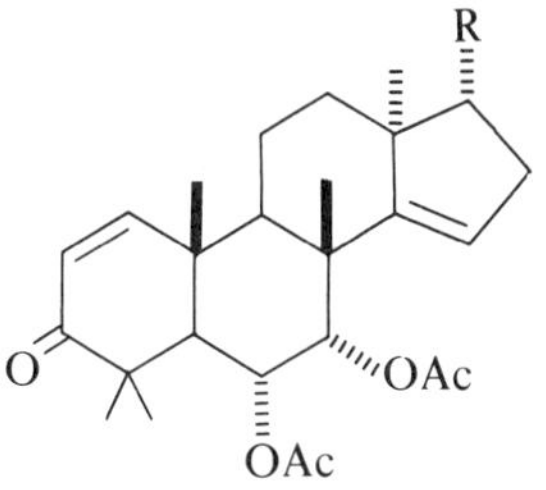

(17) R =

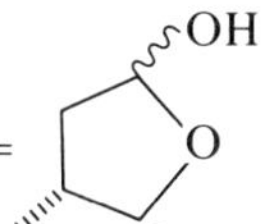

(18) R =

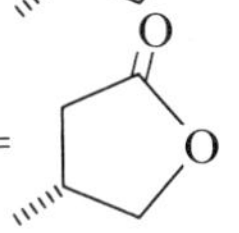

addition it is not uncommon to find compounds with a γ-hydroxybutenolide side-chain, which presumably arises by oxidation of the furan.[14,20,22,53,71,92,96]

In recent years there has been much interest in tetranortriterpenoids because of biological activity, in particular cytotoxic and antifeedant activity. 12α-Hydroxyamoorstatin **(19)** from *Aphanamixis grandifolia* is one of several cytotoxic compounds isolated by Polonsky and her colleagues.[90] Recently Nakanishi[73] has obtained several similar compounds, the trichilins [e.g. trichilin A **(20)**] from *Trichilia roka*, which exhibit interesting antifeedant activity. Kraus[53] has also made a major contribution to the isolation of tetranortriterpenoids with antifeedant activity.

In a study of the bark of *Turrea floribunda*[5] we isolated two new compounds **(21 and 22)**. While these compounds are inactive, the corresponding oxidation products **(23 and 24)** have *in vitro* cytotoxic activity.

(19) 12α-Hydroxyamoorstatin   **(20)** Trichilin A

(21)   **(22)**

(23)   (24)

V. C_{26} GROUP: RING D CLEAVED (Table 4)

Baeyer–Villiger oxidation of a 16-oxo group leads to the ring D epoxylactone found in many tetranortriterpenoids. Gedunin (25)[6] is an early example of this group. As we shall see in due course, ring D cleavage often accompanies other ring cleavages. A noteworthy reaction of this system is the "merolimonol" rearrangement. Thus base treatment of gedunin results in the formation of furfuraldehyde and merogedunol (26) which is of interest because of its similarity to the quassinoids.

(25) Gedunin   (26) Merogedunol

VI. C_{26} GROUP: RING C CLEAVED (Table 5)

Cleavage of ring C appears to be restricted to *Melia* species. Nimbin (27)[28] and nimbandiol (28)[51] are representative examples. The latter has lost an extra carbon atom and is, in effect, a pentanortriterpenoid. Kraus has isolated several compounds of this type from *Azadirachta indica*.[51] One of the most complex and most fascinating tetranortriterpenoids is azadirachtin (29)[111] from *Azadirachta indica* and *Melia azedarach*. It has potent antifeedant activity.

(27) Nimbin (R^1 = CO_2Me; R^2 = OAc)   (29) Azadirachtin
(28) Nimbandiol (R^1 = R^2 = OH)

VII. C_{26} GROUP: RING B CLEAVED (Tables 6–8)

Ring B cleavage is usually accompanied by opening of the lactone ring and dehydration as in andirobin (30).[82] However, Kraus has recently reported[54] the ring B lactone toonafolin (31), from *Toona ciliata*. Methyl angolensate (32),[28] a common representative of this group, has a $1\alpha,14\beta$ ether ring. Its structure has been confirmed by a partial synthesis[31] from 7-deacetyl-7-oxokhivorin (33) as shown in Scheme 2. An alternative method of ether ring closure has been demonstrated by Mondon and his colleagues.[36] Treatment of cneorin R (35) with sodium hydride afforded methyl ivorensate (36), the ring A lactone equivalent of 32.

(30) Andirobin   (31) Toonafolin

Recently we have investigated[23] the structure of ekebergolactones I, $C_{41}H_{52}O_{17}$, and II, $C_{37}H_{46}O_{17}$, from *Ekebergia senegalensis*. These have the same gross structure and differ only in the nature of the esterifying groups. The functionality revealed by an extensive examination of the ^1H and ^{13}C spectra can be assembled to give the biogenetically reasonable structure 37. Convincing

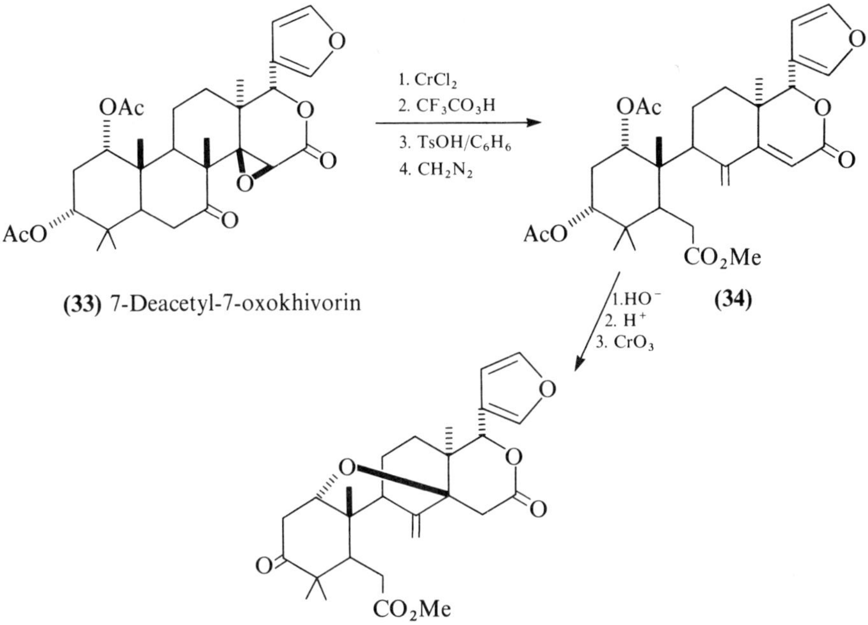

(33) 7-Deacetyl-7-oxokhivorin

1. CrCl$_2$
2. CF$_3$CO$_3$H
3. TsOH/C$_6$H$_6$
4. CH$_2$N$_2$

(34)

1. HO$^-$
2. H$^+$
3. CrO$_3$

(32) Methyl angolensate

SCHEME 2. Partial synthesis of methyl angolensate.[31]

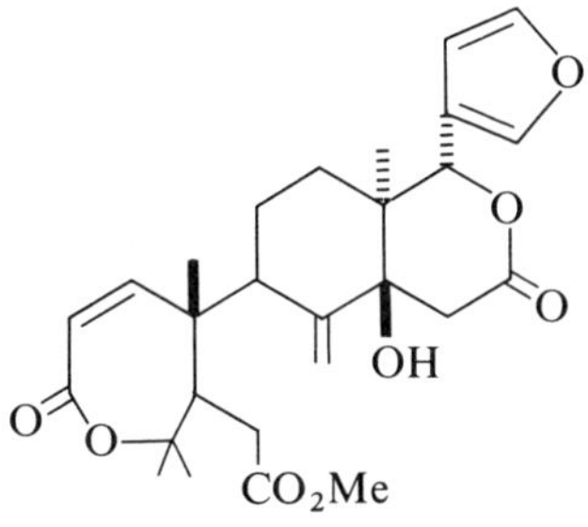

(35) Cneorin R

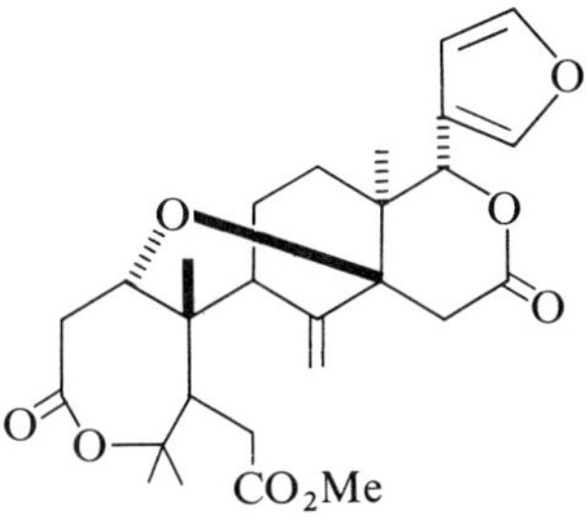

(36) Methyl ivorensate

support for this structure comes from the observation of nuclear Overhauser effects between H-5 and H-12, H-15 and the C-13 methyl group, and H-17 and H-12. The relative positions of the esters have not been determined.

Cleavage of ring B can be obscured by subsequent bond formation between C-2 and C-30 resulting in the bicyclo[3.3.1]nonane system found in

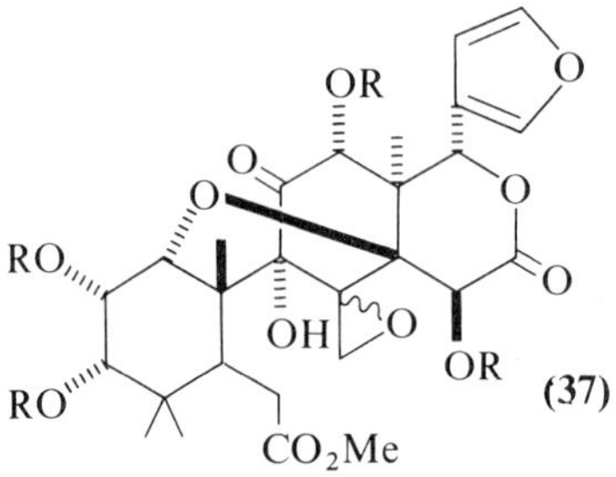

(37) Ekebergolactone

mexicanolide (38).[28] Some years ago we carried out a biomimetic partial synthesis of mexicanolide[31] from 7-deacetyl-7-oxokhivorin (33) as shown in Scheme 3. A similar synthesis, using a Beckmann rearrangement to open ring B, was reported by Ekong.[75] Further modification of the bicyclo[3.3.1]nonane system occurs in complex tetranortriterpenoids like utilin (39) and entandrophragmin (40) from *Entandrophragma* species.[45]

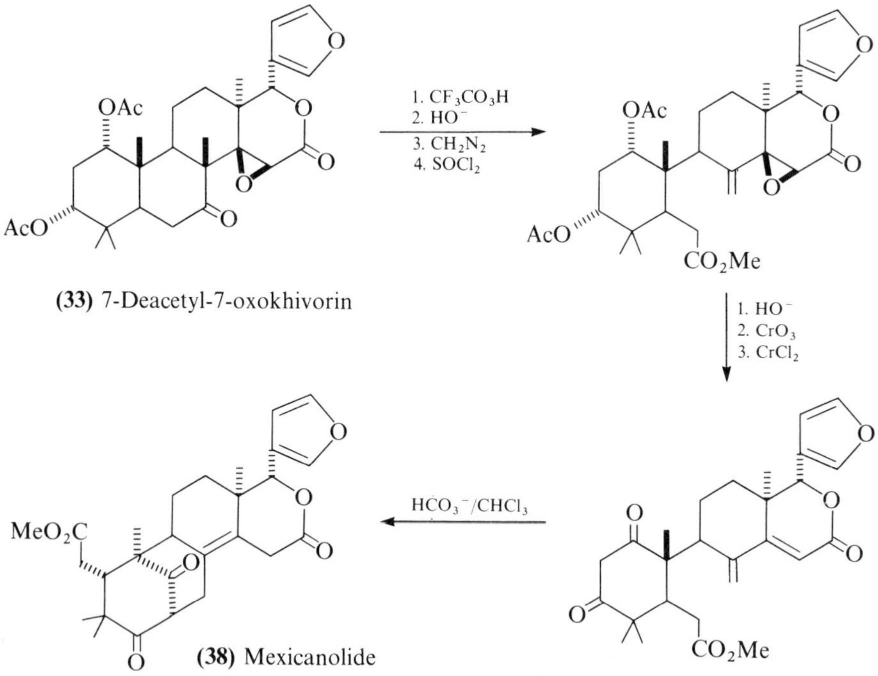

(33) 7-Deacetyl-7-oxokhivorin

1. CF_3CO_3H
2. HO^-
3. CH_2N_2
4. $SOCl_2$

1. HO^-
2. CrO_3
3. $CrCl_2$

$HCO_3^-/CHCl_3$

(38) Mexicanolide

SCHEME 3. Biomimetic partial synthesis of mexicanolide.[31]

(39) Utilin (R = Ac)

(40) Entandrophragmin (R = COCHMe$_2$)

VIII. C$_{26}$ GROUP: RING A CLEAVED (Table 9)

For many years compounds with a cleaved ring A, e.g. limonin (1), were thought to be restricted to the Rutaceae. However, this was due to an accident of isolation and many examples of ring A ε-lactones have since been obtained from the Meliaceae and the Cneoraceae. One example, tricoccin S$_{22}$ (41) from the Cneoraceae,[36] is of interest because of the presence of a hydroxyl group on C-16 and the unexpected product (42) of acid catalysed rearrangement of the epoxide.

(41) Tricoccin S$_{22}$ (42)

IX. C$_{26}$ GROUP: RINGS A AND B CLEAVED (Table 10)

Compounds with both rings A and B cleaved range from the simple bis-lactone surenolactone (43), isolated by Kraus from *Toona sureni*,[57] to the group of complex tetranortriterpenoids represented by prieurianin (44) from *Trichilia prieuriana*.[41] In solution at room temperature prieurianin suffers restricted

rotation about the C-9,C-10 bond and as a result some of the resonances in its
^{1}H and ^{13}C spectra are broad or scarcely visible. A closely related compound,
dregeanin from *T. dregeana*, was assigned structure **45**[26] with a highly strained
(virtually impossible) δ-lactone to account for the 1787 cm^{-1} carbonyl band in
its i.r. spectrum. Taylor, who published this structure "despite the scepticism of
some of his colleagues", has recently proposed the alternative structure (**46**)
with an eight-membered lactone ring.[108] This proposal was prompted by a
study of the alkaline hydrolysis of dregeanin.[21] Four products were obtained.
The first two (**47** and **48**) are epimeric at C-1 and arise by alcoholysis of the
lactone, β-elimination of the C-1 substituent, and addition of the C-11 oxygen
from the α and β faces of the molecule to the unsaturated carbonyl system. The
remaining two compounds are the bis-lactone **49** and the hemiorthoester **50** or
51. It is interesting to note that a similar hemiorthoester, hispidin A (**52**), has

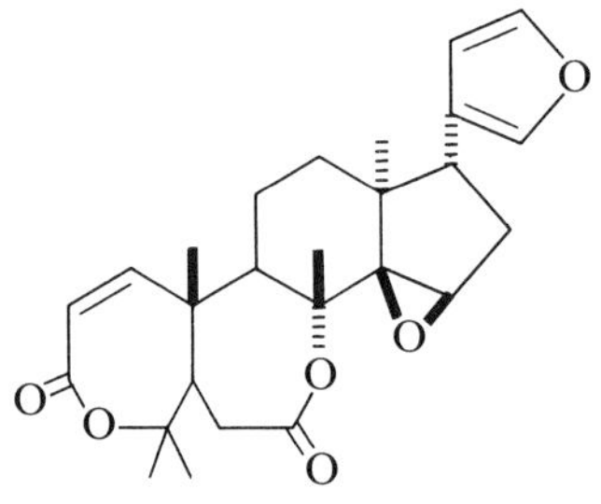

(**43**) Surenolactone

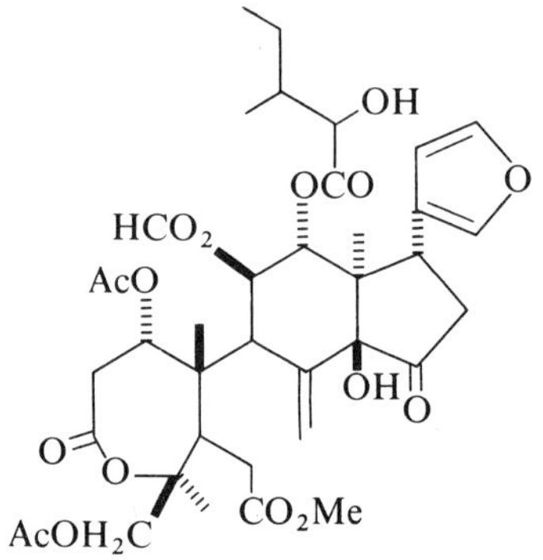

(**44**) Prieurianin

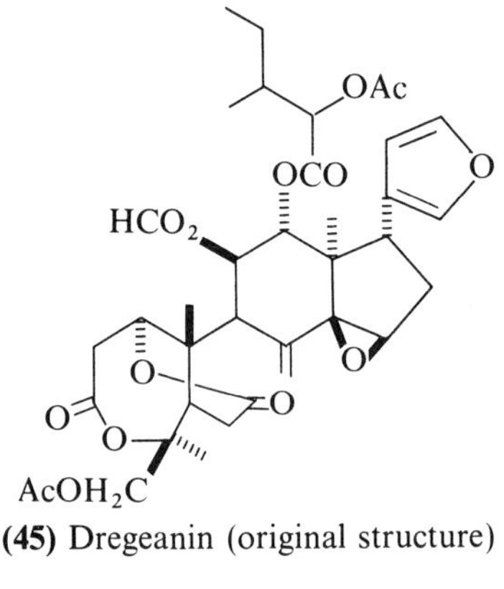

(**45**) Dregeanin (original structure)

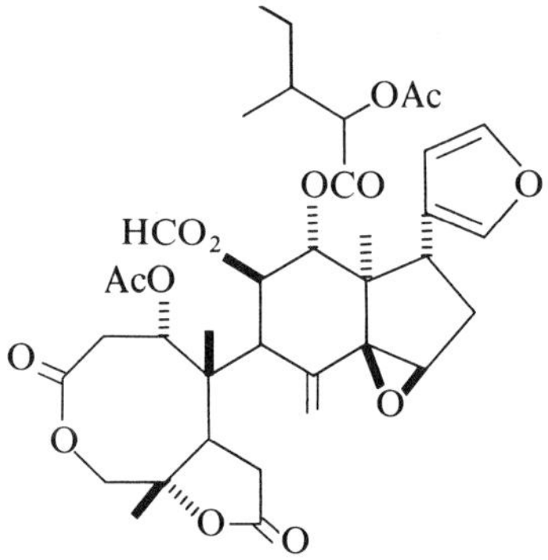

(**46**) Dregeanin (revised)

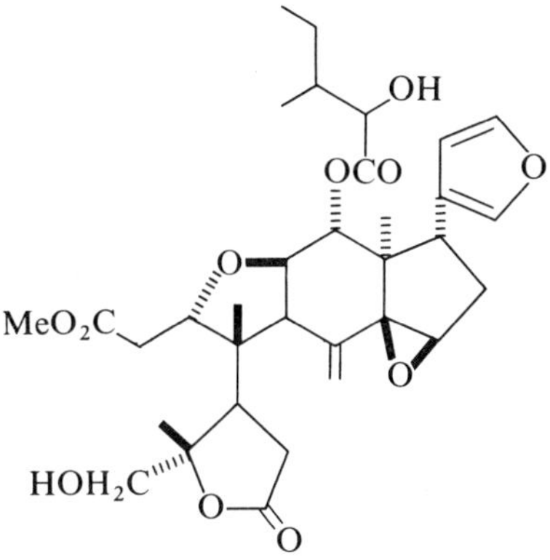

(47)

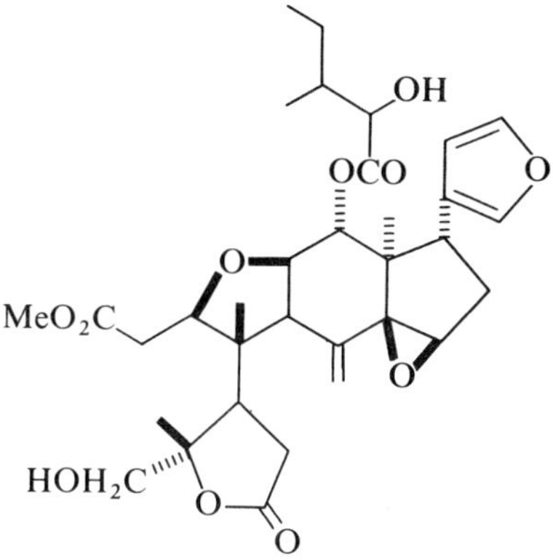

(48)

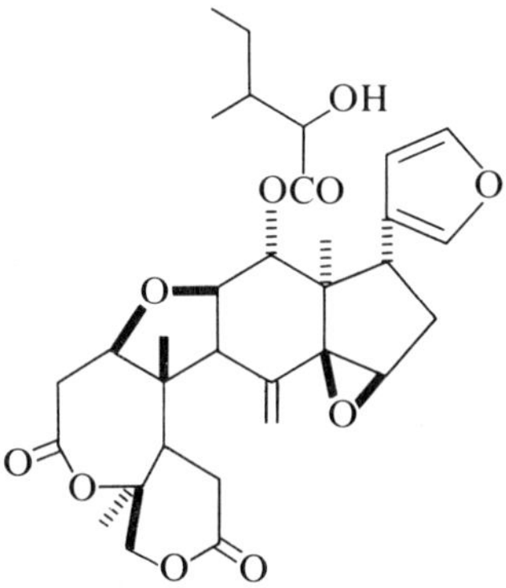

(49)

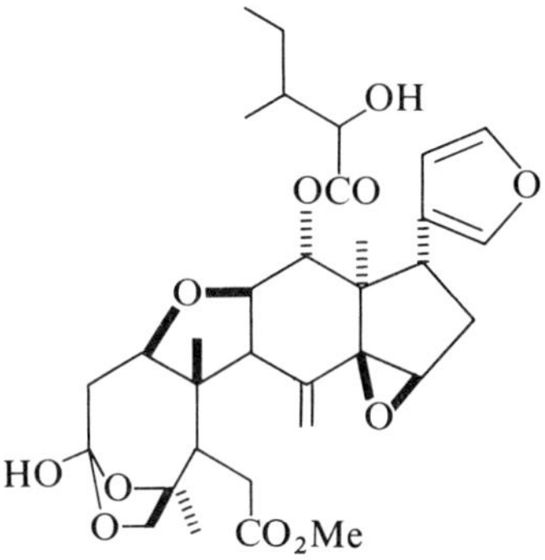

(50)

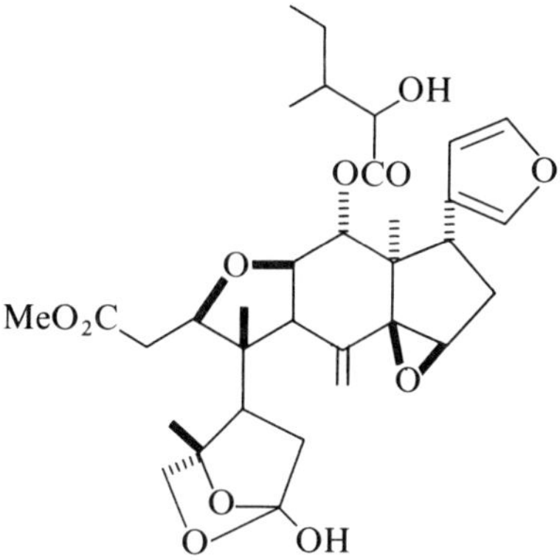

(51)

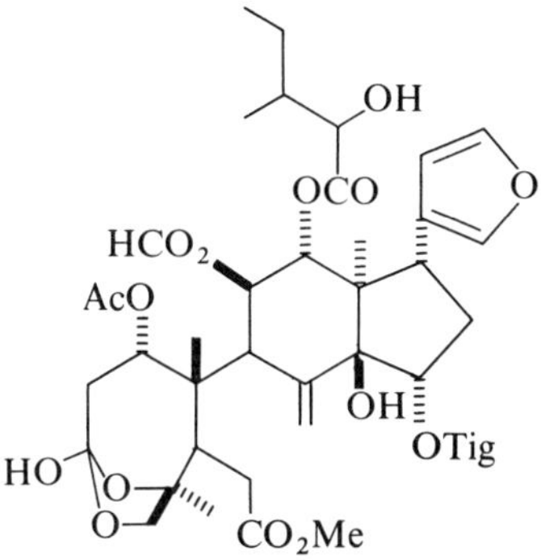

(52) Hispidin A

been isolated by Cole and his colleagues from *Trichilia hispida*.[48] In principle
the hemiorthoester **50** can open in two ways, one leading to an ε,δ-bis-lactone
(as **49**) and the other to an eight-membered γ-bis-lactone (as **46**) which is the
proposed structure for dregeanin.

X. INVESTIGATION OF *Carapa procera* AND *C. grandiflora*

Several years ago Professor Taylor and I examined the bark of *Carapa
procera* and isolated procerin (**53**)[105] and a novel spirolactone (**54**).[15] The latter
is the only example of its kind in the tetranortriterpenoids but several fungal
metabolites, e.g. andibenin (**55**), of mixed sesquiterpene–polyketide origin have
been shown to have the same spiro-lactone ring A arrangement. Extraction of
the seeds of *C. procera*, collected in the University garden in Yaoundé, afforded[9]
a new compound CP-3, $C_{26}H_{30}O_8$ (*m/z* 470). Its spectroscopic properties
revealed four tertiary methyl groups, a secondary acetate, di- and tri-substituted
double bonds, a β-substituted furan, and two lactones including the typical ring
D epoxylactone. The basic skeleton has twenty-four carbons, i.e. a
hexanortriterpenoid, and two carbons including a methyl group have been lost
from the normal tetranortriterpenoid skeleton. Decoupling experiments at
360 MHz enabled us to assemble the part structure **56**. It seemed likely that unit
(a) is ring C and unit (b) the remnants of rings A and B. Expansion of this part-
structure to the acceptable structure **57** for CP-3 is straightforward.

(53) Procerin   (54)

In biogenetic terms the loss of two carbons can be satisfactorily
accommodated in terms of the process shown in Scheme 4. The most attractive
starting point is the spiro-lactone **58** similar to **54** above. A retro-Prins reaction
leads to cleavage of ring B with formation of a methyl ketone and introduction of
a ring C disubstituted double bond. Vinylogous 1,3 dicarbonyl cleavage permits

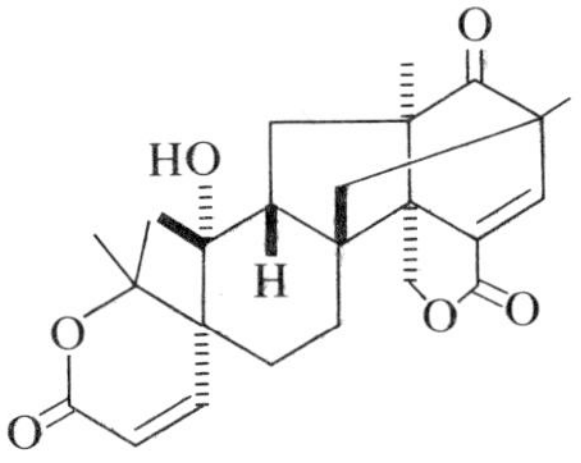

(55) Andibenin

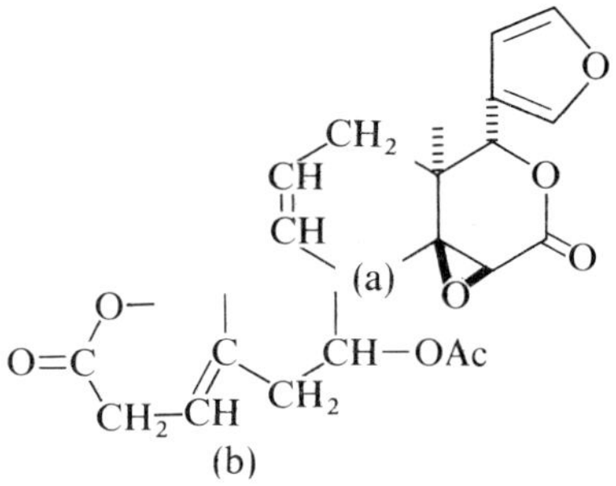

(56)

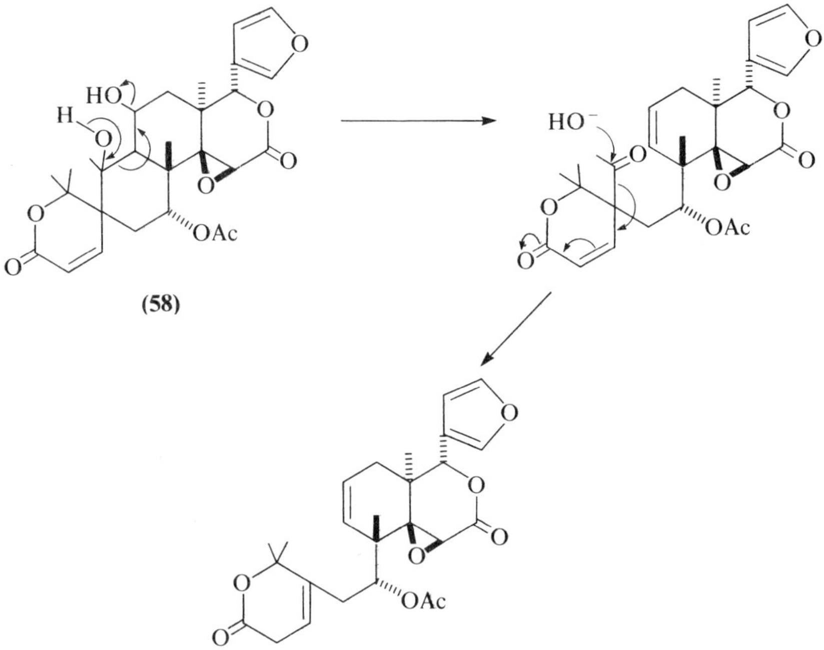

(58)

(57) CP-3

SCHEME 4. Possible route for the biogenesis of CP-3.

the loss of two carbon atoms, including a methyl group, as ethanoic acid and affords structure **57** which has all the features of CP-3.

Support for this structure came from two reactions. First, treatment of CP-3 with acidic ethanol, in an effort to move the trisubstituted double bond into conjugation with the lactonic carbonyl, yielded the diene ethyl ester **59** whose spectroscopic properties accord with its structure. Basic hydrolysis, on the other hand, resulted in loss of the acetate and the trisubstituted double bond with formation of the ether **60**.

In an effort to find more evidence we examined the minor constituents of the extract. Separation was difficult but we were gratified to obtain two compounds, in minor amounts insufficient for ^{13}C spectra, whose i.r. and ^{1}H n.m.r. spectra suggested structures **61** and **62**. The presence of the methyl ketone functions provides encouraging support for the novel cleavage of ring B proposed above and for the structure of CP-3.

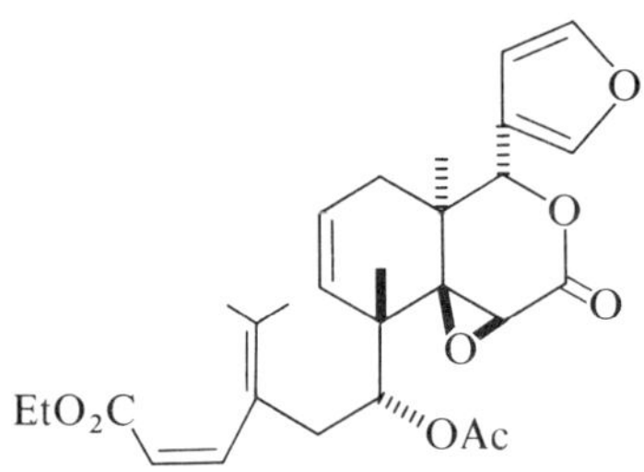

(59)

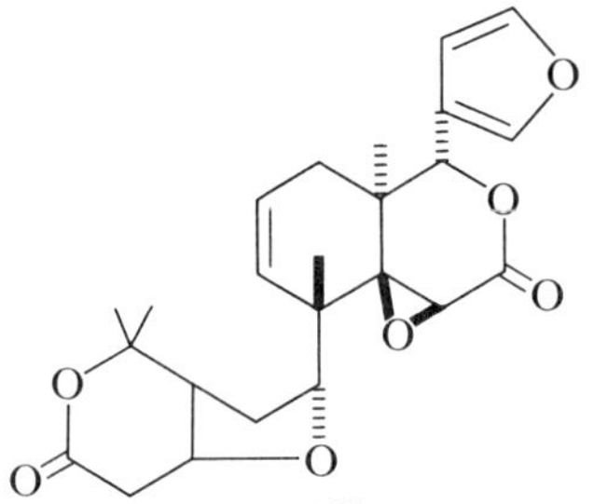

(60)

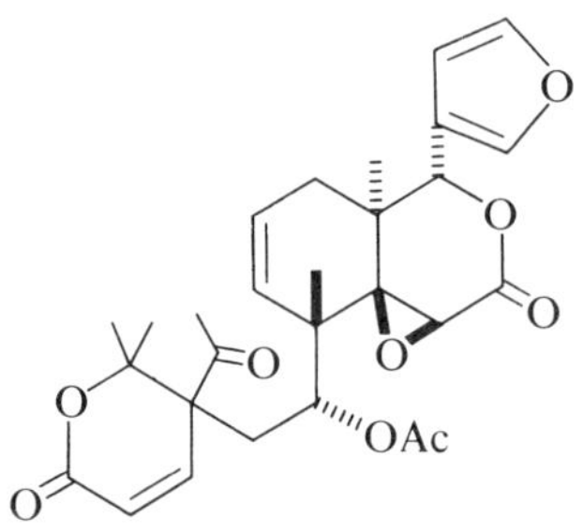

(61)

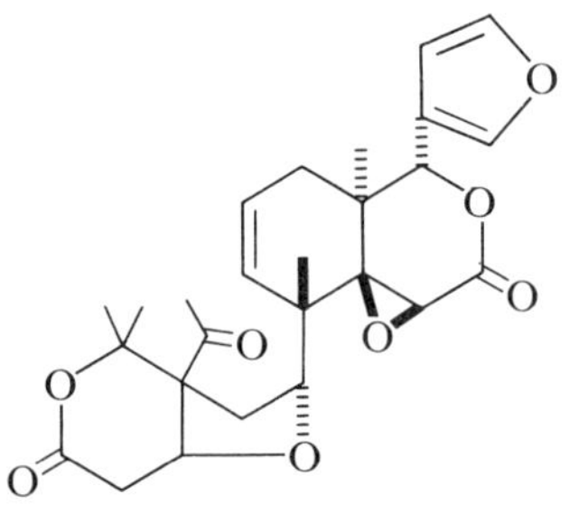

(62)

Extraction of a related species, *C. grandiflora*, afforded two compounds, CG-2 ($C_{26}H_{28}O_7$) and CG-3 ($C_{26}H_{30}O_8$), which have four and five tertiary methyl groups respectively. Both have an α,β-unsaturated δ-lactone, a unit $-C-CH_2-C(O-)H-C-$, and the same C/D ring system as CP-3 (**57**). Only CG-3 has hydroxyl absorption in the i.r. The remaining spectroscopic features of CG-2 (δ_H 4.17, 4.49 (ABq, J 2 Hz); δ_C 159.8(s), 86.3(t)) reveal the presence of vinyl ether. The ^{13}C resonance at 105.2(s) in CG-3 indicates a hemiacetal, i.e. the hydrate of the vinyl ether. Thus we arrived at the structures **63** and **64** for CG-2 and CG-3. The base peak in the mass spectrum of **63** at m/z 193, due to the ion **65**, is consistent with the proposed structure. The relationship between the two compounds was confirmed by dehydration of CG-3 with thionyl chloride to give CG-2. As expected, reaction of CG-3 with acidic methanol yielded the methyl acetal **66**.

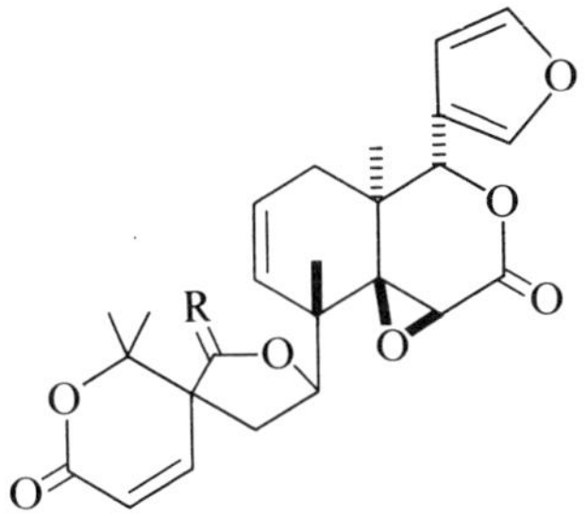

(63) CG-2 (R = CH$_2$)

(64) CG-3 (R = Me,OH)

(66)   (R = Me,OMe)

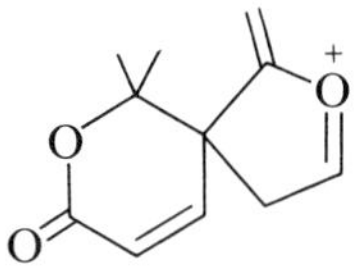

An attempt was made to open the enol ether ring of CG-3 (**64**) by treatment with TsOH in benzene. The major product, isolated by preparative t.l.c., proved to be the methyl ketone **62** identical with the compound isolated from the *C. procera* extract.

XI. C$_{25}$ GROUP: PENTANORTRITERPENOIDS OF THE CNEORACEAE (Table 11)

Discussion of the Cneoraceae has been reserved to the last since the major compounds, the C$_{25}$ pentanortriterpenoids, represent deep-seated modification of the normal tetranortriterpenoid intact skeleton. All the work in this series is due to Mondon and his colleagues and space does not permit me to do justice to their achievements. Tricoccin S$_{14}$ (**67**)[68] provides a representative example of this group. Rings A and B have been cleaved, ring D has lost a carbon atom, the C-13 methyl group is involved in a cyclopropane ring, and C-7 has become detached from C-8 and is now bonded to C-30. The detailed mechanism of these modifications is still a matter of speculation. The carbon framework **68** helps to emphasize the relationship between pentanortriterpenoids and tetranortriterpenoids.

The structural variations involve different modes of hemiacetal or acetal formation of the C-7 carbonyl group as in cneorin C (**69**),[66] K (**70**),[68] and

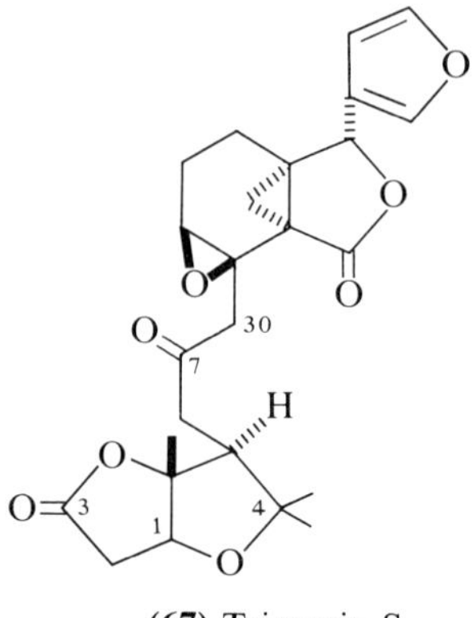

(67) Tricoccin S$_{14}$

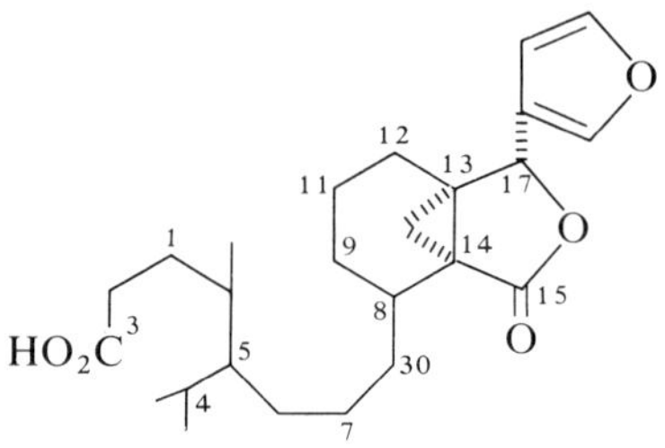

(68)

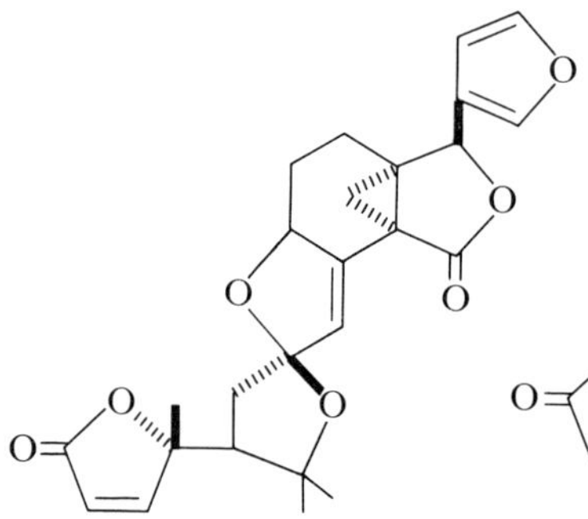

(69) Cneorin C

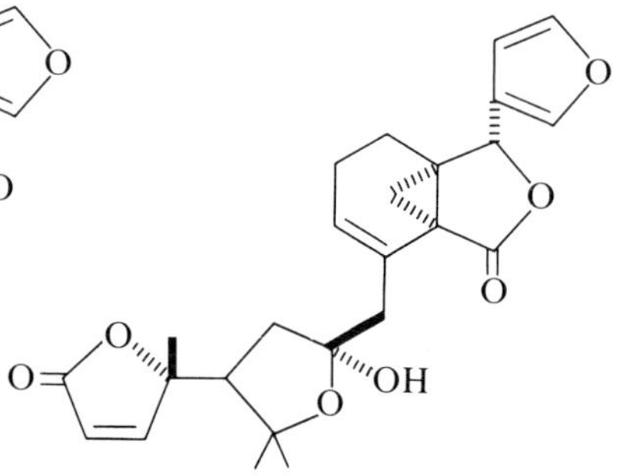

(70) Cneorin K

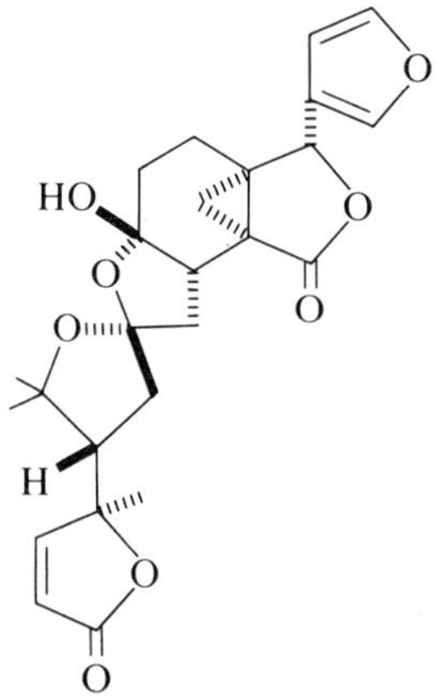

(71) Tricoccin S$_2$

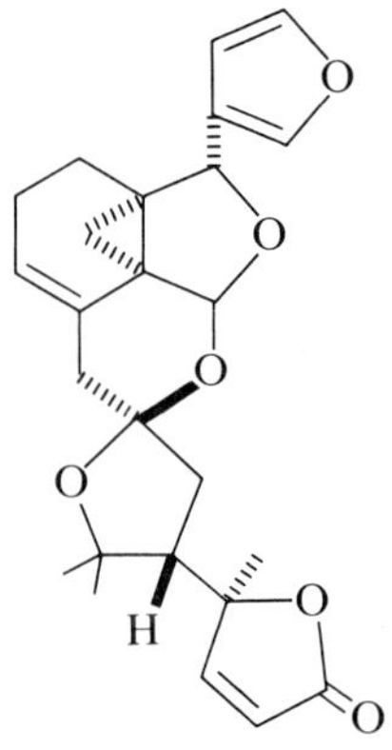

(72) Tricoccin S$_4$

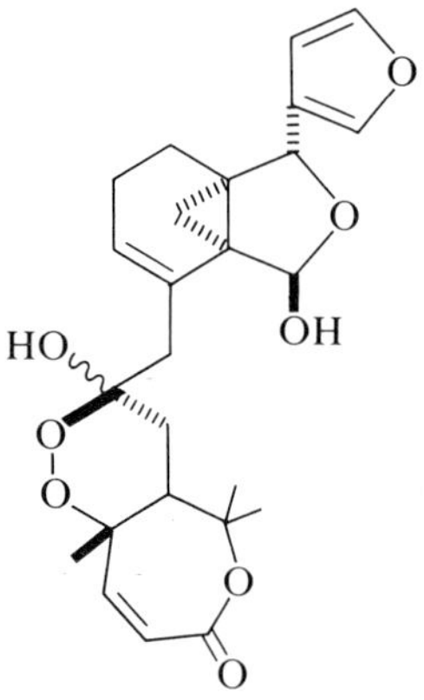

(73) Tricoccin S$_{27}$

tricoccin S_2 (71).[70] The picture becomes more complicated in tricoccin S_4 (72)[39] where the C-14 carbonyl group has been reduced and is involved in acetal formation. Recently several peroxides have been isolated.[37,38] It has been suggested that these represent an intermediate stage in the formation of the pentanortriterpenoids. Tricoccin S_{27} (73)[37] is an interesting example since it also retains the ring A ε-lactone commonly found in the tetranortriterpenoid series.

TABLE 1*

Sources of tirucallane and euphane derivatives

Compound	C=C	OH	C=O		Source
Sapelin A (7)	7	3α, 23R, 25		20αH; 21,24R-oxide	Entandrophragma cylindricum[16]
Sapelin B (8)	7	3α, 23R, 24S		20αH; 21,25-oxide	E. cylindricum[16]
Sapelin F	7	3α, 21, 23R, 24S, 25			E. cylindricum[17]
Entandrolide	7, 24			3,4-lactone	E. spp.[86]
Triol (6)	7	3α, 21, 23		20αH; 24,25-oxide	E. spp.[86]
Methyl kulonate	7, 24	16β	3	20βH; 21CO$_2$Me	Melia azedarach[18]
Sendanolactone (9)	7, 24		3, 6	20βH; 21,16β-lactone	M. azedarach[81]
3-Deoxymelianone	7	21		20αH; 21,23-oxide; 24,25-oxide	M. azedarach[98]
Cneorin NP$_{32}$	7	23α, 25	3	21,24-lactone	Neochamaelea pulverulenta[67]
Cneorin NP$_{34}$	7	24, 25	3	21,23-lactone	N. pulverulenta[67]
3,4-Secotirucalla-4(28),7,24-trien-3,21-dioic acid					Guarea cedrata[4]
3,4-Secotirucalla-4(28),7,24-trien-3,21-dioic acid 3-methyl ester					G. cedrata[4]
Hispidone	7	23R, 24S	3	20αH; 21,25-oxide	Trichilia hispida[47]

* For compounds published before 1970 see ref. 28.

TABLE 2

Sources of apotirucallane derivatives

Compound	C=C	OH	C=O	Source
16-Oxograndifoliolenone	1, 14	7αOAc, 23R, 25	3, 16 20αH; 21,24R-oxide	Khaya grandifoliola[25]
Lactone	14, 24	3β, 7α	20αH; 21,23-lactone	Melia azedarach[58]
Triacetate	14, 24	1αOAc, 3αOAc, 7αOAc	23 20αH; 21,24-oxide	K. ivorensis[3]
Sapelin C	1, 14	7α, 23R, 25	3 20αH; 21,24R-oxide	Entandrophragma cylindricum[17]
Sapelin D	14	3α, 7α, 23R 25	20αH; 21,24R-oxide	E. cylindricum[17]
Sapelin E	1, 14	7α, 23R, 24S	3 20αH; 21,25-oxide	E. cylindricum[17]
Melianin A	14	1αOAc, 3αOBz, 7αOAc, 23R, 25	20αH; 21,24-oxide	M. azedarach[83]
Melianin B	14	1αOAc, 3αOBz, 7αOAc, 23, 24	20αH; 21,25-oxide	M. azedarach[83]
Ketol	14	7α, 21αOAc	3 20αH; 21,23-oxide; 24,25-oxide	Chisocheton paniculatus[22]
Diol	14	3α, 7α, 21αOAc	20αH; 21,23-oxide; 24,25-oxide	C. paniculatus[22]
Tetrol	14	3αOAc, 7α, 21α, 24, 25	20αH; 21,23-oxide	C. paniculatus[22]
Triol	14	3αOAc, 7α 23R, 25	20αH; 21,24R-oxide	C. paniculatus[22]
Spicatin	1, 14	7αOAc, 23, 24, 25	3 20αH; 21,24-oxide	E. spicatum[29]
Glabretal (11)*		3α, 7αOAc, 21	14,18-cyclo; 20αH; 21,23-oxide; 24,25-oxide	Guarea glabra[40]

*Also 3α-(α-hydroxyvalerate), 3α-angelate, 3α-tiglate, 3α-methacrylate, and 3-ketone

TABLE 3

Sources of C_{26} group: intact carbon skeleton

Compound	C=C	OH	C=O	Source
Nimbolin A	14	1αOAc, 3αOAc, 7αOCinn	28,6α-oxide	Melia azedarach and Azadirachta indica[33]
Nimbidinin	14	1α, 3α, 7α	12 28,6α-oxide	M. indica[65]
Heudelottin C	1	7αOCOCH(OH)Pr^i, 11β, 12αOCOCH(OAc)Bu^s	3 14β,15β-oxide	Trichilia heudelottii[85]
Heudelottin E	1	7αOCOCH(OH)Pr^i, 11βOCOH, 12αOCOCH(OH)Bu^s	3 14β,15β-oxide	T. heudelottii[85]
Heudelottin F	1	7αOCOCH(OH)Pr^i, 11βOCOH, 12αOCOCH(OAc)Bu^s	3 14β,15β-oxide	T. heudelottii[85]
Cedrelone	1, 5	6	3, 7 14β,15β-oxide	Cedrela toona[19]
1,2-Dihydrocedrelone	5	6	3, 7 14β,15β-oxide	C. toona[19]
6α-Acetoxyepoxyazadiradione	1	6αOAc, 7αOAc	3, 16 14β,15β-oxide	Carapa guianensis[62]
Trifolin		1αOAc, 3α, 7αOCOCH(OH)Pr^i	14β,15β-oxide	T. trifolia[110]
Vilasinin	14	1α, 3α, 7α	28,6α-oxide	Azadirachta indica[87]
Vilasinin 1,3-diacetate	14	1αOAc, 3αOAc, 7α	28,6α-oxide	Chisocheton paniculatus[22]
Sendanin		1α, 3αOAc, 7α, 12αOAc, 29SOAc	11 14β,15β-oxide; 19,29-oxide	M. azedarach[76]
Desacetylsendanin		1α, 3αOAc, 7α, 12αOAc, 29	11 14β,15β-oxide; 19,29-oxide	M. azedarach[77]
Enol ether	5, 9(11)	6	3, 7 1,11-oxide; 14β,15β-oxide	Khaya anthotheca[42]
11α-Acetoxyazadirone	1, 14	7αOAc, 11αOAc	3	K. anthotheca[34,42]

TABLE 3—contd.

Compound	C=C	OH	C=O		Source
11β-Acetoxyazadirone	1, 14	7αOAc, 11βOAc	3		K. anthotheca[42]
Dysobinin	1, 14	6βOAc, 7αOAc	3		Dysoxylum binectari-ferum[100]
Aphanastatin		1αOAc, 2α, 3αOAc, 7α 12α, 29OCOBu^s	11	14β,15β-oxide; 19,29-oxide	Aphanamixis grandifolia[89]
Sendanal	14	1α, 3αOAc, 6α 7α, 12αOAc	28		M. azedarach[80]
17-Epiazadiradione	1, 14	7αOAc	3, 16	17αH	Azadirachta indica[50]
17β-Hydroxyazadiradione	1, 14	7αOAc, 17β	3, 16		A. indica[50]
17α-Hydroxyazadiradione	1, 14	7αOAc,17α	3, 16		A. indica[99]
Amoorstatin		1α, 3αOAc, 7α, 29	11	14β,15β-oxide; 19,29-oxide	Aphanamixis grandifolia[88]
12α-Hydroxyamoorstatin (19)		1α, 3αOAc, 7α, 12α, 29	11	14β,15β-oxide; 19,29-oxide	A. grandifolia[90]
Amoorstatone		1α, 3αOAc, 7α, 29	11, 15	14βH; 19,29-oxide	A. grandifolia[90]
6α-Acetoxyazadirone	1, 14	6αOAc, 7αOAc	3		Chisocheton paniculatus[97]
6α-Acetoxyazadiradione	1, 14	6αOAc, 7αOAc	3, 16		C. paniculatus[97]
6α-Hydroxyazadiradione	1, 14	6α, 7αOAc	3, 16		C. paniculatus[97]
17β-Hydroxy-6α-acetoxyazadiradione	1, 14	6αOAc, 7αOAc, 17β	3, 16		C. paniculatus[22]
Hemiacetal (17)					C. paniculatus[22]
γ-Lactone (18)					C. paniculatus[22]
Trichilin A (20)		1α, 2αOAc, 3αOAc, 7α, 12β, 29OCOBu^s	11	14β,15β-oxide; 19,29-oxide	Trichilia roka[73]
Trichilin B		1α, 2αOAc, 3αOAc, 7α, 12α, 29OCOBu^s	11	14β,15β-oxide; 19,29-oxide	T. roka[73]
Trichilin C		1α, 2αOAc, 3αOAc, 7α, 11β, 29OCOBu^s	12	14β,15β-oxide; 19,29-oxide	T. roka[73]
Trichilin D		1α, 2αOAc, 3αOAc, 7α, 29OCOBu^s	11	14β,15β-oxide; 19,29-oxide	T. roka[73]
Trichilin E		1αOAc, 2α, 3αOAc, 7α, 12β, 29OCOBu^s	11	14β,15β-oxide; 19,29-oxide	T. roka[73]

TABLE 3—*contd.*

Compound	C=C	OH	C=O	Source
1α-Methoxy-1,2-dihydro-epoxyazadiradione		1αOMe, 7αOAc	3, 16 14β,15β-oxide	*Azadirachta indica*[52]
1β,2β-Diepoxyazadiradione		7αOAc	3, 16 1β,2β-oxide; 14β,15β-oxide	*A. indica*[52]
7-Acetylneotrichilenone	1	7αOAc	3, 15 14αH	*A. indica*[52]
7-Desacetyl-7-benzoyl-azadiradione	1, 14	7αOBz	3, 16	*A. indica*[52]
7-Desacetyl-7-benzoyl epoxyazadiradione	1	7αOBz	3, 16 14β,15β-oxide	*A. indica*[52]

TABLE 4

Sources of C_{26} group: D-ring cleaved

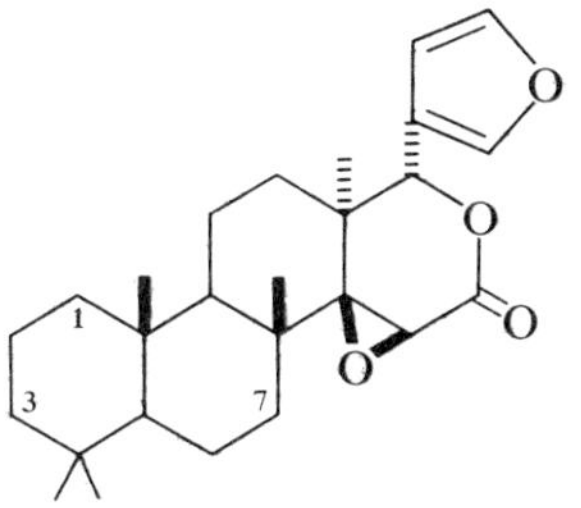

Compound	C=C	OH	C=O	Source
6α-Acetoxygedunin	1	6αOAc, 7αOAc	3	*Carapa guianensis*[62]
6α-Hydroxygedunin	1	6α, 7αOAc	3	*C. guianensis*[62]
Nyasin		1αOAc, 3αOAc, 7αOAc, 11β		*Khaya nyasica*[30]
7-Desacetyl-7-benzoyl-gedunin	1	7αOBz	3	*Azadirachta indica*[52]

J. D. CONNOLLY

TABLE 5

Sources of C_{26} group: C ring cleaved

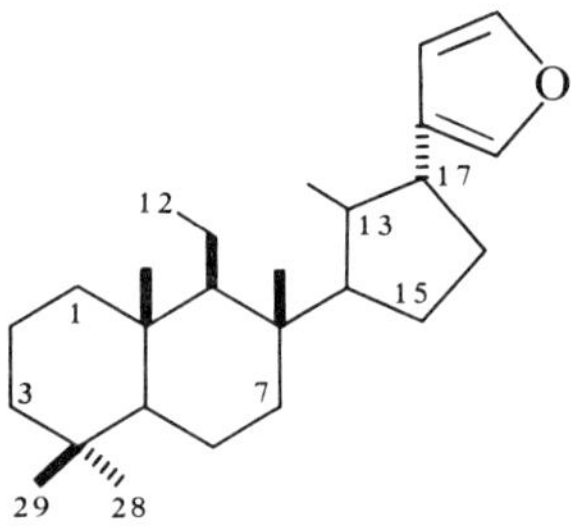

Compound	C=C	OH	C=O	CO₂Me		Source
Nimbolin B	13	1αOAc, 3αOAc, 7αOCinn, 12			12,15-oxide; 28,6α-oxide	Melia azedarach[33]
Heudebolin	13	1αOAc, 3αOAc, 7αOAc, 12			12,15-oxide; 28,6α-oxide	Trichilia heudelottii[1]
Ochinal	13	1αOBz, 3αOAc	12		7α,15β-oxide; 28,6α-oxide	M. azedarach[78]
Ochinin acetate	13	1αOCinn, 3αOAc		12	7α,15β-oxide; 28,6α-oxide	M. azedarach[78]
Nimbinene	3, 13	6αOAc	1	12	7α,15β-oxide; 28-nor	Azadirachta indica[51]
6-Deacetyl-nimbinene	3, 13	6α	1	12	7α,15β-oxide; 28-nor	A. indica[51]
Nimbandiol (28)	2, 13	4α, 6α	1	12	7α,15β-oxide; 28-nor	A. indica[51]
6-O-Acetyl-nimbandiol	2, 13	4α, 6αOAc	1	12	7α,15β-oxide; 28-nor	A. indica[51]
Ochinolide A	13	1αOAc, 3αOAc, 7αOBz			12,15α-lactone; 28,6α-oxide	M. azedarach[49,79]
Ochinolide B	13	1αOAc, 3αOAc, 7αOTig			12,15α-lactone; 28,6α-oxide	M. azedarach[49,79]
Nimbolidin A	13	1αOAc, 3αOAc, 7αOBz, 15αOAc		12	28,6α-oxide	M. azedarach[49]
Nimbolidin B	13	1αOAc, 3αOAc, 7αOTig, 15αOAc		12	28,6α-oxide	M. azedarach[49]
Nimbolinin B	13	1αOAc, 3αOAc, 7αOTig, 12			12,15α-oxide; 28,6α-oxide	M. azedarach[49]
1-Desacetyl-nimbolinin B	13	1α, 3αOAc, 7αOTig, 12α			12,15α-oxide; 28,6α-oxide	M. azedarach[49]
Azadirachtin (29)						M. azedarach and Azadirachta indica[111]

TABLE 6

Sources of C_{26} group: B-ring cleaved

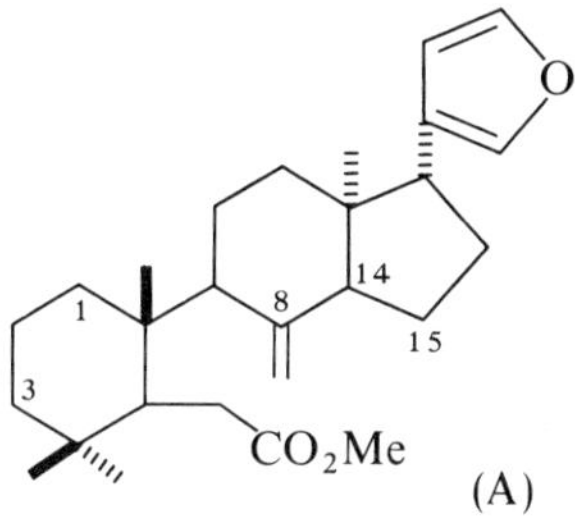

(A) 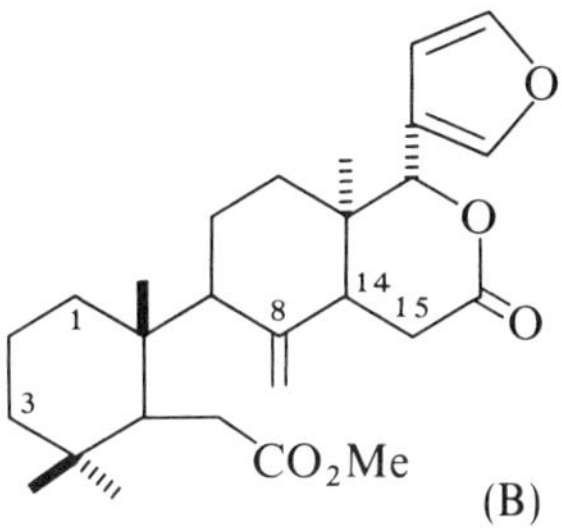   (B)

Compound	Type	C=C	OH	C=O		Source
Tricoccin S$_7$	A	1, 14		3, 16	1,8α-oxide; 8βMe	*Cneorum tricoccon*[71]
Toonacilin	A	1	11αOAc, 12αOAc	3	14β,15β-oxide	*Toona ciliata*[55]
6-Acetoxy-toonacilin	A	1	6OAc, 11αOAc, 12αOAc	3	14β,15β-oxide	*T. ciliata*[55]
Toonafolin (**31**)	A			3	7,8α-lactone; 8βMe; 1α,11α-oxide; 14β,15β-oxide	*T. ciliata*[54]
6,12α-Diacetoxy methyl angolensate	B		6OAc, 12αOAc	3	1α,14β-oxide	*Guarea thompsonii*[27]
Deoxyandirobin	B	1, 14		3		*Soymida febrifuga*[91]
Ekebergin	B		2αOCOBu^s, 3α, 15βOAc		1α,14β-oxide	*Ekebergia capense*[107]

TABLE 7

Sources of C_{26} group: B-ring cleaved and reclosed

Compound	C=C	OH	C=O		Source
Xylocarpin		3βOAc	1	8α,30α-oxide	*Xylocarpus granatum*[84]
Deoxyxylocarpin	8(30)	3βOAc	1		*X. granatum*[84]
Dihydromexicanolide	8(14)	3β	1		*Cabralea eichleriana*[112]
Angustadienolide	8(30), 14	3βOAc	1		*Cedrela angustifolia*[61,109]
2-Hydroxy-angustadienolide	8(30), 14	2, 3βOAc	1		*C. angustifolia*[61,109]
2-Hydroxyfissinolide	8(14)	2,3βOAc	1		*Khaya ivorensis*[3]
Lactone	8(14)		1	3,4-lactone	*K. ivorensis*[3]
Dihydro-2-hydroxyfissinolide		2,3βOAc	1		*K. madagascariensis*[104]
Dihydrokhayasin		3βOCOPr^i	1		*K. anthotheca*[2]
Dihydrocarapin acetate	14	3βOAc	1		*K. nyasica*[103]
Xyloccensin A	14	1,3βOCOPr^i, 30αOCOPr^i		1,8α-oxide	*Xylocarpus moluccensis*[24]
Xyloccensin B		1,3βOCOPr^i 30αOCOPr^i		1,8α-oxide	*X. moluccensis*[24]
Xyloccensin D	14	1, 2, 3βOCOPr^i, 30αOCOPr^i		1,8α-oxide	*X. moluccensis*[24]
Xyloccensin F		1, 2, 3βOCOPr^i, 30αOCOPr^i		1,8α-oxide	*X. moluccensis*[24]
Hemiacetal	14	3α	1	3β,8β-oxide	*Cedrela glaziovii*[24]
Febrifugin	8(30)	3βOTig	1		*Soymida febrifuga*[95]

TABLE 8

Sources of C_{26} group: B-ring cleaved; entandrophragmin type

Compound	Source
Utilin (39)	*Entandrophragma utile*[45]
Entandrophragmin (40)	*E. cylindricum*[43]
Bussein A (74)	*E. bussei*[44]
Bussein B (75)	*E. bussei*[44]
Candollein (76)	*E. candollei*[43]
Compound E₃ (77)	*E. cylindricum*[43]
Phragmalin (78)	*E. caudatum*[8]
Phragmalin diisobutanoate (79)	*Chukrasia tabularis*[20]
Phragmalin isobutanoate-propanoate (80)	*C. tabularis*[20]
12α-Hydroxyphragmalin diisobutanoate (81)	*C. tabularis*[20]
12α-Hydroxyphragmalin isobutanoate-propanoate (82)	*C. tabularis*[20]
Chukrasin A (83)	*C. tabularis*[11,93]
Chukrasin B (84)	*C. tabularis*[11,93]
Chukrasin C (85)	*C. tabularis*[11,93]
Chukrasin D (86)	*C. tabularis*[11,93]
Chukrasin E (87)	*C. tabularis*[11,93]
Tabularin (88)	*C. tabularis*[11]
Procerin (53)	*Carapa procera*[105]
Spicata 2 (89)	*E. spicatum*[29]
Pseudrelone B (90)	*Pseudocedrela kotschyii*[106]
Febrinin A (91)	*Soymida febrifuga*[94]
Febrinin B (92)	*S. febrifuga*[94]

(74) Bussein A (R¹ = COCHMeEt; R² = Ac)   (78) Phragmalin (R¹ = R² = R³ = H)

(75) Bussein B (R¹ = COCHMe₂; R² = Ac)   (79) R¹ = R² = COCHMe₂; R³ = H

(89) Spicata-2 (R¹ = R² = COCHMe₂)   (80) R¹ = COCHMe₂; R² = COEt; R³ = H

(81) R¹ = R² = COCHMe₂; R³ = OAc

(82) R¹ = COCHMe₂; R² = COEt; R³ = OAc

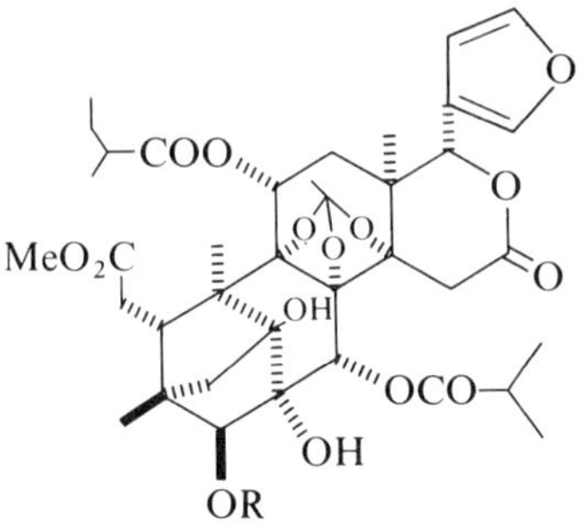

(76) Candollein (R = COCHMeEt)

(77) R = COC(OH)MeEt

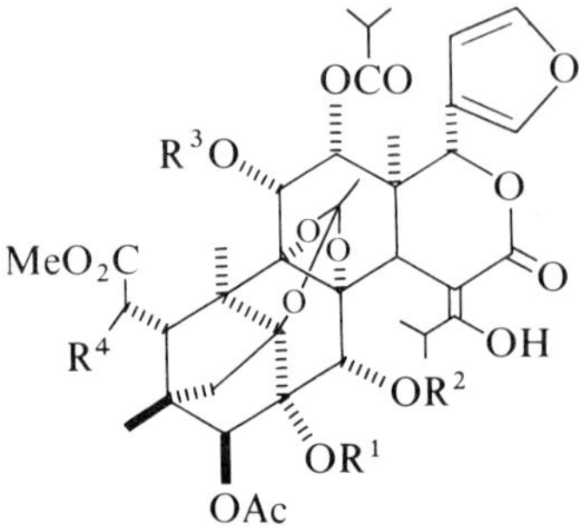

(83) Chukrasin A (R^1 = H; R^2,R^3 = Ac + COCHMe$_2$; R^4 = OH)

(84) Chukrasin B (R^1 = H; R^2 = R^3 = COCHMe$_2$; R^4 = H)

(85) Chukrasin C (R^1 = H; R^2,R^3 = Ac + COCHMe$_2$; R^4 = H)

(86) Chukrasin D (R^1 = Ac; R^2,R^3 = Ac + COCHMe$_2$; R^4 = H)

(87) Chukrasin E (R^1 = Ac; R^2 = R^3 = COCHMe$_2$; R^4 = H)

(88) Tabularin

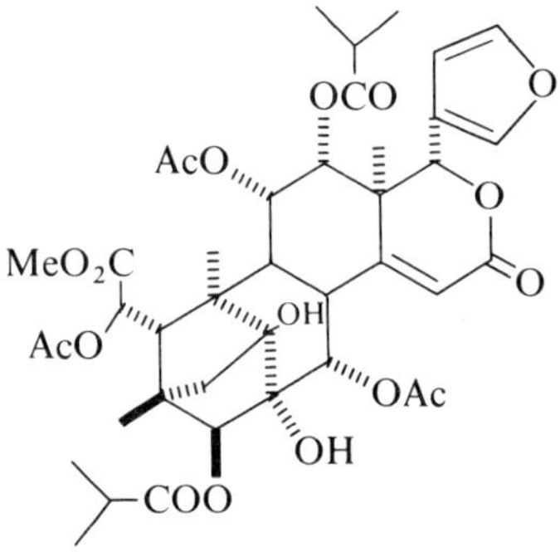

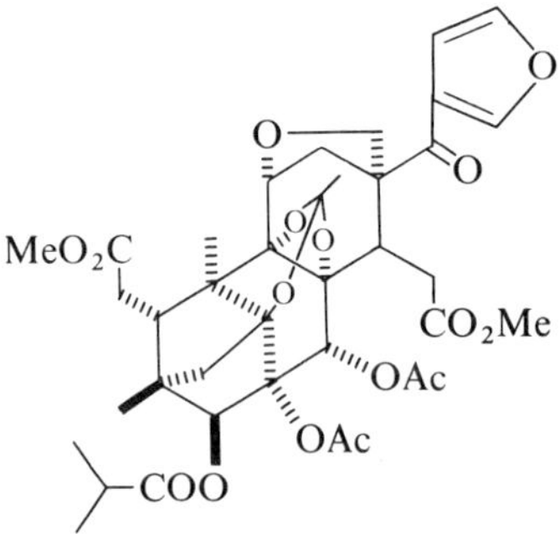

(90) Pseudrelone B

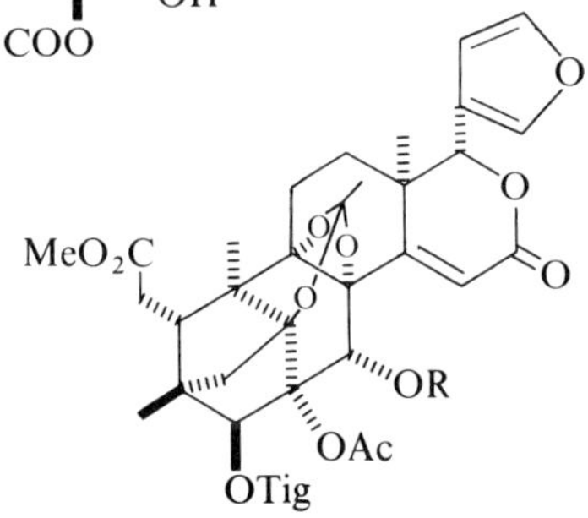

(91) Febrinin A (R = COEt)

(92) Febrinin B (R = Ac)

TABLE 9

Sources of C_{26} group: A-ring cleaved

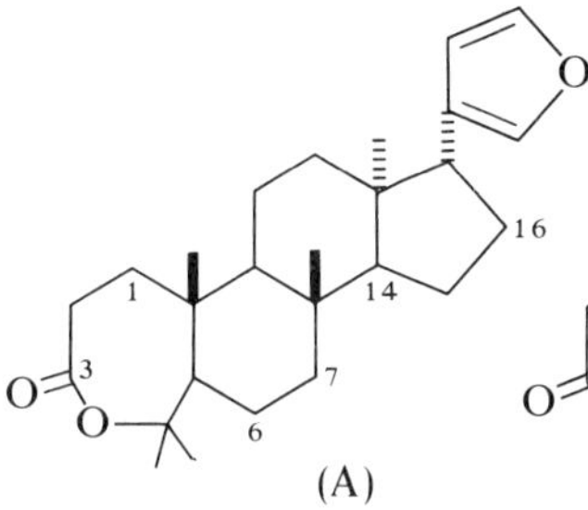

(A) 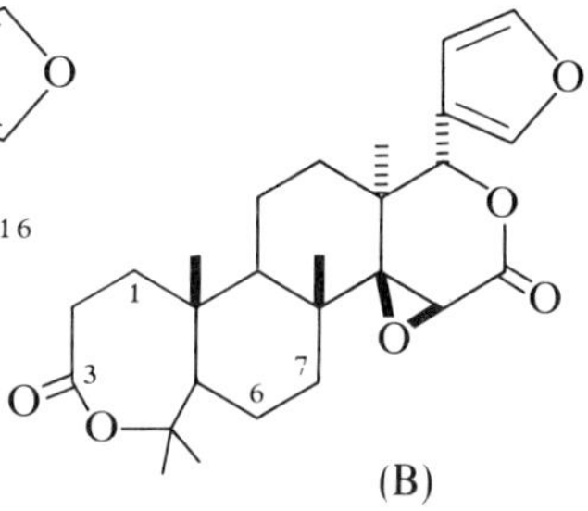   (B)

Compound	Type	C=C	OH	C=O		Source
Tricoccin S_{22}	A	1	7α, 16α		$14\beta,15\beta$-oxide	*Cneorum tricoccon*[36]
Tricoccin S_{32} triacetate	A		1αOAc, 7αOAc, 16αOAc		$14\beta,15\beta$-oxide	*C. tricoccon*[36]
Tricoccin S_{13} (93)	A					*C. tricoccon*[35]
Evodulone	A	1	7αOAc	16	$14\beta,15\beta$-oxide	*Carapa procera*[101]
Proceranone	A	1, 14	7αOAc			*C. procera*[102]
Surenone	A	1	6α	7	$14\beta,15\beta$-oxide	*Toona sureni*[56]
Surenin	A	1	6αOAc, 7αOAc		$14\beta,15\beta$-oxide	*T. sureni*[56]
6β-Acetoxy-7α-obacunol	B	1	6βOAc, 7α			*Trichilia trifolia*[110]
6β-Acetoxy-7α-obacunyl acetate	B	1	6βOAc, 7αOAc			*T. trifolia*[110]
Dihydronomilin acetate	B		1αOAc, 7αOAc			*Xylocarpus granatum*[74]

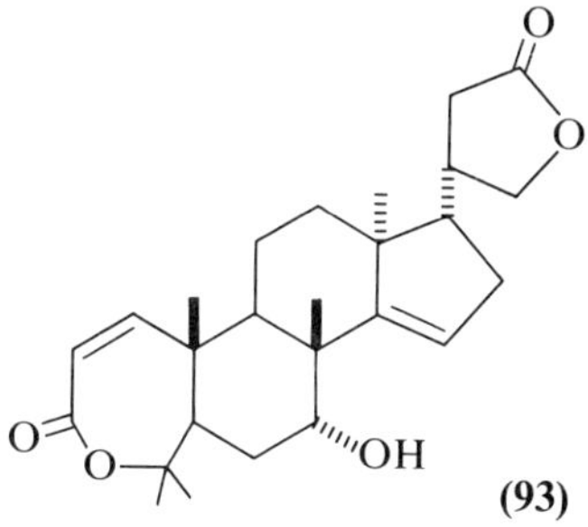

(93)

TABLE 10

Sources of C_{26} group; both A- and B-rings cleaved

Compound	Source
Surenolactone (**43**)	*Toona sureni*[57]
Prieurianin (**44**)	*Trichilia prieuriana*[41]
14,15β-Epoxyprieurianin (**94**)	*Guarea guidona*[63]
Rohitukin (**95**)	*Aphanamixis polystacha*[26]
2′-Hydroxyrohitukin (**96**)	*Guarea cedrata*[4]
Compound D-4 (**97**)	*Trichilia prieuriana*[21]
Compound D-5 (**98**)	*T. prieuriana*[21]
Compound B (**99**)	*Guarea thompsonii*[21]
Compound C (**100**)	*G. thompsonii*[21]
Dregeanin (**46**)	*Trichilia dregeana*,[26] *G. thompsonii*[108]
Deacetyldregeanin (**101**)	*T. dregeana*,[26] *T. heudelottii*[26]
Hispidin A (**52**)	*T. hispida*[48]
Hispidin B (**102**)	*T. hispida*[48]
Polystachin (**103**)	*Aphanamixis polystacha*[72]
Rohituka 1 (**104**)	*A. polystacha*[12]
Rohituka 2 (**105**)	*A. polystacha*[12]
Rohituka 3 (**106**)	*A. polystacha*[12]
Rohituka 4 (**107**)	*A. polystacha*[12]
Rohituka 5 (**108**)	*A. polystacha*[12]
Rohituka 6 (**109**)	*A. polystacha*[12]
Rohituka 7 (**110**)	*A. polystacha*[12]
Rohituka 8 (**111**)	*A. polystacha*[12]
Cneorin R (**35**)	*Neochamaelea pulverulenta*[36]
Methyl ivorensate (**36**)	*Khaya ivorensis*[3]

(**94**) R = Ac·
(**98**) R = H

(**95**) Rohitukin (R = H)
(**96**) R = OH

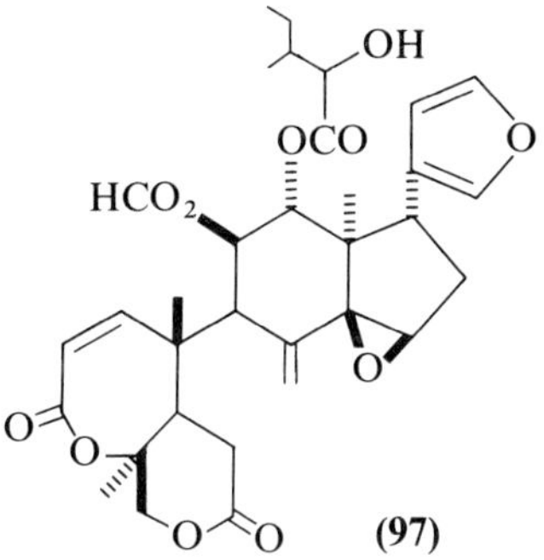

(97)

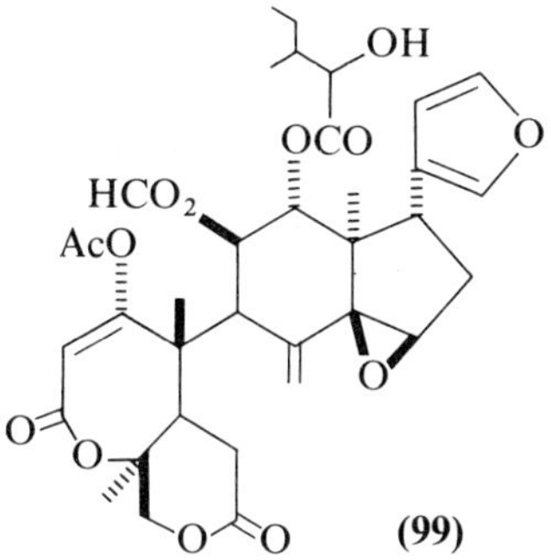

(99)

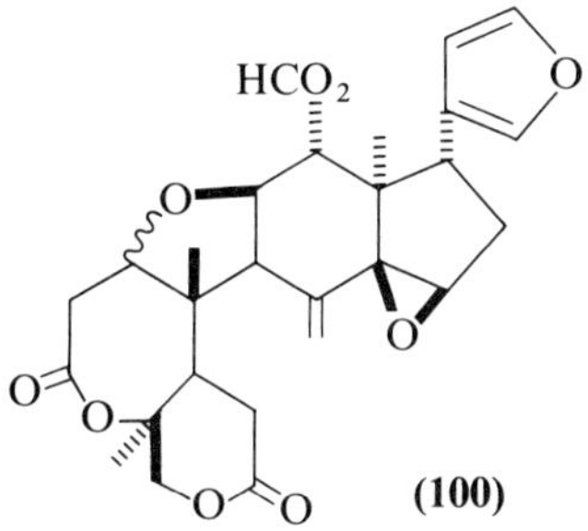

(100)

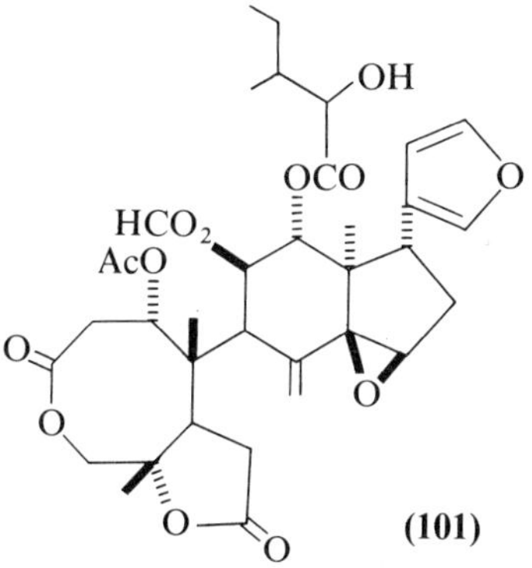

(101)

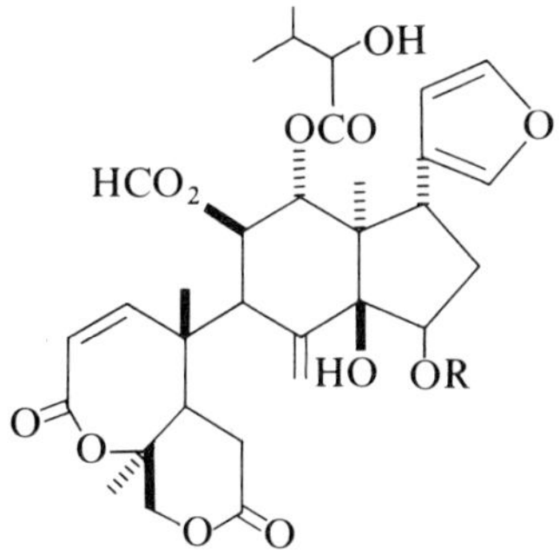

(102) Hispidin B (R = Tig)
(110) Rohituka-7 (R = Ac)

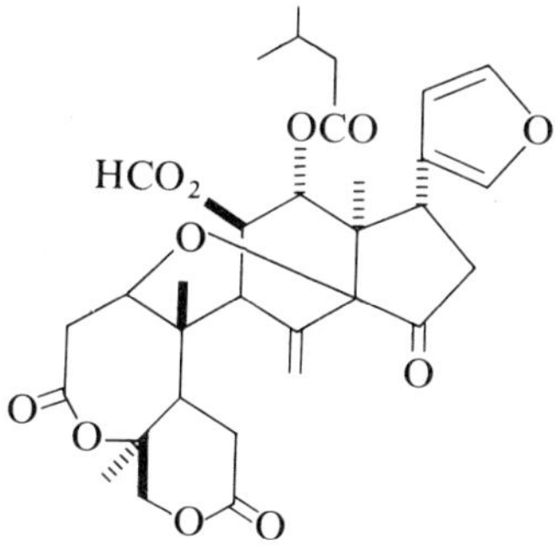

(103) Polystachin

A recent X-ray analysis has revealed that the 15-acetate of rohituka 7 is β and not α as drawn in **110**. The configuration at C-15 in related compounds may also need to be amended accordingly (Professor D. A. H. Taylor, personal communication).

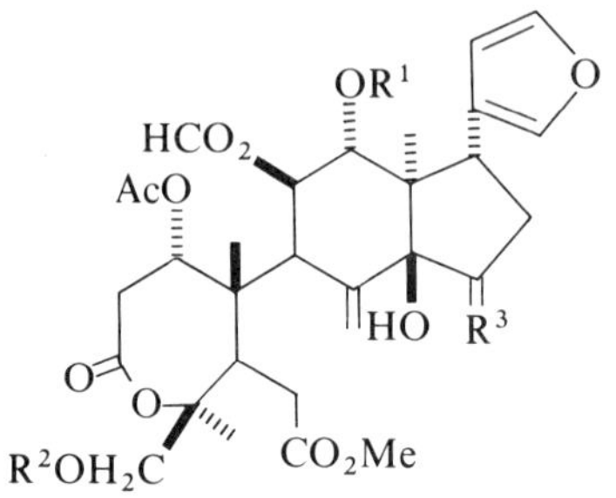

(104) Rohituka 1 (R^1 = $COCH_2CHMe_2$; R^2 = H; R^3 = H,αOAc)

(105) Rohituka 2 (R^1 = $COCH(OH)CHMe_2$; R^2 = H; R^3 = H,αOAc)

(107) Rohituka 4 (R^1 = $COCH_2CHMe_2$; R^2 = Ac; R^3 = O)

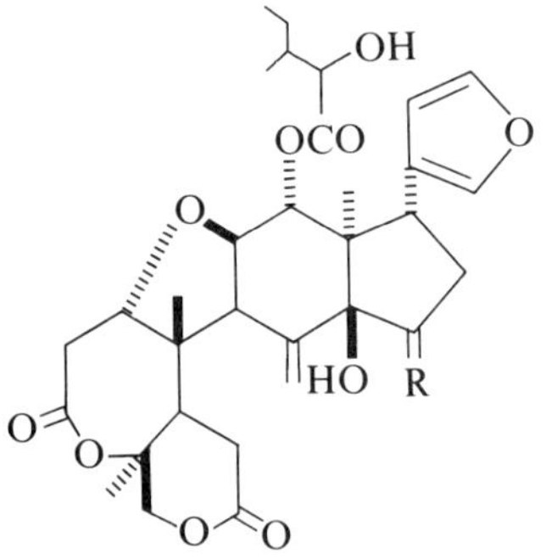

(106) Rohituka 3 (R = O)

(108) Rohituka 5 (R = H,αOAc)

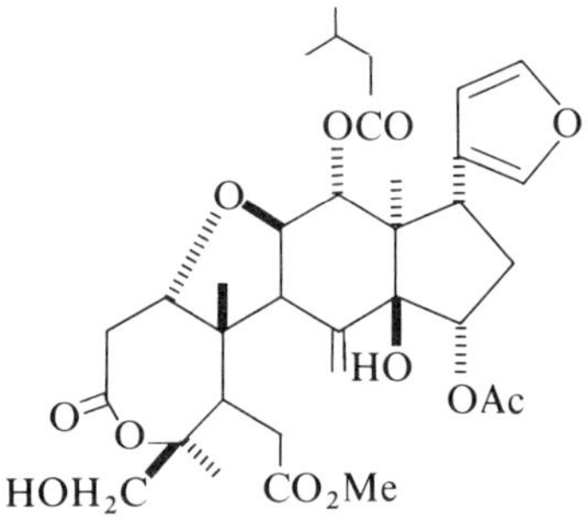

(109) Rohituka 6

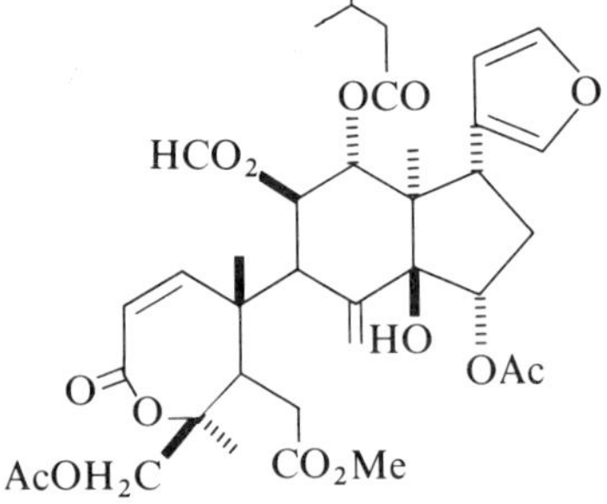

(111) Rohituka 8

TABLE 11

Sources of C_{25} group: pentanortriterpenoids of the Cneoraceae

Compound	Source
Cneorin B (112)	Neochamaelea pulverulenta[66]
Cneorin C (69)	N. pulverulenta[46]
Cneorin D (113)	N. pulverulenta[69]
Cneorin F (114)	N. pulverulenta[66]
Cneorin K (70)	N. pulverulenta[68]
Cneorin K_1 (115)	N. pulverulenta[68]
Cneorin NP_{29} (116)	Cneorum tricoccon[38]
Cneorin Q (117)	C. tricoccon[38]
Tricoccin R_1 (118)	C. tricoccon[69]
Tricoccin R_9 (119)	C. tricoccon[68]
Tricoccin R_{10} (120)	C. tricoccon[68]
Tricoccin S_1 (121)	C. tricoccon[70]
Tricoccin S_2 (71)	C. tricoccon[70]
Tricoccin S_4 (72)	C. tricoccon[39]
Tricoccin S_5 (122)	C. tricoccon[70]
Tricoccin S_{10} (123)	C. tricoccon[70]
Tricoccin S_{14} (67)	C. tricoccon[68]
Tricoccin S_{16} (124)	C. tricoccon[37]
Tricoccin S_{27} (73)	C. tricoccon[37]
Tricoccin S_{33} (125)	C. tricoccon[39]
Tricoccin S_{42} (126)	C. tricoccon[37]

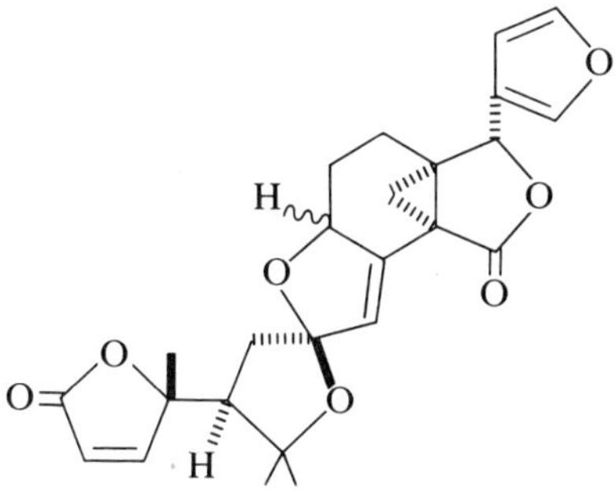

(112) Cneorin B

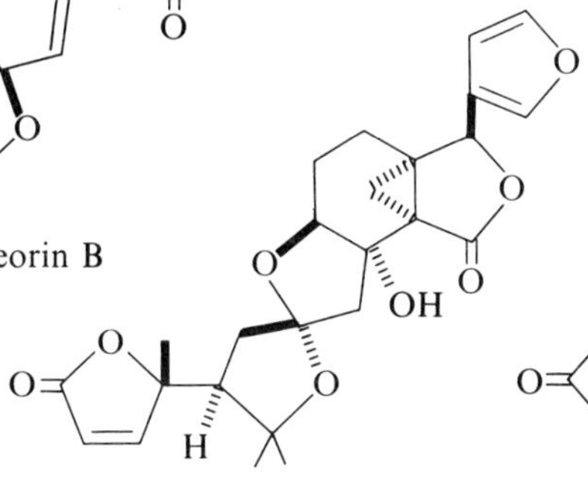

(113) Cneorin D 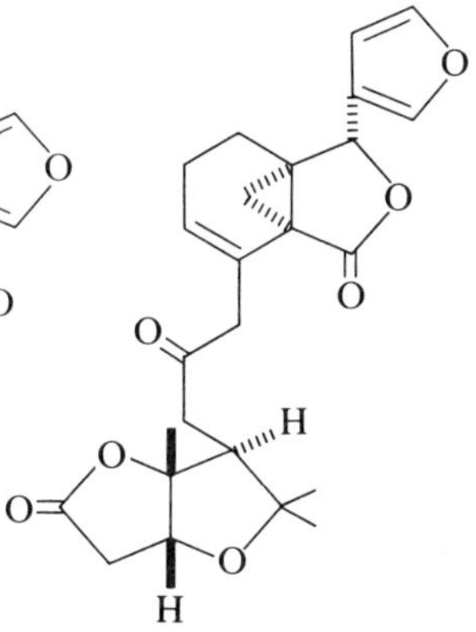   (114) Cneorin F

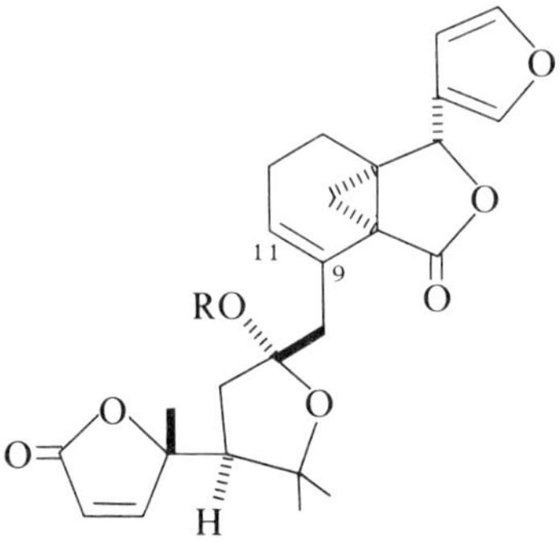

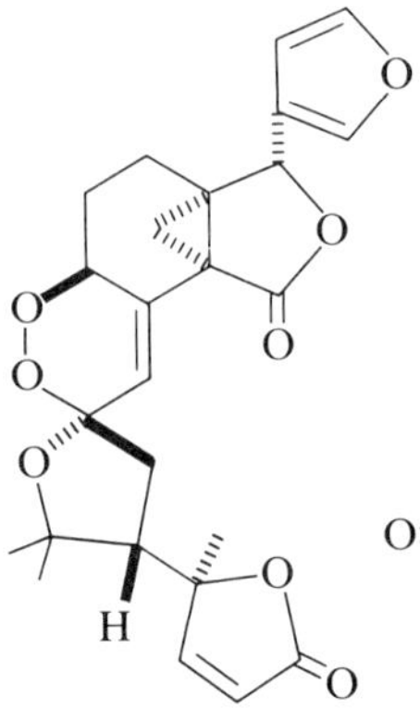

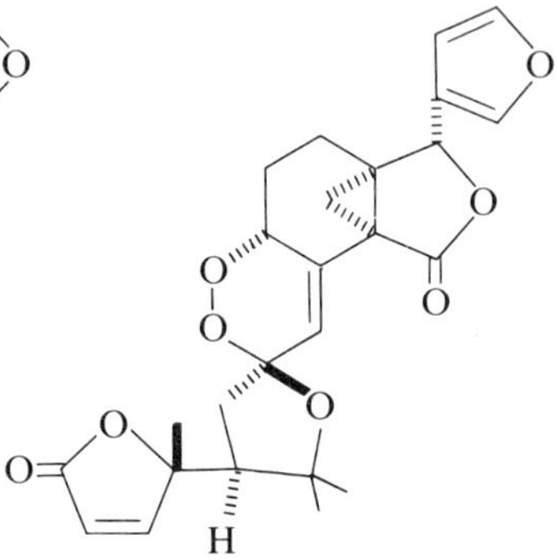

(115) Cneorin K_1 (R = Me)   (116) Cneorin NP_{29}

(120) Tricoccin R_{10} (R = H; $9\beta,11\beta$-oxide)

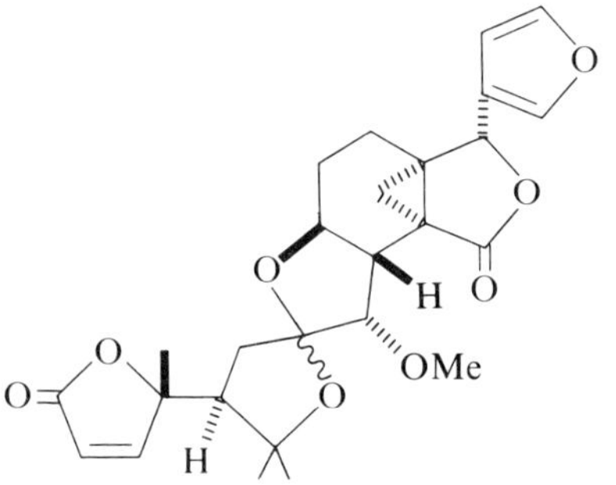

(117) Cneorin Q

(118) Tricoccin R_1

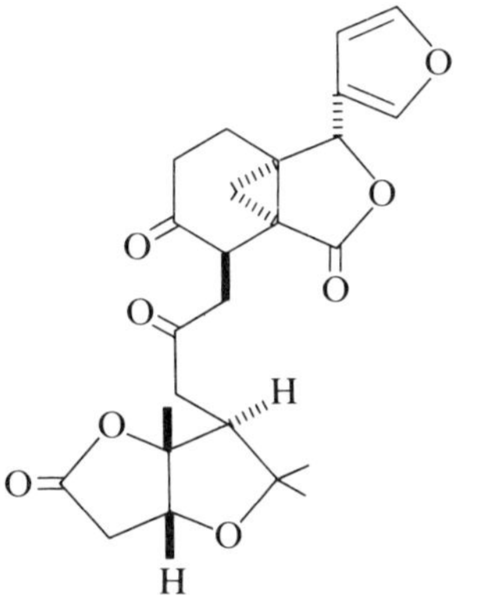

(119) Tricoccin R_9

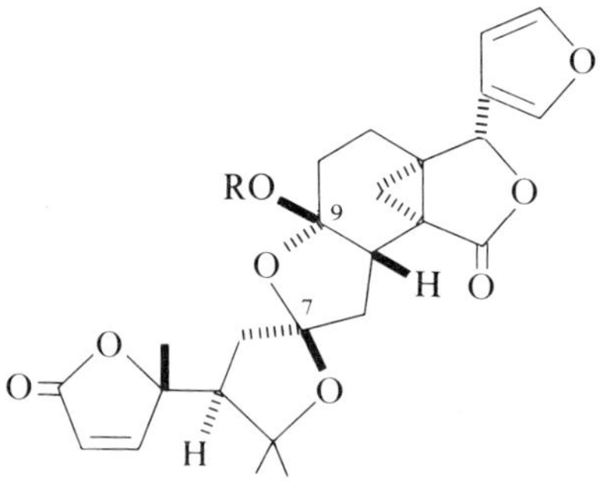

(121) Tricoccin S_1 (R = H)
(122) Tricoccin S_5 (R = H; 9-epimer)
(123) Tricoccin S_{10} (R = Me; 7-epimer)

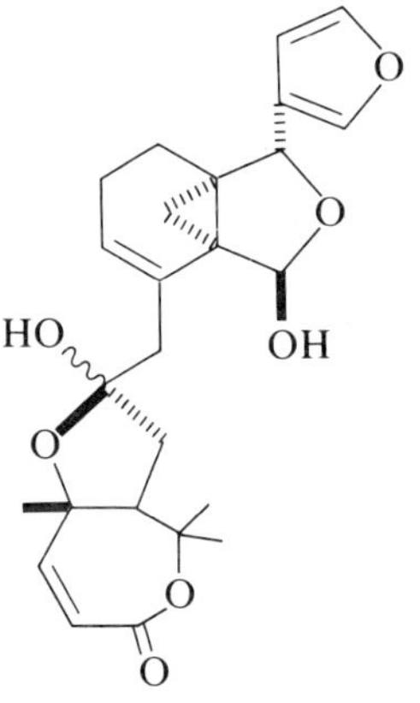

(124) Tricoccin S_{16}

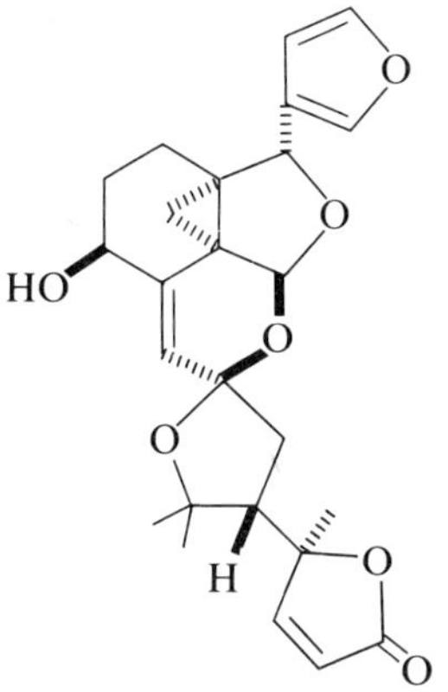

(125) Tricoccin S_{33}

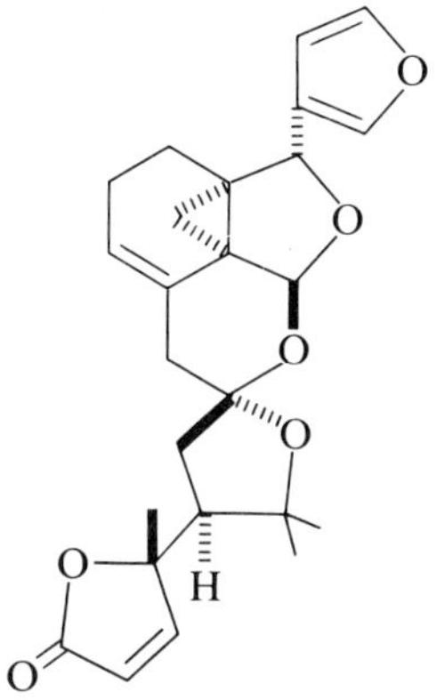

(126) Tricoccin S_{42}

ACKNOWLEDGEMENTS

In the field of tetranortriterpenoids I have had the privilege and pleasure of working with many colleagues, in particular Professor K. H. Overton in Glasgow, Professor R. McCrindle in Guelph, Professor D. A. H. Taylor in Durban, Dr. K. K. Purushothaman in Madras, and Professor B. L. Sondengam, Dr. J. F. Ayafor, and Mr. S. F. Kimbu in Yaoundé. I am deeply grateful to them all for their collaboration and friendship.

REFERENCES

1. Adesida, G. A. and Okorie, D. A. (1973). *Phytochemistry* **12**, 3007.
2. Adesogan, E. K., Okorie, D. A. and Taylor, D. A. H. (1970). *J. Chem. Soc. (C)* 205.
3. Adesogan, E. K. and Taylor, D. A. H. (1970). *J. Chem. Soc. (C)* 1710.
4. Akinniyi, J. A., Connolly, J. D., Rycroft, D. S., Sondengam, B. L. and Ifeadike, N. P. (1980). *Can. J. Chem.* **58**, 1865.
5. Akinniyi, J. A., Connolly, J. D., Rycroft, D. S., Taylor, D. A. H. and Pettit, G. R., unpublished work.
6. Akisanya, A., Bevan, C. W. L., Halsall, T. G., Powell, J. W. and Taylor, D. A. H. (1961). *J. Chem. Soc.* 3705.
7. Arigoni, D., Barton, D. H. R., Corey, E. J., Jeger, O., Caglioti, L., Sukh Dev., Ferrini, P. G., Glazier, E. R., Melera, A., Pradhan, S. K., Schaffner, K., Sternhell, S., Templeton, J. F. and Tobinga, S. (1960). *Experientia* **16**, 41.
8. Arndt, R. R. and Baarchers, W. H. (1972). *Tetrahedron* **28**, 2333.
9. Ayafor, J. F., Kimbu, S. F., Sondengam, B. L., Connolly, J. D. and Rycroft, D. S., unpublished work.

10. Bevan, C. W. L., Ekong, D. E. V., Halsall, T. G. and Toft, P. (1967). *J. Chem. Soc. (C)* 820.
11. Brown, D. A. and Taylor, D. A. H. (1978). *J. Chem. Res. (S)* 20.
12. Brown, D. A. and Taylor, D. A. H. (1978). *Phytochemistry* 17, 1995.
13. Buchanan, J. G. St. C. and Halsall, T. G. (1970). *J. Chem. Soc. (C)* 2280.
14. Burke, B. A., Chan, W. R., Magnus, K. E. and Taylor, D. R. (1969). *Tetrahedron* 25, 5007.
15. Cameron, A. F., Connolly, J. D., Maltz, A. and Taylor, D. A. H. (1979). *Tetrahedron Lett.* 967.
16. Chan, W. R., Taylor, D. R. and Yee, T. H. (1970). *J. Chem. Soc. (C)* 311.
17. Chan, W. R., Taylor, D. R. and Yee, T. H. (1971). *J. Chem. Soc. (C)* 2662.
18. Chiang, C. and Chang, G. C. (1973). *Tetrahedron* 29, 1911.
19. Chatterjee, A., Chakraborty, T. and Chandrasekharan, S. (1971). *Phytochemistry* 10, 2533.
20. Connolly, J. D., Labbe, C. and Rycroft, D. S. (1978). *J. Chem. Soc. Perkin Trans. 1* 285.
21. Connolly, J. D., Labbe, C., Rycroft, D. S., Okorie, D. A. and Taylor, D. A. H. (1979). *J. Chem. Res. (S)* 256.
22. Connolly, J. D., Labbe, C., Rycroft, D. S. and Taylor, D. A. H. (1979). *J. Chem. Soc. Perkin Trans. 1* 2959.
23. Connolly, J. D., Labbe, C., Rycroft, D. S. and Taylor, D. A. H., unpublished work.
24. Connolly, J. D., MacLellan, M. A., Okorie, D. A. and Taylor, D. A. H. (1976). *J. Chem. Soc. Perkin Trans. 1* 1993.
25. Connolly, J. D. and McCrindle, R. (1971). *J. Chem. Soc. (C)* 1715.
26. Connolly, J. D., Okorie, D. A., De Wit, D. L. and Taylor, D. A. H. (1976). *J. Chem. Soc. Chem. Comm.* 909.
27. Connolly, J. D., Okorie, D. A. and Taylor, D. A. H. (1972). *J. Chem. Soc. Perkin Trans. 1* 1145.
28. Connolly, J. D., Overton, K. H. and Polonsky, J. (1970). *In* "Progress in Phytochemistry" (Reinhold, L. and Liwschitz, Y., eds.), Vol. 2, pp. 385–455. Interscience, London.
29. Connolly, J. D., Phillips, W. R., Mulholland, D. A. and Taylor, D. A. H. (1981). *Phytochemistry* 20, 2596.
30. Connolly, J. D. and Taylor, D. A. H. (1973). *J. Chem. Soc. Perkin Trans. 1* 686.
31. Connolly, J. D., Thornton, I. M. S. and Taylor, D. A. H. (1973). *J. Chem. Soc. Perkin Trans. 1* 2407.
32. Dreyer, D. L. (1968) *Fortschr. Naturst.* 26, 190.
33. Ekong, D. E. U., Fakunle, C. O., Fasina, A. K. and Okogun, J. I. (1969). *J. Chem. Soc. Chem. Comm.* 1166.
34. Ekong, D. E. U., Okogun, J. I. and Sondengam, B. L. (1975). *J. Chem. Soc. Perkin Trans. 1* 2118.
35. Epe, B. and Mondon, A. (1978). *Tetrahedron Lett.* 3901.
36. Epe, B. and Mondon, A. (1979). *Tetrahedron Lett.* 2015.
37. Epe, B. and Mondon, A. (1979). *Tetrahedron Lett.* 4045.
38. Epe, B., Oelbermann, U., Mondon, A. and Remberg, G. (1979). *Tetrahedron Lett.* 3839.
39. Epe, B., Trautmann, D., Mondon, A. and Remberg, C. (1979). *Tetrahedron Lett.* 1365.

40. Ferguson, G., Gunn, P. A., Marsh, W. C., McCrindle, R., Restivo, R., Connolly, J. D., Fulke, J. W. B. and Henderson, M. S. (1975). *J. Chem. Soc. Perkin Trans. 1* 491.
41. Gullo, V. P., Miura, I., Nakanishi, K., Cameron, A. F., Connolly, J. D., Duncanson, F. D., Harding, A. E., McCrindle, R. and Taylor, D. A. H. (1975). *J. Chem. Soc. Chem. Comm.* 345.
42. Halsall, T. G. and Troke, J. A. (1975). *J. Chem. Soc. Perkin Trans. 1* 1758.
43. Halsall, T. G., Wragg, K., Connolly, J. D., MacLellan, M. A., Bredell, L. D. and Taylor, D. A. H. (1977). *J. Chem. Res. (S)* 154.
44. Hanni, R., Tamm, C., Gullo, V. P. and Nakanishi, K. (1975). *J. Chem. Soc. Chem. Comm.* 563.
45. Harrison, H. R., Hodder, O. J. R., Bevan, C. W. L., Taylor, D. A. H. and Halsall, T. G. (1970). *J. Chem. Soc. Chem. Comm.* 1388.
46. Henkel, G., Diercks, H., Epe, B. and Mondon, A. (1975). *Tetrahedron Lett.* 3315.
47. Jolad, S. D., Hoffmann, J. J., Cole, J. R., Tempesta, M. S. and Bates, R. B. (1980). *J. Org. Chem.* **45**, 3132.
48. Jolad, S. D., Hoffmann, J. J., Schram, K. H., Cole, J. R., Tempesta, M. S. and Bates, R. B. (1981). *J. Org. Chem.* **46**, 641.
49. Kraus, W. and Bokel, M. (1981). *Chem. Ber.* **114**, 267.
50. Kraus, W. and Cramer, R. (1978). *Tetrahedron Lett.* 2395.
51. Kraus, W. and Cramer, R. (1981). *Chem. Ber.* **114**, 2375.
52. Kraus, W., Cramer, R. and Sawitzki, G. (1981). *Phytochemistry* **20**, 117.
53. Kraus, W. and Grimminger, W. (1980). *Nouveau J. Chemie* **4**, 651.
54. Kraus, W. and Grimminger, W. (1981). *Liebigs Ann. Chem.* 1838.
55. Kraus, W., Grimminger, W. and Sawitzki, G. (1978). *Angew. Chem. (Int.)* **17**, 476.
56. Kraus, W. and Kypke, K. (1979). *Tetrahedron Lett.* 2715.
57. Kraus, W., Kypke, K., Bokel, M., Grimminger, W., Sawitzki, G. and Schwinger, G. (1982). *Liebigs Ann. Chem.* 87.
58. Lavie, D., Jain, M. K. and Kirson, I. (1967). *J. Chem. Soc. (C)* 1347.
59. Lavie, D. and Levy, E. C. (1971). *Tetrahedron* **27**, 3941.
60. Lavie, D., Levy, E. C. and Jain, M. K. (1971). *Tetrahedron* **27**, 3927.
61. Lavie, D., Levy, E. C., Rosito, C. and Zelnik, R. (1970). *Tetrahedron* **26**, 219.
62. Lavie, D., Levy, E. C. and Zelnik, R. (1972). *Bioorg. Chem.* **2**, 59.
63. Lukacova, V., Polonsky, J., Moretti, C., Pettit, G. R. and Schmidt, J. M. (1982). *J. Nat. Prod.* **45**, 288.
64. Lyons, C. W. and Taylor, D. R. (1975). *J. Chem. Soc. Chem. Comm.* 517.
65. Mitra, C. R., Garg, H. S. and Pandey, G. H. (1970). *Tetrahedron Lett.* 2761.
66. Mondon, A. and Epe, B. (1976). *Tetrahedron Lett.* 1273.
67. Mondon, A., Epe, B. and Oelbermann, U. (1981). *Tetrahedron Lett.* 4467.
68. Mondon, A., Epe, B. and Trautmann, D. (1978). *Tetrahedron Lett.* 4881.
69. Mondon, A., Trautmann, D., Epe, B. and Oelbermann, U. (1976). *Tetrahedron Lett.* 3291.
70. Mondon, A., Trautmann, D., Epe, B. and Oelbermann, U. (1976). *Tetrahedron Lett.* 3295.
71. Mondon, A., Trautmann, D., Epe, B., Oelbermann, U. and Wolff, C. (1978). *Tetrahedron Lett.* 3699.
72. Mulholland, D. A. and Taylor, D. A. H. (1979). *J. Chem. Res. (S)* 294.
73. Nakatani, M., James, J. C. and Nakanishi, K. (1981). *J. Am. Chem. Soc.* **103**, 1228.
74. Ng, A. S. and Fallis, A. G. (1979). *Can. J. Chem.* **57**, 3088.

75. Obasi, M. E., Okogun, J. I. and Ekong, D. E. U. (1972). *J. Chem. Soc. Perkin Trans. 1* 1943.
76. Ochi, H., Kotsuki, H., Hirotsu, K. and Tokoroyama, T. (1976). *Tetrahedron Lett.* 2877.
77. Ochi, M., Kotsuki, H., Ishida, H. and Tokoroyama, T. (1978). *Chem. Lett.* 99.
78. Ochi, M., Kotsuki, H., Kataoka, T., Tada, T. and Tokoroyama, T. (1978). *Chem. Lett.* 331.
79. Ochi, M., Kotsuki, H., Ido, M., Nakai, H., Shiro, M. and Tokoroyama, T. (1979). *Chem. Lett.* 1137.
80. Ochi, M., Kotsuki, H. and Tokoroyama, T. (1978) *Chem. Lett.* 621.
81. Ochi, M., Kotsuki, H., Tokoroyama, T. and Kubota, T. (1977). *Bull. Chem. Soc. Japan* **50**, 2499.
82. Ollis, W. D., Ward, A. D., Oliveira, H. M. D. and Zelnik, R. (1970). *Tetrahedron* **26**, 1637.
83. Okogun, J. I., Fakunle, C. O., Ekong, D. E. U. and Connolly, J. D. (1975). *J. Chem. Soc. Perkin Trans. 1* 1352.
84. Okorie, D. A. and Taylor, D. A. H. (1970). *J. Chem. Soc. (C)* 211.
85. Okorie, D. A. and Taylor, D. A. H. (1972). *J. Chem. Soc. Perkin Trans. 1* 1488.
86. Okorie, D. A. and Taylor, D. A. H. (1977). *Phytochemistry* **16**, 2029.
87. Pachapurkar, R. V., Kornule, P. M. and Narayaman, C. R. (1974). *Chem. Lett.* 357.
88. Polonsky, J., Varon, Z., Arnoux, B., Pascard, C., Pettit, G. R. and Schmidt, J. M. (1978). *J. Am. Chem. Soc.* **100**, 7731.
89. Polonsky, J., Varon, Z., Arnoux, B., Pascard, C., Pettit, G. R., Schmidt, J. M. and Lange, L. M. (1978). *J. Am. Chem. Soc.* **100**, 2575.
90. Polonsky, J., Varon, Z., Marazano, C., Arnoux, B., Pettit, G. R., Schmidt, J. M., Ochi, M. and Kotsuki, H. (1979). *Experientia* **35**, 987.
91. Purushothaman, K. K. and Chandrasekharan, S. (1974). *Indian J. Chem.* **12**, 207.
92. Purushothaman, K. K., Chandrasekharan, S., Connolly, J. D. and Rycroft, D. S. (1977). *J. Chem. Soc. Perkin Trans. 1* 1873.
93. Ragetti, T. and Tamm, C. (1978). *Helv. Chim. Acta* **61**, 1814.
94. Rao, M. M., Gupta, P. S., Singh, P. P. and Krishna, E. M. (1979). *Indian J. Chem.* **17B**, 158.
95. Rao, M. M., Krishna, E. M., Gupta, P. S. and Singh, P. P. (1978). *Indian J. Chem.* **16B**, 823.
96. Rao, M. M., Meshulam, H., Zelnik, R. and Lavie, D. (1975). *Phytochemistry* **14**, 1071.
97. Saika, B., Kataky, J. C. S., Mathur, R. K. and Baruah, J. N. (1978). *Indian J. Chem.* **16B**, 1042.
98. Schulte, K. E., Ruecker, G. and Matern, H. U. (1979). *Planta Med.* **35**, 76.
99. Siddiqui, S., Fuchs, S., Lucke, J. and Voelter, W. (1978). *Tetrahedron Lett.* 611.
100. Singh, S., Garg, H. S. and Khanna, N. M. (1976). *Phytochemistry* **15**, 2001.
101. Sondengam, B. L., Kamga, C. S. and Connolly, J. D. (1979). *Tetrahedron Lett.* 1357.
102. Sondengam, B. L., Kamga, C. S., Kimbu, S. F. and Connolly, J. D. (1981). *Phytochemistry* **20**, 173.
103. Taylor, D. A. H. (1969). *J. Chem. Soc. (C)* 2439.
104. Taylor, D. A. H. (1970). *J. Chem. Soc. (C)* 336.
105. Taylor, D. A. H. (1974). *J. Chem. Soc. Perkin Trans. 1* 437.
106. Taylor, D. A. H. (1979). *Phytochemistry* **18**, 1574.
107. Taylor, D. A. H. (1981). *Phytochemistry* **20**, 2263.

108. Taylor, D. A. H. (1982). *J. Chem. Res. (S)* 55.
109. Taylor, D. A. H. and Wehrli, F. W. (1973). *J. Chem. Soc. Perkin Trans. 1* 1599.
110. Taylor, D. R. (1971) *Rev. Latinoamer. Quim.* 87.
111. Zanno, P. R., Miura, I., Nakanishi, K. and Elder, D. L. (1975). *J. Am. Chem. Soc.* **97**, 1975.
112. Zelnik, R. (1971). *Phytochemistry* **11**, 1866.

CHAPTER 7

Limonoids of the Rutaceae

DAVID L. DREYER

U.S.D.A., Albany, California, U.S.A.

I. INTRODUCTION

Modification of the tetracyclic triterpene skeleton in plants through oxidation and degradation has reached its zenith in four closely related families of the order Rutales: Rutaceae, Meliaceae, Cneoraceae, and Simaroubaceae. Plants of these families produce a series of degraded triterpenes, limonoids and the related quassinoids. Limonoids are highly functionalized C_{26} tetranortriterpenes in which the side-chain has been

215

D. L. DREYER

degraded by loss of four carbons resulting in the formation of a β-substituted furan ring at C-17. Quassinoids are C_{20} and C_{19} triterpenoids which have lost C-17 and attached side-chain carbons and represent further steps down an oxidative pathway.

Each of these plant families produces a distinct and characteristic limonoid pattern. The Rutaceae produces a homogenous group of closely related, largely A- and D-ring seco limonoids while the Meliaceae produces a large number of diverse C_{26} tetranortriterpenes in which oxidation and skeletal rearrangement is the most varied.[25,31,46,94] The Cneoraceae is a very small family which produces a homogenous group of largely pentanortri-terpenoids.[106] The Simaroubaceae produces largely C_{19} and C_{20} degraded limonoids, the quassinoids.[25,91,92]

Although many additional limonoids, some even of new structural types, have been reported in the decade since the last major reviews in this area, the basic biosynthetic patterns and distribution recognized at that time have not greatly changed. Thus, limonoids in the Rutaceae do not show nearly as much structural variation as those in the Meliaceae. Moreover, limonoids from the Rutaceae tend to be dominated by the widespread occurrence of limonin itself, whereas no single limonoid occurs with such high frequency in plants of the Meliaceae. The chemistry of limonoids **1** to **11** occurring in the Rutaceae (Fig. 1) has been previously summarized.[25,31,46,94]

The gross biogenetic route from a tetracyclic triterpene, euphol or more likely Δ^7-tirucallol, leading to typical rutaceous tetranortriterpenes was advanced[7] at the time structures were proposed for obacunone (**1**) and limonin (**8**). The steps involved, illustrated in Scheme 1 must broadly be:

1. Loss of four terminal carbons from the side-chain and formation of a β-substituted furan ring at C-17.

(**1**) Obacunone (R = H)
(**22**) Zapoterin (R = OH)

(**7**) Ichangin

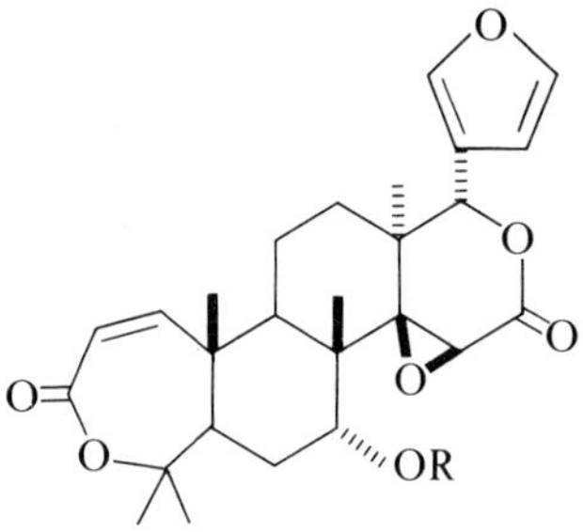

(2) 7α-Obacunol (R = H)

(16) 7α-Obacunyl acetate (R = Ac)

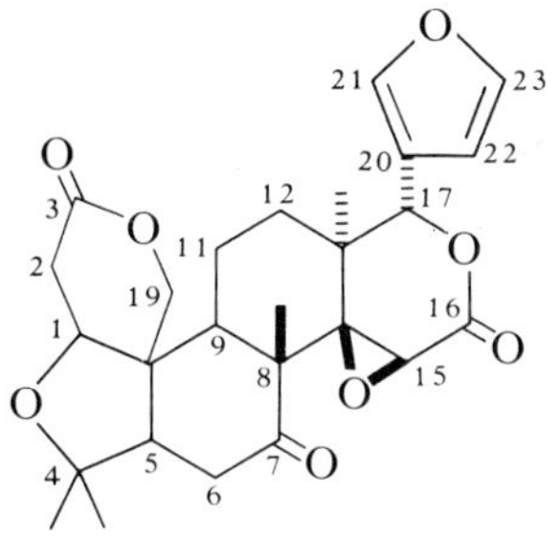

(8) Limonin

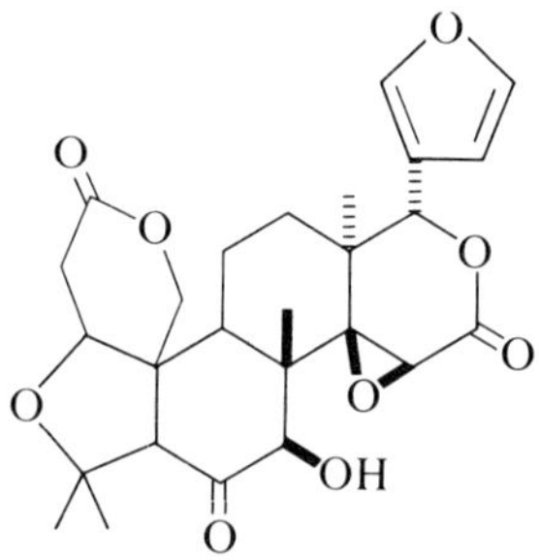

(9) Rutaevin

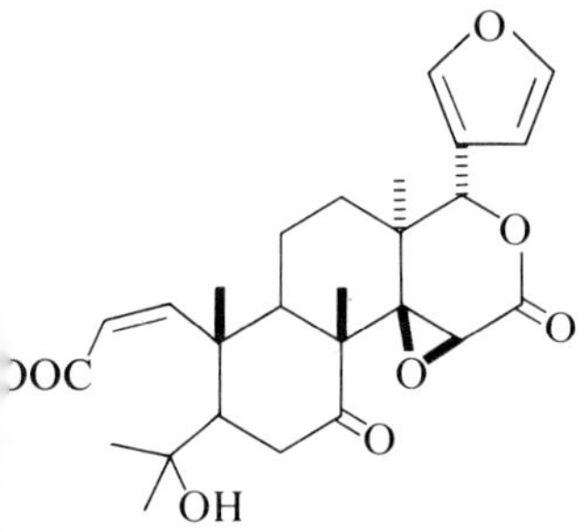

(3) Obacunoic acid

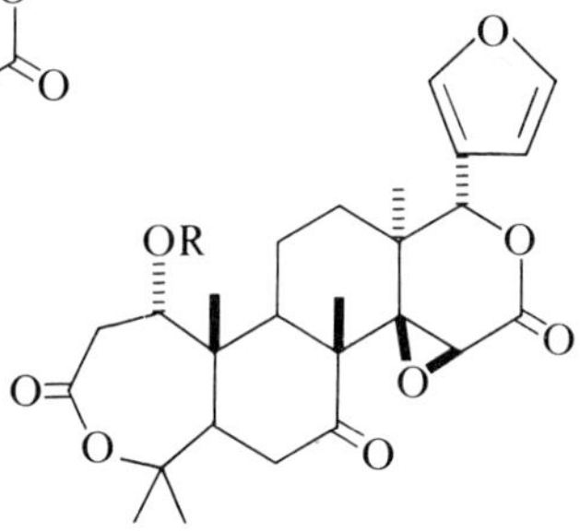

(4) Deacetylnomilin (R = H)

(5) Nomilin (R = Ac)

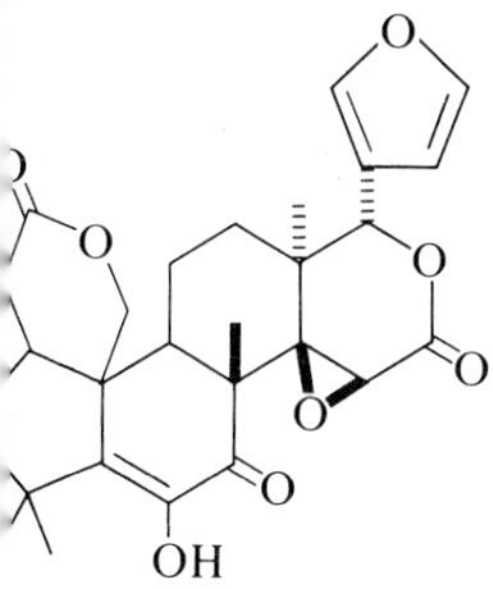

(10) Limonin diosphenol

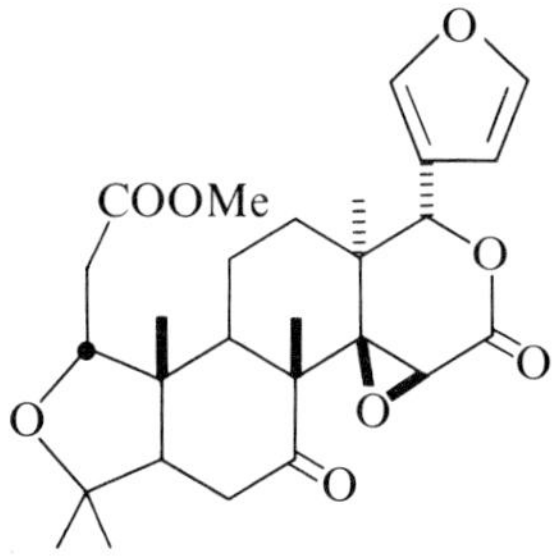

(6) Methyl epiisoobacunoate (veprisone)

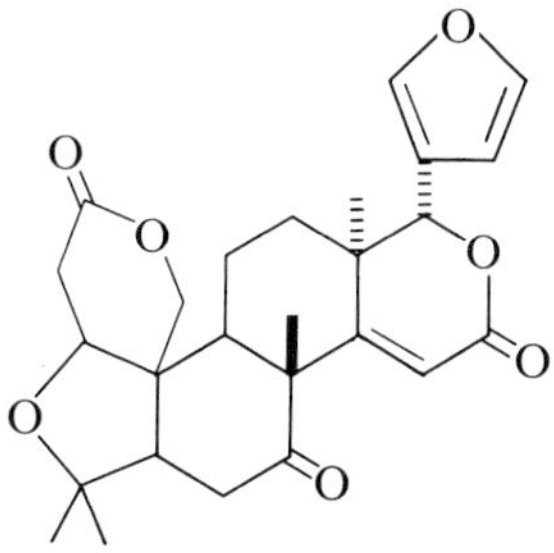

(11) Deoxylimonin

FIG. 1. Limonoids known from Rutaceae in 1968.

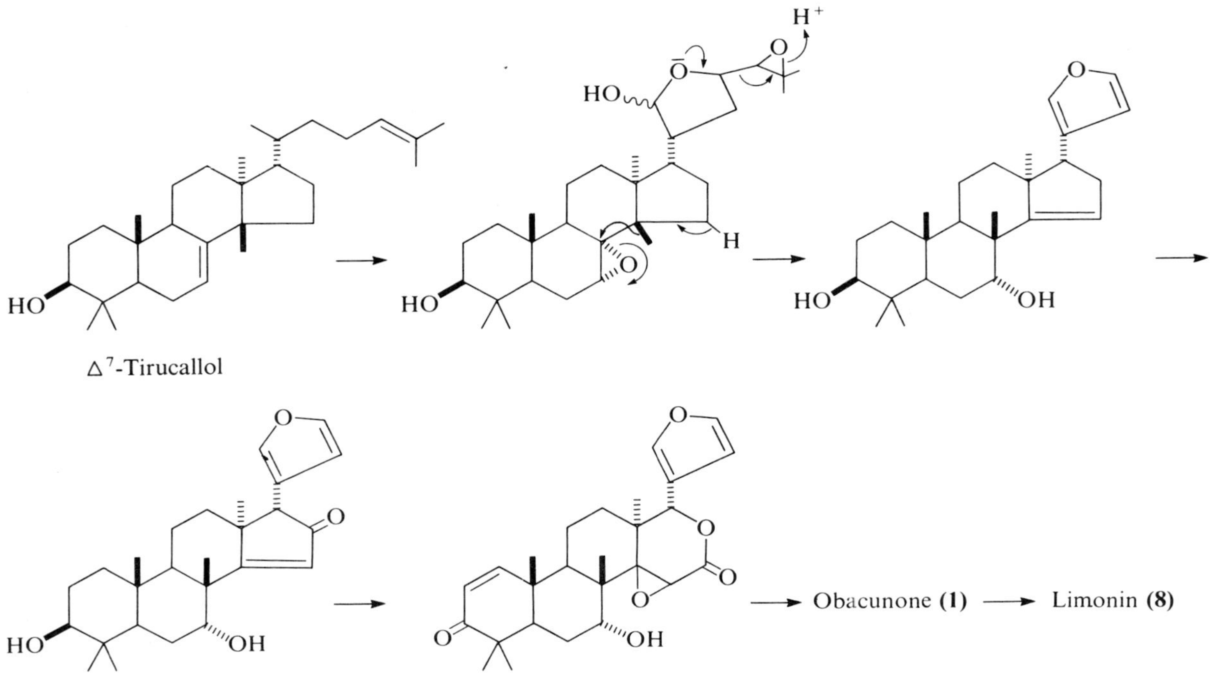

SCHEME 1. Probable biosynthetic route to limonin.

Δ⁷-Tirucallol

Obacunone (1) → Limonin (8)

2. α-Epoxidation of a 7,8 double bond as typically found in a Δ^7-euphol or Δ^7-tirucallol system.

3. Opening of the 7α-epoxide group with concurrent Wagner–Meerwein migration of the C-30 methyl to C-8 and loss of a proton to generate a apotirucallene with a 14,15 double bond. This leads to the invariable oxygen function at C-7 and accounts for its observed stereochemistry when present as a hydroxy group. This reaction also introduces the initial functionality into the D-ring.

4. Allylic oxidation of the Δ^{14}-apotirucallene leading to a 16-keto group followed by a kind of Baeyer–Villiger oxidation to generate the typical limonoid D-ring δ-lactone system.

5. Expansion of the A-ring in a similar fashion leads to the nomilin or obacunone systems.

Direct support for these biogenetic ideas through the appropriate labeling studies has been slow in appearing but the structures of many limonoids subsequently isolated, especially from the Meliaceae, have provided much indirect evidence for the route proposed. Their structures possess many of the structural features that might well be expected for intermediates in such a biogenetic scheme. On the whole, the Rutaceae appears to be poor at accumulating intermediates in limonoid biosynthesis especially when compared to the Meliaceae.

Work in the Rutaceae during the decade since the last reviews has shifted from emphasis on isolation and structure determination to more biochemical questions centering on biosynthesis, metabolism, and the biological and technological roles of limonoids. The biosynthesis and metabolism of limonin is of intrinsic interest but also of technological importance because of its role in citrus bitterness. Much chemical work on limonin (8) and related substances is motivated by the fact that they caused bitterness in processed citrus products.[74] Control of the biosynthesis and/or metabolism of limonin and congeners might allow manipulation of bitterness levels in processed citrus products.

II. Structure Determination

The ready accessibility of plant material and the role of limonoids in citrus technology has channelled much work on new rutaceous limonoids to *Citrus* species. The availability of large amounts of *Citrus* seeds, for example grapefruit, makes it profitable to look for minor limonoids in the mother liquors of such extracts after removal of limonin and other major limonoids. These minor limonoids are of special interest since they may be intermediates in the biosynthesis of limonin or represent products of limonin metabolism. Many of the new limonoids reported since the last reviews[25,31,46,94] represent simple functional group modification of previously known structures.

Spectroscopic methods have played an especially important role in structure determination of limonoids. Contributions have appeared dealing in a systematic fashion with ^{1}H NMR studies,[93] the use of lanthanide shift reagents,[14,38] chemical shifts of C-methyl groups,[59,88] ^{13}C NMR,[15,36,51,109,110] mass spectra,[6,73] and ORD/CD.[33] Scattered through the limonoid literature are occasional applications of the nuclear Overhauser effect, two-dimensional J spectroscopy, and coupled oscillator methods to structure determination.

The high degree of oxidation frequently found in limonoids results in the introduction of a large number of functional groups on the tetranortriterpene system. This causes the NMR signals to be spread out over the entire spectrum instead of being lumped together into an uninterpretable methylene envelope as is so often the case with steroids and triterpenes. As a result, detailed assignments can often be made to most or all of the signals without going to especially high magnetic fields.

The spatial proximity of many of the functional groups and their through space effect on each other with modest changes in the nature of the groups allows many conclusions to be drawn about group location and stereochemistry. Especially useful has been the fact that the position of the H-15 signal is sensitive to the nature and stereochemistry of the group at C-7. The position of the H-15 signal varies in a predictable manner depending on the presence of a 7-keto, 7α- or 7β-hydroxy group. Previously the relationship of these groups had to be determined by the base catalyzed limonol to merolimonol conversion.[80] Similarly a group at C-11 can cause diagnostic changes in the NMR signal of H-1 in the obacunone system. As with other areas of natural products chemistry, ^{13}C NMR has been useful in relating the carbon skeleton of new limonoids with previously known types.

A. NOMILINIC, DEACETYLNOMILINIC, ISOOBACUNOIC, AND EPIISOOBACUNOIC
ACIDS

These four acids were initially isolated as minor constituents from C. paradisi (grapefruit) seed extracts by Bennett.[10,68] Compounds 14 and 15 were previously known substances prepared during structure work on obac-unone.[68] The identities of 14 and 15 follow from comparison of their methyl esters with those from 12 and 13 prepared from nomilin (5) and deacetylnomilin (4) respectively. The methyl ester of 13 has also been isolated from a Citrus × Forunella hybrid.[12]

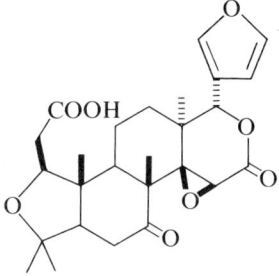

(12) Nomilinic acid (R = Ac)

(13) Deacetylnomilinic acid (R = H)

(14) Isoobacunoic acid

(15) Epiisoobacunoic acid
(epimeric at C-1)

B. 7α-OBACUNYL ACETATE, 7α-NOMILYL ACETATE, LIMONYL ACETATE, LIMONOL, DEOXYLIMONOL, AND DEOXYLIMONIC ACID

Limonoids having 7-hydroxy groups are not common in the Rutaceae to the same extent as they occur in the Meliaceae. The more frequently occurring 7-keto derivatives are more characteristic of this family especially when compared to limonoids of the Meliaceae. However, recently Bennett and Hasegawa[13] have reported 7α-obacunol (2) [previously known from seeds of *Casimiroa edulis*,[32] *Lovoa trichiliodes* (Meliaceae),[1] and as its acetate from *Cneorum tricoccon* (Cneoraceae)[41]] from *Citrus paradisi* seed extracts. In addition they[13] found 19 and 20 from the same source. Deoxylimonol (20) has also recently been found as a component of human faeces.[48] Unfortunately, nothing appears known about the diet of the individuals who provided the faeces. 7α-Obacunyl and limonyl acetates have been found in the seeds of a *Citrus* × *Poncirus* hybrid.[13]

The structures 16–20 were assigned by simple chemical conversions from known limonoids. Deoxylimonic acid (21), previously known from structure work on limonin, has also been found in *C. paradisi* seed extracts.[53]

7α-Nomilyl acetate (17) has been isolated from *Dictamnus albus* (Rutaceae)[83] as well as *Xylocarpus granatum* (Meliaceae),[2] *Cneorum tricoccon* (Cneoraceae),[82] and *Neochamaelea pulverulenta* (Cneoraceae).[82] Its structure was demonstrated by single-crystal X-ray structure determination[2] which showed that the 1-acetoxy group is in the α-configuration. Similar conclusions have recently been drawn from the NMR spectra of nomilin (5) taken at high magnetic fields.[79] Thus, those limonoids from the Rutaceae with a 1-hydroxy

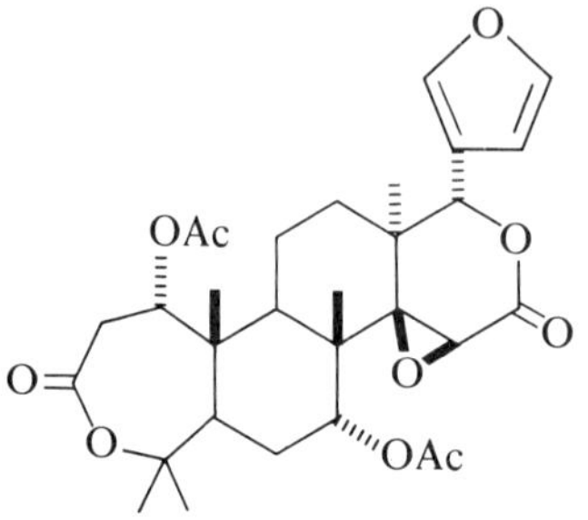

(17) 7α-Nomilyl acetate

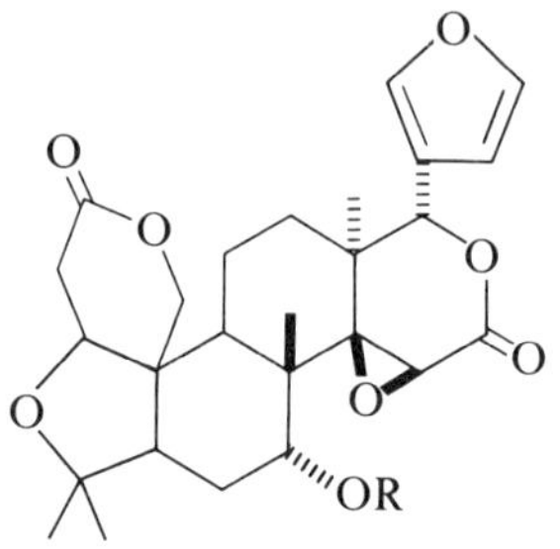

(18) Limonyl acetate (R = Ac)

(19) Limonol (R = H)

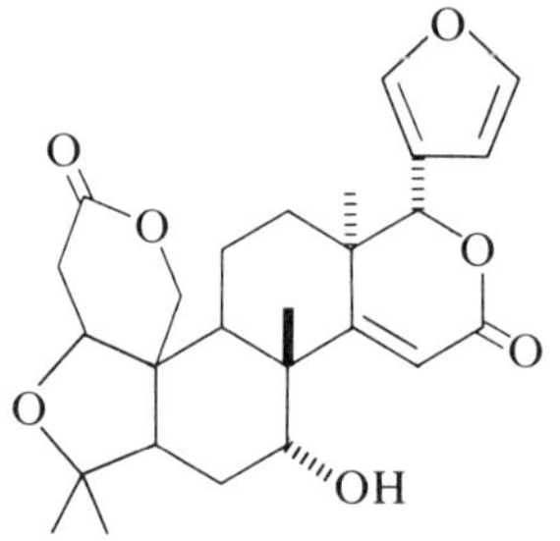

(20) Deoxylimonol

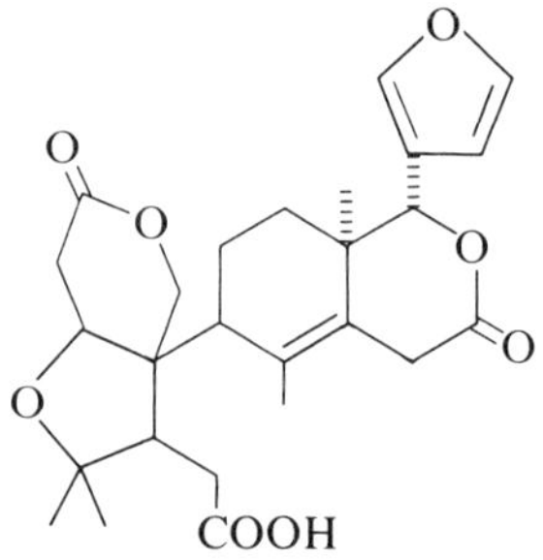

(21) Deoxylimonic acid

group and their derivatives are stereochemically homogenous with limonoids from the Meliaceae and Cneoraceae having similar structural features in the A-ring.

C. ZAPOTERIN, CLAUSENOLIDE, ATALANTOLIDE, AND SPATHELIN

Casimiroa edulis is a citrus relative with edible fruit native to Mexico. Zapoterin has long been known as a constituent of the seeds of *Casimiroa edulis*[61] where it co-occurs with obacunone (1) and 7α-obacunol (2).[32] The spectroscopic properties, formation of a monoacetate, and oxidation to a ketone showed that it was a hydroxy derivative of obacunone. The 12-hydroxyobacunone structure initially proposed[32] has been revised to 11α-hydroxyobacunone (22) on the basis of NMR, including NOE, arguments.[84]

Clausenolide (23) is a C_{25} limonoid[19] from *Clausena* which forms a monoacetate and has a hydroxy group which can be oxidized to a ketone.

Structure **23** was assigned from an X-ray crystal structure study. Clausenolide is likely to arise biogenetically from deacetylnomilinic acid **(13)** through oxidation to the 1-keto derivative and decarboxylation of the resulting β-ketoacid.

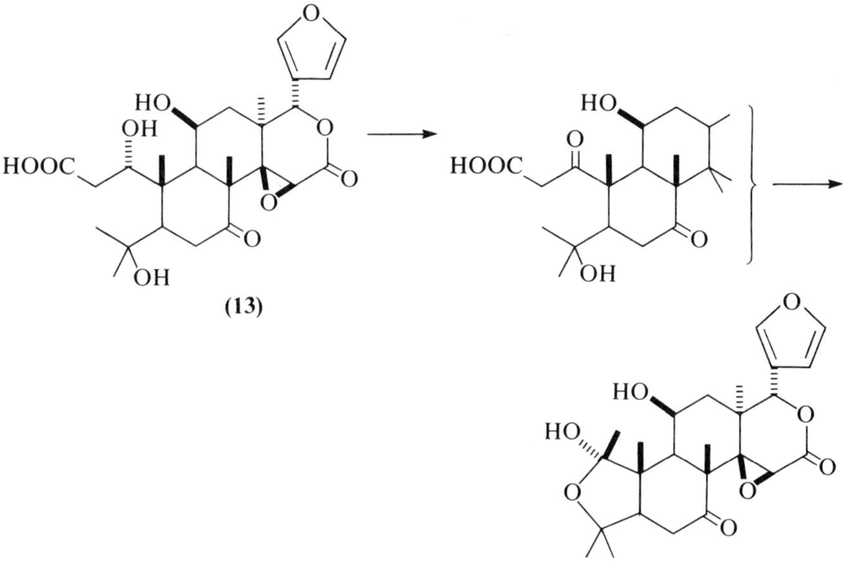

(13)

(23) Clausenolide

Atalantolide **(24)** co-occurs with atalantin **(38)** and other limonoids in *Atalantia monophylla*.[101,112] Jones oxidation of atalantolide gave initially an α-diketone, which after treatment with base was converted to the diosphenol **25**. The structure of **24** rests largely on its NMR properties and their comparison with those of the co-occurring atalantin **(38)**.[36,99]

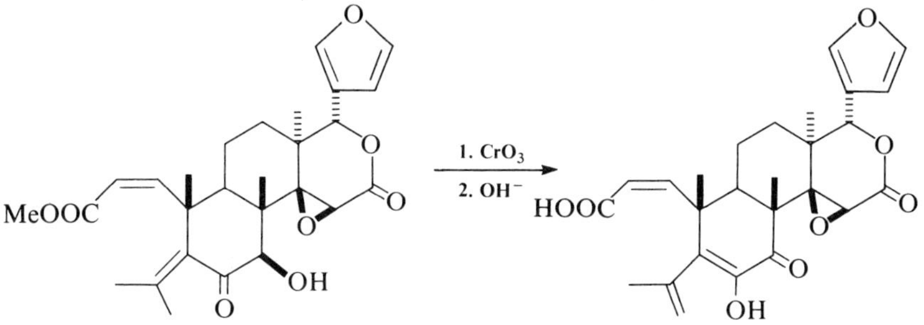

(24) Atalantolide

(25)

Spathelin (**26**) has been isolated from a plant of the genus *Spathelia* [subfamily Spathelioideae (Rutaceae)][16] which has been placed in the Simaroubaceae by Bentham and Hooker but in the Rutaceae by Engler.[39] Chromous chloride reduction of spathelin (**26**) removed two epoxide groups and the acetoxy group giving **27** (Scheme 2). Treatment of **26** with base caused a number of changes in the system including formation of a diosphenol group to give **28**. Base catalyzed isomerization of the diosphenol **28** gave a C-1 epimer

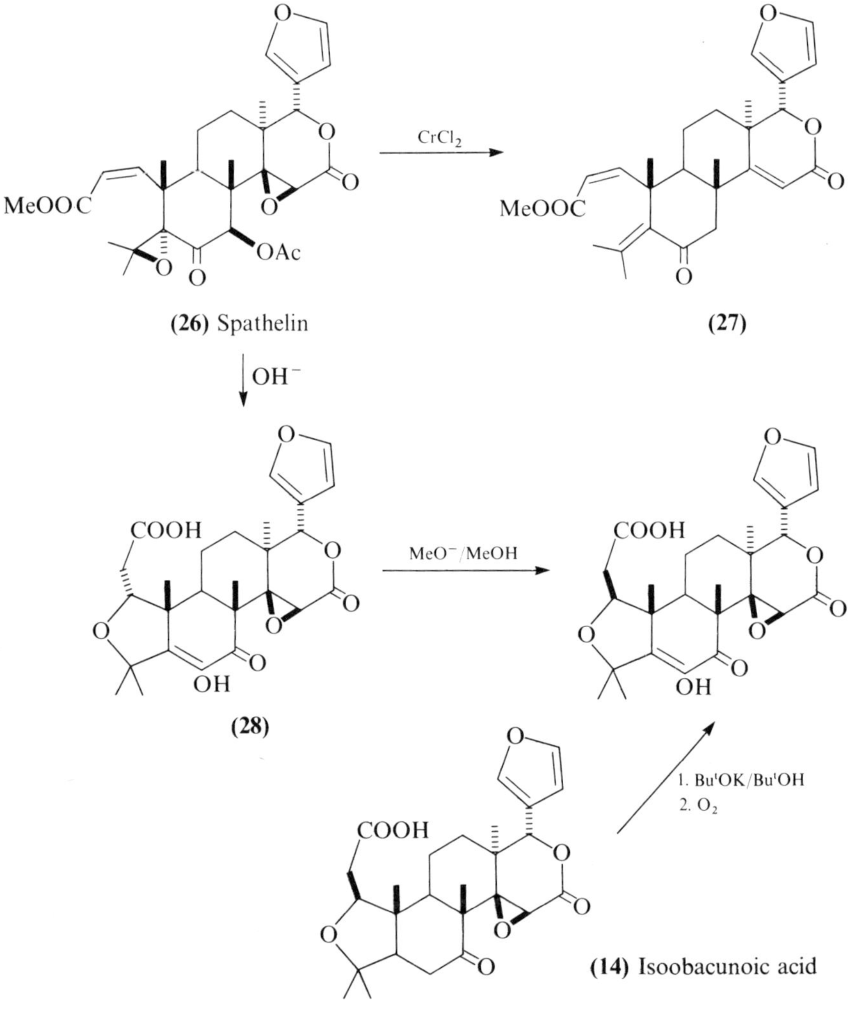

(26) Spathelin   **(27)**

(28)

(14) Isoobacunoic acid

SCHEME 2. Reactions of spathelin.

identical (via the methyl esters) with that formed by oxidation of isoobacunoic acid (14). The correlation of spathelin with 14 establishes the stereochemistry of all centres except at C-5 and C-7. The stereochemistry at C-7 in 26 was defined by the diagnostic chemical shift of the closely positioned H-15 similar to that found in rutaevin acetate (29). The α-position of the 4,5-epoxide was shown by the downfield position of the H-7 signal, again with reference to the same signal in 29. Spathelin is thus the acetyl derivative of 4,5α-epoxyatalantolide.[16]

The isolation of 26 from *Spathelia sorbifolia* argues for a relationship of the subfamily Spathelioideae closer to the Rutaceae than to the Simaroubaceae. Only one plant, *Harrisonia abyssinica*, presently classified in the Simaroubaceae is known to produce C_{26} limonoids. Obacunone (1), harrisonin (30),[69] and 12β-acetoxyharrisonin (31)[72] as well as chromones[118] occur in this plant so that there is a close parallel in the chemistry of *Harrisonia* and *Spathelia*.

(29) Rutaevin acetate

(30) Harrisonin (R = H)
(31) 12β-Acetoxyharrisonin (R = OAc)

D. CALAMIN, RETROCALAMIN, CYCLOCALAMIN, AND METHYL ISOOBACUNOATE DIOSPHENOL

These substances are all methyl esters isolated by Bennett and Hasegawa[12] from calamondin, a well known *Citrus × Fortunella* hybrid. Calamnin (32) was the major limonoid in the crude extracts. However, during isolation on silica gel column chromatography it was converted to one of the minor limonoids also present in the crude extracts, retrocalamin (33). By minimizing contact time on the column and running at 5° calamin was recovered from the extracts as the major component. The proton NMR spectroscopic properties of calamin (32) suggested that it was a hydroxy methyl deacetylnomilinate derivative. The chemical shift of H-15 indicated the presence of a 7β-hydroxy instead of the usual 7-keto group. A singlet assignable to H-5 located the keto group at C-6. The calamondin limonoids were interconverted by the reactions outlined in Scheme 3. Heating **32** with pyridine gave retrocalamin (**33**) which contained three fewer carbons. The NMR spectrum showed that **33** contained only three C-methyls suggesting loss of C-4 and its two attached C-methyl groups via a reverse aldol reaction. The most non-polar component in the extracts was identified as the previously known[30] methyl isoobacunoate diosphenol (**34**) although this was the first encounter with **34** as a natural product. The final limonoid from calamondin, cyclocalamin (**35**), was converted to **34** by Jones reagent followed by acid catalyzed enolization. This chemistry when considered with the appropriate NMR data required a 6-keto-7β-hydroxy system in **35**. Finally the chemical interconversions within this group of substances were completed by the acid catalyzed conversion of **32** to **35** as well as the 6-keto-7β-hydroxy analogue of deacetylnomilin (**36**) (Scheme 3). When the course of this reaction was monitored by carbon NMR it could be shown that **37** was an intermediate in the conversion of **32** to **35**. Detailed consideration of the ^{13}C NMR data for cyclocalamin (**35**) and rutaevin (**9**) required that both have a 5β instead of the usual 5α H. This allows formation of a less strained cis-fused 5-membered A-ring and with a chair B-ring giving more flexibility in the system.

Examination of *F. margarita* seed extracts confirmed the presence of the same compounds (**32–35**). Thus, their presence in the calamondin hybrid derives from the *Fortunella* parent.

E. ATALANTIN, DEHYDROATALANTIN, CYCLOEPIATALANTIN, AND ISOLIMONIC ACID

These limonoids all show only four C-methyl signals in the NMR and have the common structural feature that the C-19 methyl group has been oxidized and closed onto C-4 by an ether bridge. The atalantins **38**, **39**, and **40** from *Atalantia monophylla* have been chemically interrelated (Scheme 4).[9,36,99,101,112] Atalantin (**38**) and its dehydro derivative (**39**) are methyl esters. Oxidation of **38** with Jones reagent gives **39** which has the properties of an α-diketone. The usual NMR arguments based on the chemical shifts of H-15 located a β-hydroxy group at the 7-position in **38**. Since **39** is an α-diketone **38** must have a keto group at the 6-position. The proton NMR spectra of **38** and **39** indicated the presence of an α,β-unsaturated ester system and showed an AB quartet for two nonequivalent protons at C-19. These data were accommodated in terms of structures **38** and **39** for atalantin and its dehydro derivative respectively.[36,99]

Accompanying **38** and **39** in *Atalantia* was atalantolide (**24**), previously discussed, and cycloepiatalantin (**40**). Cycloepiatalantin formed a mo-noacetate and had spectroscopic properties indicating the presence of an α,β-

unsaturated cyclopentenone, an α-hydroxyketone system, and the normal D-ring δ-lactone system.[36] Treatment of **40** with base and esterification with diazomethane gave an epimer of atalantin (**41**) which on oxidation with Jones

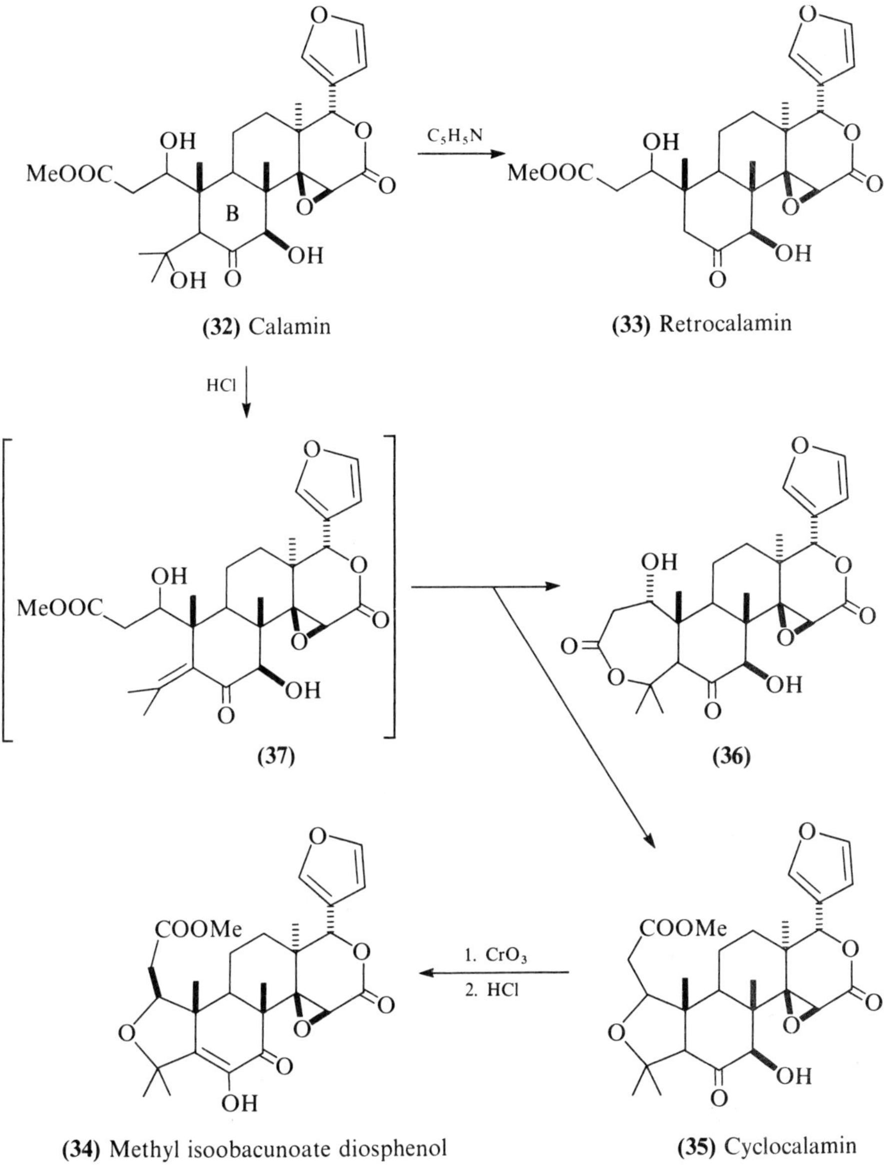
C5H5N
HCl
1. CrO3
2. HCl
(32) Calamin
(33) Retrocalamin
(37)
(36)
(34) Methyl isoobacunoate diosphenol
(35) Cyclocalamin

SCHEME 3. Interconversions of the calamondin limonoids.

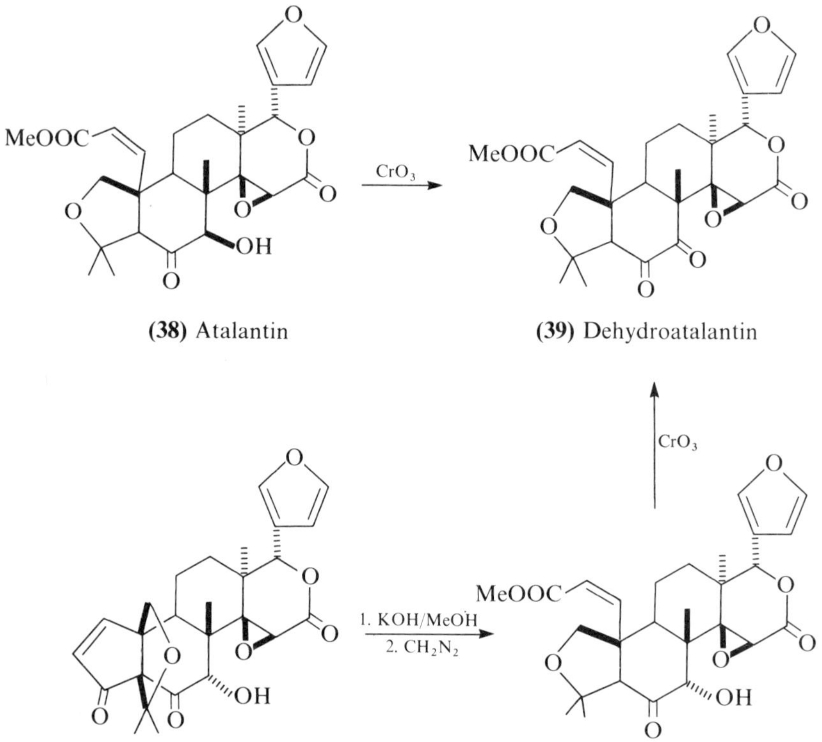

(38) Atalantin   (39) Dehydroatalantin

(40) Cycloepiatalantin   (41) Epiatalantin

SCHEME 4. Interconversions of the *Atalantia* limonoids.

reagent formed **39**. Therefore epiatalantin must have structure **41**. These data, when taken with that from detailed ^{13}C NMR studies, were interpreted in terms of structure **40** for cycloepiatalantin.[36] Unlike limonin and congeners, which have a chair B-ring, **38** and **40** have a boat B-ring.

Isolimonic acid (**42**) has been isolated as its methyl ester from seeds of several *Citrus* species by the Pasadena group.[11] Acetylation of the methyl ester of **42** gave a monoacetate. Elimination of acetic acid from the acetyl derivative with methanolic base gave, after remethylation, an α,β-unsaturated ester (**43**). The NMR spectrum of **43** showed the presence of a *trans* double bond. The appropriate NMR data on methyl isolimonate showed four *C*-methyls and by the chemical shift of H-15 that a carbonyl group was located at C-7. These data permitted structure **42** to be advanced for isolimonic acid.[11]

(42) Isolimonic acid   **(43)**

F. 7-DEACETYLAZADIRONE, 7-DEACETYLPROCERANONE, AND TECLEANIN

These three limonoids have been reported recently from *Teclea grandifolia* (Fig. 2) by Ayafor *et al.*[4] and are the first reported limonoids from the Rutaceae which lack the usual δ-lactone D-ring. These three compounds contain the structural features which might be expected for intermediates in limonin biosynthesis. The sequence of compounds involving expansion of the carbocyclic A-ring of azadirone **(47)** to a 7-membered lactone as in obacunone **(1)**, or in this case deacetylproceranone **(45)**, and finally oxidation of the C-19 methyl group and ring closure to give the limonin A- and A'-ring system as in tecleanin **(46)** are all represented. 7-Deacetylazadirone **(44)** and 7-deacetylproceranone **(45)** were known previously as their acetates **47** and **48** respectively, **47** from *Melia azadirachta*[70,102] and *Khaya anthotheca*[50] and **48** from *Carapa procera*,[103] all of the Meliaceae.

The structure of azadirone **(47)**, due to Lavie and coworkers,[50,70,102] rests on extensive chemistry including oxidation to the 16-keto derivative **(49)**, a Baeyer–Villiger ring expansion of its epoxy derivative to gedunin **(50)** and its dihydro derivative to 1,2-dihydro-7α-obacunyl acetate (dihydro derivative of **16**). By analogy with known obacunone chemistry, 7-deacetylproceranone **(45)** has been converted to the isoobacunoic acid derivative **51** (Fig. 2). In addition to the chemical conversions the structures of **44**, **45**, and especially **46** rest, to a considerable degree, on spectroscopic evidence.

G. 1-(10 → 19)-*abeo*-7α-ACETOXY-10β-HYDROXYISOOBACUNOIC ACID 3,10-
LACTONE AND 1-(10 → 19)-*abeo*-OBACUN-9(11)-EN-7α-YL ACETATE

Bennett and Hasagawa[13] have recently reported the isolation of two rearranged 7α-acetoxy limonoids (**52** and **53**) from fruit of a *Citrus × Poncirus* hybrid. The assigned structures of these materials rest on extensive

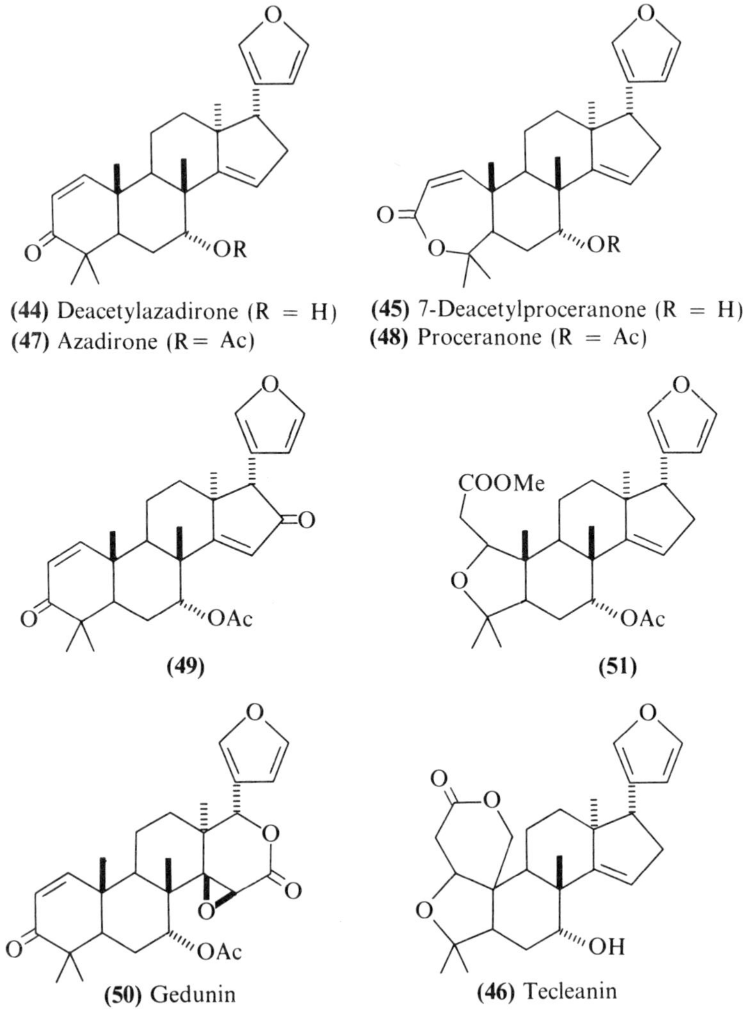

(44) Deacetylazadirone (R = H)   (45) 7-Deacetylproceranone (R = H)
(47) Azadirone (R = Ac)   (48) Proceranone (R = Ac)

(49)   (51)

(50) Gedunin   (46) Tecleanin

FIG. 2. Structures and chemistry of the *Teclea* limonoids.

spectroscopic evidence. A plausible biosynthetic route to these materials from the co-occurring 7α-limonyl acetate is outlined in Scheme 5. Models indicate that the A′-ring has enough flexibility to allow close approach of the C-1 carbonyl to the 11β-hydrogen. The abstraction of a proton from the 11β-position with migration of the *trans* H-9 to the 10-position would account for the presence of the 9(11) double bond found in 53. This biosynthetic sequence would require that H-10 of 53 possess the α-configuration and that H-10 in 53

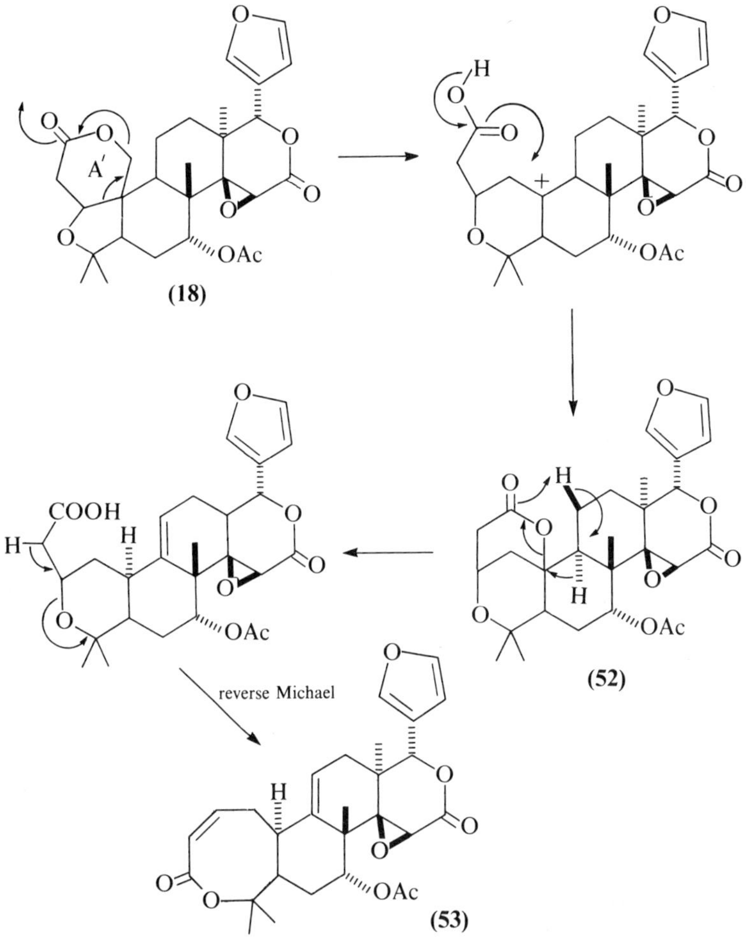

SCHEME 5. Possible biosynthetic route to **52** and **53** from 7α-limonyl acetate (**18**).

arises from H-9 in **18**. Unfortunately, the stereochemistry of this center could not be extracted from the NMR data and is still unknown.

H. 17-DEHYDROLIMONOATE A-RING LACTONE

The diketone **54** has been isolated from both lemons and oranges.[57] It is tasteless and is one of the initial metabolic products of limonin in *Citrus*. The formation of **54** accounts for the natural debittering of citrus as well as the observed decrease in limonin concentration with advancing maturity of the fruit.[121]

I. CALODENDROLIDE AND FRAXINELLONE

These two substances appear to arise biogenetically by extensive degradation of the limonoid system. Both **55**[18] and **56**[23,89,90] co-occur with limonoids and represent metabolic fragments containing the limonoid C- and D-rings. Both the relative and absolute configurations of **55** and **56** are consistent with their limonoid origin. Further, as Cassady and Liu[18] point out, calodendrolide (**55**) may well be a precursor of fraxinellone (**56**).

(54)   (55) Calodendrolide   (56) Fraxinellone

J. LIMONOIDS WITH OXIDIZED FURAN RINGS

A number of reports have appeared describing the structures of limonoids containing oxidized furan rings. The assigned structures resemble those expected from air oxidation of furans.[24] As a result, it is not always clear if these materials are artefacts of the isolation procedures. There is some suspicion that crude limonoid containing extracts on long standing are more likely to contain such furan oxidation products. Firm evidence on this point is however not available, Limonexic acid (**57**), isolated from orange extracts, has long been known[21,87] and is synthesized by dye sensitized air oxidation of limonin. The synthetic product is a mixture of isomers (**57** and **58**). More recently Makita and coworkers[77] reported limonexic acid as well as the obacunone (**59**), its acetate (**60**), and nomilin (**61**) analogues from *Citrus* seeds. Since only one isomer was obtained from the extracts it is argued that these substances **59**, **60**, and **61** must be naturally occurring.

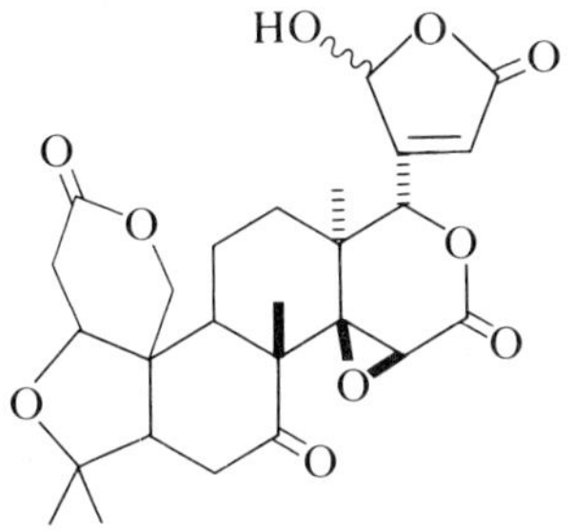

(57) Limonexic acid

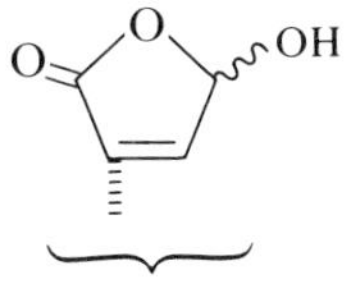

(58)

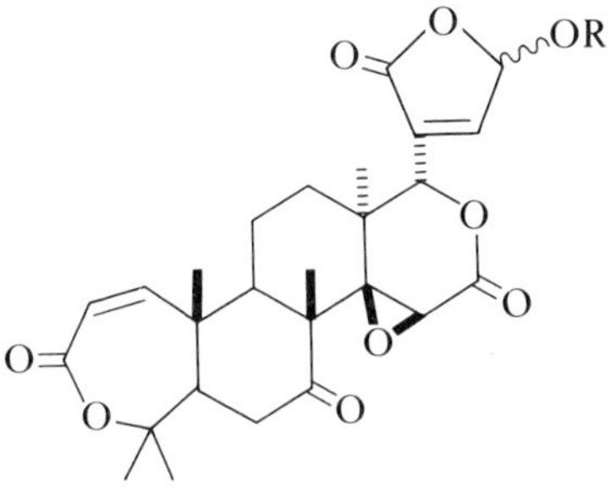

(59) R = H   (61)

(60) R = Ac

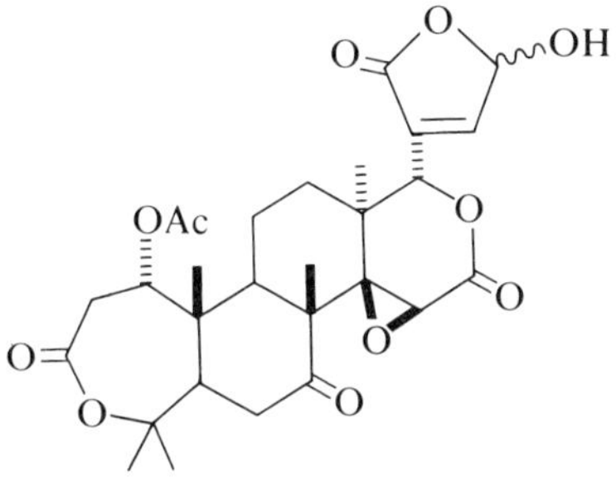

III. METABOLISM AND BIOSYNTHESIS OF LIMONIN

The Pasadena group[76] has demonstrated that limonoids are degraded by two different pathways in *Citrus* (Scheme 6). These same pathways have also been shown to occur in certain bacteria. One pathway involves conversion of limonin (8) to deoxylimonin (11) by an epoxidase. The B-ring of 11 is, in turn, opened to give deoxylimonic acid (21) by a deoxylimonin hydrolase. Finally, the A-ring lactone group of 21 is opened by a hydrolase to give a diacid (62) which was trapped as its dimethyl ester.[53] In a second degradative pathway the D-ring lactone group of limonin is hydrolytically opened and oxidation occurs at C-17 to give 17-dehydrolimonoate A-ring lactone (54). Further metabolic products beyond 62 and 54 are not known.

For the most part, the enzymes involved in these transformations appear to be relatively nonspecific. The exact nature of the A-ring, for example, in the limonin (8) to deoxylimonin (11) conversion is not critical since the same

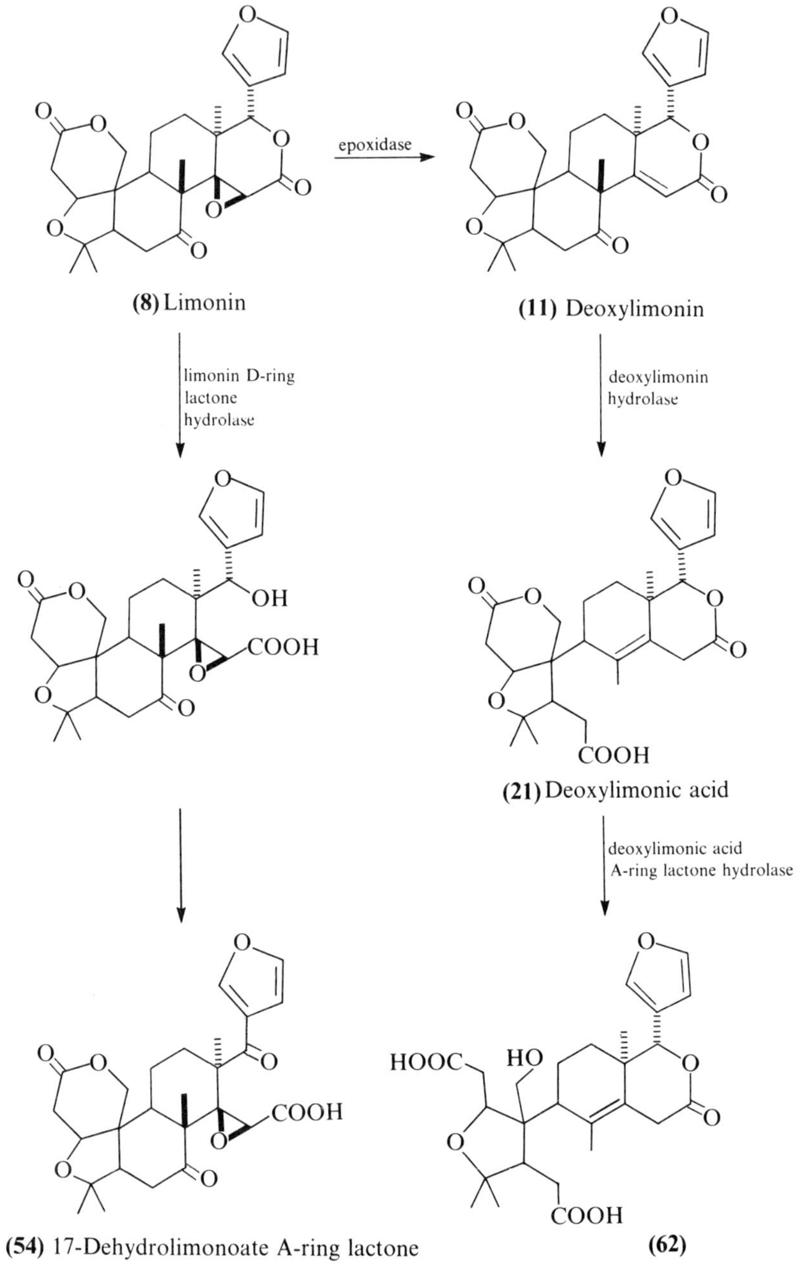

(8) Limonin

(11) Deoxylimonin

epoxidase

limonin D-ring
lactone
hydrolase

deoxylimonin
hydrolase

(21) Deoxylimonic acid

deoxylimonic acid
A-ring lactone hydrolase

(54) 17-Dehydrolimonoate A-ring lactone

(62)

SCHEME 6. Metabolism of limonin in *Citrus*.

enzyme system also converts other limonoids to their corresponding deoxy derivatives. One case of higher specificity is deoxylimonate A-ring lactone hydrolase ($21 \rightarrow 62$) which is unable to open the A-ring lactone of limonin.

Further degradation by cleavage of the 9,10 carbon–carbon bond of deoxylimonic acid (21) could lead to the formation of calodendrolide (55) or fraxinellone (56) like systems. However, the occurrence of such C/D ring fragments in *Citrus* has not yet been reported.

Limonin may be metabolized by yet another route in some rutaceous plants. This could be by the B-ring oxidation of limonin to limonin diosphenol (10) by way of rutaevin (9). Thus, Hirose and coworkers[56] found that early season fruit of *Euodia rutacarpa* contained only limonin, but in more mature, late season fruit they were able to detect only the more highly B-ring oxidized rutaevin (9) and limonin disophenol (10).

Biosynthetic studies of limonin in *Citrus* were unrewarding for many years[26] until the Pasadena workers[54] showed that limonin is biosynthesized in the leaves and translocated to the fruit and seeds, and moreover that limonin is specifically translocated from the fruit to the seeds.[52] Yokoyama and coworkers[55] have shown that the biosynthetic rate of limonin formation in *Citrus* can be decreased by the application of certain triethylamine plant bioregulators. These results are of special interest as a potential method for controlling bitterness in processed citrus products. The bioregulators probably act by inhibiting certain early steps in terpenoid biosynthesis. They had no effect on the metabolic rate of limonin.

IV. ANALYTICAL METHODS FOR LIMONOIDS

The role of limonin in causing bitterness in processed citrus products, particularly those from certain orange varieties and grapefruit, has led to much recent interest in quantitative analytical methods for limonin. An initial spectrophotometric method for limonin suffered from lack of specificity and interference from other citrus components.[121] This was followed by several TLC methods of wider utility,[20] three of which depended on Ehrlich reagent for detection.[75,81,108] The characteristic color produced with the Ehrlich reagent depends on the presence of the β-substituted furan ring and allows high selectivity for limonoids among the many other extractives present in *Citrus*, especially other terpenes and coumarins. GLC[8,65] and fluorometric[42] methods have been advanced but require lengthy sample preparation. Various HPLC methods have followed and are of increasing utility for routine analysis.[43,44,60,97] Since not all limonoids are bitter, HPLC has the advantage not only of providing separation but also of allowing an estimation of at least the bitter members. More recently, a radio-immunoassay method for limonin has appeared.[78,120] HPLC methods have been used to survey the distribution of limonoids in seeds of a number of *Citrus* species and hybrids.[98]

V. Limonoids as Insect Antifeeding Agents

A number of limonoids have shown potent antifeedant properties towards a variety of chewing insects. Some of these studies are summarized in Table 1. Frequently the test insects have been various *Spodoptera* species (for a more detailed list see ref. 116). Other chewing insects, *Locusta*, *Epilachna*, etc., have been less frequently employed.

Table 1

Insect antifeedant effects of limonoids

Compound	Test insect						References
	Acalymma	*Diabrotica*	*Epilachna*	*Locusta*	*Musca*	*Spodoptera*	
Obacunone (**1**)	−	−	×	×	×	−	67, 95
Deacetylnomilin (**4**)	−	−	×	×	×	×	95
Nomilin (**5**)	+	+	×	×	×	−	95, 115
Limonin (**8**)	−	−	×	×	×	−	95, 115
Deoxylimonin (**11**)	−	−	×	×	×	×	95
Harrisonin (**30**)	×	×	×	×	×	+	69
12β-Acetoxyharrisonin (**31**)	×	×	×	×	×	+	72
Azadirachtin (**63**)	+	+	+	+	×	+	17, 67, 95, 104
Salannin (**64**)	+	+	×	×	+	×	95, 117
Trichilin B (**65**)	×	×	+	×	×	+	86
Toonacilin (**66**)	×	×	+	×	×	×	64
6-Acetoxytoonacilin (**67**)	×	×	+	×	×	×	64

− = no effect; + = antifeedant; × = not tested

Using leaf disks treated with the test substance, azadirachtin (**63**),[122] harrisonin (**30**),[69] its 12β-acetoxy derivative (**31**),[72] and some of the trichilins [trichilin B (**65**) most active][86] have been shown to be highly deterrent towards *Spodoptera*. On the other hand, Wada and Munakata[115] reported that limonin (**8**) and nomilin (**5**) were inactive against *Spodoptera littoralis*. When limonin (**8**), deoxylimonin (**11**), obacunone (**1**), deactylnomilin (**4**), and

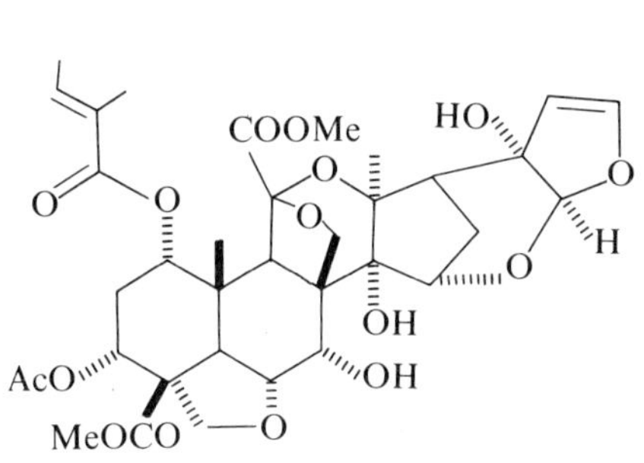

(63) Azadirachtin

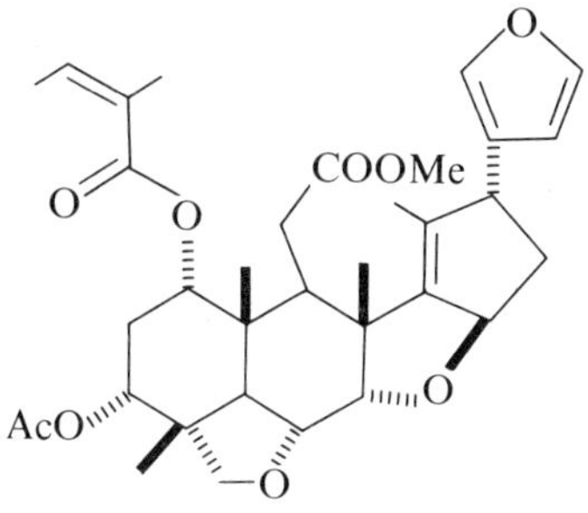

(64) Salannin

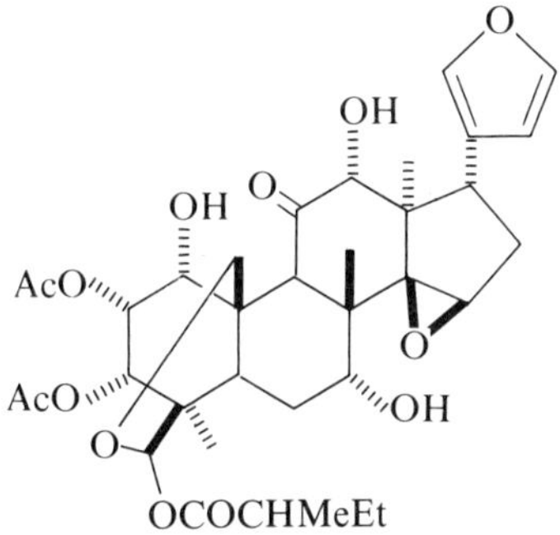

(65) Trichilin B

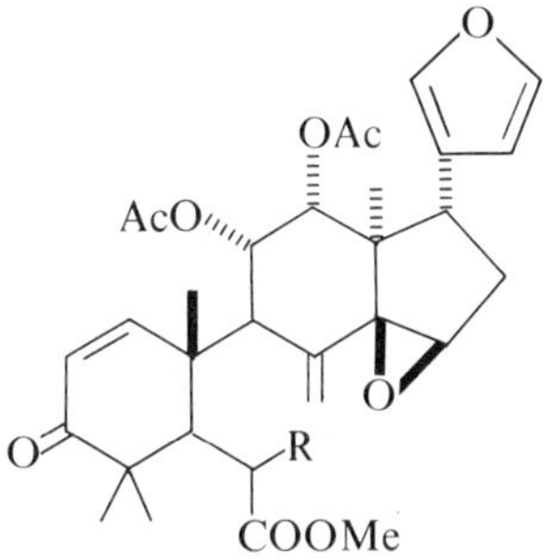

(66) Toonacilin (R = H)
(67) 6-Acetoxytoonacilin (R = OAc)

nomilin (5) were tested by the leaf disk method against the spotted cucumber beetle (*Diabrotica undecimp unctata howardi*) and the striped cucumber beetle (*Acalymma vittata*) only 5 showed modest activity as an antifeedant.[95] Toonacilin (66), and its 6-acetoxy derivative (67) and their oxidized furan derivatives, isolated by Kraus *et al.*[62,63] from *Toona ciliata* (Meliaceae), have shown activity as feeding repellents against the Mexican bean beetle (*Epliachna varivestis*).

Most interest in insect antifeedant limonoids focuses on azadirachtin (63) and other components which occur in the seeds of the widely distributed Neem tree (*Azadirachta indica*). These studies centre on the possible application of 63 in insect control by spraying Neem seed extracts on crop plants.[58] Detailed studies by Schmutterer and coworkers[104] with azadirachtin and other components of *Melia*[96] have shown that their action is not limited to just feeding deterrence. When 63 is given at low concentrations, where some feeding can occur, growth and reproductive disruptions occur in the insect. Similar results have been obtained by Kubo and Klocke[66] on a wide variety of different limonoids. It is noteworthy that most of the limonoids, nomilin (5) and the harrisonins (30, 31) excepted, which show high insect antifeeding activity are either of the C-ring seco type or have the structural features which allow opening of the C-ring. The presence of a 12α-hydroxy group could lead to cleavage of the 12,13 C–C bond with concurrent opening of the epoxide ring generating a typical C-ring seco limonoid. This same mechanism would also account for the formation of C-ring seco limonoids in the Meliaceae.

VI. CHEMOTAXONOMY OF LIMONOIDS

The limited distribution of limonoids in nature and their variation in structural complexity fills most of the main characteristics desired in chemotaxonomic markers.[49] Since the different limonoids are related to one another in a predictable biogenetic sequence this may allow correlations with the various levels of taxa to a degree which is not possible with extractives such as flavonoids and coumarins which show only limited structural variation, and which are sometimes difficult to rank in a biogenetic sequence.

A central unifying feature of the four limonoid producing families is their ability to accumulate obacunone (1) or limonoids of closely related oxidation state. Some examples which support this view are listed in Table 2. However, the presence of chromones in H. abyssinica[118] suggests that this genus should be classified with Spathelia instead of in the Simaroubaceae. On this basis one might argue that the common ancestor of the Rutaceae, Cneoraceae, and Meliaceae had the ability to degrade a tetracyclic triterpene as far as the obacunone stage. Each family then appears to have diverged to various degrees from this common structural feature to produce its own distinctive limonoid pattern. The Meliaceae has diverged to the greatest degree. Species of the Simaroubaceae have uniformly departed from modification of the intact limonoid system in favor of degradation to the C_{20} stage.

Modification of C-methyls through oxidation is an easily identifiable, unequivocal structural feature of limonoids. It may be possible that C-methyl oxidation of limonoids would correlate with their botanical distribution. To this end the occurrence of C-methyl oxidation in the five different structural positions for the four different limonoid producing families is summarized in Table 3. It is apparent that the Rutaceae is very limited in this respect, with such oxidation occurring only at the C-19 methyl. Conversely, one can find examples of compounds in the Meliaceae where, at one time or another, every C-methyl group in the system has suffered oxidation to some degree. Oxidation of C-methyls in the Cneoraceae and Simaroubaceae (excepting αC-4) is somewhat less extensive.

TABLE 2

Distribution of obacunone and related limonoids in the Rutales

	Species	Ref.
Rutaceae		
Obacunone	Widely distributed in	See Table 4
7α-Obacunol	many species and all	
7α-Obacunyl acetate	major subfamilies	
Nomilin		
7α-Nomilyl acetate		
Deacetylnomilin		
Cneoraceae		
Obacunone		82, 106
7α-Obacunol	*Neochamaelea pulverulenta*	82
7α-Obacunyl acetate	and *Cneorum tricoccon*	82, 106
7α-Nomilyl acetate		
Meliaceae		
7α-Obacunol	*Lovoa trichiliodes*	1
7α-Obacunyl acetate	*Carapa procera*	103
7α-Nomilyl acetate	*Xylocarpus granatum*	2
11β-Acetoxy-7α-nomilyl acetate	*Cedrela mexicana*	79
6β-Acetoxy-7α-obacunol	*Trichilia trifolia*	111
6β-Acetoxy-7α-obacunyl acetate	*Trichilia trifolia*	111
Simaroubaceae		
Obacunone	*Harrisonia abyssinica*	69

TABLE 3

Oxidation of C-methyl groups in limonoids-quassinoids

Family	αC-4	βC-4	C-Methyl at C-8	C-10	C-13
Rutaceae	no	no	no	yes	no
Meliaceae	yes	yes	yes	yes	yes
Cneoraceae	no	no	yes	no	yes
Simaroubaceae	always*	no	yes	no	yes

* Excepting harrisonin (**30**) and 12β-acetoxyharrisonin (**31**).

The idea was advanced[29] at one time that, within the Rutaceae itself, there was a correlation between the oxidation level of rutaceous limonoids, particularly of the C-19 methyl group, and their distribution among the three major subfamilies of the Rutaceae. Subsequent reports on the isolation of limonoids from additional genera has shown that such a simple correlation does not apply. At present there seems to be only a partial correlation of limonoid structures with the Engler classification.[39] Noteworthy is the wider than previously recognized diversity of limonoid structures in those genera placed in the Toddalioideae by Engler. In particular, in the two very closely related genera *Sargentia* and *Casimiroa*, the former contains the highly oxidized limonoids rutaevin (**9**) and limonin diosphenol (**10**)[27] while the latter contains only obacunone and derivatives with an unoxidized C-19 methyl group.[41] The recent report[4] of tecleanin (**46**), 7-deacetylproceranone (**45**), and 7-deactylazadirone (**44**), all with 5-membered carbocyclic D-rings, from *Teclea* introduces a pronounced exception to the generality that structural variation of rutaceous limonoids largely involves only the A- and B-rings.

Nevertheless limonoids in the Rutoideae and Aurantioideae still constitute homogenous groups if one allows for the fact that most reported limonoids from *Citrus* occur in minute amounts compared to limonin, nomilin and its deacetyl derivative. At present about 30 % of the genera of the subfamily Aurantioideae[34,37] have been shown to contain limonoids, while about 25 % of the Toddalioideae and only 6 % of the genera of the Rutoideae have had limonoids reported. These numbers must be considered with the fact that at least one species of 80 % of the genera of the Toddalioideae have had some definitive chemical studies reported while only about 50 % of the genera of the Rutoideae have been the subject of some definitive chemical work.

The statistical picture of limonoid distribution and structural variation is badly distorted in the Aurantioideae largely due to the intensive studies of Bennett, Hasagawa, and coworkers who have reported the isolation and structures of many minor limonoids from *Citrus* species and hybrids.

At the time of writing, the rule still holds that for any limonoid producing genus all species of that genus will contain limonoids. *Citrus* (26 sp.) has been intensively studied and every species investigated has been shown to contain limonoids.[29,37,55] Five species of the large genus *Euodia* have been studied and limonoids reported from each. Similar results have been obtained with the five *Esenbeckia* species studied. Table 4 summarizes the limonoid distribution among the various limonoid producing genera of the Rutaceae. The botanical organization used is based on that of Engler and Prantl[39] except for the Aurantioideae where that of Swingle and Reece[107] is employed.

Rouseff and Nagy[48] have recently carried out a detailed quantitative study of limonoids in seeds of 8 *Citrus* species by HPLC. Although there was some

TABLE 4

The distribution of limonoids in the Rutaceae

Subfamily Genus*	Limonoids
Rutoideae	
Euodia (120)	**(8)**[29,56,71,85]; **(9)**[29,56]; **(10)**[29,56,71]
Choisya (1)	**(8)**[37]
Dictamnus (1)	**(1)**[29]; **(3)**[31]; **(8)**[29,100,105]; **(9)**[29]; **(17)**[83]; **(56)**[23,89,90,105]
Calodendron (1)	**(8)**[30]; **(9)**[30]; **(10)**[30]; **(55)**[18]
Essenbeckia (*ca.* 30)	**(8)**[35,37]; **(9)**[35,37,114]; **(10)**[37]
Spathelioideae	
Spathelia (*ca.* 10)	**(26)**[16]
Toddalioideae	
Phellodendron (9–11)	**(1)**[31]; **(8)**[31,40]
Fagaropsis (1)	**(8)**[119]; **(9)**[119]; **(10)**[119]
Helietta (6)	**(8)**[114]
Sargentia (1)	**(9)**[27]; **(10)**[27]
Casimiroa (5)	**(1)**[61]; **(2)**[32]; **(4)**[32]; **(5)**[29]; **(22)**[32,61]
Vepris (20)	**(6)**[47]; **(8)**[5]
Teclea (25)	**(44)**[4]; **(45)**[4]; **(46)**[4]
Aurantioideae	
Clausena (23)	**(23)**[19]
Triphasia (30)	**(8)**[37]
Luvunga (12)	**(8)**[45]
Hesperethusa (1)	**(8)**[22]
Atalantia (11)	**(24)**[36,99,112]; **(38)**[9,36,99,101,112]; **(39)**[36,99,112]; **(40)**[36]
Fortunella (40)	**(1)**[29]; **(8)**[29]; **(32)**[12]; **(33)**[12]; **(34)**[12]; **(35)**[12]
Eremocitrus (1)	**(8)**[37]
Poncirus (1)	**(1)**[29]; **(4)**[29]; **(5)**[29]; **(8)**[29,113]
Microcitrus (6)	**(4)**[29]; **(8)**[29]
Citrus (16)	**(1)**[98]; **(2)**[13]; **(4)**[98]; **(4, R = Me)**[12]; **(5)**[98]; **(7)**[11]; **(8)**[98]; **(11)**[28]; **(12)**[10]; **(13)**[10]; **(14)**[10]; **(15)**[10]; **(19)**[13]; **(20)**[13]; **(21)**[55]; **(42)**[11]; **(54)**[57]
Citrus × *Poncirus*	**(16)**[13]; **(18)**[13]; **(52)**[13]; **(53)**[13]

* The number of species in each genus follows the generic name.

variation in total limonoid concentration from season to season, they found that the ratio between the various limonoids was relatively constant for fruit of the same maturity, but could vary widely from species to species. The quantitative limonoid pattern was used to argue for the possible parentage of various *Citrus* hybrids. This approach with limonoids supplements the work of Albach and Redman[3] with flavonoids from *Citrus*.

REFERENCES

1. Adesida, G. A. and Taylor, D. A. H. (1972). *Phytochemistry* **11**, 2641.
2. Ahmed, F. R., Ng, A. S. and Fallis, A. G. (1978). *Can. J. Chem.* **56**, 1020; Ng, A. S. and Fallis, A. G. (1979). *Can. J. Chem.* **57**, 3088.
3. Albach, R. F. and Redman, G. H. (1969). *Phytochemistry* **8**, 127.
4. Ayafor, J. F., Sondengam, B. L., Connolly, J. D., Rycroft, D. S. and Okogun, J. I. (1981). *J. Chem. Soc. Perkin Trans. 1* 1750.
5. Ayafor, J. F., Sondengam, B. L. and Ngadjui, B. T. (1982) *Phytochemistry* **21**, 955.
6. Baldwin, M. A., Loudon, A. G., Maccoll, A. and Bevan, C. W. L. (1967). *J. Chem. Soc. C* 1026.
7. Barton, D. H. R., Pradhan, S. K., Sternhell, S. and Templeton, J. F. (1961). *J. Chem. Soc.* 255.
8. Basker, H. B., Ben Shalom, N., Katz, M. and Sklan, D. (1973). *Lebensm. Wiss. Tech.* **6**, 34.
9. Basu, D. and Basa, S. C. (1972). *J. Org. Chem.* **37**, 3035.
10. Bennett, R. D. (1971). *Phytochemistry* **10**, 3065.
11. Bennett, R. D. and Hasegawa, S. (1980). *Phytochemistry* **19**, 2417.
12. Bennett, R. D. and Hasegawa, S. (1981). *Tetrahedron* **37**, 17.
13. Bennett, R. D. and Hasegawa, S. (1982). *Phytochemistry*, **21**, 2349.
14. Bennett, R. D. and Schuster, R. E. (1972). *Tetrahedron Lett.* 673.
15. Brown, D. A. and Taylor, D. A. H. (1978). *J. Chem. Res. (S)* 20.
16. Burke, B. A., Chan, W. R. and Taylor, D. R. (1972). *Tetrahedron* **28**, 425.
17. Butterworth, J. H. and Morgan, E. J. (1971). *J. Insect Physiol.* **17**, 969.
18. Cassady, J. M. and Liu, C. (1972). *J. Chem. Soc. Chem. Comm.* 86.
19. Chakraborty, D. P., Bhattacharyya, P., Bhattacharyya, S. P., Bordner, J., Hennessee, G. L. A. and Weinstein, B. (1979). *J. Chem. Soc. Chem. Comm.* 246.
20. Chandler, B. V. (1971) *J. Sci. Food Agric.* **22**, 473.
21. Chandler, B. V. and Kefford, J. F. (1951). *Aust. J. Sci.* **13**, 112; (1953). **15**, 28.
22. Chatterjee, A., Sarkar, S. and Schoolery, J. N. (1980). *Phytochemistry* **19**, 2219.
23. Coggon, P., McPhail, A. T., Storer, R. and Young, D. W. (1969). *J. Chem. Soc. Chem. Commun.* 828.
24. Connolly, J. D., Labbe, C. and Rycroft, D. S. (1978). *J. Chem. Soc. Perkin Trans. 1* 285.
25. Connolly, D. L., Overton, K. H. and Polonsky, J. (1970). *In* "Progress in Phytochemistry" (Reinhold, L. and Liwschitz, Y. eds.), Vol. 2, p. 285. Interscience, New York.
26. Datta, S. and Nicholas, H. J. (1968). *Phytochemistry* **7**, 955.
27. Dominguez, X. A., Butruille, D., Rudy, A., Garcia, S. G. (1977). *Rev. Latinoamer. Quim.* **8**, 47.
28. Dreyer, D. L. (1965). *J. Org. Chem.* **30**, 749.
29. Dreyer, D. L. (1966). *Phytochemistry* **5**, 367.
30. Dreyer, D. L. (1967). *J. Org. Chem.* **32**, 3442.
31. Dreyer, D. L. (1968). *Fortschr. Naturst.* **26**, 190.
32. Dreyer, D. L. (1968). *J. Org. Chem.* **33**, 3577.
33. Dreyer, D. L. (1968). *Tetrahedron*, **24**, 3273.
34. Dreyer, D. L. (1977). *Rev. Latinoamer. Quim.* **8**, 11.
35. Dreyer, D. L. (1980). *Phytochemistry* **19**, 941.
36. Dreyer, D. L., Bennett, R. D. and Basa, S. C. (1976). *Tetrahedron* **32**, 2367.

37. Dreyer, D. L., Pickering, M. V. and Cohan, P. (1972). *Phytochemistry* **11**, 705.
38. Ekong, D. E. U., Okogun, J. I. and Shok, M. (1972). *J. Chem. Soc. Perkin Trans. 1* 653.
39. Engler, A. (1931). *In* "Die Naturlichen Pflanzenfamilien" (Engler, A. and Prantl, K. eds), 2nd Edn., Vol. 19a, p. 203. Englemann, Leipzig.
40. Enina, V. (1974). *Nauka-Prakt, Farm.* 90; (1977) *Chem. Abs.* **86**, 40157.
41. Epe, B. and Mondon, A. (1979). *Tetrahedron Lett.* 2015.
42. Fisher, J. F. (1973). *J. Agric. Food Chem.* **21**, 1109.
43. Fisher, J. F. (1975). *J. Agric. Food Chem.* **23**, 1199.
44. Fisher, J. F. (1978). *J. Agric. Food Chem.* **26**, 497.
45. Ganguly, A. K., Govindachari, T. R., Manmade, A. and Mohamed, P. A. (1966). *Indian J. Chem.* **4**, 292.
46. Govindachari, T. R. (1938). *J. Indian Chem. Soc.* **45**, 1063.
47. Govindachari, T. R., Joshi, B. S. and Sundararajan, V. N. (1964). *Tetrahedron* **20**, 2985.
48. Gupta, I., Baptista, J., Furrer, R., Bruce, W. R., Krepinsky, J. J. and Yates, P. (1982). *Naturwissenschaften* **69**, 453.
49. Haensel, R. (1956). *Arch. Pharm.* **289**, 619.
50. Halsall, T. G. and Troke, J. A. (1975). *J. Chem. Soc. Perkin Trans. 1* 1758.
51. Halsall, T. G., Wragg, K., Connolly, J. D., McLellan, M. A., Bredell, L. D. and Taylor, D. A. H. (1977). *J. Chem. Res. (S)* 154.
52. Hasegawa, S., Bennett, R. D. and Verdon, C. P. (1980). *J. Agric. Food Chem.* **28**, 922.
53. Hasegawa, S., Bennett, R. D. and Verdon, C. P. (1980). *Phytochemistry* **19**, 1445.
54. Hasegawa, S. and Hoagland, J. E. (1977). *Phytochemistry* **16**, 469.
55. Hasegawa, S., Yokoyama, H. and Hoagland, J. E. (1977). *Phytochemistry* **16**, 1083.
56. Hirose, Y., Kondo, K., Arita, H. and Fujita, A. (1967). *Shoyakugaku Zasshi* **21**, 126; (1968) *Chem. Abs.* **69**, 65151.
57. Hsu, A. C., Hasegawa, S., Maier, V. P. and Bennett, R. D. (1973). *Phytochemistry* **12**, 563.
58. Jacobson, M., Reed, D. K., Crystal, M. M., Moreno, D. S. and Soderstrom, E. L. (1978). *Ent. Exp Appl.* **24**, 448.
59. Jibodu, K. O., Ohochuku, N. S. and Taylor, D. A. H. (1970). *J. Chem. Soc. (C)* 2396.
60. Johnson, E. L., Gloor, R. and Majors, R. E. (1978). *J. Chromatog.* **149**, 571.
61. Kincl, F. A., Romo, J., Rosenkranz, G. and Sondheimer, F. (1956). *J. Chem. Soc.* 4163.
62. Kraus, W. and Bokel, M. (1981). *Chem. Ber.* **114**, 267.
63. Kraus, W., Grimminger, W. and Sawitzki, G. (1978). Symposium papers. 11th IUPAC International Symposium Chem. Of Natural Products 115; (1980) *Chem. Abs.* **92**, 18787.
64. Kraus, W., Grimminger, W. and Sawitzki, G. (1978). *Angew. Chem. (Int. Ed.)* **17**, 452.
65. Kruger, A. J. and Colter, C. E. (1972). *Proc. Fla. State Hort. Soc.* **85**, 206.
66. Kubo, I. and Klocke, J. A. (1982). *Collogues de ĽI. N.R.A.*, 117.
67. Kubo, I. and Nakanishi, K. (1977). *In* "Host Plant Resistance to Pests" (Hedin, P. A., ed.), Vol. 62, p. 165. American Chemical Society Symposium Series.
68. Kubota, T., Matsuura, T., Tokoroyama, T., Kamikawa, T. and Matsumoto, T. (1961). *Tetrahedron Lett.* 325.

69. Kubo, I., Tanis, S. P., Lee, Y. Miura, I., Nakanishi, K. and Chapya, A. (1976). *Heterocycles* **5**, 485.
70. Lavie, D., Levy, E. C. and Jain, M. K. (1971). *Tetrahedron* **27**, 3927.
71. Lengyel, E. and Gellert, M. (1978). *Pharmazie* **33**, 372.
72. Liu, H., Kubo, I. and Nakanishi, K. (1982). *Heterocycles* **17**, 67.
73. Loudon, A. G. and Powell, J. W. (1970). *Org. Mass. Spect.* **3**, 321.
74. Maier, V. P., Bennett, R. D. and Hasegawa, S. (1977). *In* "Citrus Science and Technology", Vol. 1. Avi, Westport, Conn.
75. Maier, V. P. and Grant, E. R. (1970). *Agric. Food Chem.* **18**, 250.
76. Maier, V. P., Hasegawa, S., Bennett, R. D. and Echols, L. C. (1980). *In* "Nutrition and Quality of Citrus Fruits and Their Products" (Nagy, S. and Attaway, J. R. eds.), Vol. 143, p. 63. American Chemical Society Symposium Series.
77. Makita, T., Ohta, K. and Nakabayashi, T. (1980). *Agric. Biol. Chem.* **44**, 693.
78. Mansell, R. L. and Weiler, E. W. (1980). *Phytochemistry* **19**, 1403.
79. Marcelle, G. B. and Mootoo, B. S. (1981). *Tetrahedron Lett.* 505.
80. Melera, A., Schaffner, K., Arigoni, D. and Jeger, O. (1957). *Helv. Chim. Acta* **40**, 1420.
81. Misawa, Y., Matsubara, M. and Doi, N. (1971). *Nippon Shokuhin Kogyo Gakkai-Shi* **18**, 326; (1972) *Chem. Abs.* **77**, 45483.
82. Mondon, A., Trautmann, D., Epe, B., Oelbermann, U. and Wolff, Ch. (1978). *Tetrahedron Lett.* 3699.
83. Monkovic, I., Spenser, I. D. and Plunkett, A. O. (1967). *Can. J. Chem.* **45**, 1935.
84. Moss, G. P., Toube, T. P. and Murphy, J. W. (1970). *J. Chem. Soc. (C)* 694.
85. Mukherjee, J. and Roy, B. (1970). *J. Indian Chem. Soc.* **47**, 91.
86. Nakatani, M., James, J. C. and Nakanishi, K. (1981). *J. Am. Chem. Soc.* **103**, 1228.
87. Nomura, D. and Santo, T. (1965). *Bull. Fac. Agric. Yamaguchi Univ.* 635.
88. Ohochuku, N. S. and Taylor, D. A. H. (1969). *J. Chem. Soc. (C)* 864.
89. Okogum, J. I., Fakunle, C. O., Ekong, D. E. U. and Connolly, J. D. (1975). *J. Chem. Soc. Perkin Trans. 1* 1352.
90. Pailer, M., Schaden, G., Spiteller, G. and Fenzl, W. (1965). *Monatsh. Chem.* **96**, 1324.
91. Polonsky, J. (1973). *Rec. Adv. Phytochemistry* **6**, 31.
92. Polonsky, J. (1973). *Fortscht. Naturst.* **30**, 101.
93. Powell, J. W. (1966). *J. Chem. Soc. (C)* 1794.
94. Rao, M. M. (1968). *Bull. Nat. Inst. Sci. India* **37**, 205.
95. Reed, D. K., Jacobson, M., Warthen, J. D., Uebel, E. C., Tromley, N. J., Jurd, L. and Freedman, B. (1981). *U.S.D.A. Technical Bull.* 1641.
96. Rembold, H., Sharma, G. K. (1980). *Z. Pflkrankh. Pflschutz.* **87**, 290.
97. Rouseff, R. L. and Fisher, J. F. (1980). *Anal. Chem.* **52**, 1228.
98. Rouseff, R. L. and Nagy, S. (1982). *Phytochemistry* **21**, 85.
99. Sabata, B., Connolly, J. D., Labbe, C. and Rycroft, D. S. (1977). *J. Chem. Soc. Perkin Trans. 1* 1875.
100. Serkerov, S. V. and Khodzhimatov, M. (1968). *Rast. Resur.* **4**, 495; (1969). *Chem. Abs.* **70**, 109118.
101. Shringarpure, J. D. and Sabata, B. K. (1975). *Indian J. Chem.* **13**, 24.
102. Siddiqui, S., Waheed, T. N., Luecke, J. and Voelter, W. (1975). *Z. Naturforsch.* **30B**, 961.
103. Sondengam, B. L., Kamga, C. Ś., Kimbu, S. F. and Connolly, J. D. (1981). *Phytochemistry* **20**, 173.

104. Steets, R. and Schmutterer, H. (1975). *Z. Pflkrankh. Pflschutz.* **82**, 176.
105. Storer, R. and Young, D. W. (1973). *Tetrahedron* **29**, 1217.
106. Straka, H., Albers, F. and Mondon, A. (1976). *Beitr. Biol. Pflanzen* **52**, 267.
107. Swingle, W. T. and Reece, P. C. (1967). *In* "The Citrus Industry" (Rev. Edn), Vol. 1, p. 190. University of California Press.
108. Tatum, J. H. and Berry, R. E. (1973). *J. Food Sci.* **38**, 1244.
109. Taylor, D. A. H. (1974). *J. Chem. Soc. Perkin Trans. 1* 437.
110. Taylor, D. A. H. (1971). *J. Chem. Res. (S)* 2.
111. Taylor, D. R. (1971). *Rev. Latinoamer. Quim.* **2**, 87.
112. Thaker, M. R. and Sabata, B. K. (1969). *Indian J. Chem.* **7**, 870.
113. Tomimatsu, T. (1969) *Chem. Pharm. Bull.* **17**, 1723.
114. Vitagliano, J. C. and Comin, J. (1970). *An. Assoc. Quim. Argent.* **58**, 273; (1971). *Chem. Abs.* **74**, 121356.
115. Wada, K. and Munakata, K. (1971). *Agric. Biol. Chem.* **35**, 115.
116. Warthen, J. D., Jr. (1979). *U.S.D.A. Agric. Rev.* ARM-NE-4.
117. Warthen, J. D., Vebel, E. C., Dutky, S. R., Lusby, W. R. and Finegold, H. (1978). *U.S.D.A. Agric. Res. Results,* ARR-NE-2.
118. Waterman, P. G., unpublished results; Okorie, D. A. (1982). *Phytochemistry* **21**, 2424.
119. Waterman, P. G. and Khalid, S. A. (1981). *Biochem. Syst. Ecol.* **9**, 45.
120. Weiler, E. W. and Mansell, R. L. (1980). *J. Agric. Food Chem.* **28**, 543.
121. Wilson, K. W. and Crutchfield, C. A. (1968). *J. Agric. Food Chem.* **16**, 118.
122. Zanno, P., Miura, I., Nakanishi, K. and Elder, D. L. (1975). *J. Am. Chem. Soc.* **97**, 1975.

CHAPTER 8

Chemistry and Biological Activity of the Quassinoids

JUDITH POLONSKY

Institut de Chimie des Substances Naturelles, C.N.R.S., Gif-sur-Yvette, France

I. INTRODUCTION

The quassinoids are the bitter principles of simaroubaceous plants. The generic term quassinoids arises from quassin (1), the name of the first structurally identified member of this class. The structure of quassin (1) and that of its corresponding hemiketal, neoquassin (2), were established by Valenta and his coworkers.[43,44] Since then a rapid advance in structural

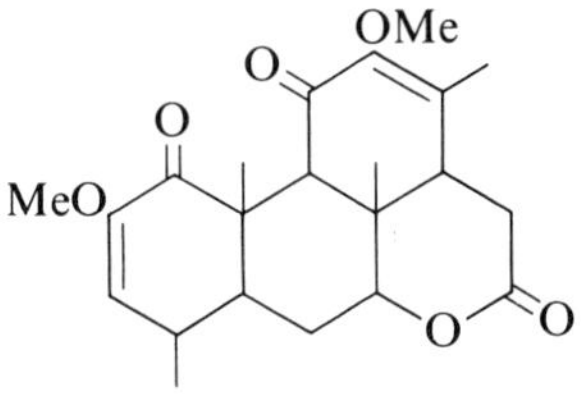

(1) Quassin 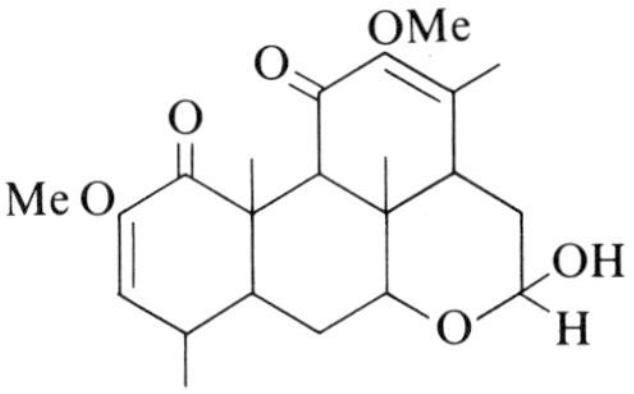   (2) Neoquassin

247

studies of the simaroubaceous bitter constituents has brought to light a group of compounds closely related to quassin, the quassinoids.

The chemistry and biogenesis of the quassinoids have been reviewed.[4,29,30] At the time when our previous review was written (1972), there were already 55 quassinoids known; ten years later the number has increased to over 75. Interest in the quassinoids has accelerated rapidly with the finding of the American National Cancer Institute in the early 1970s that these compounds display marked antileukemic activity. Since then, a wide spectrum of other biological properties for the quassinoids has been discovered and studies on chemical modifications of inactive members to yield biologically active ones have been undertaken.

It is proposed in this review to present the latest developments (1972–1982) in the quassinoid field.

II. GENERAL FEATURES OF THE QUASSINOIDS

In reviewing the essential features of the quassinoids, the new structural types discovered during the last decade will be emphasized.

The quassinoids can be divided into distinct groups according to their basic skeletons. The five skeletons observed are presented in Fig. 1. Skeletal types C and E are new. The C_{18} skeleton C differs from B by loss of one carbon atom in

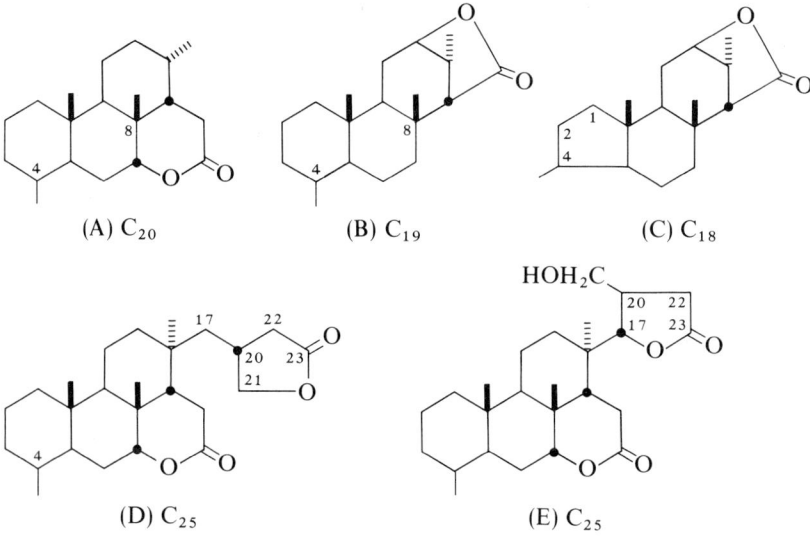

(A) C_{20} (B) C_{19} (C) C_{18}

(D) C_{25} (E) C_{25}

FIG. 1. The basic skeletons of quassinoids.

ring A, probably due to a benzylic rearrangement resulting in a ring contraction. The second new skeletal system found in the last decade is type E which differs from type D by the mode of formation of the γ-lactone. In type D this function is formed by linkage of C-23 to C-21 whereas in type E the γ-lactone is formed by linkage of C-23 to C-17 (see **6a** in Scheme 1).

The number and the positions of the methyl groups are the same on these five basic skeletons and all the quassinoids so far known have only one methyl group at C-4.

By far the majority of the numerous quassinoids known have the C_{20} basic skeleton, seven are C_{19} compounds with the basic skeleton B; three quassinoids with skeleton C, two with D, and three with E have been isolated to date. The quassinoids are heavily oxygenated lactones (δ-lactones in the C_{20} quassinoids, γ-lactones in the C_{19} and C_{18}, δ- and σ-lactones in the C_{25} compounds). They have varying numbers of different oxygen-containing groups; with the exception of carbons C-5 and C-9 and the methyl groups at C-4 and C-10, these oxygenated functions have been found on all the other carbon atoms.

III. C_{25} QUASSINOIDS

Biogenetically the quassinoids can be regarded as degraded triterpenoids. The triterpene origin of the quassinoids was proved experimentally by incorporation of specifically $^{14}C,^{3}H$-labelled mevalonolactone into one of the quassinoids, namely glaucarubinone.[29,30]

The triterpenoid precursor of the quassinoids might be, as proposed for the limonoids, Δ^{7}-euphol or more probably its 20α-isomer, Δ^{7}-tirucallol. Its conversion to the quassinoids may be schematized as follows (Scheme 1): the proto-triterpene undergoes, via the 7α-epoxide (**3**), an "apo" rearrangement during which the C-14 methyl group migrates to C-8 thus leading to the 7α-hydroxy apo compound (**4**). As proposed in the limonoid pathway, the five-membered ring is oxidatively expanded to the δ-lactone (**5**). Opening of the latter and relactonization to the 7α-hydroxyl would then give the δ-lactone (**6a**). These transformations may be accompanied by the removal of one of the methyl groups at C-4 and of four carbon atoms at the end of the side-chain with the formation of a carboxyl at C-23. The resulting intermediate (**6a**) might follow different routes for further transformations. Lactonization to the hydroxyl at C-21 and oxidation of the residual C-17 hydroxyl would lead to the C_{25} basic skeleton of simarolide (**7**) and picrasin A (**8**).[30] On the other hand, lactonization to the C-17 hydroxyl would account for the formation of

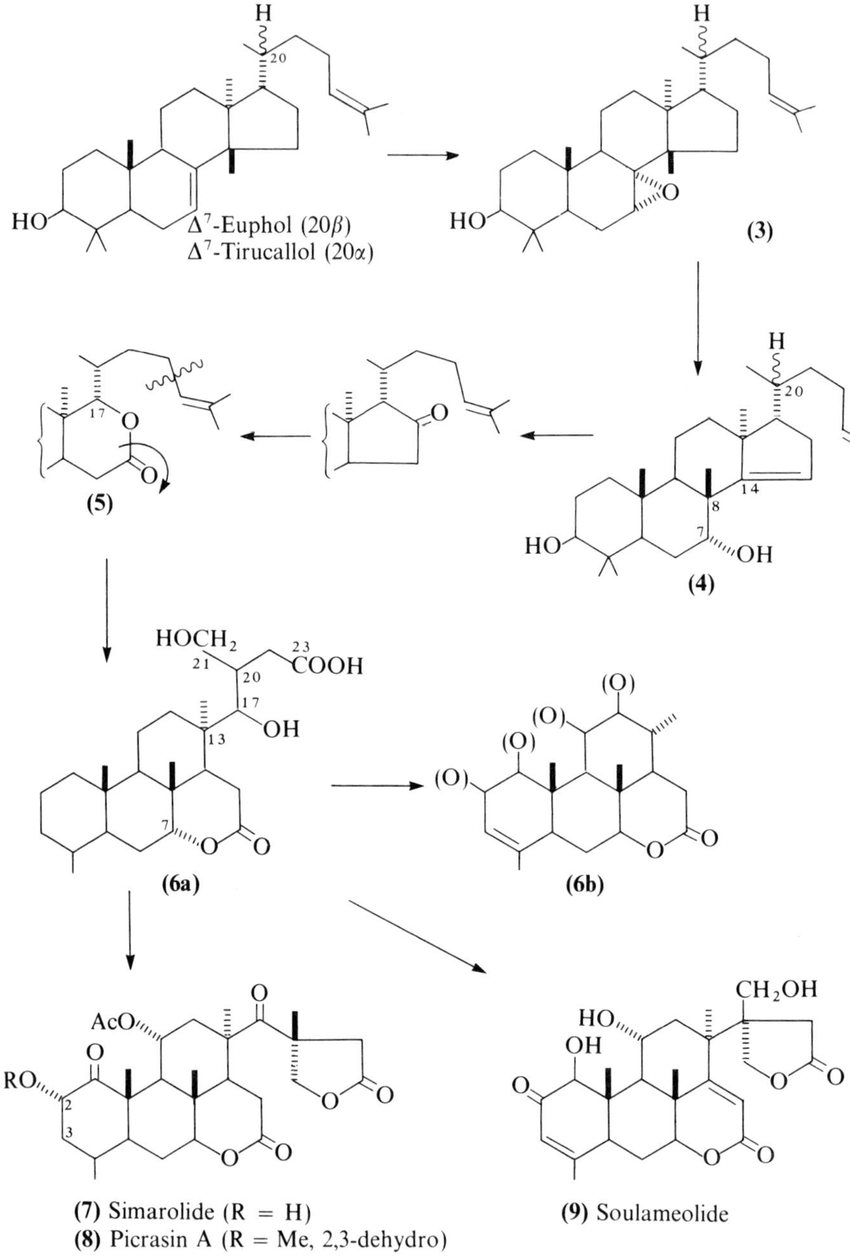

SCHEME 1. Formation of quassinoids from Δ^7-euphol and Δ^7-tirucallol.

the C_{25} basic skeleton of the three new quassinoids soulameolide (**9**).[35] simarinolide (**10**), and guanepolide (**11**).[39]

All five C_{25} quassinoids now known lack an oxygen function at position 12 but possess one at C-17. This fact further substantiates the hypothesis[11] that compounds of this type (e.g. **6a**) may be intermediates in the biosynthesis of the numerous known quassinoids with a C_{20} basic skeleton. It is clear that introduction of an oxygen function at C-12 can lead to C_{13}–C_{17} rupture (by a retro-Claisen or retro-aldol reaction) thus producing the C_{20} quassinoids (**6b**), which always have an oxygen function at C-12.

(**10**) Simarinolide   (**11**) Guanepolide

Simarinolide (**10**) and guanepolide (**11**) have A-ring structures not previously encountered among the quassinoids. Compound **11** is the first quassinoid to have two substituents at C-4, i.e. a hydroxyl and an axially oriented methyl group. The formation of the A-ring in **10** and **11** from the triterpene precursor can be explained by the following pathway:

Our previous biogenetic experiments[30] support the loss of the equatorial methyl group at C-4 in the sequence leading to **10** and hence to **11**.

Soulameolide (**9**) and guanepolide (**11**) are the only examples of natural quassinoids to have a 14,15 double bond. This suggests that the oxygenated functions at these carbon atoms, present in several quassinoids, are introduced subsequently to the formation of the δ-lactone.

IV. STRUCTURAL VARIATIONS OF THE C_{20} QUASSINOIDS

The principal variations in the structures of the quassinoids, which are basically C_{20} compounds, are as follows.

Ring A may have the structures (a), (b), (c), (d), (e), or (f). Structure (e) has so far been found only in certain quassinoids isolated from the genus *Brucea* and structure (f) only in one quassinoid, sergeolide (**36**) (see below).

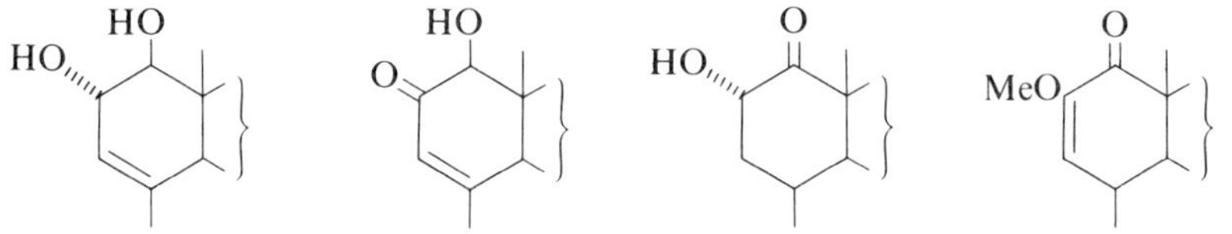

(a)   (b)   (c)   (d)

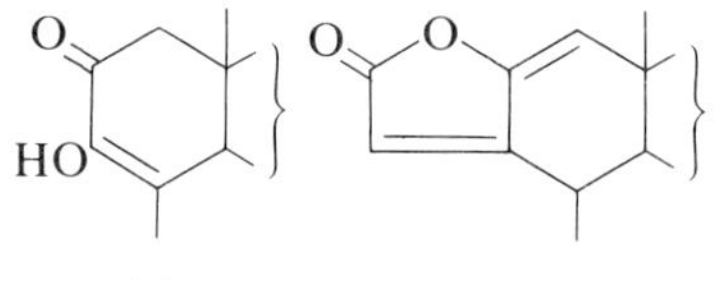

(e)   (f)

In ring C at position 8 one finds either a methyl group or a primary alcohol. The latter generally is involved in a hemiketal bridge to C-11 or an oxide bridge to C-13. The δ-lactone ring may have hydroxyl groups and in particular a β-oriented hydroxyl group at C-15 which is often found esterified with various small fatty acids. Ring B may possess at position 6 an oxygenated group.

The following structures will illustrate the observed variations with particular emphasis on compounds isolated during the last ten years.

(1) Soulameanone (**12**)[36] has a methyl group at position 8. It has the α,β-unsaturated ketol group in ring A, a structural feature common to a number of quassinoids, and a β-hydroxyl at C-15.

(2) Karinolide (13),[34] shinjulactone B (14),[7] and shinjulactone C (15)[12] are the only three quassinoids so far known which have a primary alcohol at C-8. Their structures have been established by X-ray analysis.

Karinolide (13) is the first example of a natural quassinoid to have a cyclohexadienone system and a bis-hemiketal function between C-1 and C-11. Shinjulactone B (14) is in fact a C_{19} compound which has lost a carbon atom in ring A, probably via a 1,2-dioxo compound; it has also a contracted B ring. Shinjulactone C (15), a non-bitter quassinoid, has a quite original structure with the inversed configuration at C-9, and 1,12 and 5,13 linkages.

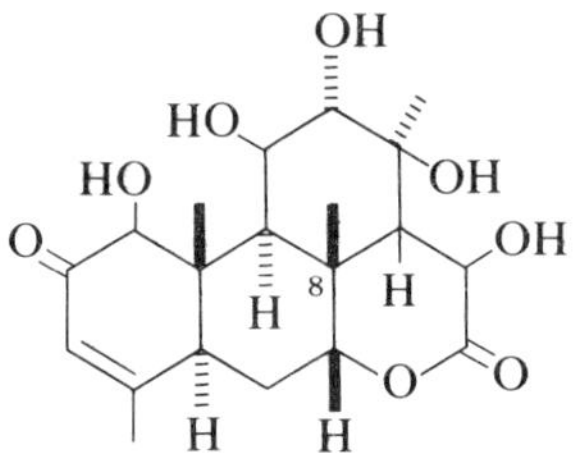

(12) Soulameanone

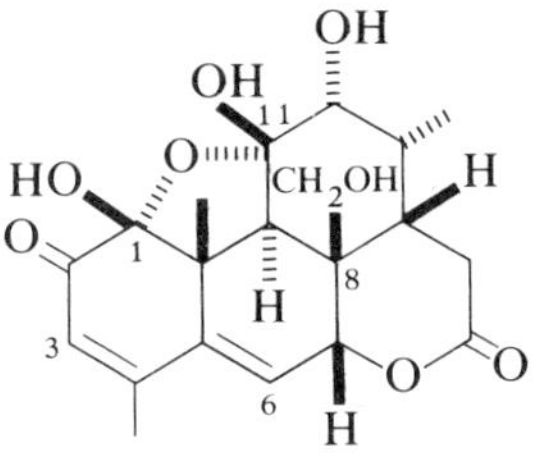

(13) Karinolide

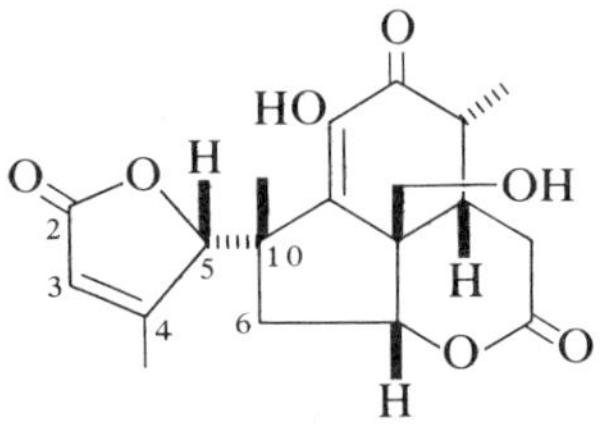

(14) Shinjulactone B

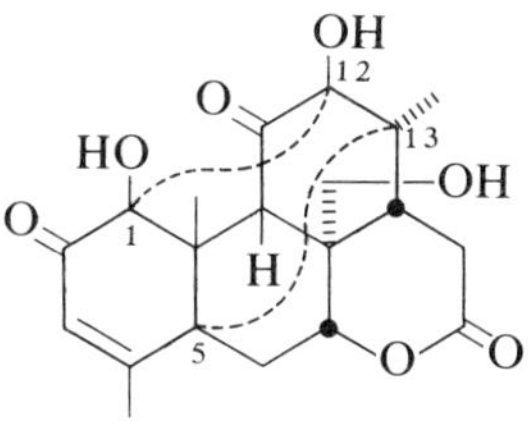

(15) Shinjulactone C

(3) Glaucarubolone (16)[30] occurs as a number of C-15 esters (17–21) in which the hydroxymethyl at C-8 forms a hemiketal bridge to C-11. Castelanone (20)[40] and soularubinone (21)[45] are new quassinoids.

These quassinoids display significant antineoplastic activity *in vivo* against the P-388 murine lymphocytic leukemia. They possess the structural requirements for this activity which are principally an α,β-unsaturated ketol group in ring A which may react with biological nucleophiles,[15] secondly the epoxymethano bridge which confers a particular conformation to ring C, and thirdly the ester chain at C-15 representing the lipophilic part of the molecule, important for the transport of the drug through the cell membranes[19] (see below).

(16) Glaucarubolone (R = H)

(17) Glaucarubinone (R = COCMeEtOH)

(18) Holocanthone (R = COMe)

(19) Ailanthinone (R = COCHMeEt)

(20) Castelanone (R = COCH$_2$CHMe$_2$)

(21) Soularubinone (R = COCH$_2$CMe$_2$OH)

(4) Until 1970 only one quassinoid was known to have a hydroxyl at position 6, namely 6-hydroxypicrasin B **(22)**.[30] This structure was later confirmed by X-ray analysis.[27] Since then the quassinoids **23**[48] and **24**,[38,41] which have the 6-hydroxyl esterified with tiglic acid and senecioic acid, respectively, have been isolated. Compounds **23** and **24** display significant antileukemic activity. This finding shows that the ester chain required for biological activity can be located either at C-15 or C-6. The quassinoids **23** and **24** are found together with inactive compounds which possess an α-glycol group instead of an α-ketol group in ring A (**23**: R^1 = OH, R^2 = H; **24**: R^1 = OH, R^2 = H).[34,38]

(22) 6-Hydroxypicrasin B

(23) R^1,R^2 = O; R^3 = CMe=CHMe
(24) R^1,R^2 = O; R^3 = CH=CMe$_2$

(25) Undulatone (R = COCMe=CHMe)

Recently another cytotoxic quassinoid, undulatone (**25**),[49] was isolated. It has ester groups at both C-15 and C-6.

(5) The bruceolide esters are the only quassinoids not to have oxygenated functions at positions 1 and 2 but at 2 and 3. They have a diosphenol group at these positions and, in ring C, the hydroxymethyl at C-8 forms an oxide bridge to C-13.

Bruceine A (**27**), B (**28**), and C (**29**), isolated in 1966 from *Brucea amarissima*, were the first C-15 esters of bruceolide (**26**) to have their structures determined.[31] These quassinoids differ in the nature of the ester group at C-15. They are esters of isovaleric, acetic, and 3,4-dimethyl-4-hydroxy-2-pentenoic acid, respectively.

Later, Kupchan *et al.*[14] isolated from *Brucea antidysenterica* several C-15 esters of bruceolide (**26**) among which were two potent antileukemic principles, bruceantin (**30**) and bruceantinol (**31**). Bruceantin is now undergoing clinical trial by the U.S. National Cancer Institute. The structure of bruceine C (**29**), with the *trans* configuration of the double bond on the ester chain, was originally attributed to bruceantinol.[14] The antileukemic activity of the bruceolide derivatives varies greatly with the nature of the ester substituent. An unambiguous structure determination of bruceine C (**29**) and bruceantinol was therefore necessary. Bruceine C proved to have the structure represented in **29**, and bruceantinol is in fact 4'-*O*-acetylbruceine C (**31**).[37]

The markedly higher antitumour activity of bruceantin and bruceantinol versus bruceine C could possibly be attributed to a greater lipophilicity in the side-chains of **30** and **31**.

(6) Four quassinoid glycosides have been isolated to date: bruceoside A (**32**),[17] bruceoside B (**33**),[17] bruceantinoside A (**34**),[26] and bruceantinoside B (**35**).[26] These quassinoids show significant antileukemic activity.

(7) Sergeolide (**36**)[20] is the first natural quassinoid to possess a butenolide function. It was isolated along with the known isobruceine B (**37**)[14] from the French Guyanan Simaroubaceae, *Picrolemma pseudocoffea*. The structure of sergeolide was established from an analysis of the 400 MHz ^{1}H NMR data and the ^{13}C NMR spectrum.

The structural similarity of **37** and **36** suggests that the latter arises from isobruceine B via the intermediacy of its 1-*O*-acetyl derivative. A Claisen type rearrangement (see below) followed by the appropriate double bond migration and lactonization would lead to **36**.

Sergeolide displays significant antileukemic activity at low dose (0.5 mg/kg). At slightly higher doses it was highly cytotoxic. It is of interest to note that the butenolide function which might be expected to act as a potent Michael acceptor should confer such a high degree of cytotoxicity to the molecule.

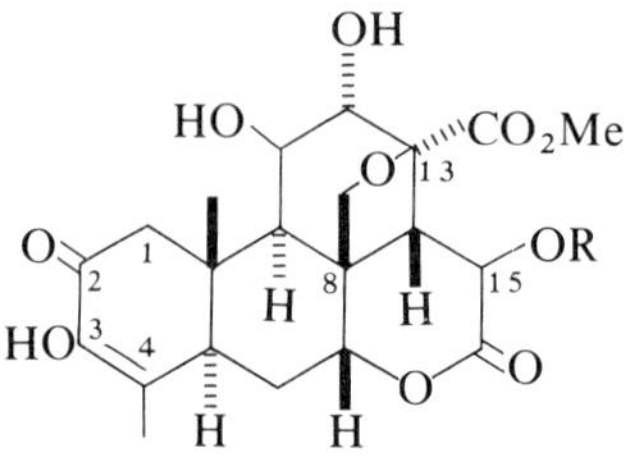

Bruceolide (R = H) (26)
Bruceine A (R = COCH$_2$CHMe$_2$) (27)
Bruceine B (R = Ac) (28)
Bruceine C (R = COCH=CMeCMe$_2$OH) (29)
Bruceantin (R = COCH=CMeCHMe$_2$) (30)
Bruceantinol (R = COCH=CMeCMe$_2$OAc) (31)

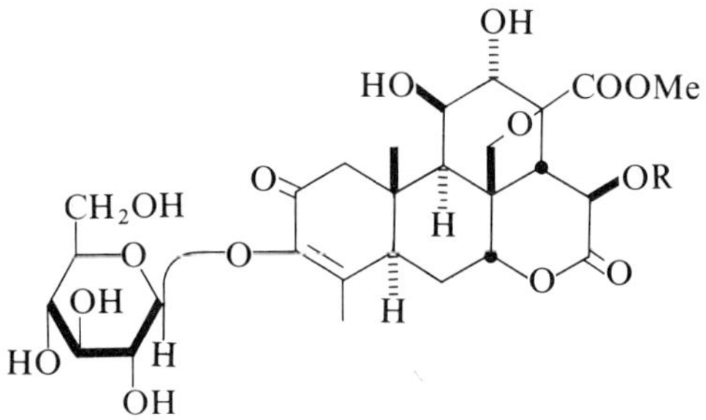

(32) Bruceoside A (R = COCH=CMe$_2$)
(34) Bruceantinoside A (R = COCH=CMeCHMe$_2$)

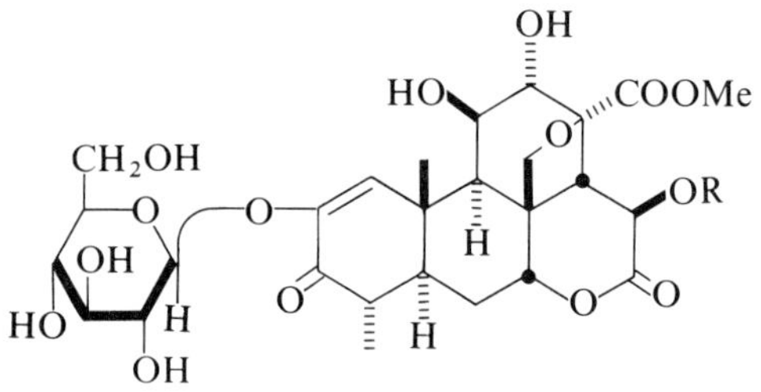

(33) Bruceoside B (R = COCH=CMe$_2$)
(35) Bruceantinoside B (R = COCH=CMeCHMe$_2$)

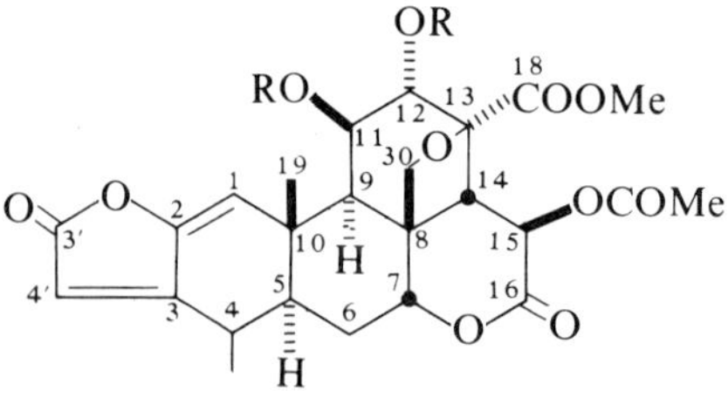

(36) Sergeolide (R = H)

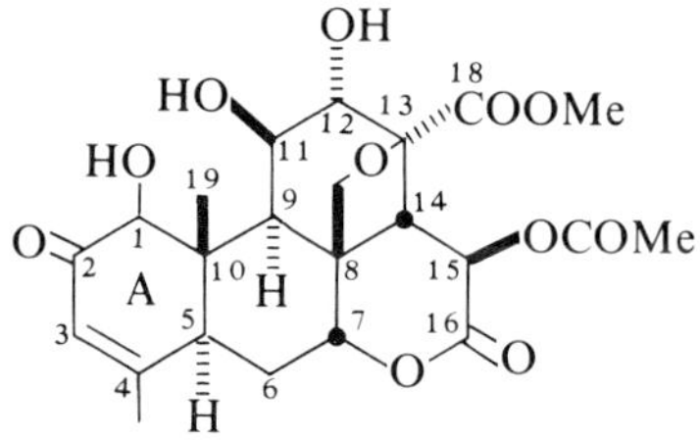

(37) Isobruceine B

V. C$_{19}$ Quassinoids

No new C$_{19}$ quassinoids have been isolated during the last decade. Six of the seven known examples which can be illustrated by the structure of samaderine C (**38**)[30] have the γ-lactone ring linked to the hydroxyl at C-12. However, there was one C$_{19}$ quassinoid, eurycomalactone, which was assigned structure **39**.[18,24] Quite recently its structure was revised to **40** as a result of an X-ray analysis.[22] Thus, all the C$_{19}$ quassinoids have their γ-lactone ring linked in the same manner.

(**38**) Samaderine C   (**39**) Eurycomalactone   (**40**) Eurycomalactone (revised structure)

VI. C$_{18}$ Quassinoids

Samaderine A (**41**),[50] laurycolactone A (**42**),[22] and laurycolactone B (**43**)[22] are the only three C$_{18}$ quassinoids so far known.

(41) Samaderine A   (42) Laurycolactone A   (43) Laurycolactone B

VII. Physical Methods

During the early structural studies 60 MHz ^{1}H NMR spectroscopy and electron impact mass spectroscopy were the most important physical tools used. Despite the great amount of information obtained by these physical methods, the structural elucidation relied heavily on chemical transformations, degradations, and correlations. Prior to 1970, of the 55 quassinoids known only two structures were solved by X-ray analysis.

In recent years higher field ^{1}H NMR spectroscopy and new mass spectroscopic methods (chemical ionization, field desorption, and "Mikes" technique) became available. Recently Baldwin et al.[1] applied these new mass spectroscopic techniques to the structural determination of some bruceolides. A detailed ^{13}C NMR study of a number of quassinoids has been published[32] and ^{13}C NMR data have since then been recorded in most of the papers dealing with constituents of Simaroubaceae. X-Ray analysis has also become more accessible and many quassinoids now have their structures established by this method.

VIII. Biological Activity

(1) Antileukemic activity. As already mentioned certain quassinoids display in vivo antileukemic activity. In Table 1 are listed some compounds which are most active against the P-388 murine lymphocytic leukemia.[3] T/C is the ratio of test group survival to control group survival in tumoured animals, expressed as a percentage. When this ratio is equal to or greater than 120, compounds are considered to be active. Bruceantin (30) was put on clinical trial in U.S.A. a few years ago, as already noted. It shows activity over a wide range of doses and in addition is active against solid tumours.

TABLE 1

Antileukemic activity

Quassinoids	Optimal dose (mg/kg)	T/C
Glaucarubinone (17)	0.25	177
Ailanthinone (19)	2.0	148
Tigloyloxy-6α-chaparrinone (24)	0.6	163
Simalikalactone D (44)[30]	1.0	198
Bruceantin (30)	0.5	220
Bruceantinol (31)	1.0	238
Isobruceine A (45)[33]	2.0	163

As to the mode of action of these compounds, Liao et al.,[19] using HeLa cells, have shown that the antitumour activity of the bruceolide (26) esters at the molecular level is due to the irreversible inhibition of protein synthesis.

Studies of the structure/activity relationships for several quassinoids[16,47] have established some of the structural requirements for optimal antineoplastic activity, particularly in the P-388 mouse leukemia. These requirements are principally the following: (a) ring A with either an α,β-unsaturated ketol group at position 1 and 2 or a diosphenol group at position 2 and 3; (b) ring C with an epoxymethano bridge between C-8 and C-11, or between C-8 and C-13; (c) the presence of a free hydroxyl group in ring A and at C-12 in addition to an ester group at C-15 and/or C-6.

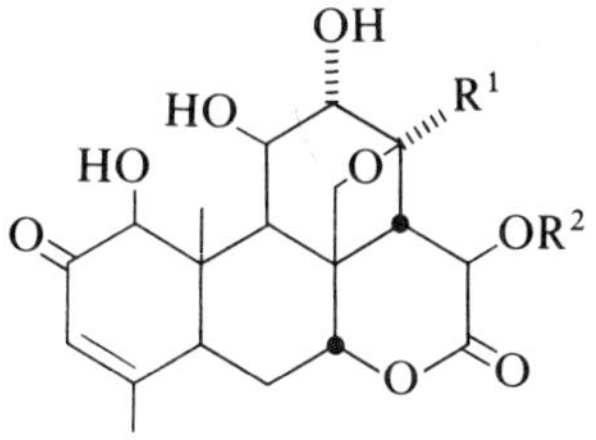

(44) Simalikalactone D (R^1 = Me; R^2 = COCHMeEt)

(45) Isobruceine A (R^1 = COOMe; R^2 = COCH$_2$CHMe$_2$)

(2) *Antiviral activity.* Certain quassinoids display *in vitro* antiviral activity, namely against the oncogenic Rous sarcoma virus.[28] This test is performed as follows: chick-embryo fibroblasts are infected by a known amount of the virus which transforms the morphology of the cells into foci. A number of quassinoids inhibit this transformation at concentrations ranging from 0.15 to 1 μg/ml, without having toxic effects on normal cells. Some of the results are shown in Table 2.

TABLE 2

Effect of some quassinoids on RSV-induced foci formation

	I_{50} (μg/ml)	I*max (μg/ml)	Inhibition (%)
Isobruceine A (**45**)	0.10	0.25	94
Simalikalactone D (**44**)	0.78	1.00	88
Chaparrinone (**46**)	0.10	0.15	76
Glaucarubolone (**16**)	0.15	0.30	96
Castelanone (**20**)	0.16	0.20	83

* Higher concentrations were toxic to normal cells.

While this inhibition does not guarantee antileukemic activity, compounds which do not display this antitransforming behaviour are invariably inactive.

(3) *Antimalarial activity.* The growth of the blood parasite *Plasmodium falciparum*, responsible for malaria, is markedly inhibited by certain quassinoids.[42] The most active compound, simalikalactone D (**44**), gave complete inhibition at 0.002 μg/ml. Glaucarubinone (**17**) and soularubinone (**21**) were equally effective at 0.006 μg/ml, whereas chaparrinone (**46**) and simarolide (**7**) had little effect even at 0.01 μg/ml. These relative activities are parallel to the antineoplastic activities of these compounds. Indeed, the two last mentioned quassinoids (**46** and **7**) do not possess the structural requirements for antileukemic activity.

(4) *Antifeedant and insecticidal properties.* A great number of quassinoids were found to act as antifeedants against the Mexican Bean Beetle (*Epilachna varvestis*) whereas only a small number, in particular simalikalactone D (**44**), isobruceine A (**45**), and glaucarubinone (**17**), proved to be active against the Southern Army Worm (*Spodoptera eridania*). Simalikalactone D (**44**) is the most potent antifeedant and is active on the level of 50 ppm.*

We have also studied the antifeedant and insecticidal behaviour of quassinoids—soulameolide (**9**), simarolide (**7**), soulameanone (**12**), chaparrinone (**46**), glaucarubinone (**17**), bruceine A (**27**), and bruceine B (**28**)—against the third stage larva of *Locusta migratoria*. The first three quassinoids do not display any antifeedant activity, whereas the others showed moderate antifeedant behaviour. On the contrary, they showed significant insecticidal activity, especially glaucarubinone (**17**) and bruceine B (**28**).[23]

(5) *Amoebicidal activity.* Extracts of many *Simarouba* species including

* These tests were performed by V. Leskinen at Professor Nakanishi's laboratories (U.S.A.), unpublished results.

Castela nicholsoni or *Chaparro amargoso* (Castamargina) have long been used by the local population in Mexico, China, etc. to treat fevers, dysentery, and amoebiasis. However, studies of the antiamoebic activity of pure quassinoids have been limited. Only two, ailanthone[6] and glaucarubin,[5] were shown to be effective amoebicides. Quite recently Gillin *et al.*[8] made an extensive study of the *in vitro* activity of 17 quassinoids against the parasite *Entamoeba histolytica*. Seven of the quassinoids tested were active; they are, in order of their decreasing activity, bruceantin, simalikalactone D, ailanthinone, glaucarubolone, glaucarubinone, ailanthone, and glaucarubin.

Apparently, there is no simple relationship between structure and activity of the seventeen quassinoids tested.

IX. CHEMICAL MODIFICATIONS

(1) Quassin (1) was modified, in five steps, by introducing at position 15 the bruceantin ester side-chain.[21] The compound obtained was found inactive against growth of HeLa cells.

(2) Okano and Lee[25] converted the quassinoid glycoside bruceoside A (32) to bruceantin (30). The methods used involved selective hydrolysis of the ester and glycosyl groups followed by esterification of the hydroxyl at position 15.

(3) Chaparrinone (46) which possesses most of the structural requirements for *in vivo* antileukemic activity (PS test system) but lacks an ester side-chain at C-15 and/or at C-6, does not display any significant antineoplastic activity. In a search for direct methods of quassinoid functionalization benzeneselininic anhydride was considered as a possible reagent for the introduction of a double bond in the lactonic ring of chaparrinone.[13] The synthetic pathway is summarized in Scheme 2.

Chapparrinone triacetate (47) on reaction with benzeneselininic anhydride afforded the desired $\Delta^{14,15}$ dehydro compound (48). In addition, two other products (49 and 50) arising, interestingly enough, from angular hydroxylation at C-5, were isolated. Deacetylation of these three products gave the free alcohols 51, 52, and 53 which however did not show any significant activity. Further preliminary experiments to introduce an oxygen function at C-15 remained unsuccessful.

(4) Chaparrin (54) is inactive and can be isolated in relatively great quantities from the Mexican Simaroubaceae. A study aimed at the development of synthetic procedures for the conversion of chaparrin into C-15 gluacarubolone esters and simple analogues in which the C-15 ester side-chain has been replaced by an alkyl group was initiated.[2] The synthetic technique is summarized in Scheme 3.

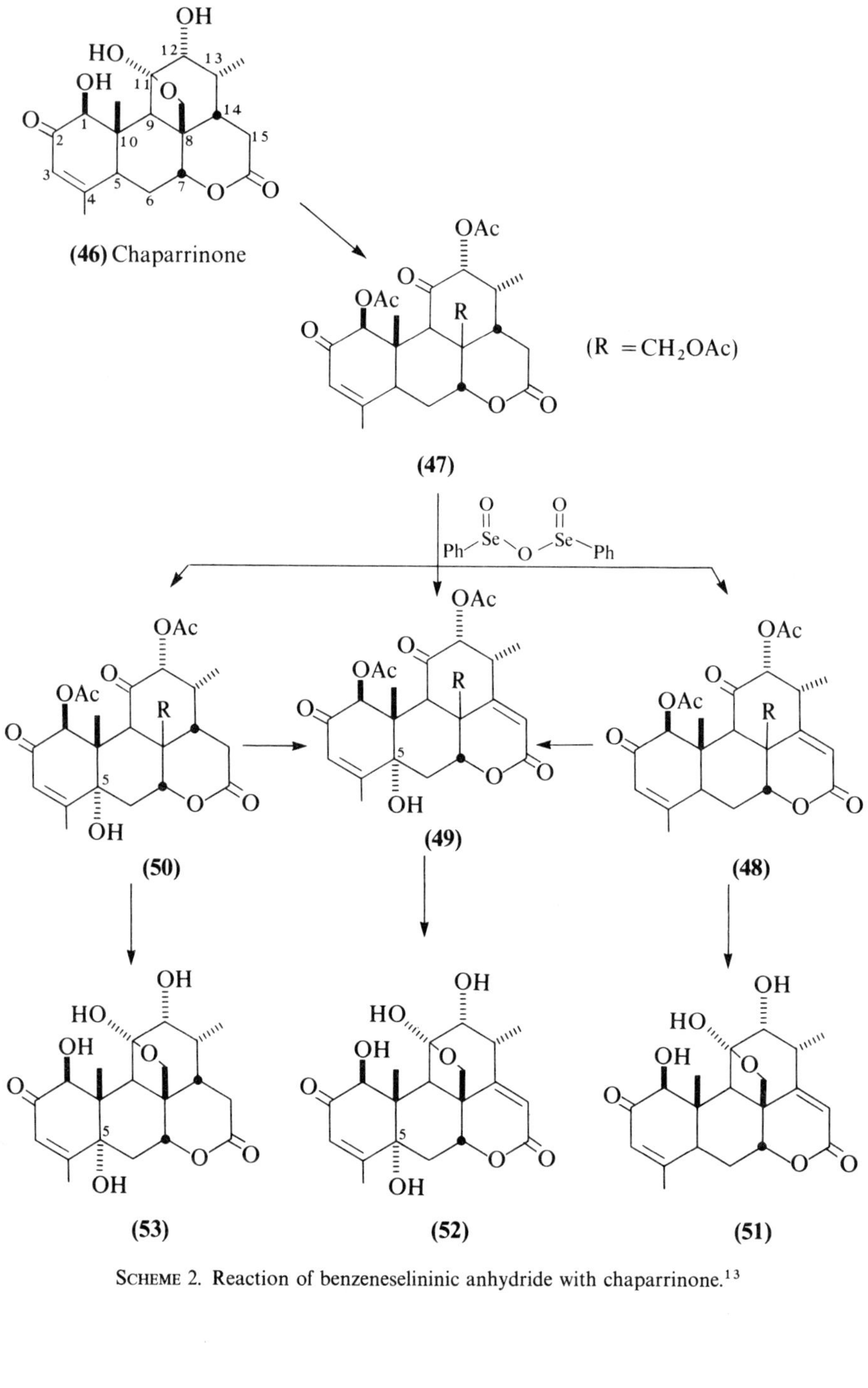

SCHEME 2. Reaction of benzeneselininic anhydride with chaparrinone.[13]

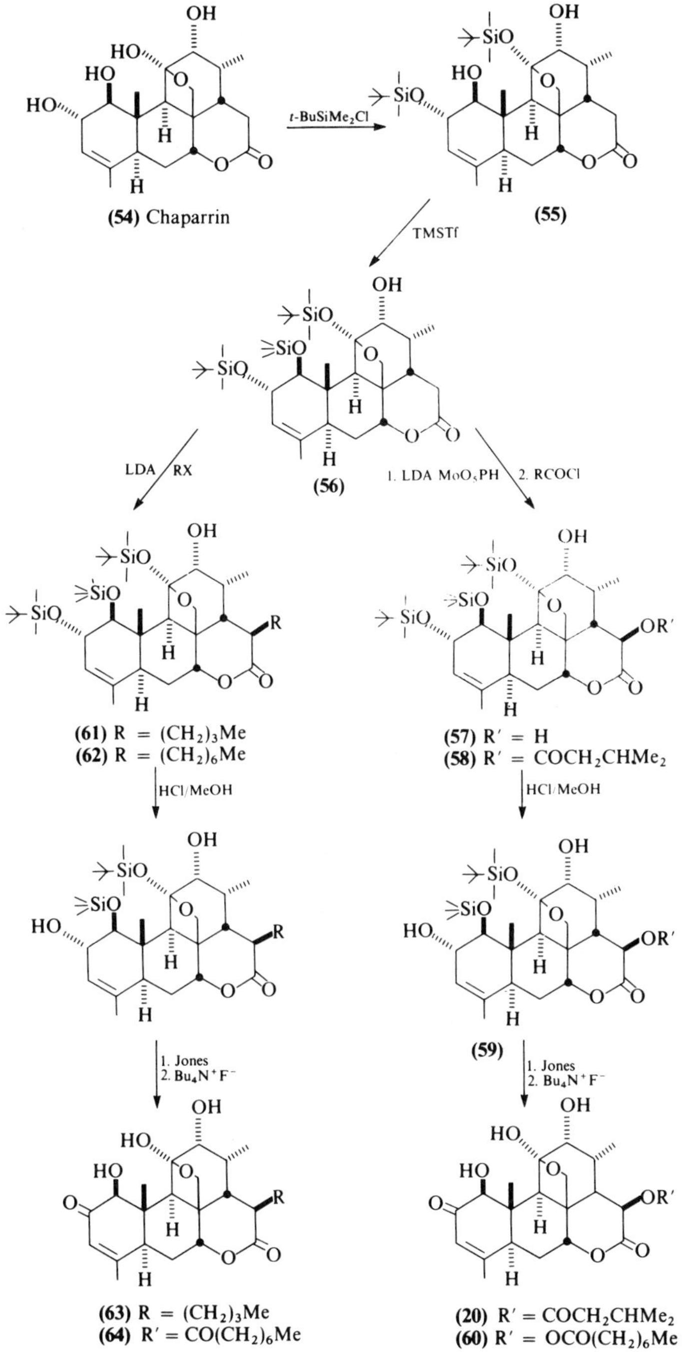

SCHEME 3. Synthetic route for the conversion of chaparrin into castelanone (20) and quassinoid analogues 60, 63, and 64.

Reaction of chaparrin (54) with t-butyldimethylsilyl chloride afforded the crystalline disilyl derivative (55). The hydroxyl function at C-1 was effectively protected using trimethylsilyl triflate to afford the trisilyl lactone (56) which upon treatment with lithium diisopropylamide (LDA) and subsequent exposure to MoO_5–pyridine–HMPA (MoO_5PH)[46] gave the required 15-hydroxy-lactone (57). Treatment of the latter with isovaleryl chloride afforded the crystalline ester 58 which was selectively desilylated to 59. Oxidation of the free allylic hydroxyl and complete desilylation of the resulting disilyl enone with tetrabutylammonium fluoride (Bu_4NF) afforded the natural cytotoxic quassinoid castelanone (20).

This reaction sequence could be applied to the synthesis of the quassinoid analogues 15-O-octanoyl glaucarubolone (60) and the 15-butyl and 15-heptyl chaparrinones (63 and 64). The alkyl analogues have been prepared by treatment of the lithium enolate derived from 56 with the corresponding alkyl halides; the alkyl lactones 61 and 62, obtained in greater than 90 % yield, were treated in an analogous manner as 58.

Compounds 60, 63, and 64 cause significant inhibition of cell transformation induced by Rous sarcoma virus at the 1 μg/ml level. Further work is under way to assess potential antineoplastic activity of these substances.

Several schools are presently involved in studies of total synthesis of quassinoids which have culminated in the recent elegant syntheses of quassin[9] and of castelanolide.[10]

REFERENCES

1. Baldwin, A. M., Carter, D. M., Darwish, F. A. and Phillipson, J. D. (1981). *Biomedical Mass Spectrometry* 8, 362.
2. Caruso, A. J., Polonsky, J. and Rodriguez, B. S. (1982). *Tetrahedron Lett.* 2567.
3. Cassady, J. M. and Suffness, M. (1980). *In* "Anticancer Agents based on Natural Product Models", pp. 207–269. Academic Press, New York.
4. Connolly, J. D., Overton, K. H. and Polonsky, J. (1970). *In* "Progress in Phytochemistry" (Reinhold, L. and Liwschitz, Y., eds.), Vol. II, pp. 385–455. Interscience.
5. Cuckler, A. C., Kuna, S. and Mushett, C. W. (1958). *Arch Int. Pharmacodyn.* 114, 307.
6. De Carneri, I. and Casinovi, C. G. (1968). *Parassitologia* 10, 215.
7. Furuno, T., Naora, H., Murae, T., Hirota, H., Tsuyuki, T., Takahashi, T., Itai, A., Iitaka, Y. and Matsushita, K. (1981). *Chem. Lett.* 1797.
8. Gillin, F. D., Reiner, D. S. and Suffness, M. (1982). *In* "Antimicrobial Agents and Chemotherapy.", 22, 342.
9. Grieco, P. A., Ferrino, S. and Vidari, G. (1980). *J. Am. Chem. Soc.* 102, 7586.

10. Grieco, P. A., Lis, R., Ferrino, S. and Jaw, J. Y. (1982). *J. Org. Chem.* **47**, 601.
11. Hikino, H., Ohta, T. and Takemoto, T. (1975). *Phytochemistry* **14**, 2473.
12. Ishibashi, M., Murae, T., Hirota, H., Tsuyuki, T., Takahashi, T., Itai, A. and Iitaka, Y. (1982). *Tetrahedron Lett.* 1205.
13. Khôi, N. and Polonsky, J. (1981). *Helv. Chim. Acta* **64**, 1540.
14. Kupchan, S., Britton, R. W., Lacadie, J. A., Ziegler, M. F. and Sigel, C. W. (1975). *J. Org. Chem.* **40**, 648.
15. Kupchan, S. M. and Lacadie, J. A. (1975). *J. Org. Chem.* **40**, 654.
16. Kupchan, S. M., Lacadie, J. A., Howie, G. A. and Sickles, B. R. (1976). *J. Med. Chem.* **19**, 1130.
17. Lee, K. H., Imakura, Y., Sumida, Y., Wu, R. Y., Hall, I. H. and Huang, H. Ch. (1979). *J. Org. Chem.* **44**, 2180.
18. Le-Van-Thoi and Nguyen-Ngoc-Suong (1970). *J. Org. Chem.* **35**, 1104.
19. Liao, L. L., Kupchan, S. M. and Horwitz, S. B. (1976). *Molec. Pharmacol.* **12**, 167.
20. Moretti, C., Polonsky, J., Vuilhorgne, M. and Prangé, T. (1982). *Tetrahedron Lett.* 647.
21. Murae, T. and Takahashi, T. (1981). *Bull. Chem. Soc. Jpn.* **54**, 941.
22. Nguyen-Ngoc-Suong, Bhatnagar, S., Polonsky, J., Vuilhorgne, M., Prangé, T. and Pascard, C. (1982). *Tetrahedron Lett.* 5159.
23. Odjo, A., Piart, J., Polonsky, J. and Roth, M. (1981). *Compt. Rend. Acad. Sci.* **293C**, 241.
24. Oei-Koch, A. and Kraus, Lj. (1980). *Sci. Pharm.* **48**, 110.
25. Okano, M. and Lee, K. H. (1981). *J. Org. Chem.* **46**, 1138.
26. Okano, M., Lee, K. H., Hall, J. H. and Boettner, F. E. (1981). *J. Nat. Prod.* **44**, 470.
27. Pascard, C., Prangé, T. and Polonsky, J. (1977). *J. Chem. Res.* (S) 324.
28. Pierré, A., Robert-Géro, M., Tempête, C. and Polonsky, J. (1980). *Biochem. Biophys. Res. Comm.* **93**, 675.
29. Polonsky, J. (1973). *In* "Recent Advances in Phytochemistry," Vol. 6, pp. 31–64. Academic Press, New York.
30. Polonsky, J. (1973). *Fortschr. Naturst.* **30**, 101.
31. Polonsky, J., Baskevitch, Z., Gaudemer, A. and Das, B. C. (1967). *Experientia* **23**, 424.
32. Polonsky, J., Baskevitch, Z., Gottlieb, H. E., Hagaman, E. W. and Wenkert, E. (1975). *J. Org. Chem.* **40**, 2499.
33. Polonsky, J., Baskevitch-Varon, Z. and Sevenet, T. (1975). *Experientia* **31**, 1113.
34. Polonsky, J., Gallas, J., Varenne, J., Prangé, T., Pascard, C., Jacquemin, M. and Moretti, C. (1982). *Tetrahedron Lett.* 869.
35. Polonsky, J., Van Tri, M., Prangé, T., Pascard, C. and Sevenet, T. (1979). *J. Chem. Soc. Chem. Comm.* 641.
36. Polonsky, J., Van Tri, M., Varon, Z., Prangé, T., Pascard, C., Sevenet, T. and Pusset, J. (1980). *Tetrahedron* **36**, 2983.
37. Polonsky, J., Varenne, J., Prangé, T. and Pascard, C. (1980). *Tetrahedron Lett.* 1853.
38. Polonsky, J., Varon, Z., Moretti, C., Pettit, G. R., Herald, C. L., Ridiout, J. A., Saha, S. B. and Khastgir, H. N. (1980). *J. Nat. Prod.* **43**, 503.
39. Polonsky, J., Varon, Z., Prangé, T., Pascard, C. and Moretti, C. (1981). *Tetrahedron Lett.* 3605.
40. Polonsky, J., Varon, Z. and Soler, E. (1979). *Compt. Rend. Acad. Sci.* **288C**, 269.
41. Seida, A., Kinghorn, A. D., Cordell, G. A. and Farnsworth, N. (1978). *J. Nat. Prod.* **41**, 584.

42. Trager, W. and Polonsky, J. (1981). *Amer. J. Trop. Med. Hygeine* **30**, 531.
43. Valenta, Z., Gray, A. M., Orr, D. E., Papadopoulos, S. and Podešva, C. (1962). *Tetrahedron* **18**, 1433.
44. Valenta, Z., Papadopoulos, S. and Podešva, C. (1961). *Tetrahedron* **15**, 100.
45. Van-Tri, M., Polonsky, J., Merrienne, C. and Sevenet, T. (1981). *J. Nat. Prod.* **44**, 279.
46. Vedejs, E., Engler, D. A. and Telschow, J. E. (1978). *J. Org. Chem.* **43**, 188.
47. Wall, M. E. and Wani, M. C. (1978). *J. Med. Chem.* **21**, 1186.
48. Wani, M. C., Taylor, H. L., Thompson, J. B. and Wall, M. E. (1978) *J. Nat. Prod.* **41**, 578.
49. Wani, M. C., Taylor, H. L., Thompson, J. B., Wall, M. E., McPhail, A. T. and Onan, K. D. (1979). *Tetrahedron* **35**, 17.
50. Wani, M. C., Taylor, H. L. and Wall, M. E. (1977). *J. Chem. Soc. Chem. Comm.* 295.

CHAPTER 9

Structural Diversity and Distribution of Lignans in the Rutales

JOHN O'SULLIVAN

School of Pharmacy, Trinity College, Dublin, Eire

I. INTRODUCTION

Lignans are a diverse group of compounds formed by the coupling of two phenylpropanoid (C_6C_3) units. Very few tracer studies have been carried out to determine the biosynthetic pathways to the lignans. However, consideration of structures of known compounds has led Birch and Liepa[14] to propose a feasible biosynthetic route. Within the Rutales they have, so far, been found only in the Rutaceae and Burseraceae. The compounds encountered may be classified into five groups: 1,4-diarylbutanes; substituted furans; dibenzylbutyrolactones; naphthalenes (tetrahydro, dihydro, aryl); 2,6-diaryl-3,7-dioxabicyclo[3.3.0]octanes. Systematic nomenclature of the lignans as a whole is complex and, in some cases, confused. In this review, therefore, the nomenclature of each group is treated separately.

II. 1,4-DIARYLBUTANES

This group represents the simplest class of lignans. To date, only one example of this type has been isolated from the Rutales, viz. 2,3-bis-(3,4-methylenedioxybenzyl)-butane-1,4-diol diacetate (1) from *Bursera arida*.[37]

267

The compound phebalarin (see Chapter 17), although with the basic structure of a lignan, is possibly an artefact.

(1)

III. SUBSTITUTED FURANS

This type has been encountered in both Rutaceae and Burseraceae (Table 1). The absolute configuration of only one of these compounds, brassilignan[54] (1-4), has been determined. In general, however, a combination of spectroscopic techniques and comparison with compounds of known stereochemistry may be useful in the assignment of absolute configuration.[44]

IV. DIBENZYLBUTYROLACTONES

This class of compound is of widespread occurrence in the plant kingdom.[43] Unlike the substituted furans (Table 1) the generally accepted numbering of the dibenzylbutyrolactones[15,51,56] assigns position 1 to the carbonyl function (Table 2). ^{1}H NMR studies[20] have been used successfully to distinguish cis/trans isomerism across positions 2 and 3 while MS can be used[51] to differentiate the benzylic group attachments at positions 2 and 3. The absolute stereochemistry, where known, of the compounds isolated from the Rutales is indicated in Table 2.

V. NAPHTHALENES

Considerable interest in the naphthalene type lignans (Tables 3–5) has been generated in recent years owing to the marked physiological actions exhibited by these compounds. A wide range of the tetrahydronaphthalene (aryltetralin) lignans have cytotoxic and/or antitumour activity.[23] The arylnaphthalenes diphyllin[53] (5-4) and justicidins A (5-5) and B[52] (5-2) are fish poisons. The naphthalene skeleton is thought to arise from a head-to-head oxidative dimerization of phenylpropanoid (C_6C_3) units followed by a further cyclization involving C–C bond formation. In view of the frequent occurrence of a *para*-hydroxy group in these precursors, Birch and Liepa[14] have suggested that this latter cyclization occurs through a spirodienone rearrangement which may generate the more highly strained *cis*-1,2-*trans*-2,3 ($1\alpha,2\alpha,3\beta$) isomer, the configuration of the more common naturally occurring tetrahydronaphthalenes (Table 3).

Several conventions for the systematic numbering of the naphthalene type lignans exist. Priority in numeration, on one hand, has been given to the point of attachment of the pendant aryl ring[13,23,37] while other authors[24] continue to use the system whereby priority is given to an oxygen function attachment in ring A. Confusion is also encountered in existing nomenclatural convention. This topic has been discussed by Dewick and Jackson.[23] They have proposed a system using α,β convention which clarifies the nomenclature of the podophyllotoxin series and other aryltetralin lignans. This convention has been adopted in Table 3 to indicate the configuration, where known, of these compounds.

The structure elucidation of the naphthalenes has been comprehensively reviewed by Ayres.[5]

VI. 2,6-DIARYL-3,7-DIOXABICYCLO[3.3.0]OCTANES

This lignan group is widely distributed in nature but within the Rutales it is, so far, confined to the Rutaceae. They are biosynthetically derived by the head-to-tail dimerization/cyclization of a methinequinone (C_6C_3) intermediate, recently discussed by Birch and Liepa.[14] The unstrained *cis* configuration of the fused heterocyclic rings has been confirmed by X-ray analysis[48,78] and by synthesis.[61] The stereochemistry of the aryl groups varies in different lignans, giving rise to three structural types: diequatorial (I), axial/equatorial (II) (Table 6); and diaxial (III) (2). No example of the diaxial isomer has yet been found in the Rutaceae. The axial/equatorial notation has been proposed by Pelter[59] to indicate the configuration of the aryl substituents attached at

(2)

positions 2 and 6. In this context "axial" describes the configuration of a substituent occupying a position approximately parallel to the axis of the molecule and "equatorial" a position approximately perpendicular (see **2**). Problems arise in the structure elucidation of compounds of type II whenever the aryl groups are differently substituted. These problems can usually be surmounted by a combination of techniques including the Gibb's test and acetylation for phenols,[57,59] ^{1}H NMR[59] and in combination with lanthanide shift reagent (LSR),[32,57] ^{13}C NMR,[59] MS,[59] and CD.[35]

The compounds encountered, thus far, are shown in Table 6.

VII. CHEMOTAXONOMY

The presently known distribution of lignans in the Rutales is shown in Table 7 which also gives the codes used for species in Tables 1–6. From the limited data available few taxonomic inferences can be drawn. Butyrolactones, arylnaphthalenes, and 2,6-diaryl-3,7-dioxabicyclo[3.3.0]octanes are also found in the Compositae,[33] tetrahydronaphthalenes in the Umbelliferae,[34] and several types in the Ranales. Within the Rutales *Bursera* has only yielded the naphthalene type and within the Rutaceae these are characteristic of the *Ruta/Haplophyllum* group. The 2,6-diaryl-3,7-dioxabicyclo[3.3.0]octanes, on the other hand, seem particularly characteristic of *Zanthoxylum*.

TABLE 1

Substituted furans

Name	Structure						Occurrence
		R^1	R^2	R^3	R^4	R^5	
1. Acanthotoxin	II:	$-CH_2-$		$-CH_2-$		H	Za-1[70]
2. Sanshodiol	III:	$-CH_2-$		Me	H	H	Za-15[2]
3. Podotoxin	II:	$-CH_2-$		Me	Me	H	Za-1[18]
4. Brassilignan (3R,4R)	I:	Me	Me	Me	Me	H	Fl-1[54]
5. Burseran	I:	$-CH_2-$		Me	Me	OMe	Bu-3[19]

TABLE 2

Dibenzylbutyrolactones

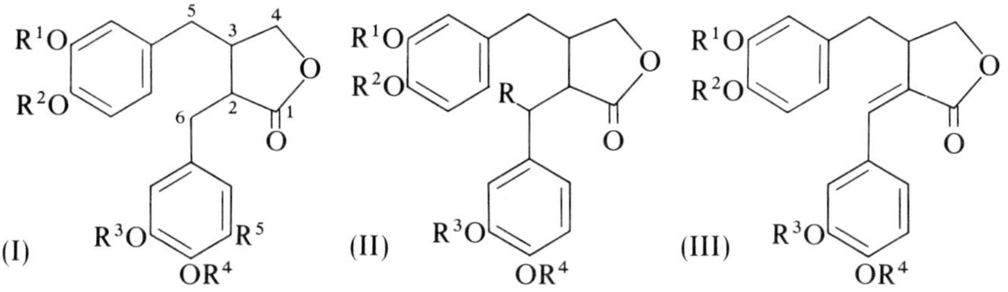

Name		Structure						Occurrence
		R	R^1	R^2	R^3	R^4	R^5	
1. Savinin (3R)	III:	H	$-CH_2-$		$-CH_2-$		H	Am-1[6], Ch-1[11], Ru-1[66], Ru-2[49] Ru-4[69], Ru-5[50] Za-17[20]
2. Hinokinin (2R,3R)	I:	H	$-CH_2-$		$-CH_2-$		H	Ch-1[11]
3. Pluviatolide (2R,3R)	I:	H	$-CH_2-$		Me	H	H	Za-17[20]
4. Suchilactone (3R)	III:	H	Me	Me	$-CH_2-$		H	Ha-7[63]
5. — (2R,3R)	I:	H	$-CH_2-$		Me	Me	H	Bu-5[51,55]
6. Sventenin	II:	OH	$-CH_2-$		$-CH_2-$		H	Am-1[6], Ru-2[30,49]
7. Artigenin methyl ether (2R,3R)	I:	H	Me	Me	Me	Me	H	Pt-1[68], Za-14[74]
8. — (2R,3R)	I:	H	$-CH_2-$		Me	Me	OMe	Bu-5[51,56]
9. Sventenin acetate	II:	OAc	$-CH_2-$		$-CH_2-$		H	Ru-4[30]

TABLE 3

Tetrahydronaphthalenes

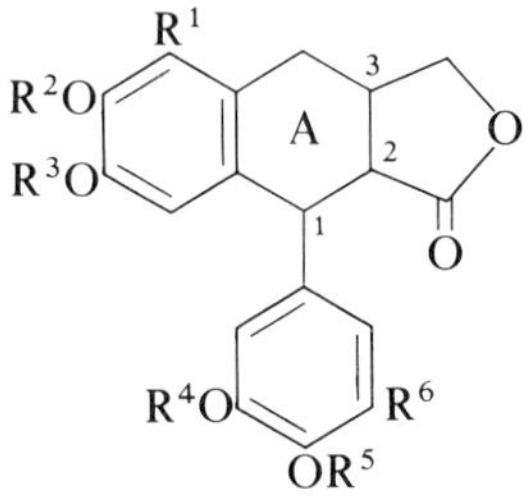

Name	R^1	R^2	R^3	R^4	R^5	R^6	Occurrence
1. Morelensin (1α,2α,3β)	H	–CH$_2$–		–CH$_2$–		H	Bu-4[41]
2. Deoxypodophyllotoxin (1α,2α,3β)	H	–CH$_2$–		–CH$_2$–		OMe	Bu-3[12], Bu-4[41]
3. — (1α,2α,3β)	OMe	–CH$_2$–		–CH$_2$–		H	Bu-1[37]
4. Austrobailignan 1 (1α,2α,3β)	H	–CH$_2$–		Me	Me	OMe	Am-1[6]
5. 5′-Demethoxy-β-peltatin A methyl ether (1α,2α,3β)	OMe	–CH$_2$–		Me	Me	H	Bu-2[7,13]
6. β-Peltatin A methyl ether (1α,2α,3β)	OMe	–CH$_2$–		Me	Me	OMe	Bu-2[13]

TABLE 4

Dihydronaphthalenes

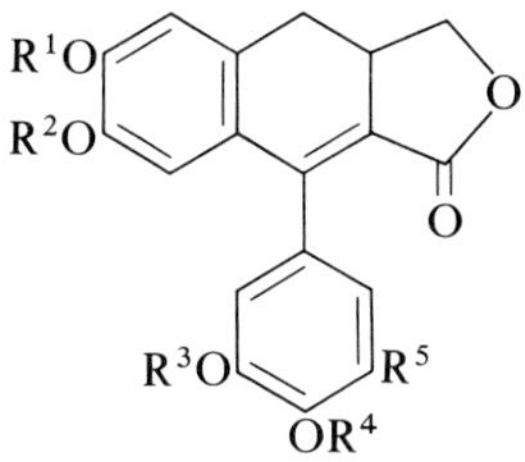

Name	Structures					Occurrence
	R^1	R^2	R^3	R^4	R^5	
1. —	–CH$_2$–		–CH$_2$–		H	Bu-1[37]
2. —	–CH$_2$–		–CH$_2$–		OMe	Bu-1[37]
3. Collinusin	Me	Me	–CH$_2$–		H	Ch-1[11]

TABLE 5

Arylnaphthalenes

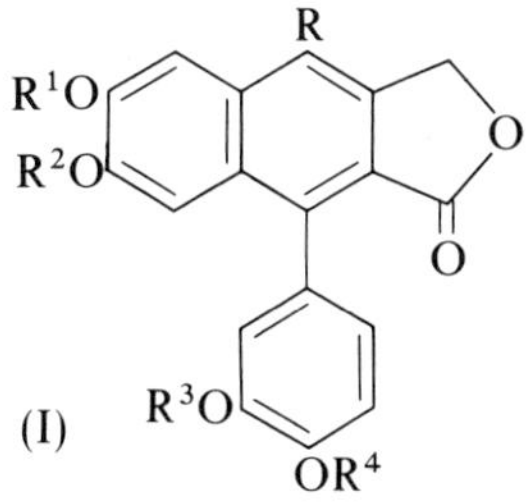

(I) 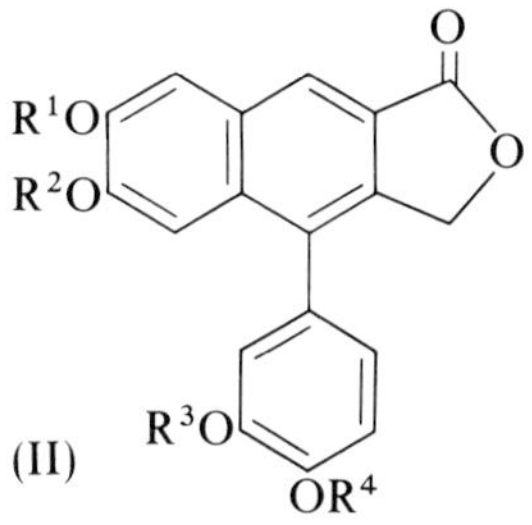   (II)

Name	Structures					Occurrence
	R	R^1	R^2	R^3	R^4	
1. Daurinol	I: H	H	H	–CH$_2$–		Ha-3[9]
2. Justicidin B	I: H	Me	Me	–CH$_2$–		Bo-1[71], Ha-3[10], Ha-5[8], Ha-7[63]
3. Helioxanthin	II: H	–CH$_2$–		Me	Me	Ru-1[67]
4. Diphyllin	I: OH	Me	Me	–CH$_2$–		Ha-4[29], Ha-5[8]
5. Justicidin A	I: OMe	Me	Me	–CH$_2$–		Cn-1[24], Ha-8[45]
6. Diphyllinin	I: OXyl	Me	Me	–CH$_2$–		Ha-4[29]

TABLE 5—*contd.*

Name	Structures				Occurrence
7. Primary acetate of diphyllinin	I: OXyl-Ac	Me	Me	$-CH_2-$	Ha-4[29]
8. Primary crotonate of diphyllinin	I: OXyl-Cr	Me	Me	$-CH_2-$	Ha-4[29]

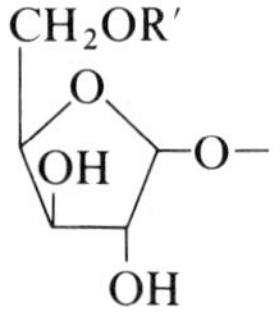

OXyl (R′ = H)

OXyl-Ac (R′ = $COCH_3$)

OXyl-Cr (R′ = COCH=CHMe)

TABLE 6

2,6-Diaryl-3,7-dioxabicyclo[3.3.0]octanes

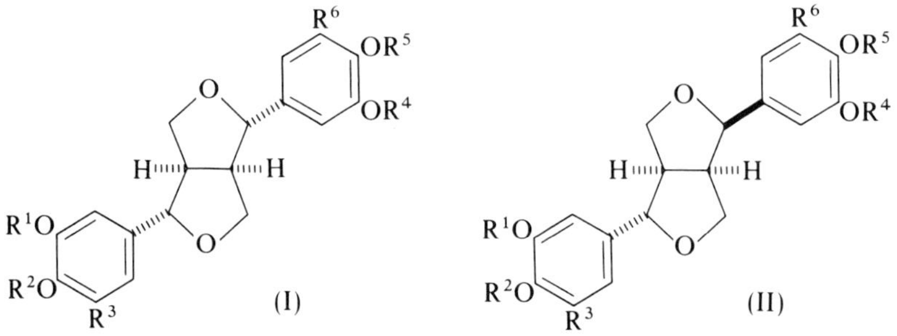

Name	Structure						Occurrence
		R^1	R^2	R^3	R^4	R^5	R^6	
1. Sesamin	I:	$-CH_2-$		H	$-CH_2-$		H	Co-1[58], Eu-2[16], Fl-2[36], Ru-3[79], Za-1[62], Za-2[22], Za-3[40], Za-4[28], Za-5[28], Za-7[47], Za-10[26], Za-11[80], Za-12[75], Za-13[76], Za-14[22], Za-15[1], Za-16[46], Za-17[20], Za-18[4], Za-19[25], Za-20[31], Za-21[4], Za-22[28], Za-23[72], Za-24[17]

TABLE 6—*contd.*

Name								Occurrence
				Structure				
2. Asarinin	II:	-CH$_2$-		H	-CH$_2$-		H	Ac-1[21], Za-1[60], Za-2[57], Za-3[39], Za-6[42], Za-9[77], Za-13[76], Za-15[3], Za-16[46], Za-23[72]
3. Piperitol	I:	Me	H	H	-CH$_2$-		H	Za-15[3]
4. Xanthoxylol	II:	Me	H	H	-CH$_2$-		H	Za-15[3,60]
5. Pluviatilol	II:	-CH$_2$-		H	Me	H	H	Ha-1[64], Za-17[20,60]
6. Methyl piperitol	I:	Me	Me	H	-CH$_2$-		H	Za-1[60]
7. Fargesin	II:	-CH$_2$-		H	Me	Me	H	Za-1[60,62], Za-2[22]
8. Eudesmin	I:	Me	Me	H	Me	Me	H	Eu-1[16], Ha-2[65], Ha-6[65], Za-1[62], Za-2[22], Za-8[73], Za-14[22]
9. Epieudesmin	II:	Me	Me	H	Me	Me	H	Za-1[60], Za-2[22], Za-8[73], Za-14[22]
10. Magnolin	I:	Me	Me	OMe	Me	Me	H	Za-2[57]
11. Syringaresinol*	I:	Me	H	OMe	Me	H	OMe	Ch-1[11], Za-1[22], Za-3[38], Za-14[22]
12. —	I:	Me	Pre	H	-CH$_2$-		H	Za-2[57], Za-13[76], Za-15[2]
13. —	II:	Me	Pre	H	-CH$_2$-		H	Za-15[2]
14. —	II:	-CH$_2$-		H	Me	Pre	H	Za-2[57]
15. —	II:	Me	Pre	H	Me	H	H	Za-2[57]
16. —	I:	Me	Me	H	Me	Pre	H	Za-2[57]
17. Lirioresinol B dimethyl ether	I:	Me	Me	OMe	Me	Me	OMe	Za-11[27]
18. —	I:	Me	Pre	H	Me	Pre	H	Za-2[57]

* See footnote to Table 7.

TABLE 7

Lignan-bearing plants from the Rutales and their contained lignans; plants belong to the Rutaceae except where marked as being from the Burseraceae (*)

Code	Plant	Lignans
Ac-1	*Acronychia muelleri*	6-2
Am-1	*Amyris pinnata*	2-1, 2-6, 3-4
Bo-1	*Boenninghausenia albiflora*	5-2

TABLE 7—contd.

Code	Plant	Lignans
Bu-1	Bursera arida*	(1), 3-3, 4-1, 4-2
Bu-2	B. fagaroides*	3-5, 3-6
Bu-3	B. microphylla*	1-5, 3-2
Bu-4	B. morelensis*	3-2
Bu-5	B. schlechtendalii*	2-5, 2-8
Ch-1	Chloroxylon swietenia	2-1, 2-2, 4-3, 6-11
Cn-1	Cneoridium dumosum	5-5
Co-1	Commiphora mukul*	6-1
Eu-1	Euodia micrococca	6-8
Eu-2	E. micrococca var. pubescens	6-1
Fl-1	Flindersia brassii	1-4
Fl-2	F. pubescens	6-1
Ha-1	Haplophyllum albert-regelli	6-5
Ha-2	H. acutifolium	6-8
Ha-3	H. dauricum	5-1, 5-2
Ha-4	H. hispanicum	5-4, 5-6, 5-7, 5-8
Ha-5	H. obtusifolium	5-2, 5-4
Ha-6	H. perforatum	6-8
Ha-7	H. popovii	2-4, 5-2
Ha-8	H. tuberculatum	5-5
Pt-1	Ptelea trifoliata	2-7
Ru-1	Ruta graveolens	2-1, 5-3
Ru-2	R. microcarpa	2-1, 2-6
Ru-3	R. montana	6-1
Ru-4	R. pinnata	2-1, 2-9
Ru-5	R. sp. 46.782	2-1
Za-1	Zanthoxylum acanthopodium	1-1, 1-3, 6-1, 6-2, 6-6, 6-7, 6-8, 6-9, 6-11
Za-2	Z. armatum	6-1, 6-2, 6-7, 6-8, 6-9, 6-10, 6-12, 6-14, 6-15, 6-16, 6-18
Za-3	Z. arnottianum	6-1, 6-2, 6-11
Za-4	Z. capense	6-1
Za-5	Z. chalybeum	6-1
Za-6	Z. clava-herculis	6-2
Za-7	Z. conspersipunctatum	6-1
Za-8	Z. culantrillo	6-8, 6-9
Za-9	Z. decaryi	6-2
Za-10	Z. dinklagei	6-1
Za-11	Z. leprieurii	6-1, 6-17
Za-12	Z. martinicense	6-1
Za-13	Z. aff. oreophyllum	6-1, 6-2, 6-12

TABLE 7—*contd.*

Za-14	*Z. oxyphyllum*	2-7, **6**-1, **6**-8, **6**-9, **6**-11
Za-15	*Z. piperitum*	1-2, **6**-1, **6**-2, **6**-3, **6**-4, **6**-12, **6**-13
Za-16	*Z. piperitum* var. *inerme*	**6**-1, **6**-2
Za-17	*Z. pluviatile*	2-1, 2-3, **6**-1, **6**-5
Za-18	*Z. rhoifolia*	**6**-1
Za-19	*Z. rubescens*	**6**-1
Za-20	*Z. simulans*	**6**-1
Za-21	*Z. tingoassuiba*	**6**-1
Za-22	*Z. viride*	**6**-1
Za-23	*Z. williamsi*	**6**-1, **6**-2
Za-23	*Z. xanthoxyloides*	**6**-1

Syringaresinol (**6**-11) has also been reported from *Z. ailanthoides* [Ishii, H., Okida, H. and Haginawa, J. (1972). *Yakugaku Zasshi* **92**, 118] and from *Z. inerme* [Ishii, H., Murakami, K., Takeishi, K., Ishikawa, T. and Haginawa, J. (1981). *Yakugaku Zasshi* **101**, 504]. These two taxa are doubtfully separable and it is probably best to regard these reports as relating to the same species.

REFERENCES

1. Abe, F., Furukawa, M., Nonaka, G., Okabe, H. and Nishioka, I. (1973). *Yakugaku Zasshi* **93**, 624.
2. Abe, F., Yahara, S., Kimiko, K., Nonaka, G., Okabe, H. and Nishioka, I. (1974). *Chem. Pharm. Bull.* **22**, 2650.
3. Abe, F., Yahara, S., Nonaka, G., Okabe, H. and Nishioka, I. (1973). *Chem. Pharm. Bull.* **21**, 1617.
4. Antonaccio, L. D. and Gottlieb, O. R. (1959). *Ann. Ass. Brasil Quim.* **18**, 183.
5. Ayres, D. C. (1978). *In* "Chemistry of the Lignans" (Rao, C. B. S., ed.) p. 123. Andhra University Press, Visakhapatnam.
6. Badawi, M. H., Seida, A. A., Kinghorn, A. D., Cordell, G. A. and Farnsworth, N. R. (1981). *J. Nat. Prod.* **44**, 331.
7. Bates, R. B. and Wood, J. B. (1972). *J. Org. Chem.* **37**, 562.
8. Batirov, E. Kh., Matkarimov, A. D. and Malikov, V. M. (1981). *Khim. Prir. Soedin.* 386.
9. Batsuren, D., Batirov, E. K., Malikov, V. H., Zemylyanskii, V. N. and Yagudaev, N. R. (1981). *Khim. Prir. Soedin.* 295.
10. Batsuren, D., Batirov, E. K. and Malikov, V. M. (1981). *Khim. Prir. Soedin.* 659.
11. Bhide, K. S., Mujumdar, R. B. and Rao, A. V. R. (1977). *Indian J. Chem.* **15B**, 440.
12. Bianchi, E., Caldwell, M. E. and Cole, J. R. (1968). *J. Pharm. Sci.* **57**, 696.
13. Bianchi, E., Sheth, K. and Cole, J. R. (1969). *Tetrahedron Lett.* 2759.
14. Birch, A. J. and Liepa, A. J. (1978). *In* "Chemistry of the Lignans" (Rao, C. B. S., ed.), p. 307. Andhra University Press, Visakhapatnam.
15. Burden, R. S., Crombie, L. and Whiting, D. A. (1969). *J. Chem. Soc.* (C) 693.

278   J. O'SULLIVAN

16. Cameron, D. W. and Sutherland, M. D. (1961). *Aust. J. Chem.* **14**, 135.
17. Carnmalm, B., Erdtman, H. and Pelchowicz, Z. (1955). *Acta Chem. Scand.* **9**, 1111.
18. Chakraborty, D. P., Roy, S., Roy, S. P. S. and Majumber, S. (1979). *Chem. Ind.* (London) 667.
19. Cole, J. R., Bianchi, E. and Trumbull, E. R. (1969). *J. Pharm. Sci.* **58**, 175.
20. Corrie, J. E. T., Green, G. H., Ritchie, E. and Taylor, W. C. (1970). *Aust. J. Chem.* **23**, 133.
21. Davenport, J. B. and Sutherland, M. D. (1954). *Aust. J. Chem.* **7**, 384.
22. Deshpande, V. H. and Shastri, R. K. (1977). *Indian J. Chem.* **15B**, 95.
23. Dewick, P. M. and Jackson, D. E. (1981). *Phytochemistry* **20**, 2277, and references cited therein.
24. Dreyer, D. L. and Lee, A. (1969). *Phytochemistry* **8**, 1499.
25. Fish, F., Gray, A. I. and Waterman, P. G. (1974). *Planta Med.* **25**, 281.
26. Fish, F., Meshal, I. A. and Waterman, P. G. (1975). *Phytochemistry* **14**, 2094.
27. Fish, F. and Waterman, P. G. (1972). *Phytochemistry* **11**, 1527.
28. Fish, F., Waterman, P. G. and Finkelstein, N. (1973). *Phytochemistry* **12**, 2553.
29. Gonzalez, A. G., Ordonez, R. M. and Luis, F. R. (1974). *An. Quim.* **70**, 234.
30. Gonzalez, A. G. and Rodriguez, L. F. (1971). *Herba. Hung.* **10**, 95.
31. Gray, A. I. and O'Sullivan, J. J. (1980). *Planta Med.* **39**, 209.
32. Groger, H. and Hofer, O. (1980). *Tetrahedron* **36**, 3551.
33. Hegnauer, R. (1977). *In* "The Biology and Chemistry of the Compositae" (Heywood, V. H., Harborne, J. B. and Turner, B. L. eds.), p. 283. Academic Press, London.
34. Hegnauer, R. (1971). *In* "The Biology and Chemistry of the Umbelliferae" (Heywood, V. H. ed.), p. 267. Academic Press, London.
35. Hofer, O. and Schölm, R. (1981). *Tetrahedron* **37**, 1181.
36. Hollis, A. F., Prager, R. H., Ritchie, E. and Taylor, W. C. (1961). *Aust. J. Chem.* **14**, 100.
37. Ionescu, F. (1975). *Diss. Abstr. Int. B* **35**, 3833.
38. Ishii, H., Hosoya, K., Ishikawa, T. and Haginawa, J. (1974). *Yakugaku Zasshi* **94**, 309.
39. Ishii, H., Hosoya, K., Ishikawa, T., Ueda, E. and Haginawa, J. (1974). *Yakugaku Zasshi* **94**, 322.
40. Ishii, H., Ishikawa, T. and Haginawa, J. (1977). *Yakugaku Zasshi* **97**, 890.
41. Joland, S. D., Wiedhopf, R. M. and Cole, J. R. (1977). *J. Pharm. Sci.* **66**, 892.
42. Karrer, W. (1958). "Konstitution and Vorkommen der Organischen Pflanzenstoffe". Birkhauser Verlag, Berlin.
43. Kato, Y. and Munakata, K. (1978). *In* "Chemistry of the Lignans" (Rao, C. B. S. ed.), p. 95. Andhra University Press, Visakhapatnam.
44. Kelm, L. H. (1978). *In* "Chemistry of the Lignans" (Rao, C. B. S., ed.), p. 175. Andhra University Press, Visakhapatnam.
45. Khalid, S. A. and Waterman, P. G. (1981). *Planta Med.* **43**, 148.
46. Komasaki, H., Kasumoto, H., Ohsuka, A. and Kotake, M. (1968). *J. Chem. Soc. Japan* **89**, 717.
47. Krajniak, E. R., Ritchie, E. and Taylor, W. C. (1973). *Aust. J. Chem.* **26**, 687.
48. Lund, E. W. (1960). *Acta Chem. Scand.* **14**, 496.
49. Martinez, E. A., Funes, J. L. B., Gonzalez, A. G. and Luis, F. R. (1969). *An. Quim.* **65**, 809.

50. Martinez, E. A., Reyes, R. E., Gonzalez, A. G. and Luis, F. R. (1967). *An. Real Soc. Espan. Fis. Quim.* **B63**, 197.
51. McDoniel, P. B. and Cole, J. R. (1972). *J. Pharm. Sci.* **61**, 1992.
52. Munakata, K., Marumo, S., Ohta, K. and Chen, Y.-L. (1965). *Tetrahedron Lett.* 4167.
53. Murakami, T. and Matsushima, A. (1961). *Yakugaku Zasshi* **81**, 1596.
54. Nimgirawath, S., Ritchie, E. and Taylor, W. C. (1977). *Aust. J. Chem.* **30**, 451.
55. Nishibe, S., private communication to Kato, Y., in ref. 43.
56. Nishibe, S., Hisada, S. and Ingaki, I. (1974). *Yakugaku Zasshi* **94**, 522.
57. O'Sullivan, J. J. and Gray, A. I., unpublished results.
58. Patil, V. D., Nayak, U. R. and Sukh Dev (1972). *Tetrahedron* **28**, 2341.
59. Pelter, A. and Ward, R. S. (1978). *In* "Chemistry of the Lignans" (Rao, C. B. S. ed.), p. 277. Andhra University Press, Visakhapatnam.
60. Pelter, A., Ward, W. S., Rao, E. V. and Sastry, K. V. (1976). *Tetrahedron* **32**, 2783.
61. Pelter, A., Ward, W. S., Watson, D. J., Collins, P. and Kay, I. T. (1979). *Tetrahedron Lett.* 2275.
62. Rao, E. V., Sastry, K. V. and Pananivelu, T. J. (1975). *Curr. Sci.* **44**, 228.
63. Razakova, D. M. and Bessonova, I. A. (1981). *Khim. Prir. Soedin.* 516.
64. Razakova, D. M. and Bessonova, I. A. (1981). *Khim. Prir. Soedin.* 673.
65. Razakova, D. M., Bessonova, I. S. and Yunusov, S. Y. (1972). *Khim. Prir. Soedin.* 665.
66. Reisch, J., Novak, I., Szendrei, K. and Minker, E. (1967). *Pharmazie* **22**, 220.
67. Reisch, J., Szendrei, K., Novak, I. and Minker, E. (1970). *Pharmazie* **25**, 435.
68. Reisch, J., Szendrei, K., Novak, I., Minker, E. and Papay, V. (1969). *Tetrahedron Lett.* 3803.
69. Reyes, E. R. and Gonzalez, A. G. (1968). *An. Quim.* **64**, 641.
70. Roy, S., Guha, R. and Chakraborty, D. P. (1977). *Chem. Ind.* (London) 231.
71. Rozsa, Zs., Reisch, J., Mester, I. and Szendrei, K. (1981). *Fitoterapia* **52**, 37.
72. Stermitz, F. R., Caola, M. and Swinehart, J. A. (1980). *Phytochemistry* **19**, 1469.
73. Swinehart, J. A. and Stermitz, F. R. (1980). *Phytochemistry* **19**, 1219.
74. Tiwari, K. P., Masood, M., Pandey, A. and Pandey, G. D. (1980). *Indian J. Chem.* **19B**, 241.
75. Tomko, J., Awad, A. T., Beal, J. L. and Doskotch, R. W. (1967). *Lloydia* **30**, 231.
76. Vaquette, J., Cavé, A. and Waterman, P. G. (1979). *Planta Med.* **35**, 42.
77. Vaquette, J., Pousset, J. L., Delaveau, P. and Paris, R. R. (1973). *Ann. Pharm. Franc.* **31**, 49.
78. Villard, A. M. and Wyart, J. (1968). *Compt. Rend. Acad. Sci.* **C266**, 1284.
79. Viquera, J. M., Sanchez, P. J. and Sepulveda, A. J. (1970). *Grasas Aceites* (Seville) **21**, 319.
80. Waterman, P. G., Gray, A. I. and Crichton, E. G. (1976). *Biochem. Syst. Ecol.* **4**, 256.

CHAPTER 10

Chemistry of the Burseraceae

SAMI A. KHALID

Faculty of Pharmacy, University of Khartoum, Sudan

I. INTRODUCTION

The family Burseraceae is widely distributed throughout tropical and subtropical regions, with major centres of radiation in arid areas, notably north-east Africa, Arabia, and tropical America.[59] Species characteristically exude resins which occur in schizogenous and schizolysigenous ducts or cavities. Mucilagenous cells are often to be found in the epidermis of the leaves and secretory canals in the phloem of leaf and stem.

The Burseraceae has usually been considered to contain about 20 genera and nearly 600 species. Many of the species are little known systematically and, accordingly, the family is in a chaotic state. Wild[58] acknowledged this problem and considered that this generic "inflation" had resulted from many botanical descriptions based on insufficient material. As a result we find, for instance, genera like *Canarium* and *Commiphora* having numerous synonym names. In view of this extensive synonymy, which created considerable problems for the author in his literature review, a list of generic synonyms for the family is included (Table 1).

Engler[19] subdivided the Burseraceae into three tribes on the basis of fruit structure:

Protieae—four genera including *Garuga* and *Protium*;
Boswellieae (Bursereae[36])—eight genera including *Aucoumea, Boswellia, Commiphora,* and *Bursera*;
Canarieae—nine genera including *Canarium, Dacryodes,* and *Icicaster*.

Many of the species are economically valuable on account of their resins, such as the gum resin of *Commiphora* which has a long ethnobotanical history in oral hygiene and treatment of skin infections.[33] Paramount among the resin combinations is Myrrh (from the Arabic "murr" = bitter) which is a product of some *Commiphora* species, notably *C. abyssinica* and *C. malmol*. More than 15 years ago Hegnauer[22] summarized the chemistry of the Burseraceae and discussed its chemotaxonomic implications. This was followed by the ethnobotanical and chemical survey of Pernet.[45] Pernet's treatise lacked any chemotaxonomic treatment and was notable for the inclusion of the tropical American rutaceous genus *Amyris* in the Burseraceae. It is my intention in this chapter to outline the chemical profile of the Burseraceae and to stress, where possible, the systematic implications of the chemistry of the family.

II. ESSENTIAL OILS

The volatile or essential oil constituents of the Burseraceae are predominantly monoterpenoid and sesquiterpenoid.

A. MONOTERPENES

Acyclic, monocyclic, and bicyclic monoterpenes have all been found widely in the family. The spectrum of acyclic monoterpenes detected or isolated covers virtually all the groups associated with this class. Myrcene (**1**) occurs

TABLE 1

Generic synonyms in the Burseraceae (the accepted name is followed by its synonyms)

Aucoumea Pierre*
Boswellia Colebr.* = *Libanotus* Stackh., *Libanus* Colebr., *Ploesslia* Endl.
Bursera Jacq. *ex* L* = *Burseria* Jacq., *Busseria* Cramer, *Elaphrium* Jacq., *Evradia*
   Adans., *Icicariba* Maza, *Russeria* Buck., *Simaruba* Boehm.
Canarium L* = *Canariastrum* Engl., *Canariellum* Engl., *Canarion* St. Lag., *Canariopsis*
   Miq. *Cenarium* L., *Colophonia* Comm. *ex* Kunth., *Libaria* Lour. *ex* Gomes,
   Mehenbethene Besl. *ex*. Gaertn., *Nanari* Adans., *Niotutt* Adans., *Pimela* Lour.,
   Sonraya Engl., *Sonzaya* Marchand, *Strania* Nor.
Commiphora Jacq.* = *Balessam* Bruce, *Balsamea* Gled., *Balsamodendron* DC,
   Balsamodendrum Kunth, *Balsamophloes* O. Berg., *Balsamus* Stackh., *Bdellium*
   Baill. *ex* Laness., *Hemprichia* Ehrenb., *Heudelotia* A. Rich., *Hizera* B. D. Jacks,
   Hitzeria Klotzsch., *Protionopsis* Blume, *Protium* Wight & Arn., *Protonopsis* Pfeiff.,
   Spondiopsis Engl.
Crepidospermum Hook. f.
Dacryodes Vahl.* = *Curtisina* Ridl., *Hemisantiria* H. J. Lam., *Pachylobus* G. Don.,
   Santiridium Pierre
Garuga Roxb.* = *Kunthia* Densst.
Haplolobus H. J. Lam.
Hemicrepidospermum Swart.
Icicaster Ridl.
Paraprotium Cuatrec
Protium Burm.* = *Dammara* Gaertn., *Icica* Aubl., *Icicopsis* Engl., *Marignia* Comm. *ex*.
   Kunth., *Tingulong* Rumph.
Santiria Blume = *Trigonochlamys* Hook.
Santiriopsis Engl. (sometimes included in *Santiria*)
Scutinanthe Thw.
Tapirocarpus Sagot.
Terebinthus Mill.
Tetragastris Gaertn. = *Caproxylon* Tussac., *Hedwigia* Sw., *Knorrea* Moc. & Sesse *ex*
   DC., *Schwaegrichenia* Reichb.
Trattinnickia Willd. = *Trattinickia* Auctt., = *Trattinickya* A. Juss.
Triomma Hook.

---

* Genera for which some phytochemical data is available.

widely, most often accompanied by the free alcohols citronellol (**2**), linalool (**3**), and geraniol (**4**), or their acetates.[1,45] Monocyclic monoterpenes are usually the dominant class within the family, with limonene (**5**) and α-phellandrene (**6**) particularly widespread.[1,5,45] *p*-Cymene has recently been reported to be the major component of *Boswellia frereana* (frankincense) oil.[49] Monocyclic derivatives also occur as aldehydes such as phellandral (**7**), ketones like carvone (**8**), and, less frequently, acids like phellandric acid (**9**).

Bicyclic monoterpenes include the thujane type [thujene (**10**)], the pinane type [α-pinene (**11**)], and the camphane type [camphene (**12**)].[45] The cooccurrence of α-pinene and verbanol (**13**) and verbanone (**14**) in the oil of olibanum, *Boswellia carteri*,[45] indicates their biogenetic relationships.

Oil from the husks of *Bursera delpechinana*[1] consisted of 35 minor and two major components, as detected by GLC. Among them 11 have been isolated by preparative GLC and identified by spectroscopic means. These included, as well as typical monoterpenes and their acetates, the furanoid *trans* (**15**) and *cis* (**16**) oxides of linalool and a mixture of two new naturally occurring esters, *cis*- and *trans*-2,6,6-trimethyl-2-vinyl-5-acetoxytetrahydropyran (**17**).

(**1**) Myrcene   (**2**) Citronellol   (**3**) Linalool   (**4**) Geraniol

(**5**) Limonene   (**6**) α-Phellandrene   (**7**) Phellandral   (**8**) Carvone

(**9**) Phellandric acid   (**10**) Thujene   (**11**) α-Pinene   (**12**) Camphene   (**13**) Verbanol

(**14**) Verbanone   (**15**)   (**16**)   (**17**)

B. SESQUITERPENES

The sesquiterpenes form the higher boiling fraction of the essential oils. Over 30 different sesquiterpenes are now recorded with wide variations in their skeletal framework. Sesquiterpenes appear to be common in *Canarium* and *Commiphora* but only one, cadinene (**18**), has been isolated from *Boswellia*, from *B. carteri*.[45]

The genus *Canarium* is a good sesquiterpene accumulator. *C. album*[31] has afforded caryophyllene (**19**) accompanied by δ-cadinene (**18**). *C. samoense*[11,24] yielded the alcohol maaliol (**20**) and the resin "black dammer" exuded by *C. strictum*[25,55] provided the ketone canarone (**21**) and the alcohols epikhusinol (**22**) and (+)-junenol (**23**). *C. strictum* is also reported[54] to contain degraded sesquiterpene hydrocarbons such as damerene ($C_{14}H_{20}O_2$). Sesquiterpenes of the elemane group, e.g. elemol (**24**), have recently been reported from *C. zeylanicum*.[5]

Eighteen sesquiterpenes have been isolated from *Commiphora erythreae*,[57] including β-bisabolene (**25**), ar-curcumene (**26**), β-santalene (**27**), and humulene (**28**). From the essential oil of *C. abyssinica*[10] nine furosesquiterpenes have been isolated, among them isofuranogermacrene (**29**), curzerenone (**30**), furanodienone (**31**), furandiene (**32**), and lindestrene (**33**). The latter type

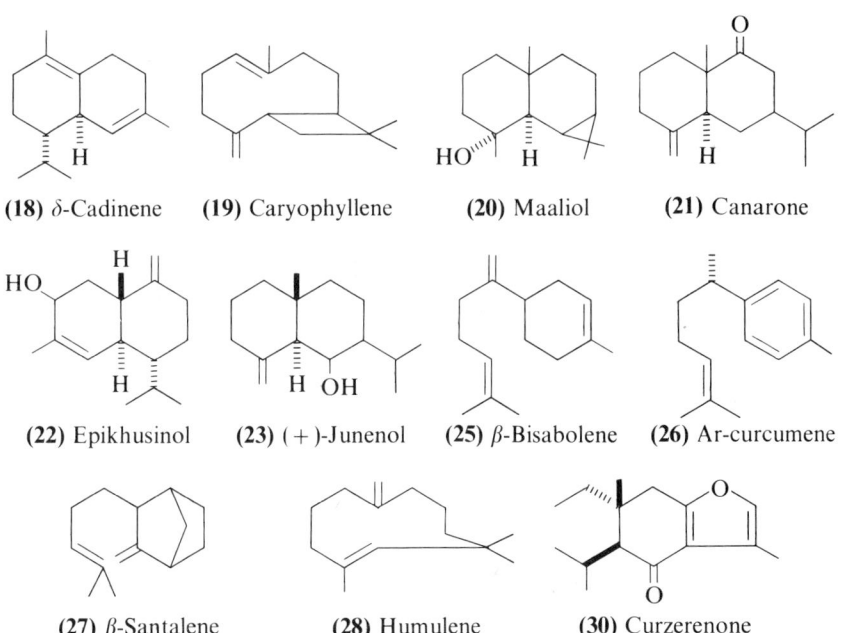

(**18**) δ-Cadinene   (**19**) Caryophyllene   (**20**) Maaliol   (**21**) Canarone

(**22**) Epikhusinol   (**23**) (+)-Junenol   (**25**) β-Bisabolene   (**26**) Ar-curcumene

(**27**) β-Santalene   (**28**) Humulene   (**30**) Curzerenone could arise from elemane type precursors by introduction of an oxygen at C-8 followed by ring closure. The co-occurrence of elemol (24) and isofuranogermacrene (29) in *C. abyssinica* are evidence in favour of such a relationship (Scheme 1).

Commiferin (34), isolated from *Commiphora myrrha*[39] together with β-eudesmol (35), still represents the sole sesquiterpene lactone of the Burseraceae. Commiferin (34) may well be derived from the co-occurring eudesmane type sesquiterpenes by oxidative ring closure (Scheme 2), a process which requires the same C-8 oxidation seen in formation of the furanosesquiterpenes (cf. Scheme 1). Two novel furanosesquiterpenes (36 and 37) have recently been reported from the gum exudates of *C. myrrha*[34] and from the opopanax oil of *C. erythreae*,[34] where they occur together with compound 31.

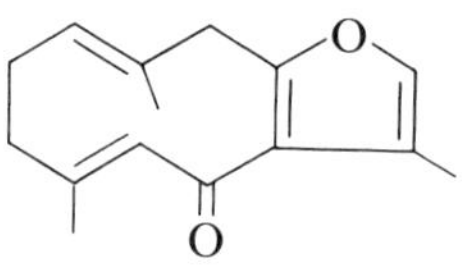

(31) Furanodienone 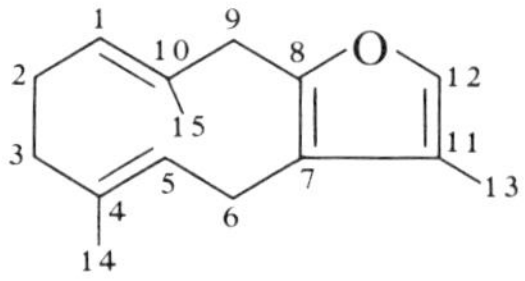   (32) Furandiene

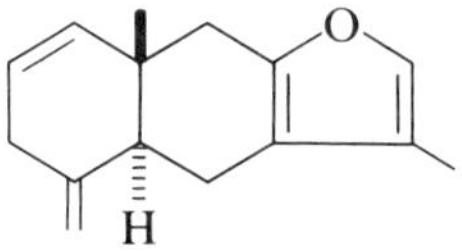

(33) Lindestrene

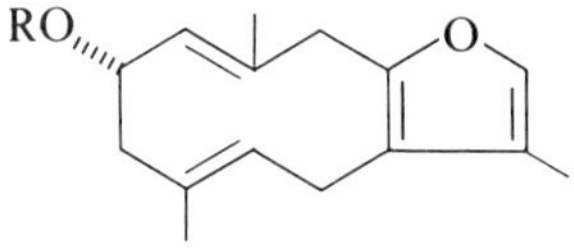

(36) R = Ac
(37) R = Me

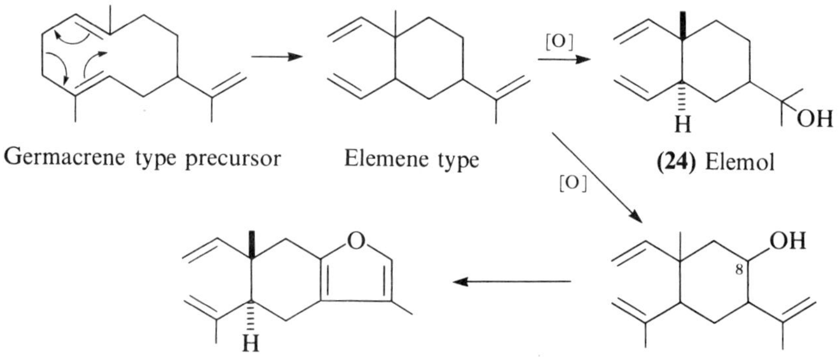

Germacrene type precursor   Elemene type   (24) Elemol

(29) Isofuranogermacrene

SCHEME 1. Possible route to elemol (24) and isofuranogermacrene (29).

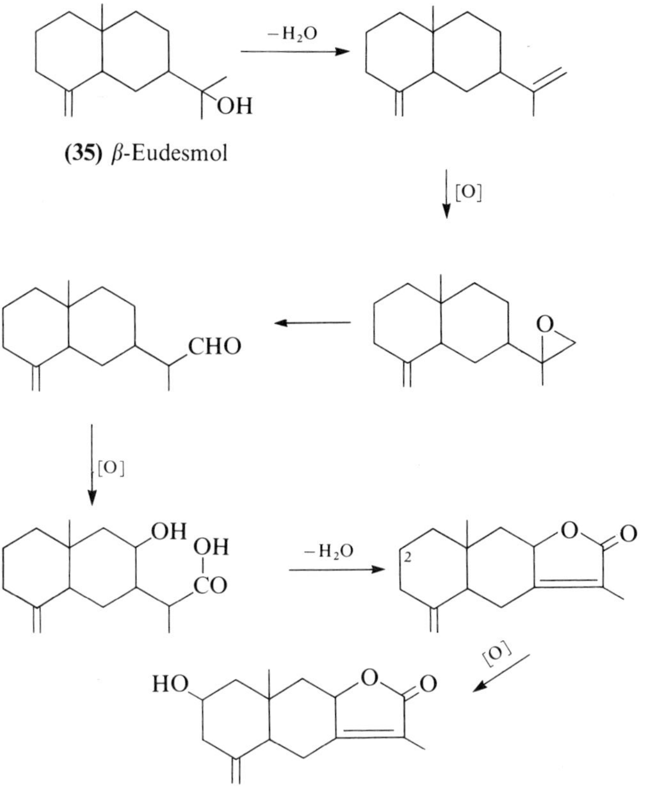

(35) β-Eudesmol

(34) Commiferin

SCHEME 2. Possible route to commiferin (34) from β-eudesmol (35) in *Commiphora myrrha*.

III. DITERPENES

The genera *Boswellia* and *Commiphora* are, to date, the source of all the diterpenes recorded in the family. They occur mainly in the gum resins and are monocyclic and macrocyclic with the exception of α-camphorene (38) from *C. mukul*. Most are based on the cembrene type, with cembrene (39) itself and the structurally related compounds 2-hydroxy-4,8,12-trimethyl-1-isopropyl-3,7,11-cyclodecatriene (40), cembrene A (41), and mukulol (42) having been reported from the gum resin of *C. mukul*.[43,47] From the frankincense produced by *Boswellia carteri*[41] incensol (43) and incensol oxide (44) have been recorded. The possible biogenetic route to 43 and 44 from cembrene type precursors is shown in Scheme 3.

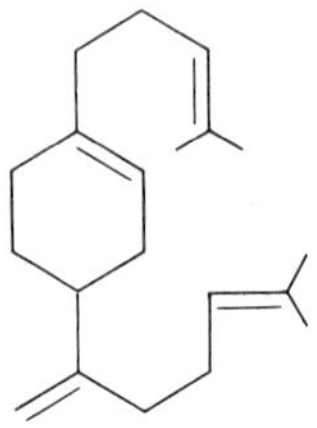

(38) α-Camphorene

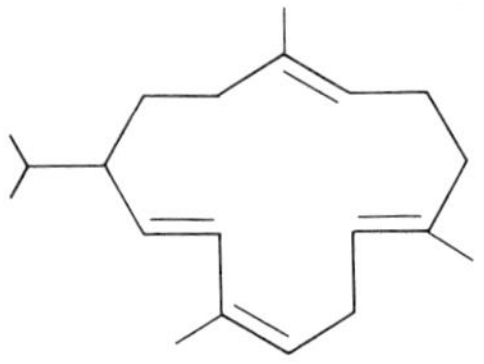

(39) Cembrene

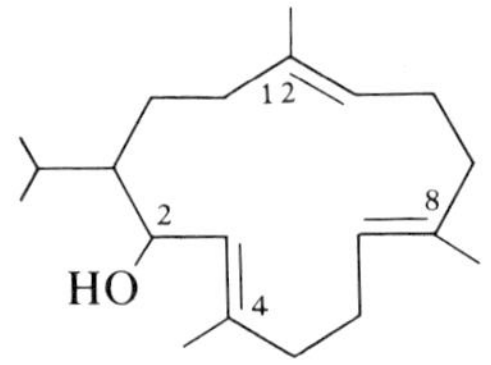

(40)

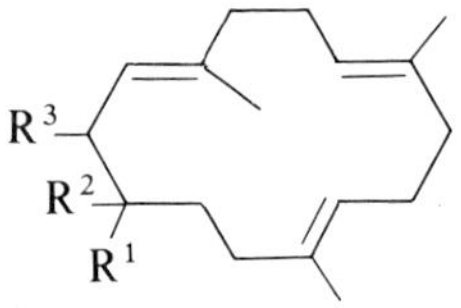

(41) Cembrene A (R^1 = CMe = CH_2; R^2 = R^3 = H)

(42) Mukulol (R^1 = H; R^2 = $CHMe_2$; R^3 = OH)

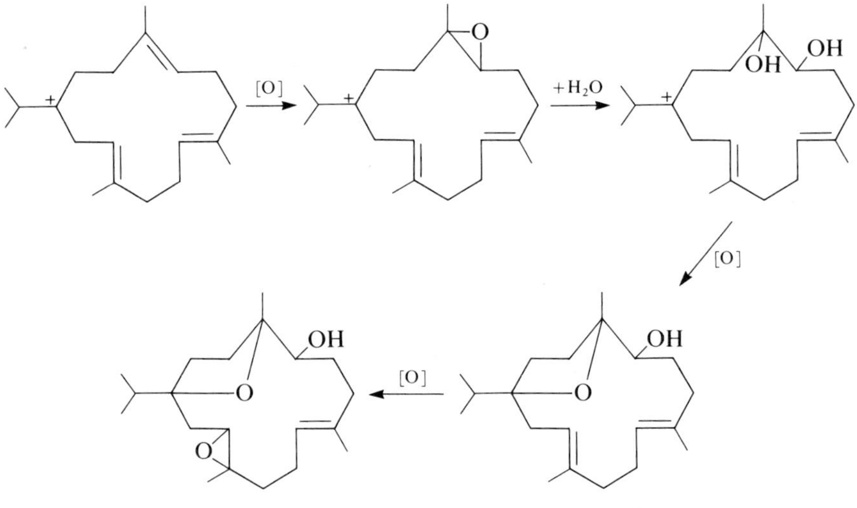

(44) Incensole oxide

(43) Incensole

SCHEME 3. Possible route to incensole (43) and its oxide (44) in *Boswellia carteri*.

IV. TRITERPENES

The Burseraceae resins accumulate both tetracyclic and pentacyclic triterpenes. They can be broadly classified into four series, the tetracyclic euphane/tirucallane type (45) and the pentacyclic lupane (46), ursane (47), and oleanane (48) types.

(45) Euphane/Tirucallane   (46) Lupane   (47) Ursane   (48) Oleanane

A. EUPHANE/TIRUCALLANE SERIES

Representatives of this class in which oxidation of the C-17 side-chain has occurred, elemolic acid (49) and isomasticadienonic acid (50), have been reported from *Canarium schweinfurthii*[45] and *Dacryodes edulis*.[18] The structurally related elemadienonic acid and elemadienolic acid and their epimers have also been isolated from *C. schweinfurthii*.[15] Compounds of this type have the capacity to undergo further modification of the side-chain leading to cyclization, as in sapelin A (51) and sapelin B (52) from *Bursera klugii*.[27] Other compounds closely allied to the sapelins include bourjotinolone A (53), flindissol (54), and melianone (55) which have been found in the allied Rutaceae and Meliaceae. Both sapelins, which also occur in *Entandrophragma*

(49) R^1 = H; R^2 = OH; R^3 = HOOC

(50) $R^1 R^2$ = O; R^3 =

(52) R^1 = H; R^2 = OH; R^3 =

(51) R^1 = H; R^2 = OH; R^3 =
(53) R^{1}R^2 = O;   R^3 =

(54) R^1 = H; R^2 = OH; R^3 = HO—

(55) R^{1}R^2 = O;   R^3 = HO—

cylindricum (Meliaceae), have shown cytotoxic activity. In both the formation of heterocyclic rings from modified C-17 side-chains and the occurrence of C-7/C-8 double bonds it is clearly possible to look upon these compounds, particularly the sapelins, as pointers toward the ability to form the limonoids which are seen widely in Rutaceae and Meliaceae.

B. LUPANE SERIES

The A-ring of lupane triterpenes occurs at three different oxidation levels in the Burseraceae (Table 2) starting from the common lupeol (56) which is widespread in *Bursera* spp. through lupeonone (57) in *Commiphora* to the very interesting 3,4-secotriterpene cannaric acid (58) found in species of *Canarium* and *Dacryodes*. Similar cleavage of the A-ring occurs widely in the limonoids of Meliaceae (Chapter 6) and to a limited extent in the Rutaceae (Chapter 7).

TABLE 2

Lupane series triterpenes

Species	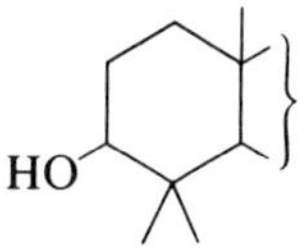 HO— (56) Lupeol	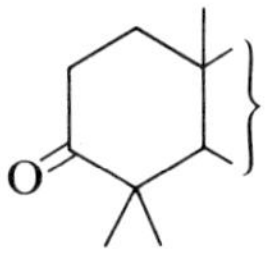 O= (57) Lupeonone	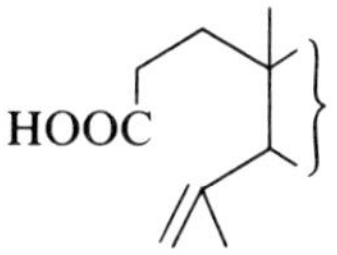 HOOC— (58) Cannaric acid
Boswellia frereana[49]	+		
Bursera delpechiana[45]	+		
B. galleotiana[22]	+		
B. gorullensis[22]	+		
B. microphylla[45]	+		
B. sessiliformis[22]	+		
Canarium muelleri[12]			+
C. zeylanicum[5]			+
Commiphora myrrha[20]		+	
Dacryodes edulis[18]			+

C. URSANE SERIES

The ursane series triterpenes show the highest degree of oxidation involving ring-A seen in the Burseraceae (Table 3). This culminates in the 1,2,3-trihydroxylation found in commic acid E (59), a feature also noted among the quassinoids of the Simaroubaceae (Chapter 8). To date ability to oxidize C-1 and C-2 has been found only in one species, *Commiphora pyracanthoides*.[50,51] Oxidation of one of the C-4 methyl groups through to the carboxylic acid is seen in compounds from both *Boswellia* spp. and *C. pyracanthoides*. Here again it is reasonable to assume that the biosynthetic change from C-4 *gem*-dimethyl groups to the single C-4 methyl found in the quassinoids will occur by initial oxidation followed by eventual decarboxylation,[26] so once again the burseraceous products can be looked upon as early stages in the reaction sequence.

Oxidation involving C-11 of ring-C has also been found, for example the occurrence of the acetates of 11β-hydroxy-α-amyrin and 11-oxo-α-amyrin in *Canarium strictum*,[25] and the methyl ester of 3-acetyl-11-hydroxy-β-boswellic acid (60) in *Boswellia serrata*.[16] Once again this is a pattern consistent with supposed early stages in the formation of quassinoids (Chapter 8) and C-ring cleaved limonoids such as nimbin (Chapter 6).

TABLE 3

Ursane series triterpenes

Species	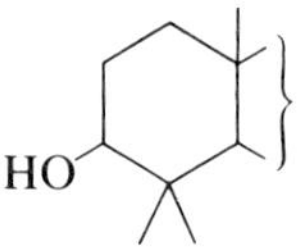 α-Amyrin	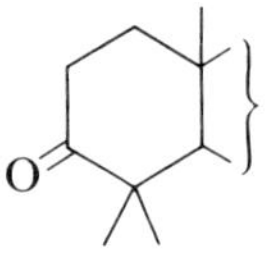 α-Amyrenone	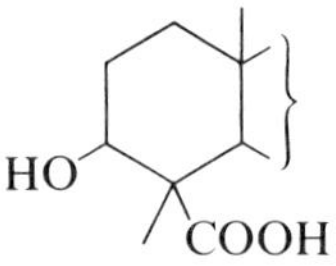 β-Boswellic acid (Commic acid B)	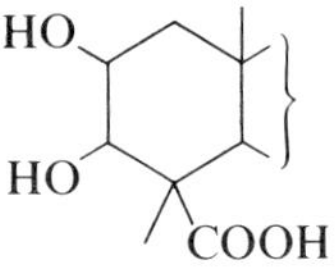 Commic acid D	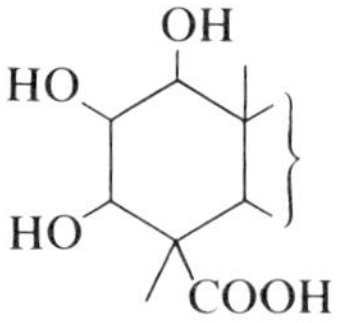 (59) Commic acid E
Aucouma klaineana[35]	+				
Boswellia carteri[45]			+		
B. dalzielli[45]			+		
B. frereana[45]			+		
Bursera					
delpechiana[45]	+				
B. galleotiana[22]	+				
B. gorullensis[45]	+				
B. schlechtendaii[35]	+				
B. sessiliflora[22]	+				
Canarium					
schweinfurthii[15]	+				
C. zeylanicum[5]	+	+			
Commiphora myrrha[22]	+				
C. pyracanthoides[20]			+	+	+
Dacryodes edulis[51]	+				

Oxidation of both A and E rings is also a feature among the ursane series. The genus *Canarium* in particular seems to specialize in oxidative modification of the E-ring with *C. strictum*[25] producing epi-ψ-taraxastanediol and its ketol and *C. luzonicum*[40] epi-ψ-taraxastanediol and brein (61).

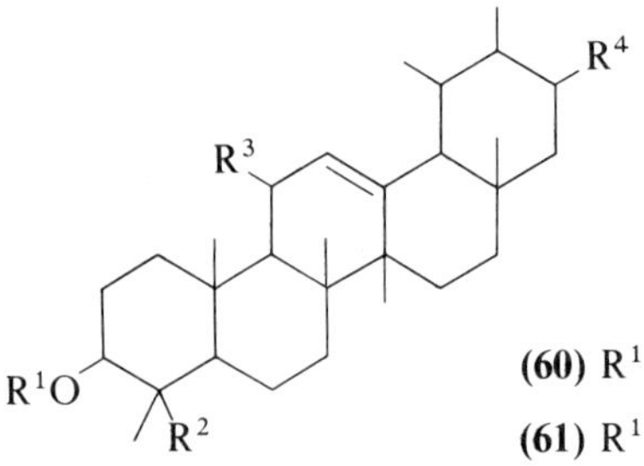

(60) R^1 = Ac; R^2 = COOMe; R^3 = OH; R^4 = H

(61) R^1 = R^3 = H; R^2 = Me; R^4 = OH

D. OLEANANE SERIES

Triterpenes of the oleanane series show similar patterns of A-ring oxidation as noted for the ursane type with the exception that C-1 oxidation has yet to be found (Table 4). Similarly compounds with A- and C-ring oxidation occur, for example olean-12-en-3,11-dione from *Canarium zeylanicum*.[5] A further modification found in this group is oxidation of both rings A and D, as in maniladiol (**62**) isolated from Manila elemi (*Canarium luzonicum*).[40]

TABLE 4

Oleanane series triterpenes

Species	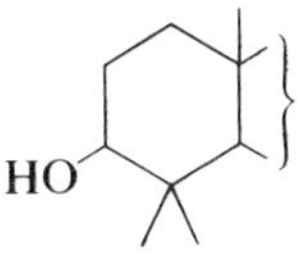 β-Amyrin	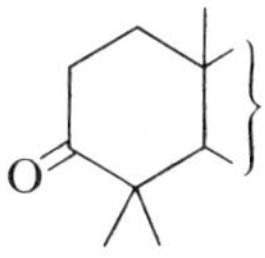 β-Amyrenone	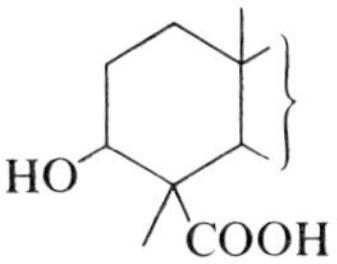 α-Boswellic acid	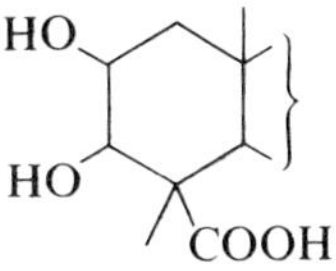 Commic acid C
Aucouma klaineana[17]	+			
Boswellia carteri[11,24]			+	
B. dalzielli[45]			+	
B. frereana[11,24]			+	
Bursera delpechiana[45]	+			
Canarium schweinfurthii[15]	+			
C. zeylanicum[5]	+	+		
Commiphora pyracanthoides[50]				+

The interesting compound 2,3-secoolean-12-en-2,3,28-tricarboxylic acid (**63**) has been reported from the heartwood of *Bursera graveolens*.[22] It is probably derived by the oxidative splitting of ring-A of the corresponding oleanic acid and is of interest as it shows oxidation of a methyl constituent other than one of those at C-4.

E. LIMONOIDS

The only limonoid so far reported to occur in the Burseraceae is cedrelone (Chapter 6, Table 3) from *Balsamodendron* (*Commiphora*) *pubescens*.[29] Cedrelone was originally reported from the meliaceous tree *Cedrela toona*. As

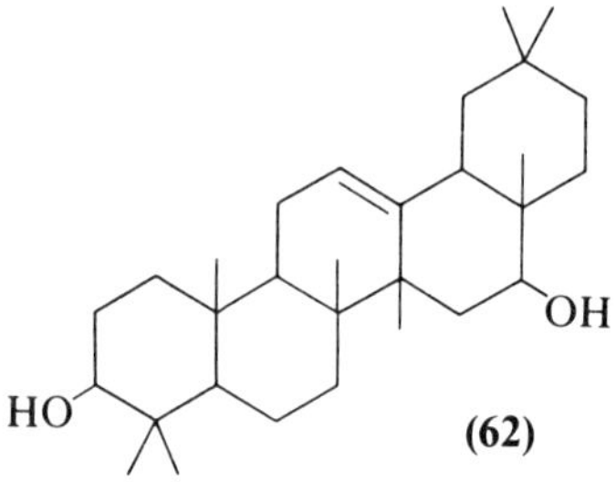

(62)

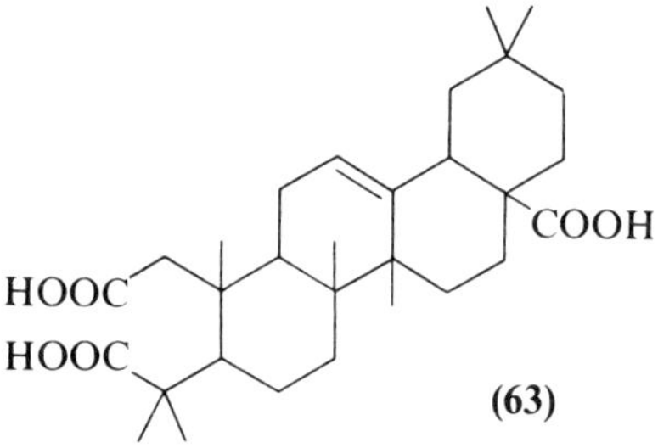

(63)

the distribution of limonoids is, to date, restricted to the three rutalean families Rutaceae, Meliaceae, and Cneoraceae, this report would appear to reinforce strongly the more tenuous evidence, from the other structurally allied triterpenes discussed previously, that Burseraceae has strong affinities with the order Rutales. However, it has to be said that there is still doubt regarding the botanical identity of the material analysed, and a re-examination of *B. pubescens* for this limonoid is essential (see below).

F. STEROIDS

Apart from the ubiquitous phytosterols a number of novel steroidal compounds have been isolated from the resin of the crude drug "guggul" (*Commiphora mukul*).[37,42,46] Among them are guggulsterols I (64) and III (65) and guggulsterones I (66) and II (67). These compounds are reported to have significant antiinflammatory activity.[4]

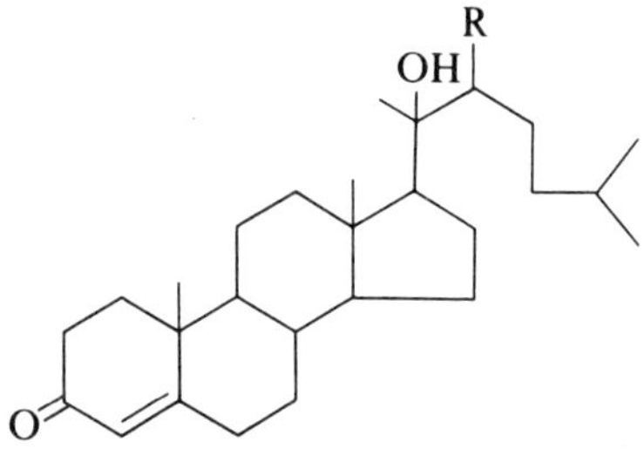

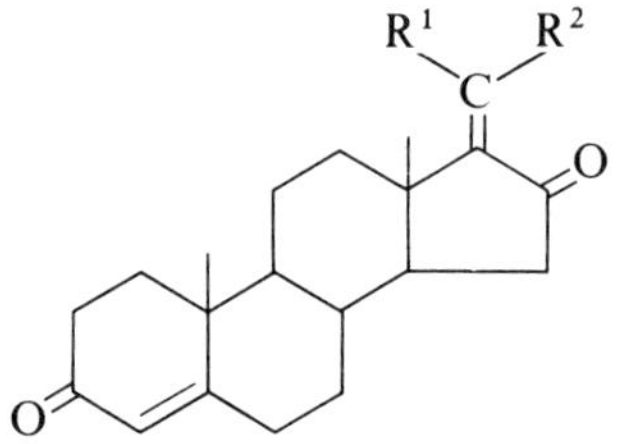

(64) Guggulsterol I (R = OH)

(65) Guggulsterol III (R = H)

(66) Guggulsterone I (R^1 = Me; R^2 = H)

(67) Guggulsterone II (R^1 = H; R^2 = Me)

V. Phenolics

A. FLAVONOIDS

Flavonoids are universal in the angiosperms and their patterns of distribution are often systematically important. Unfortunately there is still very little information on their occurrence in the Burseraceae. The common kaempferol and quercetin and their glycosides occur in *Protium*[45] and *Commiphora*[30] but of far more interest is the presence of the biflavonoid amentoflavone (68) in the leaves of *Garuga pinnata*.[3] A number of allied biflavonoids have been found in genera of the Anacardiaceae[21] but not in any of the typically rutalean families. The presence of what is usually considered to be a rather "primitive" type of biflavonoid correlates well with the suggestion that the Burseraceae are the least specialized of the rutalean families.[38]

(68) Amentoflavone

B. COUMARINS

To date only two coumarins have been reported, the coumarinolignoid propacin (Chapter 4, structure 13-10) from *Protium opacum*[60] and the 5-methylcoumarin siderin (Chapter 4, structure 1-3) from *Balsamodendron pubescens*.[29] The latter, as well as the limonoid cedrelone, are both known to occur in *Cedrela toona* and, in view of the coincidental occurrence of both compounds, the identity of the material worked on as *B. pubescens* must, at present, be considered questionable.

C. LIGNANS

To date lignans have only been recorded from the genus *Bursera* but they are, most probably, of widespread distribution. Most of the lignans so far isolated from the Burseraceae can be classified as 2,3-dibenzylbutyrolactones

(*B. schlechtendalii*[35]) or 4-aryltetrahydro- and 4-aryldihydronaphthalenes (*B. arida*,[14] *B. microphylla*,[7,13,53] *B. morelensis*,[28] and *B. fagaroides*[8]). Structures are given in Chapter 9. At present there are no records of the 2,6-diaryl-3,7-dioxabicyclo[3.3.0]octane type that are so common in the Rutaceae.

Of the types of lignans found in *Bursera* the dibenzylbutyrolactones are found widely in the Rutaceae (Chapter 9) but the arylnaphthalenes are of much more limited distribution, being found mainly in the herbaceous taxa of the Rutaceae (*Haplophyllum, Ruta, Cneoridium*) and in *Bursera*. In particular the aryltetrahydronaphthalenes are so far restricted to *Bursera* and a single report for the genus *Amyris* (Rutaceae). *Amyris* has sometimes been linked with the Burseraceae.[45] Another interesting observation is the co-occurrence of podophyllotoxin and β-peltatin type lignans in *Bursera* and Berberidaceae[52] which represents a further chemical affinity between Rutales and the ranalean complex.[23,56]

VI. Miscellaneous

A. CARBOHYDRATES

A number of species of *Boswellia* and *Commiphora* have been found to exude substantial amounts of polysaccharides. These are mainly of the arabinogalactan type and usually occur in conjunction with water insoluble resins.[48] The gum is characteristically a highly branched polysaccharide containing mainly galactose, xylose, fucose, arabinose, rhamnose, galacturonic and glucuronic acids, and their methylated counterparts. A structure for the gum of *Commiphora mukul* has been proposed by Bose and Gupta.[9] Gums of four families of the rutalean alliance have now been examined (Anacardiaceae, Burseraceae, Meliaceae, and Rutaceae) and all appear to be mainly of the arabinogalactan type.

B. FATTY ACIDS

Burseraceous fruits (both seeds and pericarp) are rich in stearic and linoleic acids. In addition to reports included in Hegnauer's review[22] the seed oils of *Canarium album*,[32] *Boswellia serrata*,[6] and *Dacryodes edulis*[18] have now been confirmed as following the general pattern. A mixture of long chain aliphatic tetrols, representing a new class of naturally occurring lipids, have been isolated from *Commiphora mukul*.[44]

C. AMINO ACIDS AND ALKALOIDS

About 15 amino acids, none of great interest, have been isolated from *Commiphora mukul*.[2] To date no alkaloids have been reported from the family.

VII. CONCLUSIONS

Little is known about the chemistry of the Burseraceae in comparison with the extensively investigated families of the Rutales. There are obvious pointers to a chemical affinity, notably in the strong abilities of the Burseraceae to perform oxidative modifications of the triterpene skeleton that may reflect early stages in limonoid and quassinoid transformations. However, to date, there are no records of alkaloids at all or of coumarins typical of the Rutaceae and its allies. The only record of a limonoid is at present open to doubt. The lignans may offer a link, notably to *Amyris* of the Rutaceae, but *Amyris* is bound to the latter by its ability to produce both alkaloids and coumarins. Clearly much more information will be required before it is possible to make a clear-cut statement regarding the chemical affinities of the Burseraceae with other elements of the Rutales. However, what is now available suggests that it is not particularly close to the small group of families that make up the Rutales *s.s.* (Rutaceae, Meliaceae, Cneoraceae, Simaroubaceae, Ptaeroxylaceae) and may have closer ties with the allied taxa of the Sapindales *s.s.*

REFERENCES

1. Adams, D. R. and Bhatnagar, S. P. (1975). *Int. Flavours Food Addit.* **6**, 185.
2. Ali, M. M. and Hassan, M. (1967). *Pakistan J. Sci. Ind. Res.* **10**, 21.
3. Ansari, F. R., Ansari, W. H. and Rahman, W. (1978). *Indian J. Chem.* **16B**, 846.
4. Arosa, R. B., Kapoor, V., Gupta, S. K. and Sharma, R. C. (1971). Indian *J. Esp. Biol.* **9**, 403.
5. Bandaranayake, W. M. (1980). *Phytochemistry* **19**, 255.
6. Bhakal, H. N. and Rao, C. V. (1967). *Curr. Sci.* **36**, 668.
7. Bianchi, E., Caldwell, M. E. and Cole, J. R. (1968). *J. Pharm. Sci.* **57**, 696.
8. Bianchi, E., Saeth, E. and Cole, J. R. (1969). *Tetrahedron Lett.* 2759.
9. Bose, S. and Gupta, K. C. (1966). *Indian J. Chem.* **4**, 87.
10. Brieskorn, C. H. and Noble, P. (1982). *Planta Med.* **44**, 87.
11. Buechi, G. and White, D. M. (1957). *J. Am. Chem. Soc.* **79**, 750; Buechi, G., Schach, V., Wittenau, M. and White, D. M. (1959). *J. Am. Chem. Soc.* **81**, 1968.
12. Carman, R. M. and Cowley, D. E. (1964). *Tetrahedron Lett.* 627.
13. Cole, J. R., Bianchi, E. and Trumbull, E. R. (1969). *J. Pharm. Sci.* **58**, 175.

14. Cole, J. R. and Wledhopf, R. M. (1978). *In* "Chemistry of Lignans" (Rao, C. B. S. ed.), pp. 39–65. Andhra University Press, India.
15. Cordoso do Vale, J., Proenca de Cunha, A. and Roque, O. R. (1979). *Bol. Fac. Farm. Coimbra* **3**, 7.
16. Corsano, S. and Iavarone, C. (1964). *Gazz. Chim. Ital.* **94**, 328.
17. Dupont, G., Dulou, R. and Vilkas, M. (1948). *Peintures, Pigments* **24**, 46.
18. Ekong, D. E. U. and Okogun, J. I. (1969). *Phytochemistry* **8**, 669.
19. Engler, A. (1931) *In* "Die Naturlichen Pflanzenfamilien" (Engler, A. and Prantl. K. eds.), Vol. 19a, pp. 405–456. Engelmann, Leipzig.
20. Enrico, M. and Carlo, I. (1972). *Chim. Ind.* (Milan) **54**, 424.
21. Geiger, H. and Quinn, C. (1975). *In* "The Flavonoids" (Harborne, J. B., Mabry, T. J. and Mabry, H. eds.), pp. 692–737. Chapman and Hall, London.
22. Hegnauer, R. (1964). *In* "Chemotaxonomie der Pflanzen", Vol. 3, pp. 310–318. Birkhauser Verlag, Basel.
23. Hegnauer, R. (1969). *In* "Perspectives in Phytochemistry" (Harborne, J. B. and Swain, T. eds.), pp. 121–138. Academic Press, London.
24. Hellyer, R. C. (1962). *Aust. J. Chem.* **15**, 157.
25. Hinge, V. K., Wagh, A. D., Paknikar, S. K. and Bhattacharyya, S. C. (1965). *Tetrahedron* **21**, 3197.
26. Holland, H. L. (1981). *Chemical Society Reviews* **10**, 435.
27. Jolad, S. D., Wiedhopf, R. M. and Cole, J. R. (1977). *J. Pharm. Sci.* **66**, 889.
28. Jolad, S. D., Wiedhopf, R. M. and Cole, J. R. (1977). *J. Pharm. Sci.* **66**, 892.
29. Joshi, B. S. and Hegde, V. R. (1979). *Proc. Indian Acad. Sci.* **88A**, 185.
30. Kakrani, H. K. (1982). *Fitoterapia* **5**, 221.
31. Kameoka, H., Cheng, Y., Miyazawa, M. and Hairao, N. (1974). *Sixth Int. Congress Essential Oils* **120**, 18.
32. Kameoka, H. and Miyazawa, M. (1976). *Yakugaku Zasshi* **25**, 561.
33. Lewis, W. H. and Elvin-Lewis, M. P. F. (1977). *In* "Medical Botany, Plants Affecting Man's Health". Wiley-Interscience, New York.
34. Maradufu, A. (1982). *Phytochemistry* **21**, 677.
35. McDoniel, P. B. and Cole, J. R. (1972). *J. Pharm. Sci.* **61**, 1992.
36. Melchior, H. (1964). *In* "Syllabus der Pflanzenfamilien", 12th Edn, Vol. 2, pp. 268–270. Borntrager, Berlin.
37. Mester, L., Mester, M. and Swarn, N. (1979). *Planta Med.* **37**, 367.
38. Metcalfe, C. R. and Chalk, L. (1950). "Anatomy of the Dicotyledons", Vol. 1, pp. 341–349. Oxford University Press, Oxford.
39. Mincione, E. and Iavarone, C. (1972). *Chim. Ind.* (Milan) **54**, 525.
40. Morice, I. M. and Simpson, J. C. E. (1941). *J. Chem. Soc.* 181.
41. Nicoletti, R. and Forcellese, V. L. (1968). *Tetrahedron* **24**, 6519.
42. Patil, V. D., Nayak, U. R. and Dev, S. (1972). *Tetrahedron* **28**, 2341.
43. Patil, V. D., Nayak, U. R. and Dev, S. (1973). *Tetrahedron* **29**, 341.
44. Patil, V. D., Nayak, U. R. and Dev, S. (1973). *Tetrahedron* **29**, 1595.
45. Pernet, R. (1972). *Lloydia* **35**, 280.
46. Purushothaman, K. K. and Chandrasekharn, S. (1976). *Indian J. Chem.* **14B**, 802.
47. Ruecker, G. (1972). *Archiv der Pharm.* **305**, 486.
48. Stephen, A. M. (1980). *In* "Encyclopedia of Plant Physiology: Secondary Plant Products" (Bell, E. A. and Charlwood, B. V. eds.), Vol. 8, pp. 556–584. Springer Verlag, Berlin.

49. Strappaghetti, G., Corsano, S., Craveiro, A. and Proietti, G. (1982). *Phytochemistry* **21**, 2114.,,
50. Thomas, A. F. and Muller, J. M. (1960). *Experientia* **16**, 62.
51. Thomas, A. F., Heusler, K. and Muller, J. M. (1961). *Tetrahedron* **16**, 264.
52. Trease, G. E. and Evans, W. C. (1972). *In* "Textbook of Pharmacognosy", 10th Edn, p. 401. Bailliere Tindall, London.
53. Trumbull, E. R. and Cole, J. R. (1969). *J. Pharm. Sci.* **58**, 176.
54. Vasisth, R. C. and Muthana, M. S. (1955). *J. Sci. Ind. Res. India* **14B**, 632; (1956). **15B**, 25.
55. Wagh, A. D., Paknikar, S. K. and Bhattacharrya, S. C. (1964). *J. Org. Chem.* **29**, 2479.
56. Waterman, P. G. and Khalid, S. A. (1981). *Biochem. Syst. Ecol.* **9**, 45.
57. Wenninger, J. A. and Yates, R. L, (1969). *J. Assoc. Off. Anal. Chem.* **52**, 1155.
58. Wild, H. (1959). *Bol. Soc. Broteriana* **33**, 67.
59. Willis, J. C. (1973). *In* "A Dictionary of the Flowering Plants and Ferns" (revised by Airy Shaw, H. K.), 8th Edn, p. 172. Cambridge University Press, Cambridge.
60. Zoghbi, M., Das Gracas, B., Roque, N. F. and Gottlieb, O. R. (1981). *Phytochemistry* **20**, 180.

CHAPTER 11

Biological Activity of Some Rutaceous Compounds

JOHN R. LEWIS

Department of Chemistry, The University, Aberdeen, U.K.

I. INTRODUCTION

The medicinal value of plants appears in all early records of human endeavour; the Chinese, 5000 years ago, the Sumerians, 4000 years ago, and succeeding generations all had their herbalists and apothecaries, and from these early beginnings the use of plant extracts in the treatment of human disease today has developed.

Rutaceous plants comprise about 150 genera and some 1600 species, and it is not surprising that some of these species are reported to have been used in herbal medicine. Rue (*Ruta graveolens*) appears in all early "pharmacopoeias" and reviews by Le Strange[32] and by Leyel[33] on herbal medicines have comprehensive listings of its use in the treatment of flatulence, amenorrhoea, pains in the joints, gout, removal of worms, wheals, and pimples, etc. Culpeper[12] surprisingly does not refer to any rutaceous plants in his "The English Physitian" of 1676 but the physicians of Myddrai in Wales, who practised through the 12th to 19th Centuries, employed rue extensively in their pharmacopoeia and one such recipe "to prevent speaking during sleep" used rue leaves or seeds pounded with vinegar and mixed with old ale, the resulting concoction being strained before it was drunk. Usher,[47] in his book "A Dictionary of Plants Used by Man" records the use of *Atalantia*, *Dictamnus*, *Glycosmis*, *Murraya*, *Ruta*, and *Zanthoxylum* species in early medicine.

A number of reviews listing the secondary metabolites found in rutaceous plants have appeared; two more recent ones by Waterman list their alkaloids[49] and coumarins[21] and discuss their taxonomic implications. Chapters 3, 4, 7, and 9 of this volume show the structural variety found in both nitrogen containing and non-nitrogenous compounds, particularly coumarins, in this plant family.

The use of plant extracts in the treatment of physiological disorders has necessitated the development of screening procedures to assess the activity of crude plant products and to monitor their purification. *In vitro* and *in vivo* screening programmes for antitumour, antiviral, antimicrobial, antifertility, uterotonic, central nervous system depressant activities, to name a few, now enable the active constituent(s) to be isolated and subsequently identified. In recent years, a number of plant extracts of rutaceous species have been found to possess real pharmacological activity which has allowed the isolation of the active principle(s) followed by their characterization. A number of such plant secondary metabolites will be discussed in detail in the following sections.

II. MISCELLANEOUS BIOLOGICAL ACTIVITIES

A. UTEROTONIC ACTIVITY

Euodia rutaecarpa. The Chinese drug Wu-Chu-Yu has been used in Chinese medicine for two millenia and it is reported to have a number of uses ranging from being a stimulant to an anthelmintic. It has been used to treat headache, abdominal pain, dysentery, postpartum haemorrhage, and amenorrhoea. The active extract is obtained by aqueous extraction of the unripe fruits of *E. rutaecarpa.* This extract has also been reported to decrease mouse litter size and to have a stimulative uterotonic action on the hamster. King and his collaborators[29] at the Chinese University of Hong Kong have isolated two active principles associated with the uterotonic activity and have shown these to be the alkaloids rutaecarpine hydrochloride (1) and dehydroevodiamine chloride (2); evodiamine (3) also isolated was not active.

(1) Rutaecarpine hydrochloride (R = H)   **(3)** Evodiamine

(2) Dehydroevodiamine chloride (R = Me)

B. ANTITHYROID ACTIVITY

Murraya paniculata. Khosa and his collaborators[28] at the Institute of Medical Science at Banaras Hindu University isolated a coumarin (**4**) from the leaves of *M. paniculata* that markedly inhibited the activity of the thyroid gland in rats when given interperitoneally at a dose of 20 μg/100 g.

C. ANTIMICROBIAL ACTIVITY

Aegle marmelos. The essential oil obtained by steam distillation of the leaves of this plant has been shown to possess a broad spectrum of antifungal activity. No specific compounds were identified in this study however.[26]

Clausena heptaphylla and *Murraya koenigii.* Chakraborty and his collaborators have studied a number of carbazole alkaloids for activity against a range of fungi.[11,13] *Microsporum gypseum* and *Trichophyton rubrum* were found to be particularly affected and their growth was inhibited at a concentration of 10 μg/ml by 3-methyl-6-hydroxycarbazole (**5**) (girinimbine) which was one of the most active of all the alkaloids tested.

Fagara xanthoxyloides. In Nigeria and other parts of West Africa tooth cleaning is done with the aid of what is called a chewing stick. It is common

HOCHCH(OMe)C(Me) = CH$_2$

(4)   **(5)** Girinimbine

practice to clean the teeth first with one of these chewing sticks prior to using one of the recognized brands of tooth paste on a brush.

These chewing sticks come from either the root or stem part of a local plant, and the appropriate stem or root is washed free of sand, trimmed of small offshoots, and cut into splinters, approx. 15–30 cm in length, which are sold in bundles in the markets. Chewing sticks impart varying taste sensations; using material made from *Fagara xanthoxyloides*, a tingling peppery taste and numbness has been reported. El Said[15] has studied the antimicrobial properties of buffered extracts of these local plants and has shown that the root of *F. xanthoxoyloides* provided an extract that possessed marked antimicrobial activity against oral flora. Following this observation Odebiyi and Sofowora[36] isolated from the root extract four compounds which possessed antimicrobial activity. Three were identified as canthin-6-one (**6**), chelerythrine chloride (**7**), and berberine chloride (**8**), while the fourth remained unidentified.

Ptelea trifoliata, commonly called the "hop tree" due to its use for making beer, has been a source for herbal remedies in North American homoeopathy since the early 19th Century. Numerous conflicting accounts of the isolation of active principles have been recorded by Bailey,[5] and a report by Lucas suggested that the leaves possessed antibiotic properties against *Mycobacterium tuberculosis*.[34] Mitscher *et al.*[35] have carefully investigated an

(**6**) Canthin-6-one (R = H)   (**7**) Chelerythrine chloride

(**8**) Berberine chloride   (**9**) Pteleatinium chloride

extract of *Ptelea trifoliata* (whole plant) and have found several alkaloids and coumarins to be present; of the secondary metabolites isolated only the quaternary alkaloid pteleatinium chloride (9) showed antimicrobial activity against *M. tuberculosis* and *Staphylococcus aureus*. The six other quaternary alkaloids isolated were not active.

III. ANTISICKLING ACTIVITY

During the investigation by Sofowora and his colleagues[40] into the antimicrobial activity of *Fagara xanthoxyloides* it was observed that blood agar plates showed retention of the red colour within the zone of inhibition of the extract. This observation led to an examination of the effect of the plant extract on haemoglobulin erythrocytes.

Sickle cell anaemia is an inherited disorder of the blood which is associated with the red blood corpuscles having a reduced capacity to bind oxygen and hence produces loss of the red colour in the erythrocyte. This disorder is apparent when the erythrocytes of patients possessing cell anaemia are examined under the microscope, the cell aggregation having a sickle shape. The reversal of this disorder can be induced by a number of agents, e.g. by testosterone, and can be induced by treating normal erythrocytes with sodium metabisulphite.

The observation that an aqueous extract of *Fagara xanthoxyloides* preserved the colour of blood agar plates[15] suggested that the extract in some way prevented the discolouration of the blood by acting on the erythrocyte so as to inhibit cell lysis. In a preliminary investigation Sofowora and Isaacs[40] demonstrated that metabisulphite crenation of human AA erythrocytes could be reversed through treatment with the root extract of *F. xanthoxyloides*. Following up this observation of an antisickling effect, Sofowora and colleagues[41] isolated an acidic fraction from the plant extract which possessed activity (Table 1) and subsequently showed that the active constituent was 2-hydroxymethylbenzoic acid (10). In a later publication Elujoba and Sofowora[16] reported that an acidic antisickling fraction could be obtained from 6 Nigerian *Fagara* species and that this fraction contained several benzoic acid derivatives of which 2-hydroxymethylbenzoic acid (10), *p*-hydroxybenzoic acid (11), and vanillic acid (12) possessed antisickling activity. Farnsworth and his colleagues[24] in an independent investigation of the extracts obtained from samples of Nigerian *Fagara xanthoxyloides* could find no antisickling activity against erythrocytes obtained from asymptomatic sickle cell anaemia patients (a different procedure to that used by Sofowora).

306   J. R. LEWIS

TABLE 1

Effect of the root extract of *F. xanthoxyloides* on crenated haemoglobin AA and sickled haemoglobin AS erythrocytes

Blood treated with	Crenated cells after 15 h (% of total erythrocytes)*	Sickled cells after 15 h (% of total erythrocytes)†
Root extract	5	40
Testosterone propionate (4 μg/ml)	40	50
T.C. 199 (control)	70	95

*Experiment performed with haemoglobin AA erythrocytes. The erythrocytes were 100% crenated at 0 h.

†Experiment performed with haemoglobin AS erythrocytes. 95%+ of the erythrocytes were sickled (metabisulphite) at 0 h.

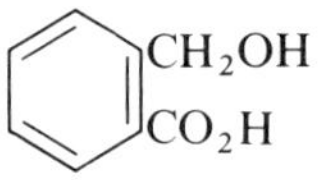

(10)    (11) 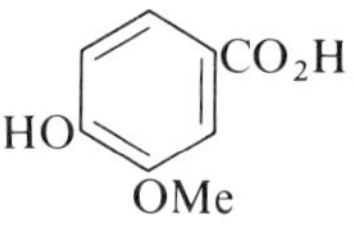   (12)

IV. ANTITUMOUR ACTIVITY

The search for a cure for cancer has led to extensive screening programmes being developed in the Eastern and Western hemispheres and resulted in both synthetic compounds and natural secondary metabolites being eagerly sought for test. In Europe individual institutes, in the U.S.A. the National Cancer Institutes as well as universities and pharmaceutical companies, have established antitumour screens. The primary test for cytostatic activity uses KB cells, subcultured cells derived from a human carcinoma of the nasopharynx. These are grown *in vitro* and the ED_{50} (the dose required to kill 50% of the cells) required for continued investigation is <2 μg/ml of substrate. Additional tests use *in vivo* PS388 and LE1210 mouse leukemias, and in the test now widely used by the National Cancer Institute (PS388) the activity of the sample must show that the ratio of survivors to control, expressed as a percentage, is >125 (i.e. T/C >125) to warrant further investigation. Thereafter active compounds are tested against a wide range of tumour systems so as to establish whether they possess specific antitumour activity.

In 1972 Bowen and Lewis[8] commenced collecting rutaceous plants from Malaysia, Hong Kong, and Sri Lanka so as to be able to assess their extracts for alkaloids, steroids/terpenoids, and antitumour activity. Table 2 shows the results obtained; although several plant extracts gave positive results for

TABLE 2

Antitumour activity of extracts of Malayan rutaceous plants[8]

Plant name	Plant part	$ED_{50}(\mu g/ml)$ KB	T/C(%) LE 1210	T/C(%) PS 388	Occurrence of alkaloids*
Acronychia porteri	S	100		90–109	+
	L	100		90–118	(+)
Atalantia roxburghiana	S	17–100		100–112	+
	L	100		87–118	+
Burkillanthus malaccensis	S	100		95–109	(+)
	L	100		100–113	−
Clausena excavata	B			105–110	−
Euodia euneura	S	100		104–127	−
Euodia glabra	S	87		100–110	(+)
	L	19	96–108	95–123	−
	B	30	97–103	86–112	
Euodia macrocarpa	S	100		95–104	(+)
	L	26	104–109	86–100	(+)
Euodia roxburghiana	L	100		95–109	(+)
Glycosmis calcicola	S	100		96–118	(+)
	B	100		95–100	(+)
Merope angulata	S	100		95–115	(+)
	L	100		95–125	(+)
Merrillia caloxylon	S			100–130	(+)
	L	23	96–99	95–118	+
	F	26	91–100	91–95	+
Paramignya lobata	S	100		90–104	(+)
	L	100	96–105	90–128	(+)
Tetractomia tetrandra	S	100		95–134	(+)
	L	18–25		95–133	(+)
Tetractomia roxburghii	S	100		100–110	(+)
	L	100		95–127	−
Triphasia trifolia	S	100	96–99	91–100	−
	L	100	99–108	91–104	+
Zanthoxylum myriacanthum	S	100		87–100	(+)
	L	100		105–110	(+)
	B	100		105–120	+

S = stem; L = leaf; F = fruit; B = bark.
* Based on TLC analysis with 4 systems; + = all TLC studies indicate alkaloids; (+) = some positive; − = all negative.

alkaloids (Dragendorff, Mayers, iodoplatinate reagents) no correlation with antitumour activity was found.

During the mid 1970s the National Cancer Institute considered that the order Sapindales, in which the Rutaceae were included, should be given a "high interest status" with respect to further investigation for antitumour activity. Barclay and Perdue[6] summarized these results (Fig. 1) and quoted, for the Rutaceae specifically, a number of alkaloids which possessed antitumour activity (Table 3) while Hartwell[23] has listed those rutaceous plants that have been used for the treatment of cancer. Many of the secondary metabolites isolated from rutaceous plants that possessed antitumour activity were alkaloids, including berberine (8), fagaronine (13), nitidine (14), 5-methoxycanthin-6-one (6, R = OMe), 4-methylthiocanthin-6-one (15), and acronycine (16); the flavone eupatorin (17) also showed cytotoxic activity.

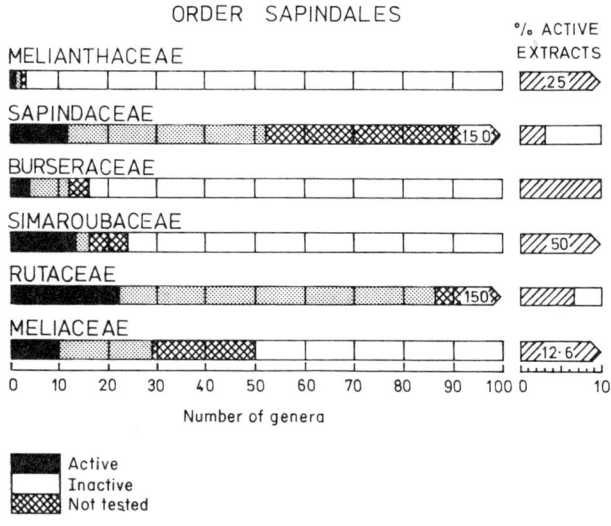

FIG. 1. Distribution of anticancer activity among extracts of some Sapindales *s.l.*[5]

Eupatorin was first isolated by Kupchan and his group from *Eupatorium* species,[31] and Adam and Lewis found it to be present in large quantities (>4%) in the juice of the fruit (Fig. 2) of the rare monotypic Malesian rutaceous plant *Merrillia caloxylon*.[2,3]

Acronycine (16) deserves special mention. It was first investigated[44] in 1966 and found to possess a broad spectrum of antitumour activity (14 out of 17 neoplasms responded). A sustained programme of research was initiated into whether it could be used clinically for the treatment of cancer. The main

TABLE 3

"High interest" anticancer compounds in the Rutaceae[6]

Compound	Source(s)
Berberine sulphate	*Zanthoxylum monophyllum*
Fagaronine	*Zanthoxylum (Fagara) xanthoxyloides*
5-Methoxycanthin-6-one	*Pentaceras australis*
8-Methoxydihydronitidine	*Zanthoxylum (Fagara) leprieurii*;
	Fagara macrophylla (= Zanthoxylum gilletii)
O-Methylfagaronine	*Z. xanthoxyloides*
4-Methylthiocanthin-6-one	*P. australis*
Nitidine chloride	very widespread in *Zanthoxylum (Fagara)*;[49]
	Toddalia asiatica
Oxynitidine	*F. macrophylla*

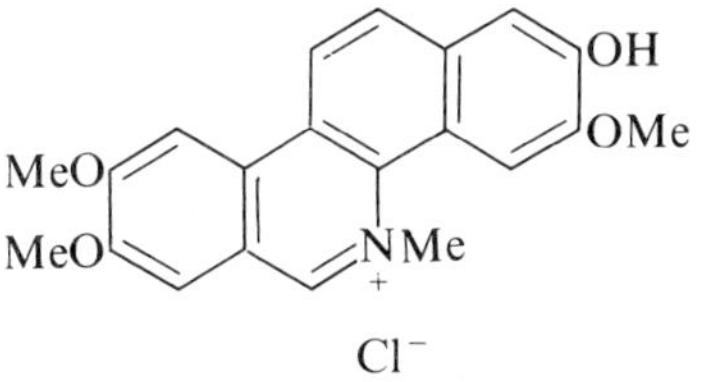

(13) Fagaronine

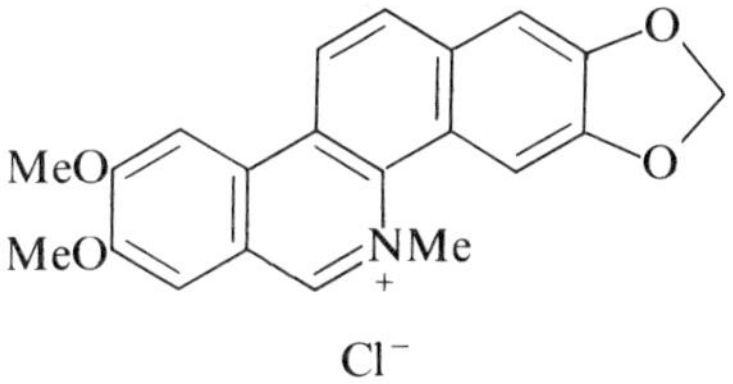

(14) Nitidine

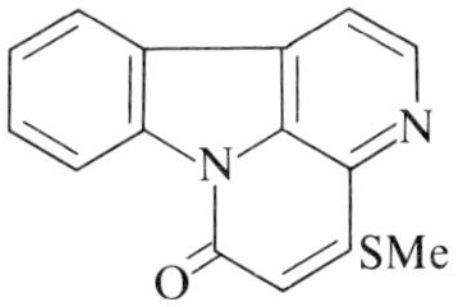

(15) 4-Methylthiocanthin-6-one

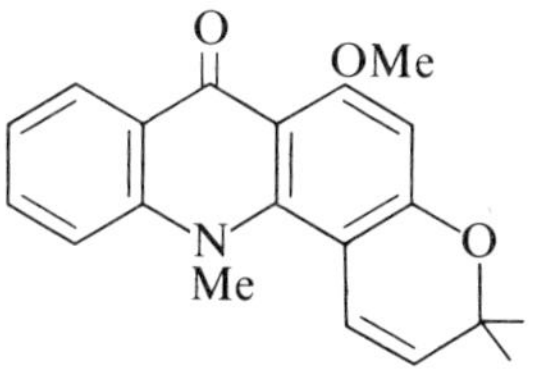

(16) Acronycine

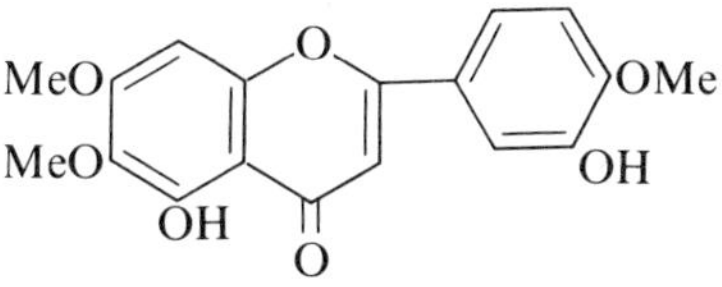

(17) Eupatorin

FIG. 2. The fruit of *Merrillia caloxylon*, showing the copious juice that contains eupatorin (**17**).[2,3]

problem to emerge in the study of acronycine as an antitumour agent was its poor solubility[45] (3 mg/l acetone, 27 mg/l chloroform, <1 mg/l water). In an attempt to increase solubility two methods have been used: (i) physical modification of the particle sizes of the drug by forming a coprecipitate with polyvinylpyrrolidone[45] which increased the solubility of the drug by two orders of magnitude, or micronizing it with starch;[46] (ii) production of a prodrug through the formation of a quaternary salt (**18**) which released acronycine on hydrolysis.[42] Repta and his collaborators[7,22,38] have studied the synthesis of acetylacroninium salts (**18**) in some detail (using various anions) and in general find that hydrolysis takes place readily, the half-life of the salt being 20 minutes in water over the pH range 0–9. The use of a complex

$$FSO_3^-$$ (**18**)

between acetylacroninium perchlorate and sodium gentisate has reduced the hydrolysis rate considerably[30] (1 % in 8 min). In all cases where solubility was increased there was a corresponding increase in inhibition of X5563 plasma cell myeloma.[45]

An alternative scheme for improving solubility and access of acronycine to tumour sites through the prodrug strategy was based on the biogenesis of acridone alkaloids (Scheme 1). In studies of the biogenesis and biomimetic synthesis of acridone alkaloids Lewis and his collaborators[1] found that the 2-amino-2'-methoxybenzophenones were converted by base into acridones at room temperature. One of their aminobenzophenones, tecleanone, was a natural product co-occurring with its acridone alkaloid (1,3-dimethoxy-10-methylacridone) (Scheme 2) in a number of rutaceous species.[49] Based on this biomimetic reaction preacronycine (Scheme 1) was synthesized[1] and found to

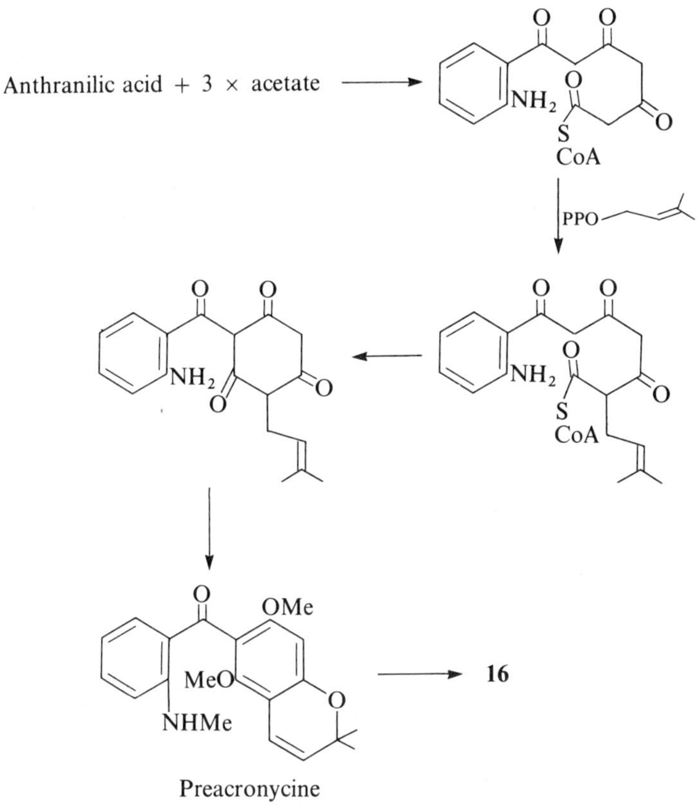

SCHEME 1. Biogenesis of the acridone alkaloid acronycine (16).[1]

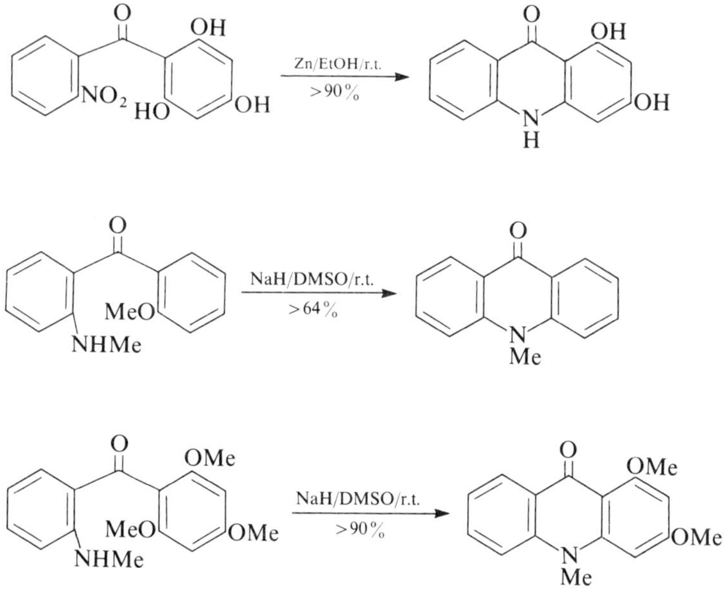

SCHEME 2. Biomimetic syntheses of acridone alkaloids.

cyclize under basic conditions to produce an equal amount of acronycine and isoacronycine. On testing with the X5563 myeloma, preacronycine was found to be only slightly active.

The mechanism of action of acronycine on cell growth appears to involve inhibition of nucleoside transport, at least in marine cells,[27] while fagaronine, a benzophenanthridine alkaloid, inhibits RNA directed DNA polymerase activity in avian myeloblastosis virus. Both fagaronine and the related nitidine inhibit viral reverse transcriptase and DMA polymerase activity in NIH Swiss mouse embryos.[37]

Cytostatic activity has also been found in a number of coumarins, furocoumarins, and pyranocoumarins. Gonzalez,[19] in his paper in 1977, quotes activities of a number of these compounds against HeLa cells (Table 4). Only rutamarin was strongly active and heliettin and xanthyletin moderately so. Recently the lignan diphyllin (19), from *Haplophyllum hispanicum*, has been found to be particularly toxic (0.5 μg/ml for ED_{50}) to HeLa cells.[18]

Cassady[10] has characterized the antitumour agent of *Micromelum integerrimum* as the coumarin micromelin (20). This epoxide showed reproducible PS388 activity (T/C 149% at 10 μg/ml) but erratic 9KB and lung carcinoma activity.

TABLE 4

Cytotoxic activity of coumarins from the
Rutaceae against HeLa cells[19]

Compound	ED_{50} ($\mu g/ml$)
Simple coumarins	
Pinnarin	100
Sabandin	100
Sabandinin	100
Furocoumarins	
Benahorin	75
Bergapten	75
Byakangelicin	100
Chalepensin	100
Furopinnarin	100
Heliettin	10
Isoimperatorin	100
Isopimpinellin	75
Pangelin	100
Psoralen	100
Rutamarin	0.1
Tederin	100
Xanthotoxin	75
Pyranocoumarins	
Luvangetin	100
Xanthyletin	10
Standard	
6-Mercaptopurine	0.1

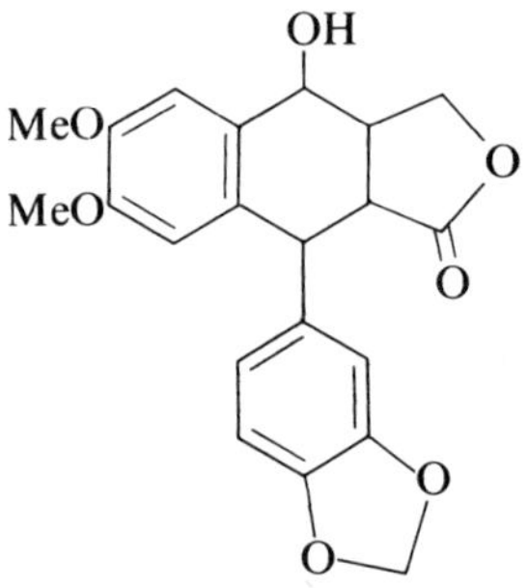

(19) Diphyllin

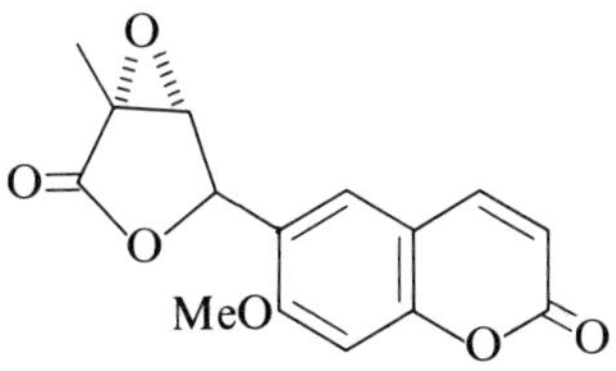

(20) Micromelin

V. Phototoxicity

It has long been known that certain plants cause ulceration and blistering of the skin when contact has been made. One of the early references to the deleterious effect of rutaceous plants on the skin was quoted by Gray[20] in his "Genera of Plants of the United States" where it is mentioned that Spanish *Ruta montana* produces blisters and ulcerous pustules when applied to the naked skin. This effect has been found to be due to the adsorption of the juice of the plant into the skin which, when followed by exposure to sunlight, can lead, after a delay of some hours, to the development of erythema.[17] The effect is one of photosensitization of the skin as no symptoms are noted in the absence of strong light. A summary of the early work in this field has been given by Brown, where the role of psoralen (21) and furocoumarins related to it are described.[9]

The ability of furocoumarins to cause photodermatitis in man is a result of the transport of the furocoumarin to cutaneous tissue whereupon a molecular complex is formed between it and native DNA. Observations as to the nature of this complex have been made by studying changes in the UV spectrum of the furocoumarins when DNA has been introduced. Decout and Lhomme[14] have made a study of the interaction between 8-alkoxypsoralens and thymine and suggest that in these model studies ring–ring stacking takes place. These complexes do not photosensitize the skin but on irradiation with light of long wavelength a further reaction occurs which results in covalent bonding between the nucleic acid bases thiamine and pyrimidine giving eventually crosslinking between DNA strands. Rapoport and his group have made a detailed study[43] of the photoreaction between pyrimidine in calf thymus DNA and 4'-hydroxymethyl-4,5',8-trimethylpsoralen and conclude that the five monoaddition products which can be isolated and characterized all have the cyclobutane ring produced as a result of a $2 + 2$ cycloaddition reaction involving the 4',5' furan ring double bond of the coumarin and the 5,6 double bond of the pyrimidine ring system or the 3,4 double bond of the pyrone ring of the coumarin and the 5,6 double bond of the pyrimidine. Their elegant spectroscopic studies on these adducts have established their stereochemistry in part and Scheme 3 indicates two of the monocycloaddition adducts. Further reaction of these monoadducts with another DNA strand would give a crosslink and account for the observed antimitotic effect of long UV and furocoumarins on cell replication.

Psoriasis is a skin disorder which shows itself as "mother of pearl" blotches on the skin surface. These can be localized or spread extensively. Psoriatic skin cells have a higher mitotic rate than normal cells and attempts have been made to attack these cells preferentially by localized treatments. Patients have been

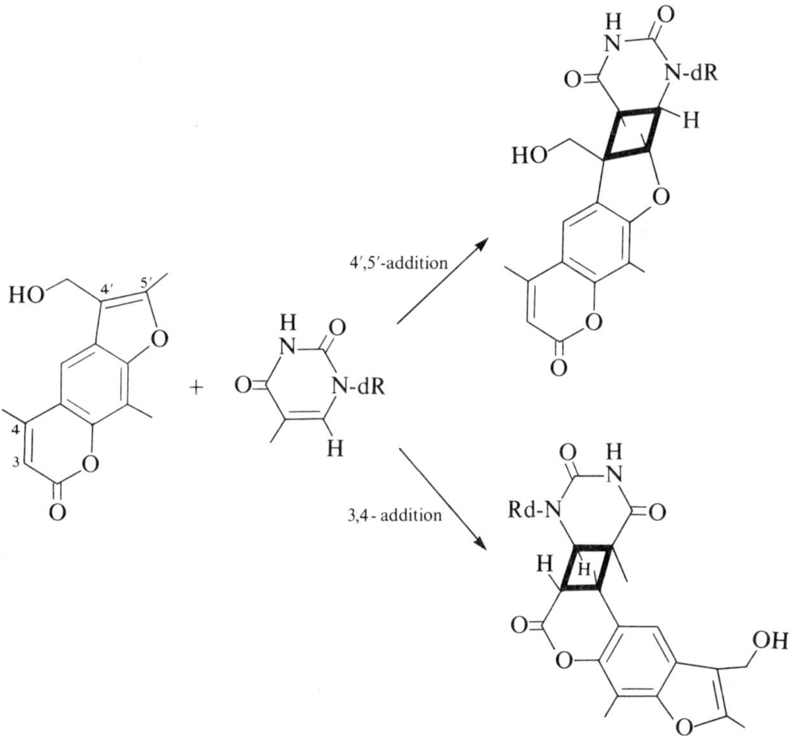

SCHEME 3. Two photo-adducts of a furocoumarin and calf thymus DNA.[43]

subjected to sunlight, UV light or X-rays in an attempt to reduce mitosis and increase pigmentation but usually this type of treatment has only given temporary relief. The use of 8-methoxypsoralen (22) followed by UV exposure has given rise to most promising results, the treatment being of a lasting nature with little or no apparent side effects.[39] Patients are given 20–50 mg doses of the drug orally 2 hours before UV irradiation at 220–390 nm (with a maximum intensity at ~365 nm). It is estimated that half of the drug is excreted within 6 hours of ingestion mainly through metabolism via its hydroxylated derivatives 23 and 24.[39]

Towers and his colleagues have examined the phototoxicity of furoquinolines and compared their effect on *E. coli* and *Saccharomyces cerevisiae* with that of 22. They found that 5 of 12 furoquinolines examined, isodictamnine (27), 7-desmethylskimmianine (28), maculosidine (29), maculine (30), and dictamnine (31), were active, but at higher concentrations than that required by 8-methoxypsoralen (22).[48]

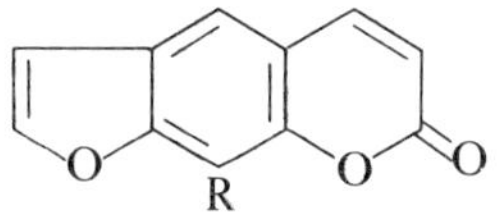

(21) Psoralen (R = H)
(22) Xanthotoxin (R = OMe)

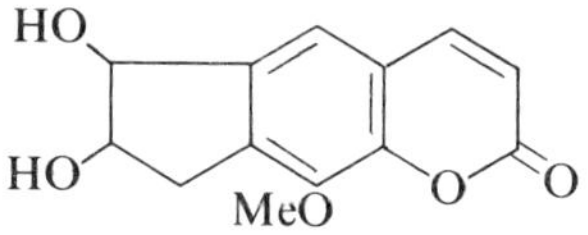

(23)

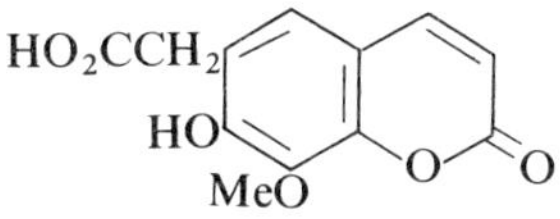

(24) 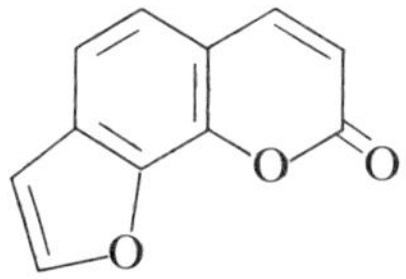   (25) Angelicin 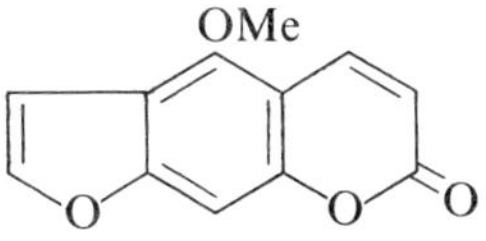   (26) Bergapten

A number of studies on furocoumarins have reported that toxic effects can be associated with their photobiology. Ashwood-Smith et al.[4] have shown that psoralen (21), angelicin (25), xanthotoxin (22), and bergapten (26) are lethal to E. coli cells (Fig. 3), but only after irradiation. Gonzalez,[19] as reported earlier, has also shown that some are weakly cytotoxic to HeLa cells. Because of these observations some concern has been expressed about suntan preparations which contain psoralen derivatives (for inducing melanin formation), while raw and cooked parsnip roots have been shown to contain 21, 22, and 26,[25] and their presence, albeit in low concentrations, may be thought to be a hazard.

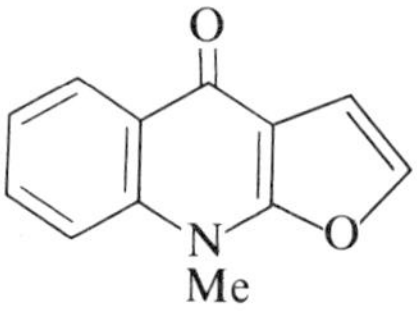

(27) Isodictamnine 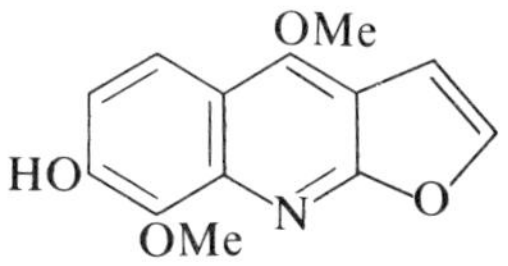   (28) 7-Demethylskimmianine 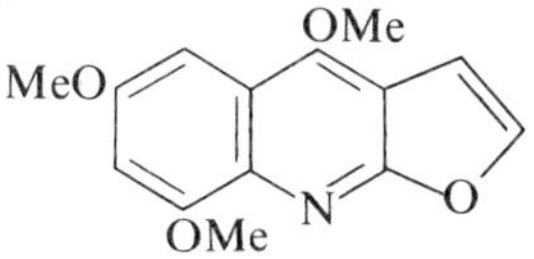   (29) Maculosidine

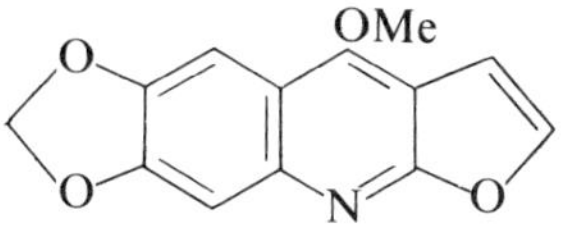

(30) Maculine 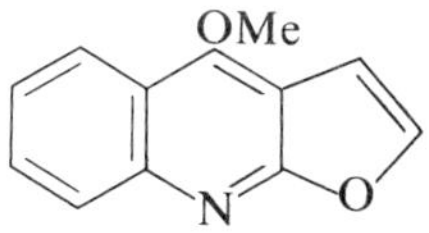   (31) Dictamnine

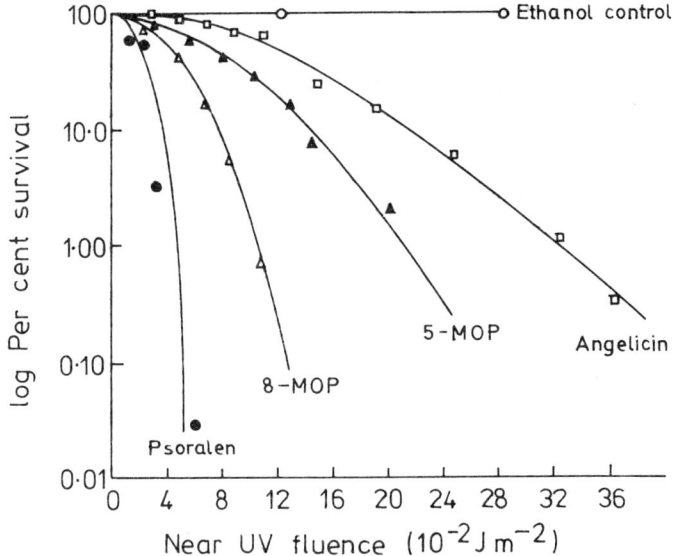

FIG. 3. Phototoxicity of furocoumarins towards *E. coli*.[4] 8-MOP = xanthotoxin; 5-MOP = bergapten.

REFERENCES

1. Adams, J. H., Brown, P. M., Gupta, P., Khan, M. S. and Lewis, J. R. (1981). *Tetrahedron* **37**, 209.
2. Adams, J. H. and Lewis, J. R. (1977). *Planta Med.* **32**, 86.
3. Adams, J. H. and Lewis, J. R. (1978). *Malays. J. Sci.* **5B**, 173.
4. Ashwood-Smith, M. J., Poulton, G. A., Barker, M. and Mildenberger, M. (1980). *Nature (London)* **285**, 407.
5. Bailey, V. L. (1960). *Econ. Bot.* **14**, 180.
6. Barclay, A. S. and Perdue Jr, R. E. (1976). *Cancer Treatment Rep.* **60**, 1081.
7. Bourne, D. W. A., Higuchi, T. and Repta, A. J. (1977). *J. Pharm. Sci.* **66**, 628.
8. Bowen, I. H. and Lewis, J. R. (1978). *Planta Med.* **34**, 128.
9. Brown, S. A. (1979). *In* "Biochemistry of Plant Phenolics" (Swain, T., Harborne, J. B. and Van Sumere, C. F. eds.), pp. 265–275. Plenum Press, New York.
10. Cassady, J. M., Ojima, N., Chang, C.-J. and McLaughlin, J. L. (1979). *Lloydia* **42**, 275.
11. Chakraborty, D. P., Das, K., Das, B. P. and Chowdhury, B. K. (1975). *Trans. Bose Res. Inst. Calcutta* **38**, 1; *Chem. Abs.* (1977). **86**, 51029z.
12. Culpeper, N. (1676). "The English Physitian" (Enlarged). George Sawbridge, London.
13. Das, K. C., Chakraborty, D. P. and Bose, P. K. (1965). *Experientia* **21**, 340.
14. Ducout, J. L. and Lhomme, J. (1981). *Tetrahedron Lett.* **22**, 1247.

15. El-Said, F., Fadulu, S. O., Kuye, J. O. and Sofowora, E. A. (1971). *Lloydia* **34**, 172.
16. Elujoba, A. A. and Sofowora, E. A. (1977). *Planta Med.* **32**, 54.
17. Fitzpatrick, T. B. (1974). *In* "Sunlight and Man" (Musago, L. ed.), pp. 369–390. University of Tokyo Press, Tokyo.
18. Gonzalez, A. G., Darias, V. and Alonso, G. (1979). *Planta Med.* **36**, 200.
19. Gonzalez, A. G., Darias, V., Alonso, G., Boada, J. N. and Rodrigue-Luis, F. (1977). *Planta Med.* **31**, 352.
20. Gray, A. (1849). "Genera of Plants of the United States".
21. Gray, A. I. and Waterman, P. G. (1978). *Phytochemistry* **17**, 845.
22. Hansen, A. B., Kreilgard, B., Huang, C.-H. and Repta, A. J. (1978). *J. Pharm. Sci.* **67**, 237.
23. Hartwell, J. L. (1971). *Lloydia* **34**, 153.
24. Honig, G. R., Farnsworth, N. R., Ferenc, C. and Loyda, N. V. (1975). *Lloydia* **38**, 387.
25. Ivie, G. W., Holt, D. L. and Ivey, M. C. (1981). *Science* **213**, 909.
26. Jain, N. K. (1977). *Indian Drugs Pharm. Ind.* **12**, 55; *Chem. Abs.* (1977). **87**, 130449.
27. Kessel, D. (1977). *Biochem. Pharmacol.* **26**, 1077.
28. Khosa, R. L. and Siddiqui, M. Z. (1973). *Indian J. Pharm.* **35**, 32; *Chem. Abs.* (1973). **78**, 143714g.
29. King, C. L., Kong, Y. C., Wong, N. S., Yeung, H. W., Fong, H. H. S. and Sankowa, U. (1980). *Lloydia* **43**, 577.
30. Kreilgard, B. (1978). *Arch. Pharm.* (*Chem. Ed.*) **6**, 116; *Chem. Abs.* (1978). **89**, 117658.
31. Kupchan, S. M., Sigel, C. W., Hemingway, R. J., Knox, J. R. and Udayamurthy, M. S. (1969). *Tetrahedron* **25**, 1603.
32. Le Strange, R. (1973). "A History of Herbal Plants", p. 216. Angus and Robertson, London.
33. Leyel, C. F. (1926). "The Magic of Herbs". Cape, London.
34. Lucas, E. H., Lickfeldt, A., Gottshall, R. Y. and Jennings, J. C. (1951). *Bull. Torrey Bot. Club.* **78**, 310.
35. Mitscher, L. A., Bathala, M. S., Clark, G. W. and Beal, J. L. (1975). *Lloydia* **38**, 109.
36. Odebiyi, O. O. and Sofowora, E. A. (1979). *Planta Med.* **36**, 204.
37. Phillips, S. D. and Castle, R. N. (1980). *J. Heterocyclic Chem.* **17**, 1489.
38. Repta, A. J., Dimmock, J. R., Kreilgard, B. and Kominski, J. J. (1977). *J. Pharm. Sci.* **66**, 1501.
39. Rusciani, A. (1980). *J. Ital. Dermatol. Venereal.* **115**, 381.
40. Sofowora, E. A. and Isaacs, W. A. (1971). *Lloydia* **34**, 383.
41. Sofowora, E. A., Isaac-Sodeye, W. A. and Ogunkoya, L. O. (1975). *Lloydia* **38**, 169.
42. Smethwick, E. L. U.S. Patent 3,843,658; *Chem. Abs.* (1975). **82**, 31430.
43. Straub, K., Kanne, D., Hearst, J. E. and Rapoport, H. (1981). *J. Am. Chem. Soc.* **103**, 2347.
44. Svoboda, G. H., Poore, G. A., Simpson, P. J. and Boder, G. B. (1966). *J. Pharm. Sci.* **55**, 758,
45. Svoboda, G. H., Sweeney, M. J. and Walkling, W. D. (1971). *J. Pharm. Sci.* **60**, 333.
46. Svoboda, G. H. U.S. Patent 3,374,588; *Chem. Abs.* (1976). **84**, 126778.
47. Usher, G. (1974). "A Dictionary of Plants Used by Man". Constable, London.
48. Towers, G. H. N., Graham, E. A., Spenser, I. D. and Abromowski, Z. (1981). *Planta Med.* **41**, 136.
49. Waterman, P. G. (1975). *Biochem. Syst. Ecol.* **3**, 149.

CHAPTER 12

Chemistry and Significance of Selected Citrus Limonoids and Flavonoids

VINCENT P. MAIER

Fruit and Vegetable Chemistry Laboratory, U.S.D.A., Pasadena, California, U.S.A.

I. OVERVIEW OF THE CITRUS INDUSTRY

The genus *Citrus* contains a number of species that produce a fruit crop that represents the main horticultural crop of the Rutaceae and of the Rutales. Citrus fruits are grown in tropical and subtropical climates in a belt extending around the world from roughly 35° north latitude to 35° south latitude.[9] During the 1979–80 season the world production of citrus fruits was approximately 47 million metric tons, of which 55% was marketed as fresh fruit. The remaining 45% of the crop was converted into a variety of processed products.[3]

The top five major producing countries are the United States, Brazil, Japan, Spain, and Italy, which together produce almost 75% of the world crop (Table 1). The sweet orange (*Citrus sinensis*) is the main fruit produced. It makes up 66% of the crop, followed by the tangerine (*C. reticulata*) 15.4%, the grapefruit

320   V. P. MAIER

TABLE 1

Major citrus fruit producing countries; data for 1979–80 crop year.[3]

Country	Production (million metric tons)	Percentage of world production
United States	15.0	32.1
Brazil	9.8	20.9
Japan	4.3	9.2
Spain	3.0	6.4
Italy	2.9	6.2
World total	46.8	

(*C. paradisi*) 8.5%, and the lemon (*C. limon*) 6.8%. The remaining 3.3% is made up of the lime (*C. aurantifolia*), the citron (*C. medica*), the sour orange (*C. aurantium*), and certain hybrids.[3]

In the United States, the citrus crop ranks about seventh in value at $1.6 billion, preceded only by corn, soybeans, hay, wheat, cotton, and tobacco (Table 2). In terms of the overall U.S. fruit and vegetable crop, citrus ranks first (Table 3), followed by potatoes, grapes, tomatoes, apples, and lettuce.[2]

Citrus fruits are grown commercially in four states in the United States, Florida with 73.3% of U.S. production, California with 18.9%, Texas with 4.8%, and Arizona with 3.0%. California produces 80% of the total lemon crop and 64% of the fresh orange crop while Florida produces 86% of all fruit processed.[7] Overall about 25% of the U.S. citrus crop is marketed as fresh fruit. The other 75% is converted into processed products, primarily juice products.[7] On the order of 5.5 million metric tons of peel, pulp, and seed

TABLE 2

Major crops of the United States; data for 1978 crop year.[2]

Crop	Value of production (billion $)	Area harvested (million hectares)
Corn grain	14.9	28.3
Soybeans	12.2	25.5
Hay	6.8	24.9
Wheat	5.3	23.0
Cotton	3.1	5.0
Tobacco	2.7	—
Citrus	1.6	0.49

TABLE 3

Major fruit and vegetable crops of the United States; data for 1978 crop year.[2]

Crop	Value of production (billion $)	Area harvested (million hectares)
Citrus	1.60	0.49
Potatoes	1.27	0.55
Grapes	1.00	—
Tomatoes	0.85	0.17
Apples	0.78	—
Lettuce	0.65	0.10

residues are generated as a result of the processing operations. By-products such as pectin, essential oils, flavonoids, citrus molasses, etc. are extracted from portions of these residues, and the remainder, which makes up the bulk, is used for animal feed in both the wet and dried forms.

The size, diversity, and technological sophistication of the citrus industry has been a stimulus for a great deal of research on the chemistry and properties of citrus fruit constituents. This work has shown that a broad array of primary and secondary metabolites contribute to the distinctive characteristics that set citrus fruits apart from other fruits and from each other. Two comprehensive reviews of citrus fruit composition, constituents, and properties have appeared in recent years.[61,62] The present chapter will focus on aspects of the chemistry, biochemistry, and technology of two classes of secondary metabolites, the flavonoids and limonoids, of importance to the citrus industry because of their taste properties.

II. THE FLAVONOID AND LIMONOID BITTER SUBSTANCES

A. COMPARATIVE ASPECTS

Bitterness of citrus fruits and/or juices is caused primarily by bitter members of two widely different classes of compounds, the flavonoids and the limonoids. Limonoid bitterness is characterized by the gradual development of bitterness in citrus juices following expression of the juice from the fruit. This is sometimes referred to as delayed bitterness and is caused by bitter limonoids, primarily limonin (1) and to a lesser extent nomilin (2).

The absence of bitterness in the intact fruit and the delay in the onset of bitterness after juicing differentiates limonoid bitterness from that due to the

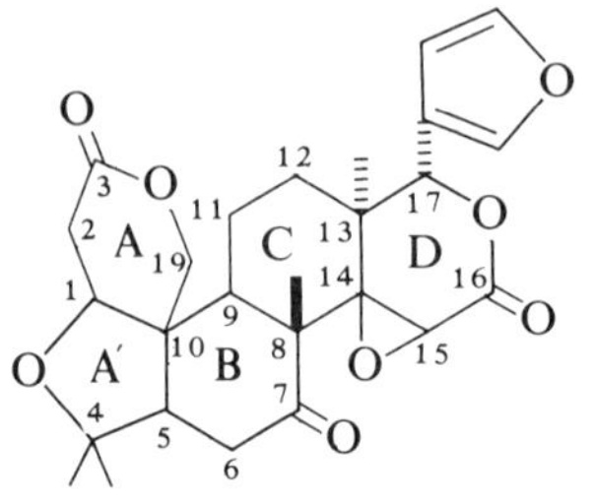

(1) Limonin

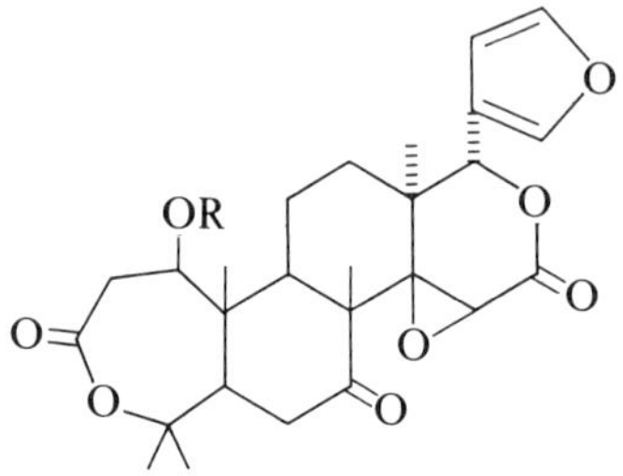

(2) Nomilin (R = Ac)

(14) Deacetylnomilin (R = H)

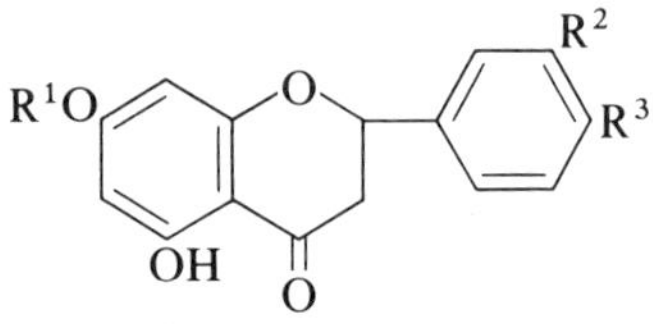

(3) Naringin (R^1 = β-neohesperidosyl; R^2 = H; R^3 = OH)

(4) Neohesperidin (R^1 = β-neohesperidosyl; R^2 = OH; R^3 = OMe)

(10) Hesperidin (R^1 = β-rutinosyl; R^2 = OH; R^3 = OMe)

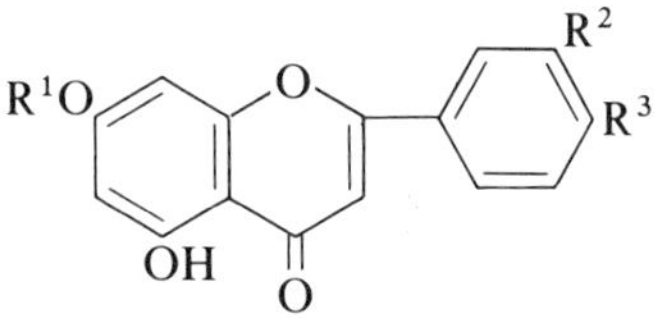

(5) Rhoifolin (R^1 = β-neohesperidosyl; R^2 = H; R^3 = OH)

(6) Neodiosmin (R^1 = β-neohesperidosyl; R^2 = OH; R^3 = OMe)

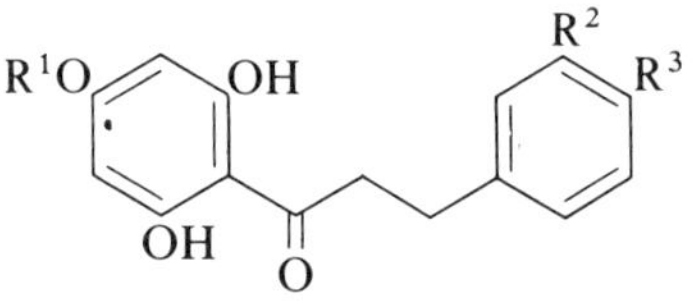

(7) Naringin dihydrochalcone (R^1 = β-neohesperidosyl; R^2 = H; R^3 = OH)

(8) Neohesperidin dihydrochalcone (R^1 = β-neohesperidosyl; R^2 = OH; R^3 = OMe)

(9) Hesperetin dihydrochalcone glucoside (R^1 = β-D-glucosyl; R^2 = OH; R^3 = OMe)

flavanone neohesperidosides, such as the naringin (**3**) of grapefruit and the neohesperidin (**4**) of the sour orange (*C. aurantium*). The presence of flavanone neohesperidosides causes the intact fruit to be bitter and also imparts immediate bitterness to the freshly prepared juice. These bitter flavonoids occur only in a few *Citrus* species, namely the grapefruit, pummelo (*C. grandis*), sour orange, and Ponderosa lemon (a hybrid). They do not occur in the sweet orange, lemon, lime, tangerine, and citron (*C. medica*).[1,51] In contrast, **1** appears to occur in all *Citrus*,[17] although it may not be present in sufficient quantities at maturity to cause delayed bitterness of the juice.[55]

Delayed bitterness is most noticeable in the juices of the navel, Murcott, and Shamouti oranges. It also occurs in the juice of grapefruit and Natsudaidai, but its contribution to the taste properties is not as obvious because of the presence of the bitter flavonoids. Juices of lemon and lime also exhibit delayed bitterness but these juices are rarely consumed undiluted because of their high acidities, consequently bitterness is not commonly noticed. Circumstances can arise whereby it is possible for limonin to occur in sufficient quantities to cause bitterness in almost any citrus juice. In addition, in cases where the intact fruit is injured, such as through frost damage before harvest or by bruising, the fresh fruit itself can exhibit limonoid bitterness. The chemistry of the delayed bitterness phenomena will be discussed in a later section.

B. TASTE PROPERTIES

While the taste properties of the bitter flavonoids and limonoids have been known from the time of their first isolation in pure crystalline form, very little quantitative information about their taste threshold concentrations was available until the work of Guadagni and associates in the 1970s.

1. *Bitterness* vs *Acceptability*

Guadagni[20] studied the relationship between **3** content, bitterness rating, and acceptability of grapefruit juice using a panel of 27 judges. Very close relationships were found to exist between **3** level and acceptability, and between bitterness rating and acceptability. Over a range of increasing **3** levels (from 120 to 730 ppm) the acceptability of grapefruit juice dropped and the bitterness rating increased. There was a linear relationship between bitterness rating and acceptability rating, with a correlation coefficient of $+0.99$. This showed that there is a close relationship between perceived bitterness and acceptability. This held over the wide range of age groups represented on the panel (20 to 61 years). Thus, while individuals vary in their ability to detect bitterness (Table 4), their assessment of acceptability at a given degree of perceived bitterness appears to be in close agreement. Similar results have been obtained with a consumer type panel of 72 persons where both limonin and naringin were added to grapefruit juice.[19] Studies have also shown that limonin bitterness reduces juice acceptability.[10,14,22]

TABLE 4

Range of individual thresholds for limonin and naringin in water and citrus juice and group thresholds in water[22]

Liquid medium	Individual thresholds[a]	
	limonin (ppm)	naringin (ppm)
Water	0.075 to 5.0	1.5 to 50
Orange juice	0.50 to 32	—
	Group thresholds[b]	
	limonin (ppm)	naringin (ppm)
Water	1.0	20
Orange juice	6.0	—

[a] Range among individuals of a 27 member panel.
[b] Concentration at which 75% of panel could detect bitterness.

2. Effects of Other Flavonoids

As shown in Table 4, the threshold concentration (the minimum concentration detected by human subjects by taste) of **1** in orange juice is 6-fold higher than it is in pure water. Guadagni et al.[22] showed that juice constituents such as sucrose and citric acid and factors such as pH and total soluble solids/total acidity could account for some of this increase in the threshold. In addition, Guadagni and associates studied the influence of certain citrus and citrus derived flavonoids on the bitterness of **1** and **3**.

(a) Flavone neohesperidosides. The bitter flavanone neohesperidosides **3** and **4** are converted to their tasteless flavone analogs rhoifolin (**5**) and neodiosmin (**6**) by dehydrogenation of the 2,3 region.[52] The close structural relationship between the flavanones and flavones led Horowitz and Gentili[45,47] to the finding that **5** had the ability to suppress the bitterness of **3**, presumably by competing for the **3** taste receptor sites but without causing a taste response of its own. At the suggestion of Horowitz and Gentili, Guadagni and associates carried out a comprehensive study using the more readily soluble **6** prepared by oxidation of **4**.[52]

The effect of **6** on the thresholds of a variety of bitter substances is summarized in Table 5. As was originally observed for **5**, **6** also has the ability to suppress the bitterness of the flavanone neohesperidoside **3**. At the same time, when 20 ppm **4** was added to a grapefruit juice containing 380 ppm **3** and 5.1 ppm **1** 82 % of a 36 judge panel preferred the **6** containing juice over the control juice. These workers showed that **6** also has the ability to suppress the bitterness of aqueous solutions of such structurally unrelated substances as **1**, caffeine, quinine sulfate (Table 5), and the bitter aftertaste of saccharin (Table 6). In addition, **6** was effective in suppressing **1** bitterness in orange juice (Table 5), and in reducing bitterness of suprathreshold levels of **1** (10 ppm) when

TABLE 5

Effect of neodiosmin on thresholds of bitter substances

Neodiosmin concentration (ppm)	Liquid medium and bitter substance	Taste threshold of bitter substance (ppm)	Ref.
0	water + naringin	20	20
10	water + naringin	65	20
0	water + limonin	1.0	24
10	water + limonin	4.0	24
0	orange juice + limonin	3.4	24
10	orange juice + limonin	5.2	24
0	water + caffeine	128	21
10	water + caffeine	230	21
0	water + quinine sulfate	4.0	21
10	water + quinine sulfate	9.5	21

TABLE 6

Effect of neodiosmin on the bitter after-taste of aqueous solutions of saccharin[21]

Neodiosmin concentration (ppm)	Medium	% of 40 judges choosing sample having the least after-taste
0	water + 20 ppm saccharin[a]	25
10	water + 20 ppm saccharin[a]	75
0	water + 40 ppm saccharin[b]	25
10	water + 40 ppm saccharin[b]	75

[a] Equivalent in sweetness to a 1 % sucrose solution.
[b] Equivalent in sweetness to a 3 % sucrose solution.

compared against control juices. In a similar manner it reduced the bitterness of suprathreshold levels of aqueous solutions of caffeine (200 ppm) and quinine sulfate (8 ppm).[21] On the basis of these results, processes for reducing bitterness by adding 10 to 150 ppm **6** to citrus juices and to other orally ingested materials (beverages, foods, pharmaceuticals, etc.) containing bitter substances were patented.[21]

(b) Dihydrochalcones. Another group of citrus derived flavonoids that have the ability to influence bitterness are the dihydrochalcone (DHC) sweeteners. The DHC sweeteners were discovered by Horowitz and Gentili as a result of their studies of the structure–taste relations of citrus flavanones.[46] The many DHCs and their properties as sweeteners have been reviewed in detail.[50] Of interest in this discussion are studies of three DHC sweeteners that are derived from three commonly occurring citrus flavanone glycosides, namely naringin DHC (**7**) from **3**, neohesperidin DHC (**8**) from **4**, and hesperetin DHC 4′-β-D-glucoside (**9**) from hesperidin (**10**) (L-rhamnose removed by hydrolysis). The sweetness thresholds of these DHCs in aqueous solution are 4.5, 0.7, and 4.5 ppm, respectively, as compared with 1300 ppm for sucrose.[23] The chalcones of **3** and **4** are also sweet, but they are unstable except in alkaline solution.

In studies of the effect of DHCs on bitterness, Guadagni *et al.*[23] found that low concentrations of both **8** and **9** raised the bitterness thresholds of **1** and **3** significantly in aqueous solutions (Table 7). These effects were greater than those caused by sucrose of equivalent sweetness, suggesting that the DHCs exert a bitterness suppressing effect in addition to the masking effect due to

TABLE 7

Effect of dihydrochalcone sweeteners on limonin and naringin bitterness in aqueous solutions[23]

Compound added and concentration	Sucrose sweetness equivalent (%)	Threshold concentration of bitter substance (ppm)	
		limonin	naringin
None	0	1.0	20
Sucrose, 10,000 ppm	1	1.0	25
Neohesperidin DHC, 17 ppm	1	1.4	33
Hesperitin DHC glucoside, 80 ppm	1	3.2	33
Sucrose, 30,000 ppm	3	1.1	30
Neohesperidin DHC, 90 ppm	3	2.7	40
Hesperitin DHC glucoside, 300 ppm	3	3.5	45

their sweetness. In further studies, **8** at 12 ppm was found to increase from 33 % to 76 % the proportion of a taste panel who indicated that they liked a given grapefruit juice. At the same time, 84 % of the panel judged the control juice to be more bitter than the **8** containing juice and gave the latter a lower bitterness rating.

C. USES OF CITRUS FLAVONOIDS

The taste properties of the aforementioned citrus flavonoids and their derivatives make them potentially useful commercially. Peel from citrus fruit processing plants is an excellent raw material from which they can be derived.

The flavanone neohesperidosides are useful as bittering agents. Compound **3** from grapefruit peel has been in use as a bittering agent for beverages for many years and offers an alternative to the use of other bitter substances such as caffeine and quinine. Bittering agents are in demand for use in tonic and other carbonated and noncarbonated beverages, and for use in confections and marmalades. Only a small portion of the peel available from grapefruit processed in the U.S. is currently used for production of **3** because the market is still small.

The sweet DHCs derived from citrus flavone glycosides have potential as sweetening agents. Because of their high sweetness intensity (**8** is 7-fold sweeter than saccharin on a weight basis) they can be considered noncaloric sweeteners.[45] Their sweetness has been characterized as "pleasant, somewhat slow in onset and of varying though usually long duration". There is no bitter aftertaste but some people note a cooling or licorice like sensation.[51] The DHCs, **8** in particular, have been widely tested for use in beverages, foods, confections, pharmaceuticals, tooth pastes, mouth washes, etc. While not yet approved by the FDA in the United States, **8** has recently been approved by the health ministry of Belgium.

The bitterness suppressing properties of the flavanone neohesperidosides **5** and **6** are being explored by the food, beverage, and pharmaceutical industries. Compound **6** is a more attractive candidate than **5** because of its greater water solubility. While both **5** and **6** occur at low levels in citrus peel, their production by dehydrogenation of the appropriate flavanone neohesperidosides is probably a more convenient approach.[51]

The availability of abundant amounts of peel from processed grapefruit, pummelo, and Natsudaidai is a potential source of substantial quantities of **3**. In the United States alone about 1.4 million metric tons of grapefruit are processed annually,[7] from which an estimated 1.4 to 2.8 million kilos of **3** could be extracted from the peel and pulp residues. In addition to its direct use as a bittering agent, **3** can also serve as a starting material for production of the other useful flavonoid neohesperidosides (Fig. 1). Compound **7** can be prepared in high yield by hydrogenation of **3** in alkaline solution; dehydrogenation of **3** yields **5**, **3** can also be converted to **4** via alkaline cleavage to yield phloracetophenone neohesperidoside which can then be condensed with isovanillin to give **4**. From **4**, **8** and **6** can be produced.[52] Neohesperidin can also be isolated directly from the peel of the sour orange and Ponderosa lemon; however, the supply of these fruits is limited.

The other flavonoid sweetener **9** is produced from **10** by either acid- or enzyme-catalyzed hydrolytic removal of the rhamnose from hesperidin dihydrochalcone[48,49] Large amounts of **10** are potentially available from the vast quantities of oranges, mandarins, and lemons processed each year.

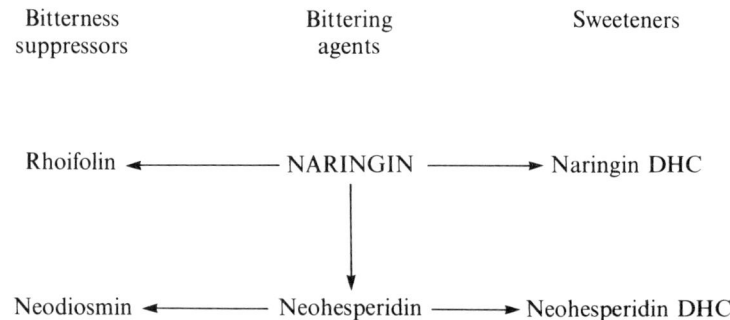

FIG. 1. Naringin as a common starting material for the synthesis of various flavonoid neohesperidosides with useful taste properties.

III. CHEMISTRY AND BIOCHEMISTRY OF LIMONOID BITTERNESS

A. DELAYED BITTERNESS

As mentioned earlier, limonoids are responsible for the delayed bitterness that develops following expression of the juice from the fruit. Commercially this is a serious problem because bitterness lowers juice quality and value. While some 29 limonoids have been isolated from the seeds of *Citrus* species and their hybrids,[6,30,58] only 4 are reported to be bitter, namely **1**, **2**, **11**, and **12**.[55] Obacunoic acid (**13**) is also a bitter limonoid but has not been isolated from *Citrus*.[5,55] Limonin (**1**) is the most abundant limonoid present in citrus juices and is the primary cause of delayed bitterness. Recently the **2** content of fruit tissues and citrus juices has been determined.[44,63] Although **2** occurs at levels lower than **1**, it undoubtedly also contributes to delayed bitterness. The bitterness of **2** has been reported to be about twice that of **1** on a weight basis.[44] In addition, other limonoids also occur in citrus fruit tissues. Minor amounts of deacetylnomilin (**14**), **2**, obacunone (**15**), deacetylnomilinic acid (**16**), and **12** have been identified in extracts of navel oranges, and 17-dehydrolimonoate A-ring lactone (**17**) was isolated from the peel and juice of navel oranges.[53]

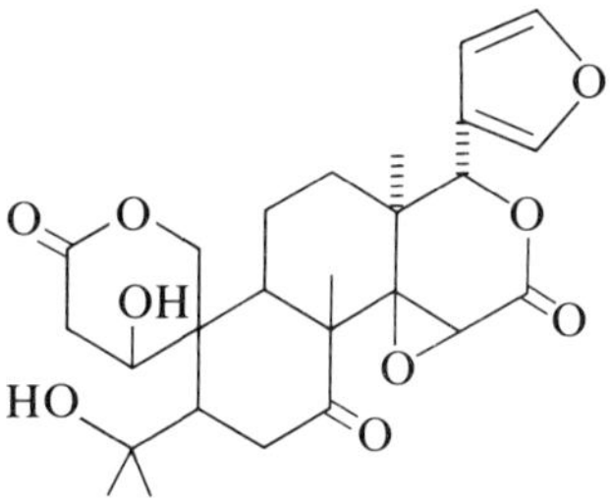

(**11**) Ichangin

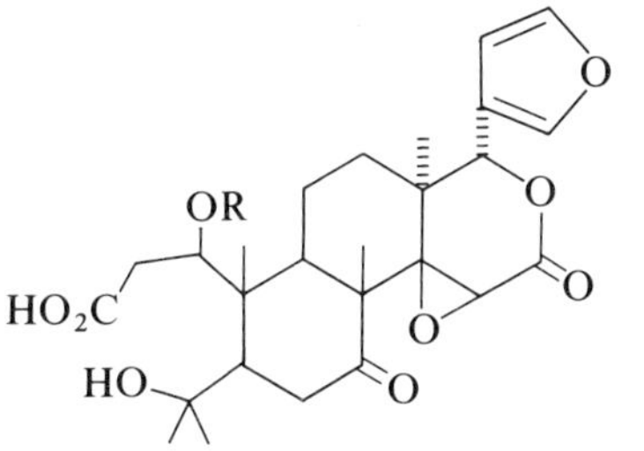

(**12**) Nomilinic acid (R = Ac)

(**16**) Deacetylnomilinic acid (R = H)

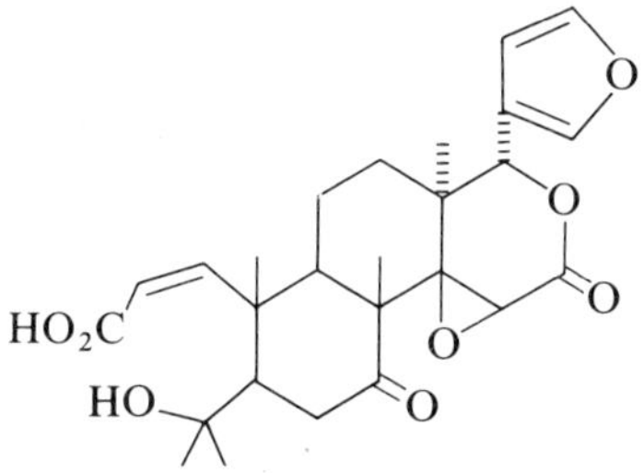

(**13**) Obacunoic acid

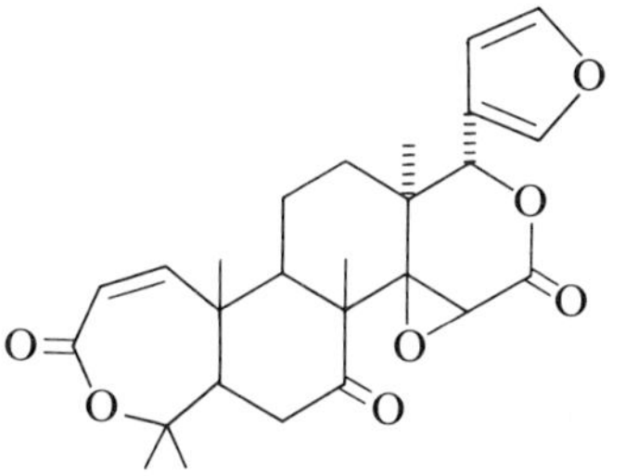

(**15**) Obacunone

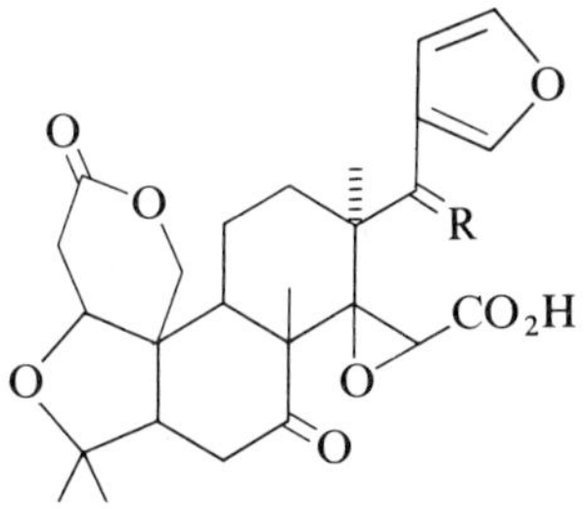

(**17**) 17-Dehydrolimonoic acid
A-ring lactone (R = O)

(**18**) Limonoic acid A-ring
lactone (R = H,OH)

The chemistry of **1** has been reviewed previously.[13,18,55,58] With regard to the delayed bitterness phenomenon the key reactions are the opening and closing of the lactone rings. In citrus tissues the naturally occurring form is a salt of limonoic acid A-ring lactone (**18**), in which the A-ring is closed and the D-ring is open.[56,60] This tasteless compound is only stable in the salt form. Under acidic conditions, and more rapidly in the presence of citrus limonoate D-ring lactone hydrolase, the D-ring closes to form **1**.[59]

It is of interest with regard to the identification of **18** as the naturally occurring form of **1** in *Citrus* tissues to review the chemistry of the A- and D-lactone rings of **1**. Both lactone rings can be hydrolyzed by refluxing with excess dilute NaOH for several hours. Crystalline limonoic acid D-ring lactone (**19**) can be prepared by acidification of concentrated solutions of the salt of limonoic acid (**20**) with strong acid. Treatment with CH_2N_2 in MeOH at 0° yields the methyl ester of **19**.[60] Crystalline **18** (or similar limonoids) cannot be prepared because of the greater facility with which the D-ring lactonizes as compared to the A-ring.[30] However, aqueous mixtures of the two monolactones can be prepared by partial hydrolysis of **1** or partial acid catalyzed lactonization of **20** at pH 5. Partitioning acidified aqueous mixtures of the two monolactones with ethyl acetate, ethyl ether, chloroform, etc. results

(**19**) Limonoic acid D-ring lactone

(**20**) Limonoic acid

(**21**) *trans*-19-Hydroxyobacunoic acid

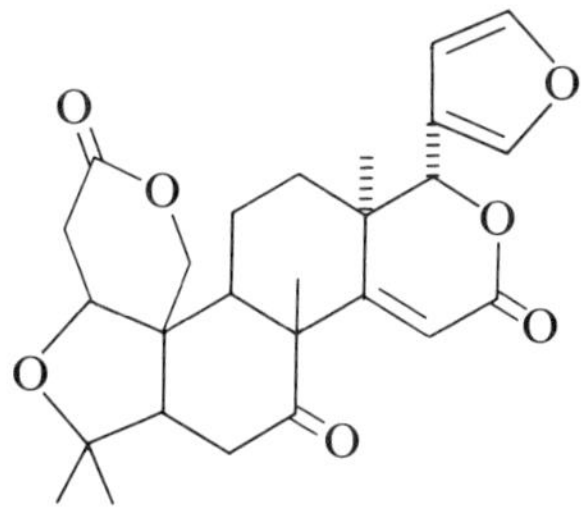

(22) Deoxylimonin

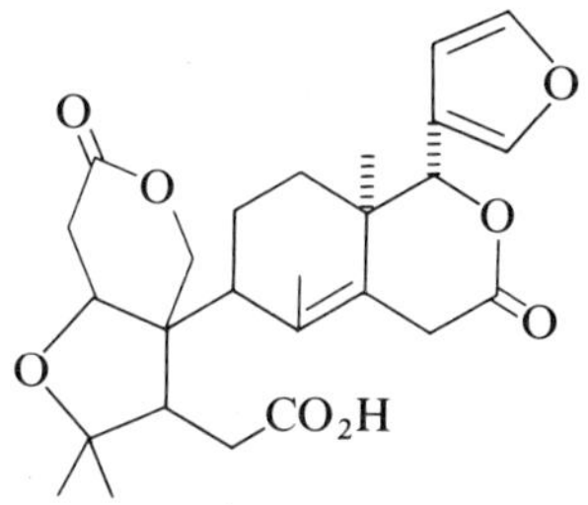

(23) Deoxylimonic acid

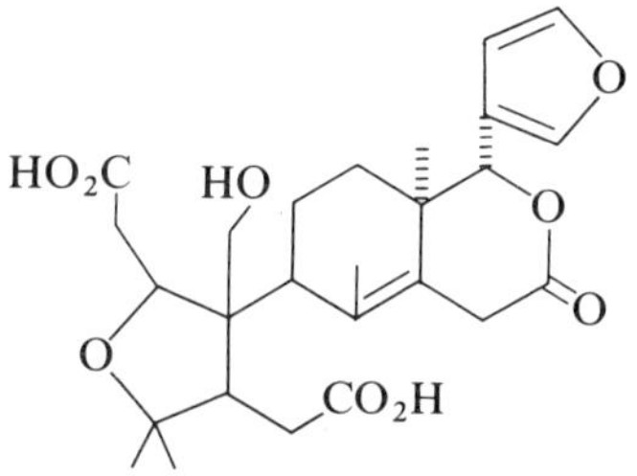

(24) Deoxylimonic acid D-ring lactone

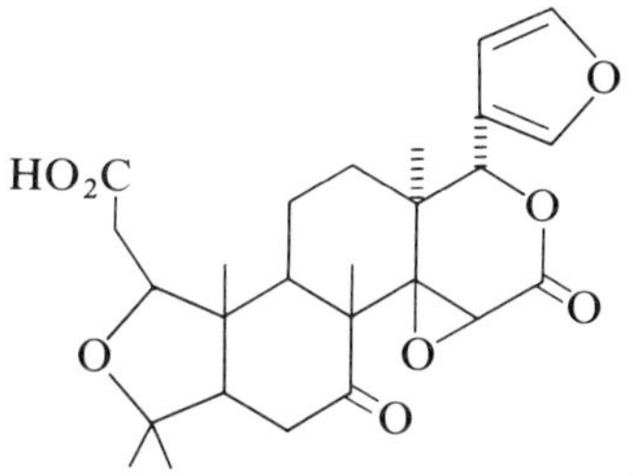

(25) Isoobacunoic acid

in rapid lactonization of the D-ring and the recovery of 1 under these conditions. Consequently, it is not possible to use partitioning systems to isolate 18 from citrus tissues or juices.

Since the two carboxyl groups of limonoic acid have widely different pK_a values (pK_a of C-3 = 4.7; pK_a of C-16 = 2.7) high voltage paper electrophoresis (PE) at pH 3.1 and 5.7 allows separations of all four forms of 1. Thus, at pH 3.1, 1 and 19 have zero charge and do not migrate whereas 20 and 18 have a charge of -1 and do migrate. At pH 5.7, 1 with zero charge does not migrate; 18 and 19 each have a charge of -1 and migrate the same distance; and 20 has a charge of -2 and migrates twice the distance of the monolactones. Thin-layer chromatography (TLC) on microcrystalline cellulose using isopropanol, ammonia, water (9:1:1) also is an effective method of separating the four forms of 1. The hydroxy-acid functions of the open-ring forms of 1 are stable during the brief duration and mild conditions of both the PE and TLC methods.[60]

A very specific method for detecting the presence of limonoids with an open D-ring such as 18 in aqueous tissue extracts (or fruit juices) is through the use of bacterial limonoate dehydrogenase. This enzyme attacks only the open D-ring of limonoids and converts them into the corresponding 17-dehydro derivatives. The open A-ring of **20** is not attacked. By immediate addition of this enzyme to freshly prepared aqueous extracts of citrus fruit tissues and to freshly expressed citrus juices Hasegawa et al.[31] and Brewster et al.[8] showed that **17** was formed in amounts equivalent to total limonin formed following acid catalyzed lactonization of companion samples. This confirmed **18** as the naturally occurring form of **1** in the fruit tissues.

In attempting to determine what form of **1** or other limonoid occurs naturally in plant tissues the method of extraction is also important because of the instability of the open-ring forms. (A simple test for the presence of **1** is its bitter taste.) Cold aqueous extracts at about pH 5 provide a suitable medium such that PE and TLC can be carried out before noticeable conversion takes place. In this work, low temperatures and rapid work-up are important because of the presence of lactonizing enzymes, whereas heat treatment to inactivate enzymes is inappropriate because it catalyzes lactonization.

It is also essential to use fresh, undamaged tissues because once cells are damaged limonoids and acids are released from their respective compartments and mix and lactonization occurs. While such considerations are not of importance where the purpose is to identify the parent limonoids, they are important in studies of the biosynthesis, metabolism, and transport of limonoids and in citrus juice technology. Some workers in these areas tend to ignore the instability problem or assume that no change in the form of the limonoids present occurs during extraction and detection.

In the case of citrus fruits Maier and coworkers identified **18** as the naturally occurring form of **1** in the fruit tissues.[56,60] This was aided by the observation that the highest concentration of limonoids in the fruit (except the seeds) occurs in the carpellary membranes, albedo and central vascular bundle, as opposed to the juice vesicles. Dissection to separate these tissues from the juice vesicles (which contain substantial quantities of sugars and citric acid) reduced the soluble solids content of subsequent aqueous extracts and gave extracts of lower acidity (pH 5.5 vs. 3.5 for juice vesicles) and higher limonoid levels. Identification of the presence of **18** and the absence of **20**, **19**, and **1** (traces) was done by PE, TLC, and the use of limonoate dehydrogenase as described above.

The absence of bitterness in the intact fruit is explained by the presence of nonbitter **18** and the absence of **1**. The **18** appears to be located in a compartment of the cell in which the pH is neutral or alkaline, probably the cytoplasm, as opposed to the vacuoles which contain the organic acids. When the fruit tissues and cells are disrupted during juice extraction, **18** is mixed with the acidic juice and gradual lactonization to **1** occurs. The rate of lactonization is accelerated by pasteurization of the juice and by the presence of limonoate

D-ring lactone hydrolase. Also, if the intact fruit (either pre- or post-harvest) is damaged sufficiently (such as by bruising or frost) to disrupt cells and thereby bring **18** in contact with acids, conversion of **18** to **1** will occur and the intact fruit itself will be bitter.

B. METABOLISM

1. *Bacteria*

Because of the difficulty of studying **1** metabolism in citrus fruit systems, Hasegawa and coworkers and other investigators have studied bacterial systems. These studies were also prompted by the desire to find enzymes that might be used to debitter citrus juices. A number of bacteria have now been reported that have the ability to metabolize limonoids. Hasegawa[35] has recently reviewed this work which is summarized in Table 8.

Five species of bacteria have now been reported that have the ability to metabolize limonoids, based on studies using **20** and **1**. Hasegawa and coworkers have shown the presence of three metabolic pathways (Table 9), namely the 17-dehydrolimonoid, the deoxylimonoid, and the *trans*-19-hydroxyobacunoic acid (**21**) pathways.[35] Limonin D-ring lactone hydrolase plays a key role in limonoid metabolism in that it is the link between the dehydrolimonoid pathway which requires an open D-ring and the deoxylimonoid pathway which requires a closed D-ring. Limonin D-ring lactone hydrolase has high intramolecular specificity, in that it does not attack the A-ring lactone of limonoids. However, it shows broad specificity in attacking the D-ring of a wide variety of limonoids such as **2**, **11**, **14**, and **15**.[25] However, the presence of the 14,15-epoxide group is essential, as the enzyme does not attack deoxylimonin (**22**) or deoxylimonic acid (**23**). This is in keeping with the finding that deoxylimonin hydrolase, the second enzyme on the deoxylimonoid pathway, appears not to attack limonoids with an open D-ring.[37] (Limonin epoxidase, the first enzyme of the pathway has not yet been isolated.) Consequently, in those organisms such as *Pseudomonas* and Bacterium 342-152-1 which have both the 17-dehydrolimonoid and deoxylimonoid pathways, limonoate D-ring lactone hydrolase provides the connecting link. The pH stability of **1** vs. **18** suggests that the pathways may occur in different cell compartments.

The **21** pathway in *C. fascians* is a recent discovery. Activity of limonate A'-ring transeliminase has been detected but the enzyme has not yet been isolated.[27,35] This pathway differs from the others in that the initial attack is on the A'-ring rather than the D-ring. *C. fascians* is unique among the

TABLE 8

Limonoid metabolizing bacteria, metabolites produced, and enzymes identified

Bacteria	Major metabolites	Pathways identified	Enzymes identified	Enzyme status	Ref.
Arthrobacter globiformis	$17^{a,b}$	17-Dehydrolimonoid	Limonoate dehydrogenase (Limonoate-NAD oxidoreductase)	Isolated	28
			Limonoate D-ring lactone hydrolase	Cell-free	39
Pseudomonas 321-18	22 (trace),^{c} 23^c	17-Dehydrolimonoid	Limonoate dehydrogenase (Limonoate-NAD(P) oxidoreductase)	Isolated	38
			Limonin D-ring lactone hydrolase	Isolated	25
		Deoxylimonoid	Limonin epoxidase	Postulated	25, 37
			Deoxylimonin hydrolase	Isolated	37
Bacterium 342-152-1	17^a, 22^a, 23^a	17-Dehydrolimonoid	Limonoate dehydrogenase (Limonoate-NAD oxidoreductase)	Isolated	33, 42
		Deoxylimonoid			
Corynebacterium fascians	17^a, $21^{a,d}$	17-Dehydrolimonoid	Limonoate dehydrogenase (Limonoate-NAD oxidoreductase)	Isolated	26, 35
		trans-19-Hydroxy-obacunoic acid	Limonoate A'-ring *trans*-eliminase (lyase)	Postulated	27. 35
Acinetobacter sp.	22^e, 23^e	Deoxylimonoid	Deoxylimonin hydrolase	Isolated	64

^{a} From limonoate. ^{b} 17-Dehydrolimonoate A-ring lactone from limonin. ^{c} From limonoate or limonin. ^{d} Open D-ring form. ^{e} From limonin.

TABLE 9

Limonoid enzymes and their status in *Citrus*

Enzyme	Reaction in *Citrus*	Source of enzyme or location of activity	Enzyme status	General remarks	Ref.
Limonin D-ring lactone hydrolase	1 ⇌ **18**	Orange and grapefruit seeds. Also present in fruit and leaves	Isolated and purified	Attacks most citrus limonoids with D-ring like limonin. Does not attack deoxylimonin or deoxylimonate. Limonoate A-ring lactone is present in fruit, leaves, seeds	25, 59
Limonoate dehydrogenase	**18** ⇌ **17**	Navel orange fruit albedo tissues	Activity demonstrated using tissue slices and ^{14}C-labeled substrate	Requires open D-ring. Bacterial enzyme substrate specificity as above. 17-Dehydrolimonoate A-ring lactone is present in fruit tissues, juice, seeds, and seedlings	30, 36, 53
Limonin epoxidase	1 → **22**	Calamondin leaves (*C. microcarpa*)	Activity demonstrated using detached young leaves and ^{14}C-labeled substrate	Attacks closed D-ring. Deoxylimonin is present in grapefruit seeds	15, 30
Deoxylimonin hydrolase	**22**^{a} → **23**	Not yet demonstrated in *Citrus*	Enzyme isolated from *Pseudomonas* 321-18 and *Acinetobacter* sp.	Appears to require closed D-ring. Deoxylimonin and deoxylimonic acid are present in grapefruit seeds	15, 30, 37, 64

contd.

TABLE 9—*contd.*

Enzyme	Reaction in *Citrus*	Source of enzyme or location of activity	Enzyme status	General remarks	Ref.
Deoxylimonic acid A-ring lactone hydrolase	**23 → 24**	Grapefruit seeds	Crude enzyme isolated	Enzyme specific for deoxylimonic acid; does not hydrolyze A-ring of limonin	30
Limonate A′-ring transeliminase	**20**[a] **→ 21**	Not yet demonstrated in *Citrus*	Enzyme postulated	*Corynebacterium fascians* produces *trans*-19-hydroxyobacunoic acid from limonoate	27, 35

[a] Reaction in bacteria.

limonoid metabolizing bacteria in that it produces limonoid enzymes constitutively. The other bacteria must be induced to produce limonoid enzymes by the presence of a limonoid compound in the medium.

2. Citrus

Bitterness of citrus juices due to **1** is most pronounced with juices extracted from fruit harvested soon after commercial maturity is reached. The degree of bitterness decreases as the fruit matures further on the tree, which suggests that metabolism of **18** exceeds accumulation. Maier *et al.*[56,60] showed this to be the case for **18**. Since decreases in **18** (or total limonoid) also occur during postharvest storage of citrus fruits, metabolism rather than transport of limonoids out of the fruit is involved. Accelerated postharvest metabolism can be triggered by brief exposure of the fruit to ethylene gas.[57] Treatment of the fruit for 3 hours with 20 ppm ethylene followed by storage for 3–5 days in air is as effective as continuous exposure of the fruit to 20 ppm ethylene for 3–5 days. This process has potential commercial utility in reducing juice bitterness. The 3-hour exposure is preferred because it does not lead to accelerated respiration and the development of off-flavors as does continuous exposure to ethylene.

Table 9 lists the known **1** metabolizing enzymes and the state of knowledge with regard to their presence in *Citrus* tissues. Most of the knowledge of limonoid enzymes was made possible by advances in bacterial enzymes and metabolism, although the first limonoid enzyme isolated was limonin D-ring lactone hydrolase from *Citrus* seeds.[59] This is still the only limonoid enzyme that has been isolated and purified from *Citrus* sources. Deoxylimonic acid A-ring lactone hydrolase, which catalyzes the conversion of **23** to deoxylimonic acid D-ring lactone (**24**), has recently been extracted from grapefruit seeds as a crude enzyme preparation.[30] Through a combination of ^{14}C feeding studies that demonstrated specific enzyme activities, and identification of key metabolites, there is strong evidence in support of the presence of both the dehydrolimonoid and deoxylimonoid pathways in *Citrus*.

C. PRACTICAL CONSIDERATIONS

As mentioned earlier, one purpose of the studies of the bacterial metabolism of limonoids was to find enzymes that could be used in the production of citrus juices free of limonoid bitterness. The addition of limonoate dehydrogenases isolated from *A. globiformis* and *Pseudomonas* 321-18 directly to freshly prepared citrus juices has been shown to prevent **1** formation by converting **18** to its stable nonbitter 17-dehydro derivative.[8,31] While successful in preventing juice bitterness, the direct addition of the dehydrogenases to juice has drawbacks. First, the acidic nature of citrus juices means that the enzymes are not at optimum activity and, second, because the dehydrogenases do not attack **1** itself they are not effective on juice that is already bitter.

Two novel approaches using intact bacterial cells rather than extracted enzymes have been used recently to circumvent the aforementioned difficulties. Hasegawa et al.[39,41] have demonstrated debittering of navel orange juice with *A. globiformis* cells immobilized in beads of acrylamide gel. Natural juice (pH 3.6) was pumped upward through a column packed with the beads. Since **1** was converted to **17**, both limonoate D-ring lactone hydrolase and linonoate dehydrogenase of the bacteria were operative. Also, since the hydrolase catalyzes the net conversion of **1** to **18** only at pH 7 and above, it appears that in the immobilized cells the limonoid enzymes are functioning in a compartment where the pH is more favorable than that of the juice. The enzyme system was very stable and allowed the immobilized cells to be used many times without loss of effectiveness. The system was equally as effective in debittering juice containing **2**.[40] In view of the wide variety of limonoids attacked by the hydrolase-dehydrogenase enzymes,[25] the immobilized cell system also offers a convenient and specific method for the synthesis of 17-dehydrolimonoids from their parent lactones.

Vaks and Lifshitz[64] reported the use of cells of *Acinetobacter* sp. entrapped in a dialysis sac for debittering orange juice adjusted to pH 4.5 with NaOH. Living cells of the bacteria were placed in a short length of dialysis tubing along with 0.1 M phosphate buffer, pH 6.0. The dialysis sac was placed in a batch of pH adjusted juice for 2 hours. Five consecutive transfers using fresh juice were achieved before activity fell to 50 % of the initial. Studies with a **1** model system showed that the bacteria converted **1** to **22** and ultimately **23**. Both of these limonoids were found to be nonbitter at 200 ppm.

D. ACCUMULATION AND SITES OF BIOSYNTHESIS IN *Citrus* TISSUES

In *Citrus*, limonoids have been shown to occur in the leaves, twigs, fruit, and seeds.[32] In general, the limonoid concentration is highest in the earliest stages of growth of leaves and fruit and decreases gradually thereafter (Tables 10, 11), although the total limonoid content per leaf or fruit tends to increase throughout the growth and enlargement stages and decreases during maturation.[11,29,44] On the other hand, both the concentration and total amount per seed increases steadily during fruit growth and maturation.[44] Hasegawa et al.[29] found that in immature lemon leaves, fruits, and seeds the ratio of **2** to **1** is 1:1 whereas in the mature tissues the ratios are 1:20, 1:15, and

TABLE 10

Limonoid content of lemon leaves during early growth[32]

Leaf size (mg/leaf)	Limonoic acid A-ring lactone content[a]	
	(μg/leaf)	(ppm)
50	100	1800
100	145	1250
200	190	900
250	215	800

[a] Determined as total limonin by thin-layer chromatography after acid catalyzed lactonization.

TABLE 11

Limonoid content of navel oranges during early growth[32]

Fruit size (g/fruit)	Limonoic acid A-ring lactone content (mg/fruit)[a]
5	0.7
10	1.5
20	3.0
30	4.4
35	5.1

[a] Determined as total limonin by thin-layer chromatography after acid catalyzed lactonization.

1:1.3, respectively. Hashinaga et al.[44] obtained generally similar results with Natsudaidai.

Hasegawa et al.[32] using ^{14}C-acetate and ^{14}C-mevalonate showed that limonoids were actively synthesized in young lemon leaves. No evidence of limonoid biosynthesis in fruit or seed tissues was found.[29] Labelled 18 fed to leaves was transported to the fruit. In addition, the salt of the 3-methyl-^{14}C-ester of 23 fed to lemon fruit and the salt of the labelled methyl ester of 16 fed to calamondin fruit were translocated from the fruit tissue into the seeds.[29] Thus, limonoid biosynthesis occurs in the leaves and limonoids are transported to the fruit and the seeds. The high levels of limonoids in seeds (grapefruit seeds 1.6 % limonoids on a wet weight basis, Valencia orange seeds 1.2 %, and lemon seeds 1.2 %), and the fact that limonoid levels do not decrease with maturity as they do in the fruit, indicates that the seeds act as a storage tissue for limonoids as suggested previously by Kefford and Chandler.[54] The fact that the insoluble dilactone **1** rather than the water soluble salt of **18** predominates in seeds[56] is in agreement with this point.

Hasegawa et al.[43] have shown that limonoid biosynthesis in lemon leaves is markedly inhibited by low concentrations of either 2-(3,4-dimethylphenoxy)-triethylamine or 2-(4-ethylphenoxy)triethylamine whereas limonoid metabolism is generally unaffected. Casas and coworkers[12] reported that the first of these compounds and 2-(3,4-dichlorophenylthio)triethylamine each lowered the limonoid content of navel orange fruit.

Direct studies of **1** biosynthesis through feeding of labeled precursors have not yet been reported. In another chapter of this volume, Dreyer discusses the steps involved in a pathway leading to **15** and **1** put forth at the time structures were proposed for these compounds.[4] Based on limonoids found in grapefruit seeds Bennett[5] proposed **16** as a central intermediate in **1** biosynthesis. He suggested that **1** could be synthesized from **16** through two routes, one via **11** and the other via isoobacunoic acid (**25**). Dreyer had previously suggested that **25**[18] and **11**[16] might be immediate precursors of **1** by alternative routes.

REFERENCES

1. Albach, R. F. and Redman, G. H. (1969). *Phytochemistry* **8**, 127.
2. Anonymous (1979). "Agricultural Statistics", pp. 435–437, U.S. Dept. Agric., U.S. Printing Office, Washington, D.C.
3. Anonymous (1981). *Foreign Agriculture Circular*, FCF 4-81, July, pp. 1–26, U.S. Dept. Agric., Washington, D.C.
4. Barton, D. H. R., Pradhan, S. K., Sternhell, S. and Templeton, S. F. (1961). *J. Chem. Soc.*, 255.
5. Bennett, R. D. (1971). *Phytochemistry* **10**, 3065.
6. Bennett, R. D. and Hasegawa, S. (1981). *Tetrahedron* **37**, 17.
7. Berry, R. E. (1979) *In* "Tropical Foods: Chemistry and Nutrition" (Inglett, G. E. and Charalambous, G., eds.), Vol. 1, p. 96. Academic Press, New York.
8. Brewster, L. C., Hasegawa, S. and Maier, V. P. (1976). *J. Agric. Food Chem.* **24**, 21.
9. Burke, J. H. (1967). *In* "The Citrus Industry" (Reuther, W., Webber, H. J. and Batchelor, L. D., eds.), Vol. 1, p. 41. University of California.
10. Carter, R. D., Buslig, B. S. and Cornell, J. A. (1975). *Proc. Florida State Hort. Soc.* **88**, 358.
11. Casas, A. and Rodrigo, M. I. (1981). *J. Sci. Food Agric.* **32**, 252.
12. Casas, A., Rodrigo, M. I. and Mallent, D. (1980). *Rev. Agroquim. Tecnol. Ailment.* **19**, 513.
13. Connolly, J. D., Overton, K. H. and Polonsky, J. (1970). *In* "Progress in Phytochemistry" (Reinhold, L. and Liwschitz, Y., eds.), Vol. 2, p. 385. John Wiley and Sons, New York.

14. Cornell, J. O. (1974). Abstracts of 25th Annual Citrus Processors Meeting, p. 3, Lake Alfred, Florida.
15. Dreyer, D. L. (1965). *J. Org. Chem.* **30**, 749.
16. Dreyer, D. L. (1966). *J. Org. Chem.* **31**, 2279.
17. Dreyer, D. L. (1966). *Phytochemistry* **5**, 367.
18. Dreyer, D. L. (1968). *Fortschr. Naturst.* **26**, 190.
19. Fellers, P. J. (1973). Abstracts of 24th Annual Citrus Processors Meeting, p. 6, Lake Alfred, Florida.
20. Guadagni, D. G. (1972). Private communication, Berkeley, California.
21. Guadagni, D. G., Horowitz, R. M., Gentili, B. and Maier, V. P. (1979). U.S. Patent, 4,154,862, May 15.
22. Guadagni, D. G., Maier, V. P. and Turnbaugh, J. G. (1973). *J. Sci. Food Agric.* **24**, 1277.
23. Guadagni, D. G., Maier, V. P. and Turnbaugh, J. G. (1974). *J. Sci. Food Agric.* **25**, 1199.
24. Guadagni, D. G., Maier, V. P. and Turnbaugh, J. G. (1976). *J. Food Sci.* **41**, 681.
25. Hasegawa, S. (1976). *J. Agric. Food Chem.* **24**, 24.
26. Hasegawa, S. (1981). Private communication, Pasadena, California.
27. Hasegawa, S. and Bennett, R. D. (1981). Private communication, Pasadena, California.
28. Hasegawa, S., Bennett, R. D., Maier, V. P. and King, A. D. Jr. (1972). *J. Agric. Food Chem.* **20**, 1031.
29. Hasegawa, S., Bennett, R. D. and Verdon, C. P. (1980). *J. Agric. Food Chem.* **28**, 922.
30. Hasegawa, S., Bennett, R. D. and Verdon, C. P. (1980). *Phytochemistry* **19**, 1445.
31. Hasegawa, S., Brewster, L. C. and Maier, V. P. (1973). *J. Food Sci.* **38**, 1153.
32. Hasegawa, S. and Hoagland, J. E. (1977). *Phytochemistry* **16**, 469.
33. Hasegawa, S. and Kim, K. S. (1975). Abstracts of the Citrus Research Conference, p. 16, U.S.D.A., Pasadena, California.
34. Hasegawa, S. and King, A. D. Jr. (1982). Private communication, Pasadena, California.
35. Hasegawa, S. and Maier, V. P. (1981). *Proc. Int. Soc. Citriculture*, in press.
36. Hasegawa, S., Maier, V. P. and Bennett, R. D. (1974). *Phytochemistry* **13**, 103.
37. Hasegawa, S., Maier, V. P., Border, S. N. and Bennett, R. D. (1974). *J. Agric. Food Chem.* **22**, 1093.
38. Hasegawa, S., Maier, V. P. and King, A. D. Jr. (1974). *J. Agric. Food Chem.* **22**, 523.
39. Hasegawa, S., Patel, M. N. and Snyder, R. C. (1982). *J. Agric. Food Chem.* **30**, 509.
40. Hasegawa, S. and Pelton, V. (1983). *J. Agric. Food Chem.* **31**, 178.
41. Hasegawa, S., Pelton, V. A. and Snyder, R. C. (1981). Abstracts of the Citrus Research Conference, pp. 1–2, U.S.D.A., Pasadena, California.
42. Hasegawa, S., Verdon, C. P. and Schroeder, M. F. (1980). Abstracts of the Citrus Research Conference, p. 16, U.S.D.A., Pasadena, California.
43. Hasegawa, S., Yokoyama, H. and Hoagland, J. E. (1977). *Phytochemistry* **16**, 1083.
44. Hashinaga, F., Ejima, H., Nagahama, H. and Itoo, S. (1977). *Bull. Fac. Agric.* **27**, 171, Kagoshim Univ., Japan.
45. Horowitz, R. M. (1964). *In* "Biochemistry of Phenolic Compounds" (Harborne, J. B., ed.), pp. 545–571. Academic Press, New York.
46. Horowitz, R. M. and Gentili, B. (1963). U.S. Patent 3,087,821, April 30.
47. Horowitz, R. M. and Gentili, B. (1969). *J. Agric. Food Chem.* **17**, 698.
48. Horowitz, R. M. and Gentili, B. (1969). U.S. Patent 3,429,873, Feb. 25.

49. Horowtiz, R. M. and Gentili, B. (1971). U.S. Patent 3,583,894, June 8.
50. Horowitz, R. M. and Gentili, B. (1974). *In* "Symposium: Sweeteners" (Inglett, G. E., ed.), pp. 182–193. AVI Publishing Co., Westport, Connecticut.
51. Horowitz, R. M. and Gentili, B. (1977). *In* "Citrus Science and Technology" (Nagy, S., Shaw, P. E. and Veldhuis, M. K., eds.), Vol. 1, pp. 397–426. AVI Publishing Co., Westport, Connecticut.
52. Horowtiz, R. M. and Gentili, B. (1977). *Proc. Int. Soc. Citriculture* **3**, 743.
53. Hsu, A. C., Hasegawa, S., Maier, V. P. and Bennett, R. D. (1973). *Phytochemistry* **12**, 563.
54. Kefford, J. F. and Chandler, B. V. (1970) *In* "Advances in Food Research" (Chichester, C. O., Mrak, E. M. and Steward, G. F., eds.), Suppl. 2, pp. 150–164. Academic Press, New York.
55. Maier, V. P., Bennett, R. D., Hasegawa, S. (1977). In "Citrus Science and Technology" (Nagy, S., Shaw, P. E., and Veldhuis, M. K., eds.), Vol. 1, pp. 355–396. AVI Publishing Co., Westport, Connecticut.
56. Maier, V. P. and Beverly, G. D. (1968). *J. Food Sci.* **33**, 488.
57. Maier, V. P., Brewster, L. C. and Hsu, A. C. (1973). *J. Agric. Food Chem.* **21**, 490.
58. Maier, V. P., Hasegawa, S., Bennett, R. D. and Echols, L. C. (1980). *In* "Citrus Nutrition and Quality" (Nagy, S. and Attaway, J. A., eds.), pp. 63–82. ACS Symposium Series 143, American Chemical Society, Washington, D.C.
59. Maier, V. P., Hasegawa, S. and Hera, E. (1969). *Phytochemistry*, **8**, 405.
60. Maier, V. P. and Margileth, D. A. (1969). *Phytochemistry* **8**, 243.
61. Nagy, S. and Attaway, J. A., eds. (1980). *In* "Citrus Nutrition and Quality", ACS Symposium Series 143, pp. 1–456. American Chemical Society, Washington, D.C.
62. Nagy, S., Shaw, P. E. and Veldhuis, M. K., eds. (1977). "Citrus Science and Technology", Vol. 1, pp. 1–531. AVI Publishing Co., Westport, Connecticut.
63. Rouseff, R. L. (1982). *J. Agric. Food Chem.* **30**, 504.
64. Vaks, B. and Lifshitz, A. (1981). *J. Agric. Food Chem.* **29**, 1258.

CHAPTER 13

Chemotaxonomy of the Genus *Citrus*

RAINER W. SCORA and J. KUMAMOTO

Department of Botany and Plant Sciences, University of California, Riverside, California, U.S.A.

I. INTRODUCTION

In ordinary plants, embryos are formed from a paternal and maternal complement, and the resulting offspring carry the characteristics of both parents. In the genus *Citrus*, however, embryos may also be formed from the nucellar tissue of the mother tree alone. Since no sexual reproduction is involved, the resulting offspring will be exact duplicates of the maternal parent. This vegetative reproduction or apomixis through seeds is called nucellar embryony. It has enabled hybrids and budsprouts to perpetuate themselves almost indefinitely.

These stabilized biotypes have added greatly to the complexity of *Citrus* taxonomy and this is reflected in the extreme positions taken by Tanaka,[13] who recognized 162 species in 1977, and Malik[5] who advocated just one. The most widely accepted view however is a modification of Swingle's[12] system which recognized 16 species in the genus *Citrus*. Of these, 10 species which are

343

edible are in the subgenus *Citrus*, and 6 species are in the subgenus *Papeda*. The latter 6 are inedible due to the presence of acrid oil droplets in the juice vesicles.

Citrus is indigenous to the sub-Himalayan areas, Malaysia, and China. The citron (*C. medica*) was the first citrus to reach Europe with the Persian campaign of Alexander the Macedonian. The Roman empire added the lemon (*C. limon*) and sour orange (*C. aurantium*), and later the Arabs brought lime (*C. aurantifolia*) and pummelo (*C. grandis*) to North Africa and Spain.

It was from this sphere that Christopher Columbus brought citrus seeds to the Americas on his second voyage, November 1493. Once the Portuguese succeeded in sailing around Africa they brought sweet oranges (*C. sinensis*) from the Far East, and these then were dispersed to the Americas and subsequently to Australia. The mandarin (*C. reticulata*) of Chinese origin arrived late, in 1805, at Kew and was dispersed from there to the Mediterranean area and subsequently to other parts of the world.

II. Enzymic Chemotaxonomic Markers

Living matter can be characterized by proteins, and some of these, the enzymes, catalyze specific reactions. Enzymes are nearly direct gene products and are thus ideally suited as taxonomic markers. Gel electrophoresis has made it possible to identify large numbers of gene loci and to enumerate the proportion that show variation, or are polymorphic in having more than one allele. Further, we can easily find the proportion of gene loci that are heterozygous in single individuals. In a gene/enzyme system with a monomeric structure, a homozygote will form only one band which, adopting the terminology of Torres,[16] is called slow or fast depending upon its migration. A heterozygote will thus have two bands, one fast and one slow. In a dimeric structure the homozygote also produces subunits of only one kind which homodimerize to form only one type of band, either slow slow (SS) or fast fast (FF). A heterozygote produces two kinds of subunits; these produce three kinds of bands, the two homodimers and an intermediately migrating heterodimer (FS). In addition to these basic alleles others are known in *Citrus* which were named after the biotype where first discovered. The zymograms from the major citrus biotypes are summarized in Table 1.

Oxidase browning of young shoot tissue can also be used as a diagnostic marker in citrus taxonomy. Browning will occur in the presence of an enzyme and a suitable substrate. The mandarin, *C. reticulata*, possesses both, enzyme and substrate. The acid group *C. limon*, *C. aurantifolia*, and *C. medica*, as well as *C. grandis*, the subgenus Papeda, and the genera *Poncirus* and *Fortunella* do

TABLE 1
Summary of genotypes and other taxonomic characters

Species and biotypes	Isozymes				Browning		Amylase bands 1 and 3	Total protein electrophoresis		Leaf petiole	Sabinene content of leaf oil (%)	Embryony	Pollen tectation (SEM)
	Got-1	Got-2	Pgi-1	Pgm	enzyme	substrate		35,500 daltons	33,700 daltons				
C. aurantifolia	FF	SM	SS	FM	–	–	–	++	–	sub-alate	0.1	poly	semi-tectate
C. aurantium	SS	MM	WS	FS	+	+	–	+++	+++	alate	0.2	poly	tectate perforate
C. grandis	FF	MM	SS	SS	–	–	+	+	–	alate	5.9	mono	tectate microperforate
C. jambhiri	FS	FS	FS	FI	+	+	+	++	++	non-alate	20.1	poly	semi-tectate
C. limon	FS	SM	WS	FS	–	–	–	+	++	non-alate	1.9	poly	semi-tectate
C. medica	FF	SS	SS	FF	–	–	–	++	–	non-alate	0.2	mono	semi-tectate
C. paradisi	FS	MM	SS	SS	+	+	+	++	++	alate	35.8	poly	tectate microperforate
C. reticulata	SS	FM	FF	FF	+	+	+	++	+++	sub-alate	0.2–40.0	mono/poly	semi-tectate
C. sinensis	SS	MM	FS	FS	+	+	+	+	++	sub-alate	38.0	poly	tectate perforate

Got = glutamic oxalacetic transaminase; Pgi = phosphoglucose isomerase; Pgm = phosphoglucose mutase.
S = a slow migrating allele; F = fast; M and I = intermediate migration. W = an ultra-fast migrating allele of Pgi-1.

not possess either enzyme or substrate.[3] All the hybrids with mandarin parentage that were tested were browning, and it was concluded that this character is indicative of mandarin parentage. The grapefruit possesses the substrate only, but not the enzyme. However, we have found that the expression of the enzyme is recessive in the grapefruit, because browning reappeared in 12 out of 67 inbred segregate grapefruit trees.

III. Origin of the Grapefruit

All *Citrus* taxa originated in Asia with the exception of the grapefruit (*C. paradisi*) which originated in the New World on Barbados. Since this was a rather recent event, and the grapefruit was cultivated for the export trade to Europe, let us start looking at species interrelationships with this taxon.

We notice (Table 1) that the grapefruit, *C. paradisi*, is homozygous in phosphoglucose mutase, designated SS. This requires that each parent must have at least one S allele to contribute. Eliminating those that cannot, namely *C. reticulata* (mandarin), *C. jambhiri* (rough lemon), *C. medica* (citron), *C. aurantifolia* (lime), and *C. ichangensis* (ichang), only four possible choices remain as immediate parents for the grapefruit, namely *C. grandis* (pummelo), *C. aurantium* (sour orange), *C. sinensis* (sweet orange), and *C. limon* (lemon).

According to Sloane,[10] these four were present on the island of Barbados by 1692, and they will give us six binary combinations: *C. grandis* × *C. sinensis*, *C. grandis* × *C. aurantium*, *C. grandis* × *C. limon*, *C. sinensis* × *C. limon*, *C. aurantium* × *C. limon*, and *C. aurantium* × *C. sinensis*. The last combination *C. aurantium* × *C. sinensis* can be eliminated because glutamic oxalacetic transaminase-1 in *C. paradisi* is heterozygous designated FS and the F allele is absent in this combination. Morphological characteristics also rule out this combination.

In the zymograms of amylase,[4] we are primarily concerned with bands 1 and 3, at 6 and 9.5 mm from the origin. They appear to be diagnostic and are present in *C. paradisi*, *C. sinensis*, *C. aurantium*, *C. reticulata*, and *C. jambhiri*. They are missing in *C. grandis*, *C. limon*, *C. medica*, and *C. aurantifolia*. Since *C. grandis* and *C. limon* do not possess them, this pair is unlikely to have been the parental combination. Neither is *C. grandis* crossed with any of the other biotypes like *C. medica* and *C. aurantifolia*, which have previously been eliminated by Pgm. On the other hand, a combination of *C. grandis* × *C. sinensis* is a possibility as well as *C. grandis* × *C. aurantium*, but *C. reticulata* and *C. jambhiri*, which would also have been a possibility, have previously been eliminated by Pgm as possible immediate parents.

Citrus foliage and fruits are rich in essential oils. The oil chambers form schizogenously, at first through the separation of the walls of the central cells. Further expansion results in wall degradation which is suggestive of lysigeny.[15] We sometimes are able to tap the essential oil directly from a fruit oil gland with a short-needled syringe.

Some of the mono- or sesqui-terpene components may change because of various factors. Rootstocks, for instance, may influence the fraction of a given component or even double the total oil yield.[2] Since all *Citrus* biotypes are usually grafted, this is an important consideration for comparative studies. Maturation of plant tissue also produces compositional changes. Even physiological changes, like breaking dormancy, are accompanied by changes in the composition of Valencia (*C. sinensis*) rind oil. While some components change dramatically, others change more uniformly or are relatively stable.[7]

A study of oil inheritance shows that hybrids tend to express parental characters. In the extreme cases, intermediate values for a given oil component are present in the hybrids. In cases where the parents are similar, the hybrids tend to be similar.[8]

Leaf oil sabinene (**1**), an early eluting monoterpene, can be used as a diagnostic marker in the ancestry of the grapefruit. *C. paradisi* leaf oil has a sabinene content of about 35%, *C. sinensis* 38–42%, *C. grandis* about 6%, while the other possible parents have less than 2% (Table 1). Thus low sabinene crosses of *C. grandis* × *C. aurantium* and *C. aurantium* × *C. limon* are very unlikely to be the parental combinations, as is *C. grandis* × *C. limon*, which we have already eliminated by amylase. Further support for these eliminations comes from pollen tectation and oxidase browning.

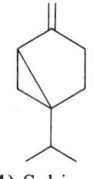

(**1**) Sabinene

Of the remaining two combinations, consideration of pollen tectation, of leaf and fruit morphology, and breeding, make a *C. sinensis* × *C. limon* combination also unlikely (Table 1).

If these above assumptions are acceptable to us, we are left with *C. grandis* × *C. sinensis* as the only combination satisfying morphological as well as chemical characteristics. They become the only possible immediate parents for the grapefruit. We can conclude that the pollen self-incompatible *C. grandis* must have been the maternal parent with the apomictic *C. sinensis* the pollen parent.[9]

IV. SWEET ORANGE

A natural question arises; where did the sweet orange, *C. sinensis,* get its high sabinene content? Our studies show that the polymorphic mandarin, *C. reticulata,* has high and low sabinene members, and it was most likely derived from a high sabinene mandarin. A cross of high sabinene *C. reticulata* (SS in glutamic oxalacetic transaminase-1, Got-1) with an old-line pummelo, *C. grandis,* for example Kao Pan (FS in Got-1), will carry the necessary genes to make a high sabinene sweet orange, *C. sinensis* (SS in Got-1). We need this old-line pummelo in order to obtain the second S allele, because all other pummelos tested were FF in Got-1. If any other pummelo had been the parent, then the F_1 would have been FS in Got-1. This would necessitate a backcross to the mandarin, or an inbred F_2 in order to satisfy the homozygous requirement.

V. SOUR ORANGE

Another cross gave rise to the sour orange, *C. aurantium.* This time a low sabinene *C. reticulata* crossed with a *C. grandis* to give rise to the sour orange, *C. aurantium.* However, the low sabinene mandarin must also carry the W allele in Pgi-1, because the sour orange, is WS in Pgi-1, and this requires a W allele, if it is to be an F_1 product of *C. grandis* (SS in Pgi-1) × *C. reticulata* (FW in Pgi-1). The pummelo parent must again be an old-line heterozygous FS in Got-1 like Kao Pan. Furthermore the alate leaf petiole of the sour orange points to the pummelo as one of the parents.

When the tropical *C. grandis* expanded to the north and the Chinese *C. reticulata* expanded southward, they hybridized and gave us *C. sinensis* and *C. aurantium.* The latter had reached Europe already during Roman times, and the *C. sinensis* was brought by the Portuguese navigators.

VI. NEW ZEALAND GRAPEFRUIT

The major export fruit cultivated in New Zealand is the "New Zealand grapefruit". However, the low sabinene content of its leaves and different isozyme patterns suggest that it could not have originated from the regular *C. sinensis* and *C. grandis* cross, and this implies that it should not be classified as *C. paradisi,* which it presently is. Winged petioles and green cotyledons in the seed point to a *C. grandis* and *C. reticulata* involvement.

VII. TROYER CITRANGE

The Troyer citrange is a man-made hybrid. It is an intergeneric cross between the Bahia Navel sweet orange (*C. sinensis*) and *Poncirus trifoliata*. Originally produced as a cold-hardy fruit, it has come into common use as a tristeza virus resistant rootstock. Essential oils, isozymes, and oxidase browning are consistent with what is expected of these man-made combinations where the parents are known, and we use such data as a check for our other assumptions.

VIII. ROUGH LEMON AND THE ACID GROUP

Still another biotype of importance as a rootstock is the wild rough lemon or *C. jambhiri* of northern India, of which several strains exist. Its position on our taxonomic chart of species relationships would be between *C. reticulata* and the acid member group, which consists of *C. medica*, *C. limon*, and *C. aurantifolia*. These rough lemons are characterized by their rough and bumpy rind, hollow fruit axis, and loosely adherent albedo. They have a relatively high sabinene content of 20% in their leaf oils, are tristeza virus resistant, and possess enzyme and substrate resulting in oxidase browning. These last three characteristics are most likely inherited from the mandarin. The true lemons have a smooth rind, a solid fruit core, and a tight albedo. They show tristeza virus susceptibility, lack the substrate and the enzyme responsible for oxidase browning, have a small amount of leaf oil sabinene, and much more citral, consisting of the stereoisomers neral and geranial. Thus, the rough lemons are quite distinct from the true lemons.[7] While one of their parents is most probably the mandarin, the other may turn out to be the true lemon or the citron. Protein patterns from the total leaf proteins place the rough lemons closer to the true lemons than to the citrons.

The acid member group consisting of *C. limon*, *C. aurantifolia*, and *C. medica* is a difficult group because these biotypes have many characters in common, suggesting introgression. All have similar terpene patterns, amylase patterns, are nonbrowning, and are tristeza virus susceptible. However, the W allele in Pgi-1 sets *C. limon* apart from *C. aurantifolia* and *C. medica*, and the M allele in isocitrate dehydrogenase sets *C. medica* apart from the other two. Within this group, *C. medica* seems to be a good basic species. It is monoembryonic or sexually reproducing, and except for three fruit forms which are either bumpy, smooth, or fingered, it is highly homogeneous.

C. limon, however, is of a hybrid nature. Crossing two strains of lemons gave us F_1 progeny which expressed the characteristics of *C. limon*, *C. jambhiri*, *C.*

aurantifolia, and *C. medica*. Three factor analyses of rind oil showed that the hybrids clustered around the parents and acid group members. Analysis of leaf oils showed that the hybrids clustered around the parents, but that a sufficient number were outside of the parental and acid member cluster, suggesting an introgression with an additional, as yet unidentified, genepool. The involvement of *C. medica* in the ancestry of *C. limon* is almost certain.[6]

C. aurantifolia is similarly assumed to be of hybrid nature. Tanaka[14] thought it to be the primary *Citrus* species, having arisen from the Papeda gene pool because of similarities of flower and fruit vesicle structure and close geographical association with the Papedas throughout most of its range. Recent numerical analysis of 145 morphological character states by Barret and Rhodes[1] points also to *C. grandis* and the related genus, *Microcitrus*, as possible parental species. If so, *C. aurantifolia* might be a trihybrid.

IX. *CITRUS HALIMII*

A recently described wild *Citrus*, indigenous to Thailand and Malaysia, *C. halimii*, has a leaf and fruit morphology intermediate between the genera *Citrus* and *Fortunella*.[11] We regard this monoembryonic taxon as a good and distinct species of the subgenus *Citrus*. What, however, is its association with the other biotypes?

The high limonene content of its rind oil is similar to that of the mandarin and sweet orange. The low sabinene content of the leaf oils, the lack of oxidase browning reaction, virus susceptibility, and amylase pattern exclude it from this group. Although these last four characteristics and the acidity of the fruit suggest an association with the acid member group, total leaf protein fractions and isozyme patterns in glutamic oxalacetic transaminase-1 and in phosphoglucose mutase seem to exclude it from this acid group.

X. CONCLUSIONS

In summary, we propose four good species for the subgenus *Citrus*, namely the tropical *C. halimii* and *C. grandis* and the subtropical *C. medica* and *C. reticulata*.

All other biotypes, no matter how well established in our minds, as well as in the commercial citrus industry, can, despite their species standing, be identified as agamic complexes.

REFERENCES

1. Barrett, H. C. and Rhodes, A. M. (1976). *Syst. Bot.* **1**, 105.
2. Bitters, W. P. and Scora, R. W. (1970). *Bot. Gaz.* **131**, 105.
3. Esen, A. and Scora, R. W. (1975). *Am. J. Bot.* **62**, 1078.
4. Esen, A. and Scora, R. W. (1977). *Am. J. Bot.* **64**, 305.
5. Malik, N. M. (1973). *Pak. J. Sci. Res.* **25**, 268.
6. Malik, N. M., Scora, R. W. and Soost, R. K. (1974). *Hilgardia* **42**, 361.
7. Scora, R. W. (1975). *Int. Flav.* **6**, 342.
8. Scora, R. W., Esen, A. and Kumamoto, J. (1976). *Euphytica* **25**, 201.
9. Scora, R. W., Kumamoto, J., Soost, R. K. and Nauer, E. M. (1982). *Syst. Bot.* **7**, 600.
10. Sloane, H. (1696). "Catalogus plantarum quae in insulae Jamaica sponte proveniunt". Brown, London.
11. Stone, B. C., Lowry, J. B., Scora, R. W. and Jong, K. (1973). *Biotropica* **5**, 102.
12. Swingle, W. T. (1943). *In* "The Citrus Industry" (Webber, H. J. and Batchelor, L. D. eds), Vol. I, pp. 129–474. Univ. California Press, Berkeley.
13. Tanaka, T. (1977). *Studia Citrologica* **14**, 1.
14. Tanaka, T. (1969). *Bull. Univ. Osaka Pref.* Ser. B. **21**, 133.
15. Thomson, W. W., Platt-Aloia, K. A. and Endress, A. G. (1976). *Bot. Gaz.* **137**, 330.
16. Torres, A. M., Soost, R. K. and Diedenhofen, U. (1978). *Am. J. Bot.* **7**, 869.

CHAPTER 14

Biogenesis, Distribution, and Systematic Significance of Limonoids in the Meliaceae, Cneoraceae, and Allied Taxa

DAVID A. H. TAYLOR

Department of Chemistry, University of Natal, Durban, South Africa

I. INTRODUCTION

The limonoids are a group of oxidized triterpenes, only known to occur in species of the families Rutaceae, Meliaceae, Cneoraceae, and in *Harrisonia abyssinica* (Simaroubaceae). A recent report[22] of the occurrence of a limonoid in *Balsamodendron pubescens* (Burseraceae) should be confirmed, as details of the botanical characterization are not available. The limonoids are defined as altered tetracyclic triterpenes containing a furan ring attached as side-chain at C-17 to a cyclopentenophenanthrene nucleus, which may have undergone oxidative opening of one or more rings. To these are added the protolimonoids, a group of intact triterpenes oxidized in the side-chain in such a way as to suggest that they are biogenetic precursors of the limonoids. This relationship has not been proved by tracer experiments but the circumstantial evidence is strong enough to leave little doubt.

The biological distribution of the limonoids and protolimonoids is such that they can be used both as markers of family relationship within the Rutales, and also of closer relationship at a generic level. To this end it is convenient to divide the limonoids into chemically classified groups on the basis of what rings in the nucleus have been oxidized, either to lactones or to ring-opened compounds (see Chapter 6). A secondary classification can sometimes be obtained on the basis of particular patterns or positions of oxidation, not leading to ring opening. In this way the limonoids and protolimonoids can be divided into ten groups with the compounds related to cneorin, which are allied but very distinctive, forming a further group 11. The characteristic oxidation patterns of these are shown in Table 1. This classification differs from one given earlier[45] by the separation of groups 4 and 5, although these are the same in oxidation level.

TABLE 1

Division of the meliaceous limonoids into chemical groups based on oxygenation patterns

Group	Example	Ring A	Ring B	Ring C	Ring D	Side-chain
1	Turreanthin (1)	+	+	+	+	+
2	Havanensin (2)	+	+	+	+	Furan
3	Khivorin (3)	+	+	+	Lactone	Furan
4	Andirobin (4)	+	Opened	+	Lactone	Furan
5	Mexicanolide (5)	+	Opened and recyclized	+	Lactone	Furan
6	Phragmalin (6)	Bridged	Opened and recyclized	+	Lactone	Furan
7	Evodulone (7)	Lactone	+	+	+	Furan
8	Prieurianin (8)	Lactone	Opened	+	+	Furan
9	Nimbin (9)	+	+	Lactone or opened	+	Furan or further oxidized
10	Obacunone (10)	Lactone	+	+	Lactone	Furan
11	Cneorin B (11)	Extensively altered		+	Oxidized	Furan

+ = remains intact

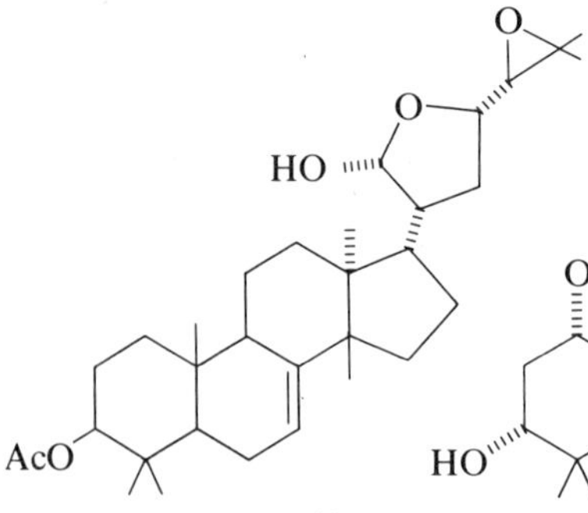

(1) Turreanthin 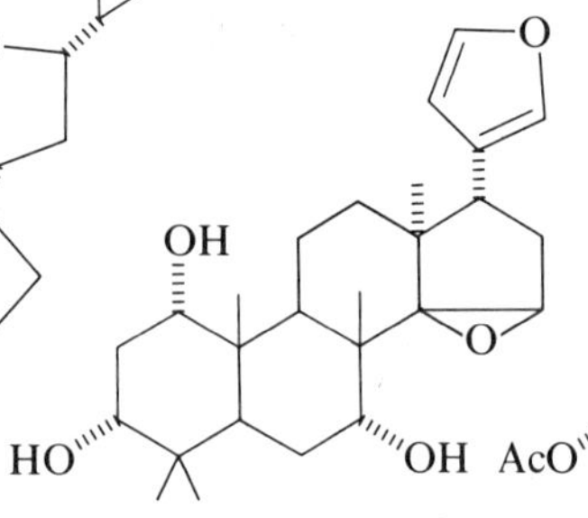   (2) Havanensin 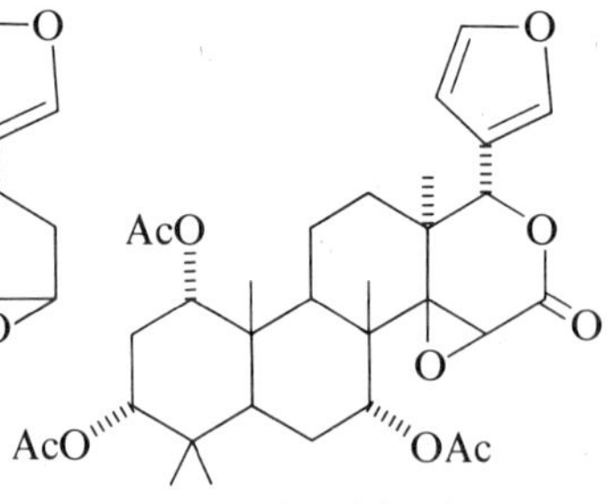   (3) Khivorin

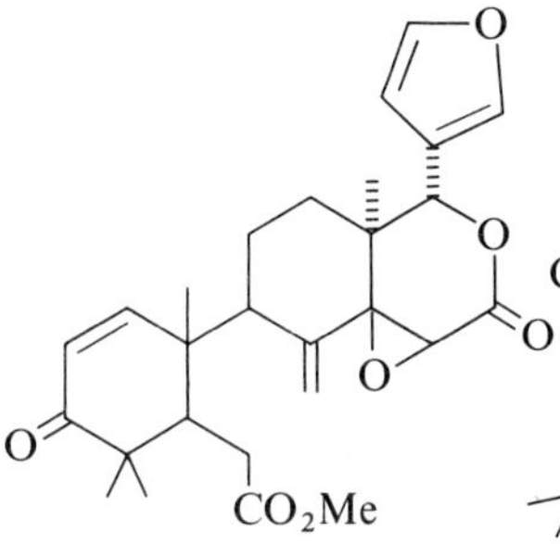

(4) Andirobin

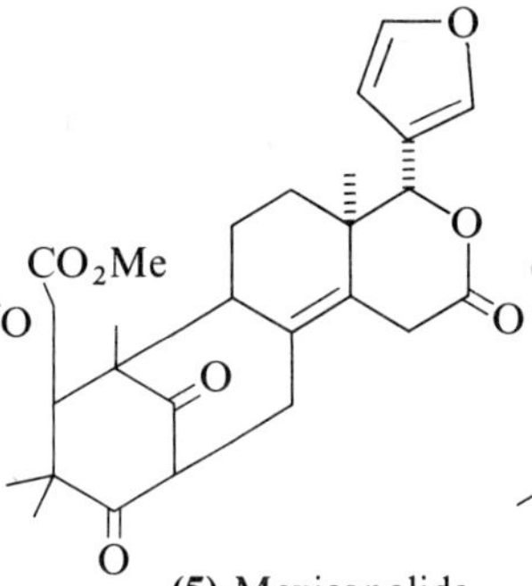

(5) Mexicanolide

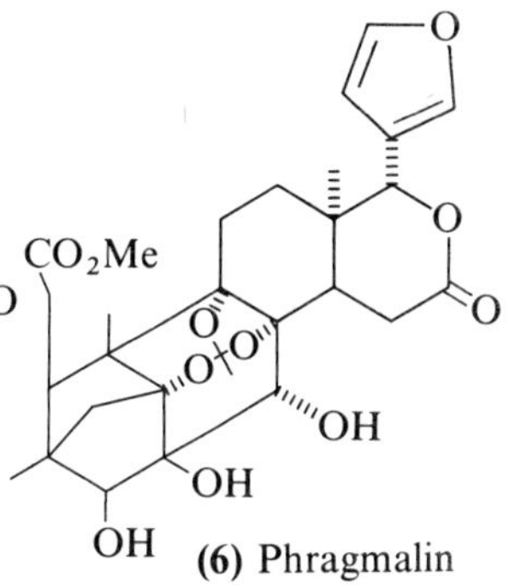

(6) Phragmalin

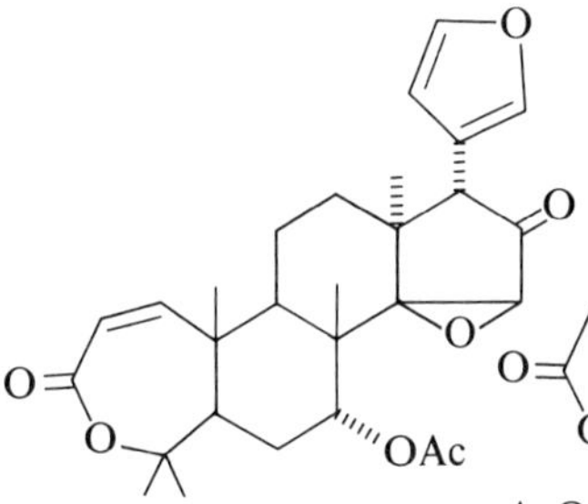

(7) Evodulone

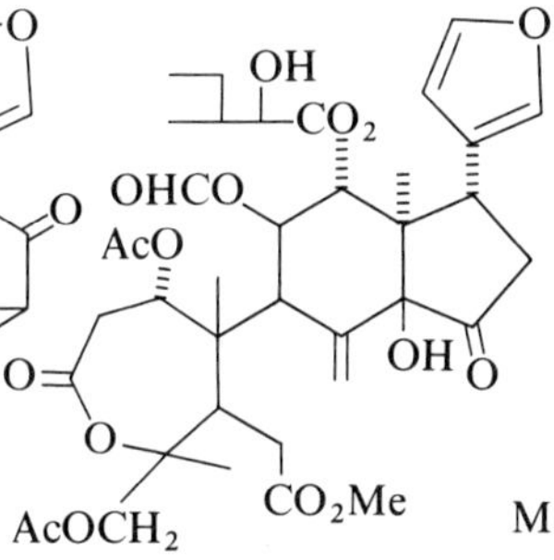

(8) Prieurianin

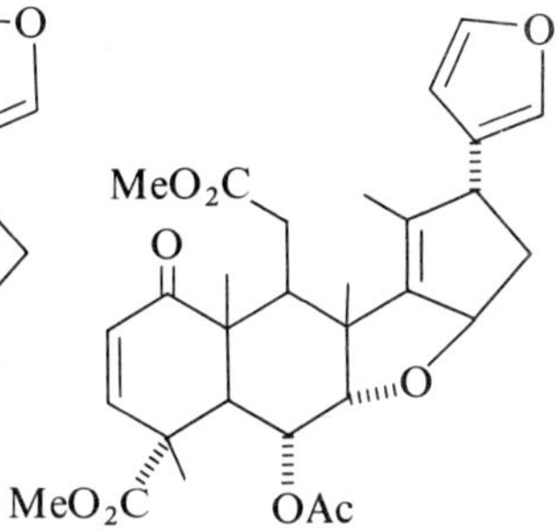

(9) Nimbin

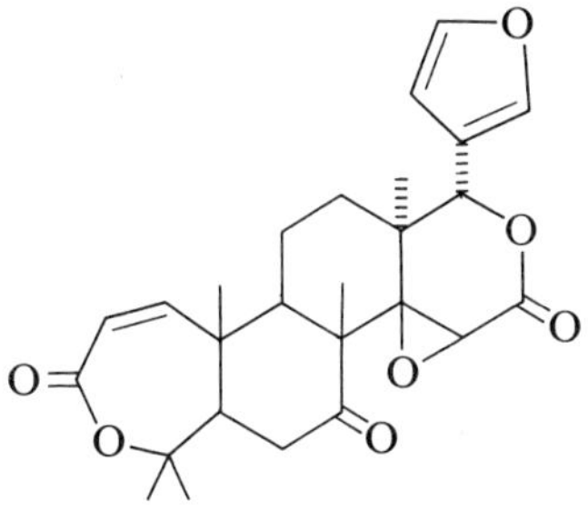

(10) Obacunone

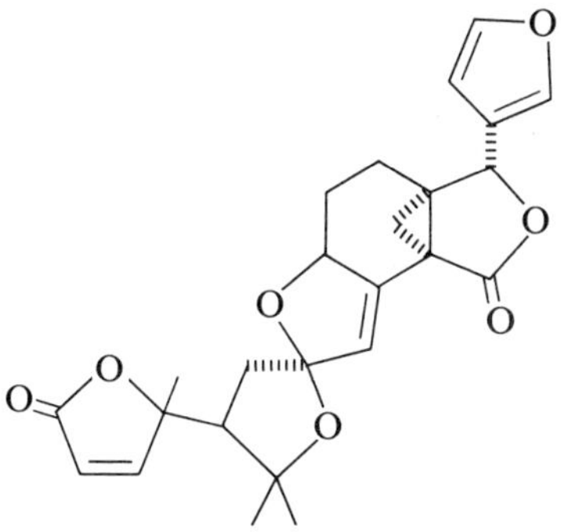

(11) Cneorin B

II. Biogenesis

Little tracer work has been done on the limonoids, and ideas on the biogenesis are derived from hypothetical chemical relations between the compounds which have been isolated. In all cases the biosynthetic process is believed to be based on peroxide oxidation, followed by rearrangement or recyclization of reactive intermediates. The supposed relationships between the groups are shown in Scheme 1; the chemistry underlying these relationships will now be examined.

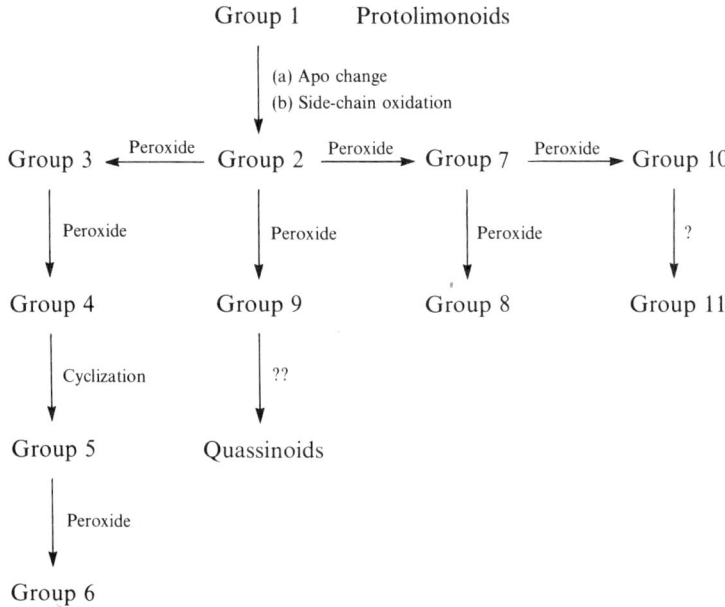

SCHEME 1. Hypothetical relationships between groups of limonoids.

Group 2. In the formation of these compounds from the protolimonoids two changes occur. In one, the so-called apo change, the $\Delta^{7,8}$-14-methyl nucleus occurring in turreanthin (**1**) and similar compounds is converted into the 7α-hydroxy-8-methyl-$\Delta^{14,15}$ structure found in grandifoliolenone (**12**). In the other the side-chain is degraded to a furan ring. In all known cases the apo change precedes the side-chain oxidation, and until contrary evidence turns up this is assumed to be universal.

The apo change has been carried out in the laboratory by Halsall[14] (Scheme 2); the method involved the oxidation of the original double bond to an epoxide which opened on treatment with a Lewis acid with concurrent carbonium ion rearrangement to give the required product. The biosynthetic route is probably similar although no 7,8-epoxides have yet been isolated.

The side-chain oxidation has also been duplicated by Halsall[7] by the periodate oxidation of turreanthin (Scheme 3). However, an alternative

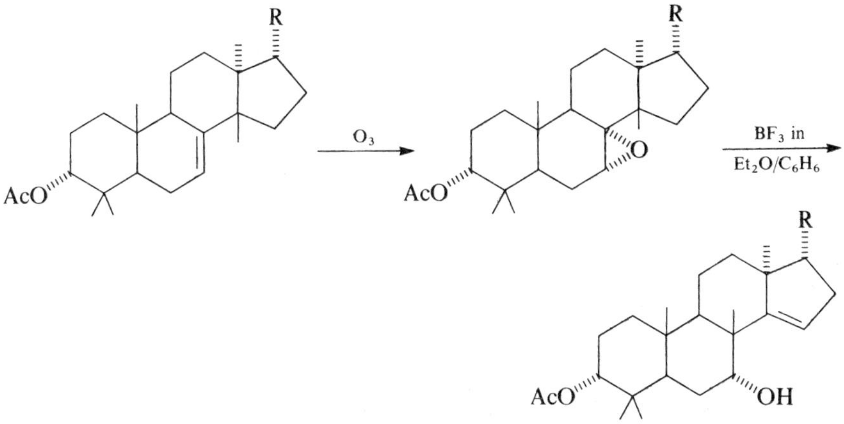

SCHEME 2. The apo change.[14]

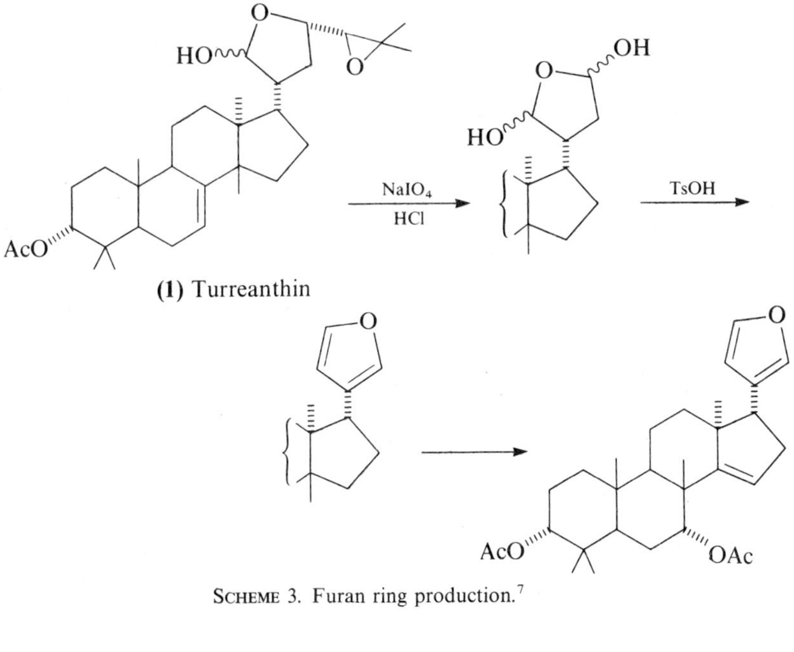

(1) Turreanthin

SCHEME 3. Furan ring production.[7]

mechanism involving the loss of the extra carbon atoms in a fragmentation reaction related to the origin of furocoumarins and furoquinolines[18] may be the true biological mechanism. The application of the apo change and side chain oxidation to turreanthin leads to the nuclear structure present in deoxyhavanensin (13), which has an additional hydroxyl group at C-1. This compound is the simplest known true limonoid.

(12) Grandifoliolenone   (13) Deoxyhavanensin

Group 3. Gedunin (14) and khivorin (3), the two simplest compounds of this group, have been partially synthesized by Lavie[30] and Halsall[6] by similar routes (Scheme 4). All the intermediates are known compounds, and it appears very likely that the sequence is natural.

Group 4. Andirobin (4) and similar compounds are readily obtained by peracid Baeyer–Villiger oxidation of 7-oxo limonoids,[13] and this probably represents the natural process.

Group 5. A suitably substituted limonoid of group 4 cyclizes spontaneously to mexicanolide (5), which has been synthesized in this way.[13] Similar compounds cyclize spontaneously to methyl angolensate (29) which is one of the most widely distributed of limonoids (Scheme 5). Again this is probably closely analogous to the natural process, although the order of the stages between 7-oxo khivorin and mexicanolide remains unknown.

Group 6. The simplest member of this rather complex group is phragmalin (6). There is little evidence of the biosynthetic route though it is perhaps noteworthy that the ring-A bridge is so far always accompanied by 8α,9α hydroxylation. Only in *Xylocarpus moluccensis*,[11,46] which contains phragmalin triacetate and a group of closely related substances of which xyloccensin D (15) is typical, has a limonoid of group 6 been found in company

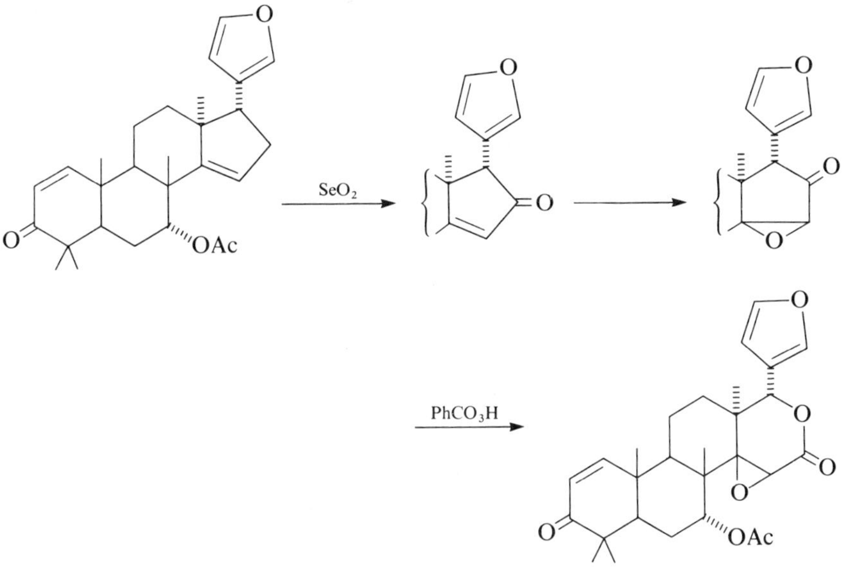

SCHEME 4. Partial synthesis of gedunin (**14**).[30]

with simpler substances. This may suggest that **15**, or an analogue, is an intermediate in the biosynthesis of phragmalin (Scheme 7).

Formation of the characteristic ring-A bridge is formally an isomerization, since oxidation of the 4α-methyl group is accompanied by reduction of the original C-1 carbonyl. Such a reaction seems in principle unlikely, since the only way of substituting an unactivated methyl group is oxidation, commonly by an oxygen radical formed from a peroxide and a suitably situated hydroxyl group, while peroxide oxidations are the basis of all limonoid phytochemistry.

It seems probable therefore that this isomerization represents a concealed oxidation followed by a rearrangement. The eventual site at which the oxidized group appears may then be C-9, which as previously mentioned is always oxidized in known phragmalin derivatives, and not in any other limonoids. The hydroxyl at C-1 in xyloccensin D (**15**) is in a suitable location to oxidize the 4α-methyl group, and on this basis the biosynthetic route shown in Schemes 6 and 7 is based.[46] So far the only evidence for this is the occurrence of the hypothetical intermediates in *Xylocarpus moluccensis*, but the proposal is being tested in the laboratory.

No compounds more complex than these are known in this biosynthetic chain.

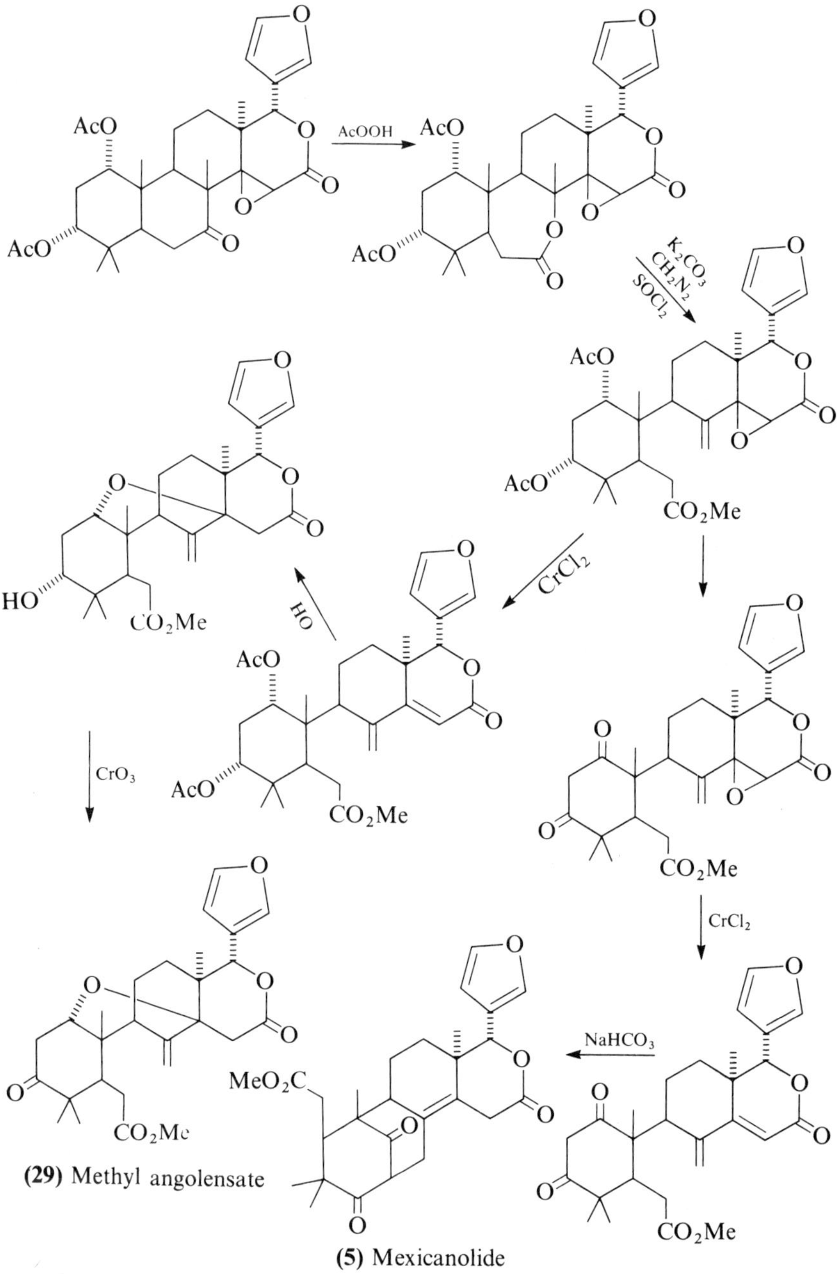

SCHEME 5. Partial synthesis of mexicanolide (5) and methyl angolensate (29).[13]

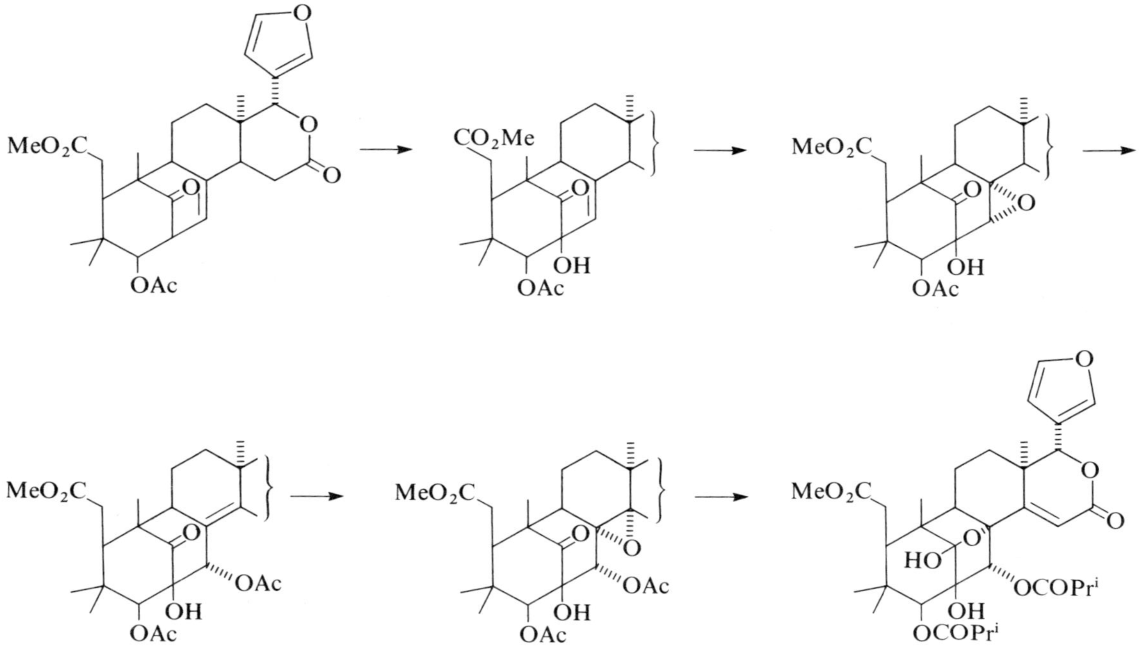

(15) Xyloccensin D

SCHEME 6. Projected biosynthetic pathway to xyloccensin D(**15**).

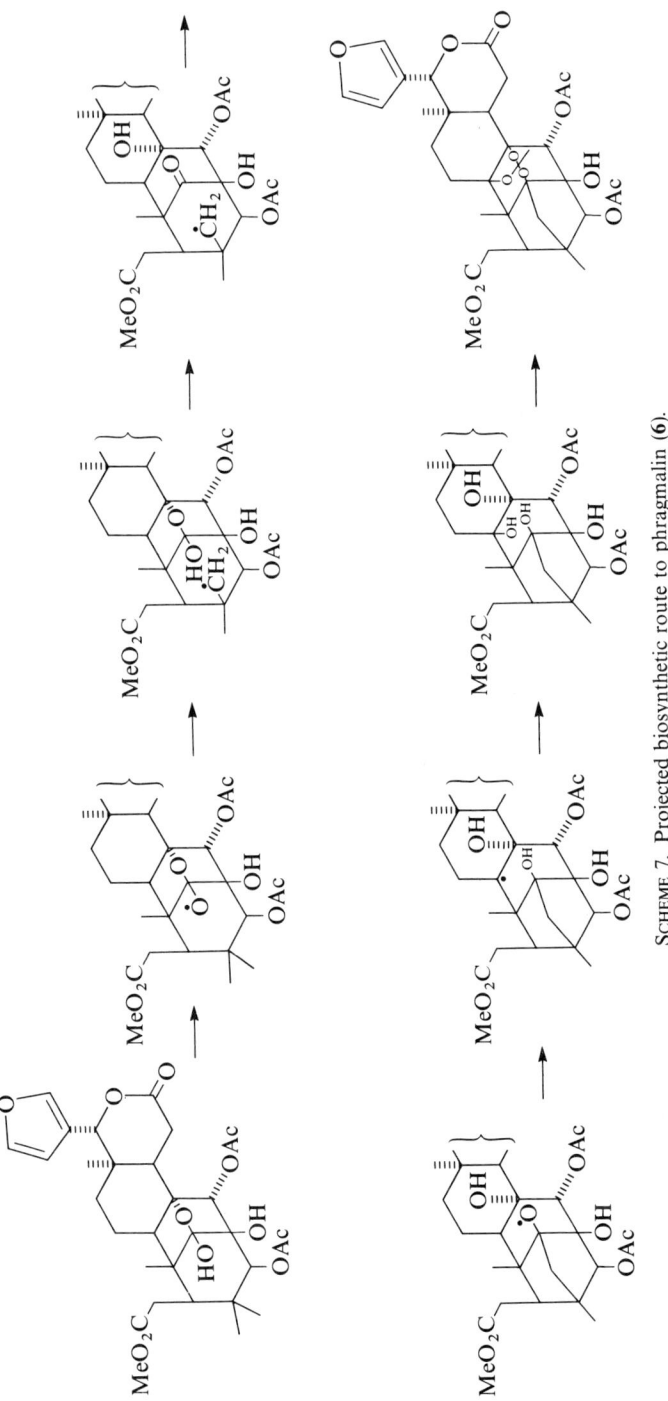

SCHEME 7. Projected biosynthetic route to phragmalin (6).

Group 7. The existence of compounds of this sort was predicted on the basis of biosynthetic theory as precursors of prieurianin (**8**). A search was made for them, and they have now been found by four groups of workers.[17,27,32,42] They presumably arise by the Baeyer–Villiger oxidation of compounds similar to heudelottin E (**16**) or more probably its unknown 1-hydroxy-dihydro derivative, although nothing is known of the details. No similar reaction has been carried out in the laboratory but it offers no obvious problems.

(**16**) Heudelottin E

Group 8. The preparation of prieurianin (**8**) and similar compounds by the ring-B Baeyer–Villiger oxidation of compounds of group 7 appears analogous to the preparation of andirobin, although it has not been carried out *in vitro.* Compounds of this group are very frequently oxidized in ring C to give an 11β,12α-diol, acylated at C-11 with formic acid and at C-12 with 2-hydroxy-3-methylvaleric acid, and oxidized in ring D to give a 14β-hydroxy-15-ketone or 15β-acyloxy group.[5] This ring-D substitution presumably derives from the more usual 14,15β-epoxide; the mechanism is not known.

Group 9. The limonoids related to nimbin (**9**) may be derived by Baeyer–Villiger oxidation of a 12-keto limonoid, though no suitable candidate compounds for this process are known. These compounds are characteristically very complex, and often display oxidation of angular methyl groups and of the furan ring.

Group 10. The *Citrus* species in which obacunone (**10**) was first found are lacking in simpler compounds, and until recently there was no evidence to indicate whether ring A or ring D was first oxidized. Obacunone and related compounds have now been found in Cneoraceae species together with compounds of group 7,[43] and it appears that in this case at least the oxidation of ring A occurs first. It is not known whether this is so in the case of *Citrus* species.

Group 11. The very complex compounds related to cneorin B (**11**) occur together with compounds of types 7 and 10, and it therefore appears that they are formed by further transformation of compounds similar to obacunone. The details are obscure and undoubtedly complex.[43]

The Quassinoids. The quassinoids, which occur in the related family Simaroubaceae, are usually considered as further transformation products of the limonoids because of the chemical and biological relationships. Alkaline hydrolysis of khivorin and similar ring-D lactones of group 3 brings about the so-called mero change, and mero-khivorin (**17**) has a lactone similar to that of quassin. However, the similarity does not extend beyond the new lactone ring, and the oxidation pattern of the quassin nucleus is quite unlike that normally found in limonoids of group 3. A second possibility for the loss of the side-chain in quassin would be a retro-aldol reaction in a 12-keto limonoid. This idea is supported by the fact that all C_{19} and C_{20} quassinoids do have a 12-keto group and by the fact that 12-substituted ring-D lactones are extremely rare in the Meliaceae. This raises the possibility of a relationship between quassin and nimbin, since the nimbin group is the only one in which ring-C opening, depending on a 12-ketone, occurs. Further, the secondary oxidation pattern of the nimbin group of limonoids and of the quassinoids shows a remarkable similarity. Both commonly lose one of the methyl groups from C-4; both commonly are oxidized on the other angular methyl groups, especially that at C-8, which is not oxidized in any other limonoid. Moreover, these secondary oxidation patterns extend to the quassinoids which have not lost the side-chain. Simarolide (**18**), for example, has a 1-oxo-4-nor structure similar to nimbin. It has no C-12 carbonyl group which may explain why the side-chain is not lost. On this hypothesis, the major distinction between the nimbin group and the quassinoids is that the latter possess the mechanism for opening ring D to a lactone, which is missing in nimbin, and which is necessary for the loss of the side-chain.

In the absence of evidence either from tracer studies or from the isolation of minor quantities of limonoids accompanying quassinoids, it is not at present possible to evaluate this proposal further.

Toona *Compounds.* Kraus has recently isolated a series of compounds from *Toona* species[28] in which only ring B is opened, giving either a lactone or a

(**17**) Mero-khivorin   (**18**) Simarolide ring-opened ester. Formally these represent a new group of limonoids. In fact they probably belong phytochemically to group 4 (rings B and D opened) and represent cases where the mechanism for oxidizing ring D has failed. This failure to carry out a stage of oxidation which might normally be expected appears to be quite common; its taxonomic significance will be discussed later.

III. DISTRIBUTION OF LIMONOIDS

As already mentioned, the limonoids are known to occur in the Rutaceae and Meliaceae, in Cneoraceae, and in *Harrisonia abyssinica*,[29] some botanical groupings being characterized by certain groups of limonoids. Although taxa such as *Khaya ivorensis* yield predominantly one limonoid,[4] this is not the normal pattern, and a mixture of limonoids of varying oxidation levels is present in very many species. This presumably implies that the biochemical utilization of intermediates is not very efficient. Usually different limonoids are present in seeds, bark, and timber; quite often limonoids are present in only one of these. Thus *Swietenia* contains limonoids only in the seed,[36] while the closely related *Khaya* contains limonoids in seed and timber,[1] and *Entandrophragma* only in timber.[3] The leaves have not been so extensively investigated. The occurrence of the various groups of limonoids is shown in Table 2. Minor occurrences are not listed; thus protolimonoids and other biochemical intermediates may be found in many genera if they are sought with enough determination.

IV. CHEMOTAXONOMY

In applying the distribution of limonoids to the chemotaxonomic study of the Meliaceae, only the most highly oxidized limonoids found in a given species can be regarded as significant. Many species in a genus, or even plants individually in a species, may fail to produce limonoids, or if they do, to oxidize them to the expected level. It appears that the loss of one or more of the necessary enzymes occurs readily and rather haphazardly. Thus in the species *Entandrophragma angolense* many West African individuals produce gedunin (group 3), while others produce methyl angolensate (group 4), and East African specimens contain no limonoid. This distribution was examined taxonomically in a collection of specimens, and no correlation could be found between the chemistry and the morphology. *Khaya senegalensis* is even more widely variable, though it does not usually fail to produce limonoids, in either East or West Africa. Other species, such as *Khaya ivorensis*, appear to give very

TABLE 2
Distribution of limonoid groups in the Meliaceae and allied taxa, arranged in tribes and genera

Limonoid group	Major constituent in: tribes of the Meliaceae	genera of the Meliaceae and other families	Notes
1	Melieae Aglaieae Guareeae Swietenieae	*Aphanamixis* *Chisocheton* *Entandrophragma* *Khaya* *Melia* *Turreanthus*	Frequent minor constituents in other genera
2	Turraeeae Melieae Aglaieae Trichilieae Cedreleae Swietenieae	*Aphanamixis* *Azadirachta* *Khaya anthotheca* *Toona* *Trichilia* *Turrea*	Frequent minor constituents in other genera
3	Melieae Trichilieae Guareeae Swietenieae Xylocarpeae	*Azadirachta* *Cabralea* *Chisocheton* *Entandrophragma* *Guarea thompsonii* *Khaya* *Trichilia* *Xylocarpus granatum*	Gedunin, but not khivorin, is a very frequent minor constituent in other genera
4	Trichilieae Guareeae Cedreleae Swietenieae Xylocarpeae	*Cabralea* *Carapa* *Cedrela* *Ekebergia* *Entandrophragma* *Khaya* *Soymida*	Methyl angolensate is one of the most widely distributed of all limonoids
5	Guareeae Cedreleae Swietenieae Xylocarpeae	*Cedrela* *Entandrophragma* *Guarea* *Khaya* *Pseudocedrela* *Swietenia* *Xylocarpus*	Found in *Guarea trichiliodes*; not other species
6	Swietenieae Xylocarpeae	*Carapa* *Chukrasia* *Entandrophragma* *Psudocedrela* *Xylocarpus*	

TABLE 2—contd.

Limonoid group	Major constituent in:		Notes
	tribes of the Meliaceae	genera of the Meliaceae and other families	
7	Cneoraceae Trichilieae Cedreleae Xylocarpeae		Minor constituent only in Cneoraceae, *Trichilia dregeana*, *Carapa procera*, and *Toona sureni*
8	Turraeeae Trichilieae Aglaieae Guareeae	*Aphanamixis* *Guarea* *Nymania* *Trichilia*	
9	Melieae	*Azadirachta* *Melia*	Accompanied by many minor compounds in *Azadirachta*
10	Cneoraceae Swietenieae	*Lovoa trichiliodes*	Also in *Harrisonia abyssinica* and Rutaceae
11	—	—	Cneoraceae

much more reproducible results. This is why the lack of ring-D oxidation in the *Toona* derivatives mentioned above occasions little surprise.

Although the study of minor constituents may be extremely valuable from a biosynthetic viewpoint, it usually seems to have little taxonomic significance. There are a few cases where this may not be so; an example is *Xylocarpus moluccensis*, which rather consistently produces a wide range of limonoids, of which the most highly oxidized member, phragmalin triacetate, is only a very minor constituent, yet is probably the most taxonomically significant.[11,46]

Although most plants which contain limonoids do contain mixtures if examined closely enough, some genera of which *Azadirachta*, *Melia*, and the members of the family Cneoraceae are outstanding, produce much more complex mixtures than the others. In these the mixture obtained, though complex, often appears to be more consistent from plant to plant than in species such as *E. angolense* which produce only a few limonoids.

The Rutaceae commonly produce mixtures of limonoids but are much more consistent throughout a genus, and indeed throughout the family, in the oxidation level reached, implying a more efficient use of the enzymes which are present.

At the family level, the chemotaxonomic distinctions are very clear. Rutaceae contains only limonoids of group 10;[16] Cneoraceae contains complex mixtures of limonoids of groups 7, 10, and 11, corresponding to one biosynthetic chain, and the equally significant 2-methylchromones related to ptaeroxylin;[43] the Meliaceae normally contains limonoids of groups 1–9. This distinction is so clear-cut that any exceptions are to be regarded with suspicion. One is *Lovoa trichilioides* which contains no limonoids in the seed or timber, but a small quantity of obacunol in the bark. There is no reason to doubt that *Lovoa* is correctly attached to the Meliaceae, but it is recognized as a problem genus,[37] and it seems that this is an apt description. The specific name is interesting, as the only other known occurrence of obacunol in the Meliaceae is in *Trichilia trifolia*.[47] Possibly *Lovoa* in some way represents a junction species.

A more serious problem is provided by *Harrisonia abyssinica* (Simaroubaceae), which was found by Nakanishi to contain limonoids of group 10.[29] A re-examination by Okorie in Ibadan confirmed this occurrence, and also found chromones of the ptaeroxylin group.[35] This combination of chromones and limonoids is very rare, the only other examples being *Spathelia* (Rutaceae)[48] and the Cneoraceae which is held separate by the presence of cneorins.[43] The chromones of the ptaeroxylin group do not occur elsewhere in the Rutaceae and are otherwise only known in Ptaeroxylaceae.[15] Morphologically, *Spathelia* and *Harrisonia* are very similar and have been combined by some botanists. It thus seems clear that *Harrisonia* does not belong in the Simaroubaceae, but in the tribe Spathelioideae, which itself belongs in some rather unclear junction area between Rutaceae, Cneoraceae, and Ptaeroxylaceae.

At the level of tribes in the Meliaceae there are also some fairly clear distinctions, although certain limonoids such as gedunin and methyl angolensate have a very wide distribution. The recent generic monograph of the Meliaceae by Styles and Pennington[38] allocates the genera in subfamilies as shown in Table 3.

Of these, subfamilies I (Melioideae) and IV (Swietenioideae) are clearly distinguished, while subfamilies II and III have unfortunately not been investigated. The Swietenioideae contain characteristically limonoids of the first biosynthetic chain, of groups 1, 2, 3, 4, 5, and 6 with, so far as is known, no members of any other group with the exception of *Lovoa* which contains obacunol as noted above. The genera in this subfamily form a progression, with *Entandrophragma*, *Chukrasia*, and *Pseudocedrela* being the most advanced. Subfamily I (Melioideae) is much more mixed. First, the tribe Melieae is distinguished from all the others. *Melia* and *Azadirachta* are very similar, and characteristically contain the ring-C oxidized limonoids of the

TABLE 3
Botanical arrangement of genera in the Meliaceae

Subfamily Tribe	Genera
Melioideae	
Turraeeae	*Munronia, Naregamia, Turraea, Humbertioturraea,* *Calodecaryia, Nymania*
Melieae	*Melia, Azadirachta*
Vavaeeae	*Vavaea*
Trichilieae	*Trichilia, Pseudobersama, Pterorhachis, Walsura,* *Lepidotrichilia, Malleastrum, Ekebergia, Astrotrichilia,* *Owenia, Cipadessa*
Aglaieae	*Aglaia, Lansium, Aphanamixis, Reinwardtiodendron, Sphaero-* *sacme*
Guareeae	*Heckeldora, Cabralea, Ruagea, Turraeanthus, Guarea,* *Chisocheton, Megaphyllaea, Synoum, Anthocarapa, Pseudo-* *carapa, Dysoxylum*
Sandoriceae	*Sandoricum*
Quivisianthoideae	*Quivisianthe*
Capuronianthoideae	*Capuronianthus*
Swietenioideae	
Cedreleae	*Cedrela, Toona*
Swietenieae	*Khaya, Neobeguea, Soymida, Entandrophragma, Chukrasia,* *Pseudocedrela, Schmardeae, Swietenia, Lovoa*
Xylocarpeae	*Carapa, Xylocarpus*

nimbin class. This both separates these two genera from the rest of the family, and also suggests a possible relationship to the Simaroubaceae, which has been discussed above. This is not a completely clear distinction, for heudebolin (19) a nimbin-type limonoid, has been found by Adesida and Okorie in the bark of *Trichilia heudelottii*, together with the more usual prieurianin-type compounds.[2] The other tribes, so far as they have been investigated, mainly contain limonoids of the prieurianin group (group 8),[9,19,31] which have been found in several genera in tribes 1, 4, 5, and 6. *Vavaea* (tribe 3) has not been investigated, while *Sandoricum* (tribe 7) contains completely unrelated pentacyclic triterpenes.[23] Gedunin and methyl angolensate are widespread in the subfamily,[10,21,26,41,50] and it seems that they have little taxonomic significance within the Meliaceae.

At the generic level, the distribution of limonoids also shows some interesting taxonomic features. *Melia* and *Azadirachta* are chemically very similar, containing many compounds in common. However, seeds of

Azadirachta contain a large range of compounds biochemically related to gedunin,[26] which are lacking in *Melia*.[24] Moreover, and rather in contrast, the very highly oxidized compounds such as azadirachtin (**20**), apparently only occur in *Azadirachta*.[49] The genus *Ekebergia* contains ekebergin (**21**)[44] which has a unique oxidation pattern, being oxidized at C-2, which is unusual, and at C-15 in a ring-D lactone. This is probably a secondary variation on a methyl angolensate type molecule, and synthetic analogues are known.[13] Compounds similar to aphanastatin (**22**) are also apparently common, having been found in *Melia*,[34] *Aphanamixis*,[40] and *Trichilia*.[33] The most interesting feature is the oxidation of the ring-A angular methyl group.

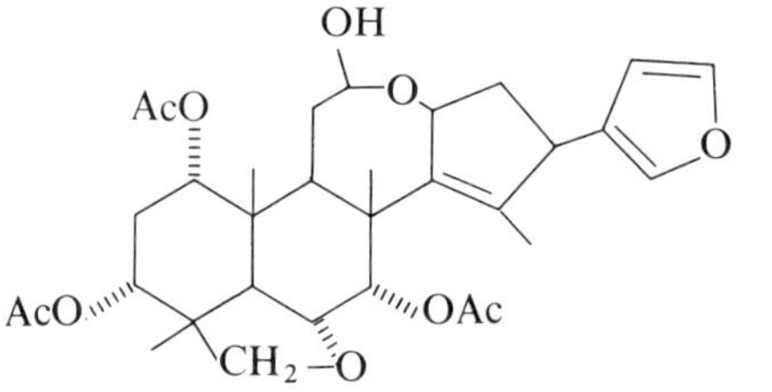

(**19**) Heudebolin

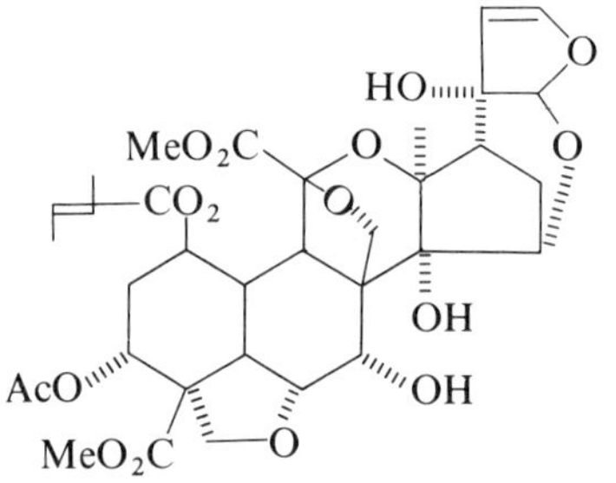

(**20**) Azadirachtin

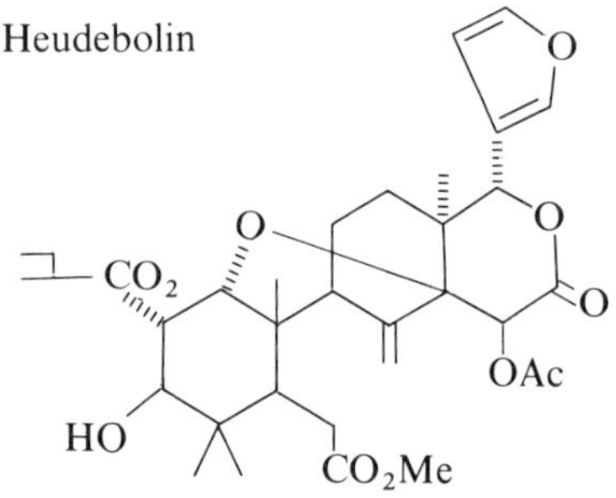

(**21**) Ekebergin

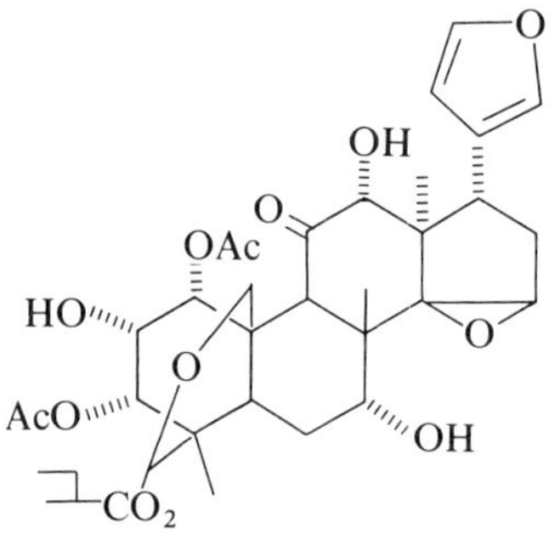

(**22**) Aphanastatin

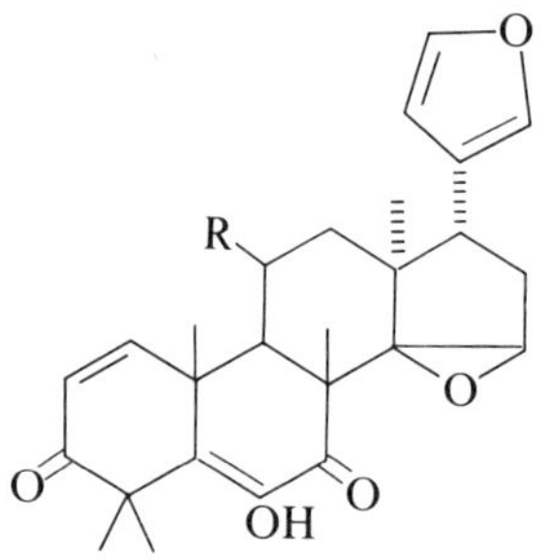

(**23**) Cedrelone (R = H)
(**26**) Anthothecol (R = OAc)

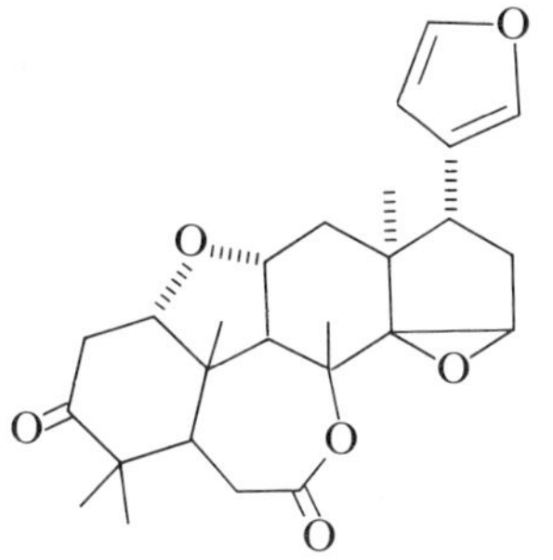

(24) Toonafolin

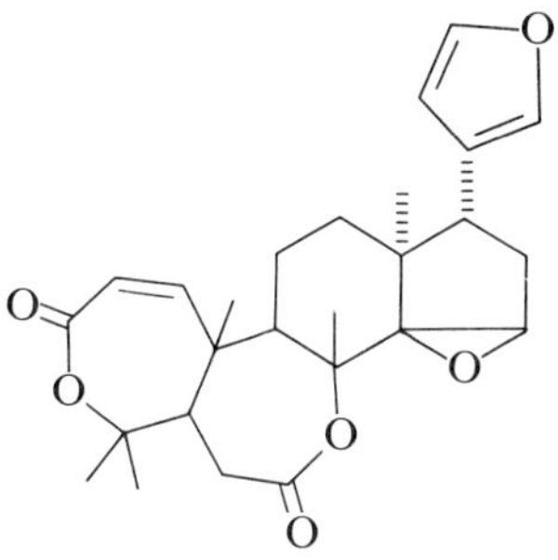

(25) Surenolactone

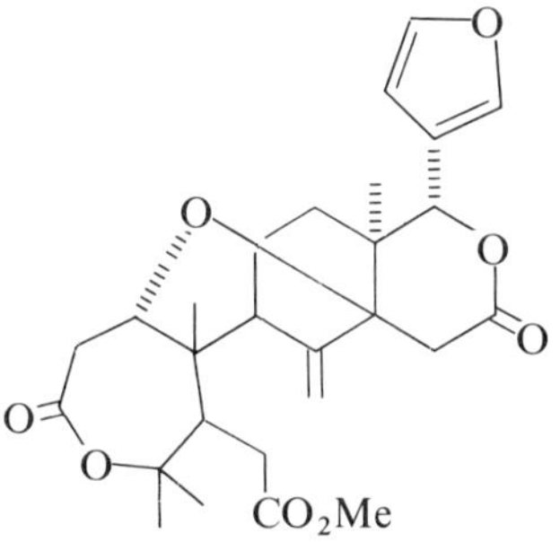

(27) Methyl ivorensate

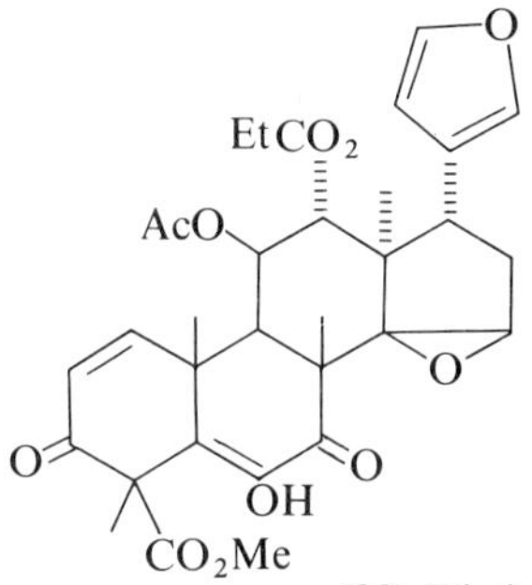

(28) Hirtin

The tribe Cedreleae is of interest. This contains the two genera *Cedrela* and *Toona*, which are said by Styles and Pennington to be distinct from all other genera in the Swietenioideae, and most in the Meliaceae, and possibly to have distinct evolutionary histories.[39] This is borne out to some extent by the chemotaxonomy. Chemically, *Cedrela*, giving mainly mexicanolide (5),[12] is entirely characteristic of the Swietenioideae, but *Toona* is not. The timber of *Toona* contains the rather unfortunately named cedrelone (23) which is apparently of the havanensin group (group 2);[20] the leaves have recently been shown to contain toonafolin (24),[25] a ring-B lactone, and surenolactone (25),[28] a ring-A/ring-B dilactone. The taxonomic significance of this is not yet clear. However, it seems probable that they are actually compounds of the mexicanolide group in which the opening mechanism for ring D has failed. An analogy is to be found in the chemistry of *Khaya anthotheca*. The genus *Khaya* is in the main a chemotaxonomically straightforward one, giving khivorin, mexicanolide, and closely related compounds.[1] However, the species *K. anthotheca*, which is morphologically extremely close to *K. ivorensis*, is very different chemically, containing havanensin (**2**) and anthothecol (**26**) which is very close to cedrelone (**23**). In view of the great morphological similarity this cannot be because of a basic dissimilarity in the biochemistry, and must depend on the lack of the ring-D opening enzyme, which is probably a very minor genetic variation. Similar enzyme failures occur fairly commonly, but not usually so dramatically as in this case.

In view of this, it does not seem too strange to suggest that the production of cedrelone is due to a similar enzyme lack in a plant which would normally be expected to produce substances related to mexicanolide. In **24** and **25** the ring-B opening enzyme system is working, while the oxidation of ring A in **25**, though unusual, is not unique. Thus *Khaya ivorensis* contains small amounts of methyl ivorensate (**27**), which is closely similar except for the failure to oxidize ring D. It seems not impossible that an intensive search might reveal in *Khaya anthotheca* similar compounds to surenolactone.

The ring-B diosphenol grouping in cedrelone and anthothecol is unusual, and its biosynthesis is unknown. However, it is not confined to these two compounds, as hirtin (**28**) from *Trichilia hirta* contains a similar system.[8] In this case, the presence of the five-membered ring D is not remarkable, since many *Trichilia* species contain this feature.

The taxonomic conclusions that may be drawn from these observations are as follows.

(1)   Gedunin, methyl angolensate, 7-oxogedunin, and simpler related compounds may occur anywhere in the Meliaceae, and have no known taxonomic significance. There may possibly be two divisions of the genus *Trichilia*, one containing these compounds and the other prieurianin derivatives, although these two types occur together in *Guarea* species.

(2)   The Swietenioideae are characterized by the possession of complex ring-D lactones, the most advanced of which occur in the genera *Chukrasia*, *Entandrophragma*, and *Pseudocedrela*.

(3)   *Lovoa trichiliodes* is unique, and may be a bridge species between the two main subfamilies.

(4)   *Toona* is unique in possession of ring-B oxidized five-membered ring-C compounds which sharply distinguish it from *Cedrela*. These differences probably result from loss of one enzyme system, and are paralleled by *Khaya anthotheca*.

(5)   The Melioideae are characterized by the possession of ring-D carbocyclic limonoids extensively oxidized elsewhere in the molecule. Oxidation of angular methyl groups is a feature of this subfamily. It is possible that the Melioideae separated from the Swietenioideae before the ring-D oxidation mechanism evolved, and that it is a separate development in the genera *Guarea*, *Trichilia*, *Ekebergia*, and

Azadirachta. There is an urgent need for the examination of *Quivisianthe* and *Capuronianthus*, regarded by Pennington and Styles as intermediate between these two major subfamilies. Unfortunately these plants, native to Madagascar, are very difficult to obtain at present.

(6)   The Melieae are distinguished by ring-C oxidized compounds. They are distinct from the rest of the Melioideae, but related by the presence of aphanastatin and similar compounds.

(7)   The limonoids of *Ekebergia* are not closely related to any others known, though presence of the 2-substituent is characteristic of Melioideae.

(8)   The limonoids of the Cneoraceae are quite distinct. The Cneoraceae are also distinct in the mixture of extractives they contain, including chromones otherwise characteristic of *Ptaeroxylon* and *Cedrelopsis*, which are considered sufficiently distinct to form a separate family, limonoids similar to the Rutaceae, and the characteristic cneorins. The Cneoraceae seem therefore very much a junction family, related to the Rutaceae and to the Ptaeroxylaceae, but distinct from either.

(9)   The position of *Harrisonia abyssinica*, discussed above, is noteworthy. It is closely related to the Spathelioideae, which has been placed in the Rutaceae and in the Simaroubaceae. The group probably belongs in a junction position, related to Cneoraceae.

(10) The relationship between the Simaroubaceae and the Meliaceae/Rutaceae is interesting.

A possible mechanism for the loss of the side-chain in quassinoids characteristic of the Simaroubaceae is by a reversed aldol reaction occurring in a 12-oxo ring-D lactone. In this case, the typical Simaroubaceae differs from the Meliaceae by containing the enzymic systems necessary both for the oxidation of ring D to a lactone and also for oxidizing C-12 to a ketone. This would suggest that the link between the families may be through the Melioideae in which oxidation at C-12 is common.

REFERENCES

1. Adesida, G. A., Adesogan, E. K., Okorie, D. A., Taylor, D. A. H. and Styles, B. T. (1971). *Phytochemistry* **10**, 1845.
2. Adesida, G. A. and Okorie, D. A. (1973). *Phytochemistry* **12**, 3007.
3. Adesida, G. A. and Taylor, D. A. H. (1967). *Phytochemistry* **6**, 1429.
4. Adesogan, E. K. and Taylor, D. A. H. (1970). *J. Chem. Soc.* (C) 1710.
5. Brown, D. A. and Taylor, D. A. H. (1978). *Phytochemistry* **17**, 1995; King, T. J. and Taylor, D. A. H. (1983). *Phytochemistry* **22**, 307.
6. Buchanan, J. G. St C. and Halsall, T. G. (1969). *J. Chem. Soc. Chem. Comm.* 1493.
7. Buchanan, J. G. St C. and Halsall, T. G. (1970). *J. Chem. Soc.* (C) 2280.
8. Chan, W. R. and Taylor, D. R. (1966). *J. Chem. Soc. Chem. Comm.* 206.

9. Connolly, J. D., Labbé, C., Rycroft, D. S., Okorie, D. A. and Taylor, D. A. H. (1979). *J. Chem. Res. (S)* 256, (*M*) 2858.
10. Connolly, J. D., Labbé, C., Rycroft, D. S. and Taylor, D. A. H. (1979). *J. Chem. Soc. Perkin Trans. 1* 2959.
11. Connolly, J. D., MacLellan, M., Okorie, D. A., and Taylor, D. A. H. (1976). *J. Chem. Soc. Perkin Trans. 1* 1993.
12. Connolly, J. D., McCrindle, R. and Overton, K. H. (1965). *J. Chem. Soc. Chem. Comm.* 162.
13. Connolly, J. D., Thornton, I. M. S. and Taylor, D. A. H. (1973). *J. Chem. Soc. Perkin Trans. 1* 2407.
14. Cotterell, G. R., Halsall, T. G. and Wriglesworth, M. J. (1970). *J. Chem. Soc. (C)* 1503.
15. Dean, F. M. and Taylor, D. A. H. (1966). *J. Chem. Soc. (C)* 114.
16. Dryer, D. L. (1983). Chapter 7.
17. Epe, B. and Mondon, A. (1979). *Tetrahedron Lett.* 2015.
18. Grundon, M. F. (1983). Chapter 2.
19. Gullo, V. P., Miura, I., Nakanishi, K., Cameron, A. F., Connolly, J. D., Duncanson, F. D., Harding, A. E., McCrindle, R. and Taylor, D. A. H. (1975). *J. Chem. Soc. Chem. Comm.* 345.
20. Hodges, R., McGeachin, S. G. and Raphael, R. A. (1963). *J. Chem. Soc.* 2515.
21. Housley, J. R., King, F. E., King, T. J. and Taylor, P. R. (1962). *J. Chem. Soc.* 5095.
22. Joshi, B. S. and Hegde, V. R. (1979). *Proc. Indian Acad. Sci.*, **88A**, 185.
23. King, F. E. and Morgan, J. W. W. (1960). *J. Chem. Soc.* 4738.
24. Kraus, W. and Bokel, M. (1981). *Chem. Ber.* **114**, 267.
25. Kraus, W. and Grimminger, W. (1981). *Liebigs Ann. Chem.* 1838.
26. Kraus, W., Cramer, R. and Sawitzka, G. (1981). *Phytochemistry* **20**, 117.
27. Kraus, W. and Kypke, K. (1979). *Tetrahedron Lett.* 2715.
28. Kraus, W., Kypke, K., Bokel, M., Grimminger, W., Sawitzki, G. and Schwinger, G. (1982). *Liebigs Ann. Chem.* 87.
29. Kubo, I., Tanis, S. P., Lee, Y., Miura, I., Nakanishi, K. and Chapya, A. (1976). *Heterocycles* **5**, 485.
30. Lavie, D. and Levy, E. C. (1970). *Tetrahedron Lett.* 1315.
31. MacLachlan, L. K. and Taylor, D. A. H. (1982). *Phytochemistry* **21**, 1701.
32. Mulholland, D. A. and Taylor, D. A. H. (1980). *Phytochemistry* **19**, 2421.
33. Nakatani, M., James, J. C. and Nakanishi, K. (1982). *J. Am. Chem. Soc.* **103**, 1228.
34. Ochi, M., Kotsuki, H., Hirotsu, K. and Tokoroyama, T. (1976). *Tetrahedron Lett.* 2877.
35. Okorie, D. A. (1982). *Phytochemistry* **21**, 2424.
36. Okorie, D. A. and Taylor, D. A. H. (1971). *Phytochemistry* **10**, 469.
37. Pennington, T. D. and Styles, B. T. (1975). *Blumea* **22**, 436.
38. Pennington, T. D. and Styles, B. T. (1975). *Blumea* **22**, 446.
39. Pennington, T. D. and Styles, B. T. (1975). *Blumea* **22**, 513.
40. Polonsky, J., Varon, Z., Marazano, C., Arnoux, B., Pettit, G. R., Schmid, J. M., Ochi, M. and Kotsuki, H. (1979). *Experientia* **35**, 987.
41. Rao, M. M., Meshulam, H., Zelnik, R. and Lavie, D. (1975). *Phytochemistry* **14**, 1071.
42. Sondengam, B. L., Kamga, C. S. and Connolly, J. D. (1979). *Tetrahedron Lett.* 1357.
43. Straka, H., Albers, F. and Mondon, A. (1976). *Beitrage Zur Biologie der Pflanzen* **52**, 267.

44. Taylor, D. A. H. (1981). *Phytochemistry* **20**, 2263.
45. Taylor, D. A. H. (1982). *In* "Flora Neotropica", Monograph 28, Meliaceae, pp. 450–459. New York Botanical Garden.
46. Taylor, D. A. H., unpublished work.
47. Taylor, D. R. (1971). *Rev. Latinoam. Quim.* **2**, 87.
48. Taylor, D. R. and Wright, J. A. (1971). *Rev. Latinoam. Quim.* **2** 84.
49. Zanno, P. R., Miura, I., Nakanishi, K. and Elder, D. L. (1975). *J. Am. Chem. Soc.* **97**, 1975.
50. Zelnik, R. and Rosito, C. M. (1966). *Tetrahedron Lett.* 6441.

CHAPTER 15

Phylogenetic Implications of the Distribution of Secondary Metabolites within the Rutales

PETER G. WATERMAN

Phytochemistry Research Laboratory, University of Strathclyde, Glasgow, U.K.

I. INTRODUCTION

It is my purpose in this chapter to bring together as many as possible of the observations recorded earlier in this volume concerning variation and distribution of secondary metabolites within the Rutales, and to assess them for their potential value as chemotaxonomic markers that can be related to an understanding of the phylogeny of the Order. In doing this I will largely confine myself to discussion of the five families Cneoraceae, Meliaceae, Ptaeroxylaceae, Rutaceae (including Flindersiaceae), and Simaroubaceae, as these clearly seem to be bound together by their ability to produce at least some of the range of secondary metabolites that can be regarded as typically

377

rutalean. In agreement with Hegnauer (Chapter 16) it seems more appropriate to treat the Burseraceae, a family usually associated with the Rutales *s.s.*, together with other families sometimes associated with the Order as members of the Sapindales *s.s.*

For the chemical taxonomist the Rutales is a gold-mine, overflowing with types of secondary metabolite which, by virtue of their limited distributions and reasonably defined biogenesis, are well suited to taxonomic analysis. The following groups appear to be the most important.

(1) *Alkaloids* (Chapters 2 and 3). The most numerous group of alkaloids found in the Rutales are the direct derivatives of anthranilic acid, of which the furoquinolines (including pyranoquinolines), acridones, and probably the carbazoles are either completely or very largely restricted to the Rutaceae.[44] Simple derivatives of tryptamine are less common but include β-indoloquinazolines, canthinones, and β-carbolines that occur in either Rutaceae or Simaroubaceae, or both. Derivatives of tyrosine and phenylalanine include the rare oxazoles and derivatives of 1-benzyltetrahydroisoquinolines (1-btiq). The former may well be restricted to the Rutaceae while the latter are well known as typical alkaloids of several of the large families of the Ranales.[23,26] Finally, imidazole alkaloids were for long known only from *Pilocarpus* (Rutaceae) but have now been recorded in *Casimiroa*, another genus of Rutaceae, and in *Cynometra* species (Caesalpiniaceae).[45]

(2) *Coumarins* (Chapters 2 and 4). Simple coumarins occur widely in the plant kingdom but the proliferation of a wide range of complex furo- and pyrano-coumarins is a feature that is largely confined to the Rutaceae[19] and to the Umbelliferae.[30]

(3) *Limonoids* (Chapters 6, 7, and 14) and *quassinoids* (Chapter 8). These triterpene derivatives that have usually undergone extensive oxidation and ring modification appear to be unique to the Rutales and, together with some of the anthranilate derived alkaloids, qualify as good taxonomic markers for the Order. The simplest limonoids are found in the Rutaceae. They occur more extensively and in increasing complexity through the Meliaceae and the Cneoraceae. The quassinoids represent a further development of the limonoid pathway and are the typical chemical markers of many genera of Simaroubaceae.

(4) *Flavonoids* (Chapter 5). At present flavonoids have not been widely investigated in the Rutales. The polymethoxy flavones of *Citrus*[7] and some related genera may be of significance, and recent studies in the genus *Acmadenia* (Rutaceae)[46] suggest that, in the future, comprehensive surveys of flavonoid constituents will be worthwhile.

(5) *Chromones* (Chapter 4). In terms of numbers of sources these are very minor constituents of the Rutales. However, they may have great taxonomic implications (see Chapter 14 and Section V of this chapter).

II. The Taxonomic Significance of the 1-btiq Alkaloids:
The Proto-Rutaceae

To date 1-btiq alkaloids have been reliably reported from only four genera, *Zanthoxylum* (in which I include *Fagara*[18]), *Phellodendron*, *Toddalia*, and *Fagaropsis*. The alkaloids isolated include many of the most typical 1-btiq groups (Fig. 1), e.g. aporphines (**1**), berberines (**2**), tetrahydroprotoberberines (**3**), protopines (**4**), and, most commonly, benzophenanthridines (**5**). In all cases the alkaloids isolated are typical of their class, and although there is little

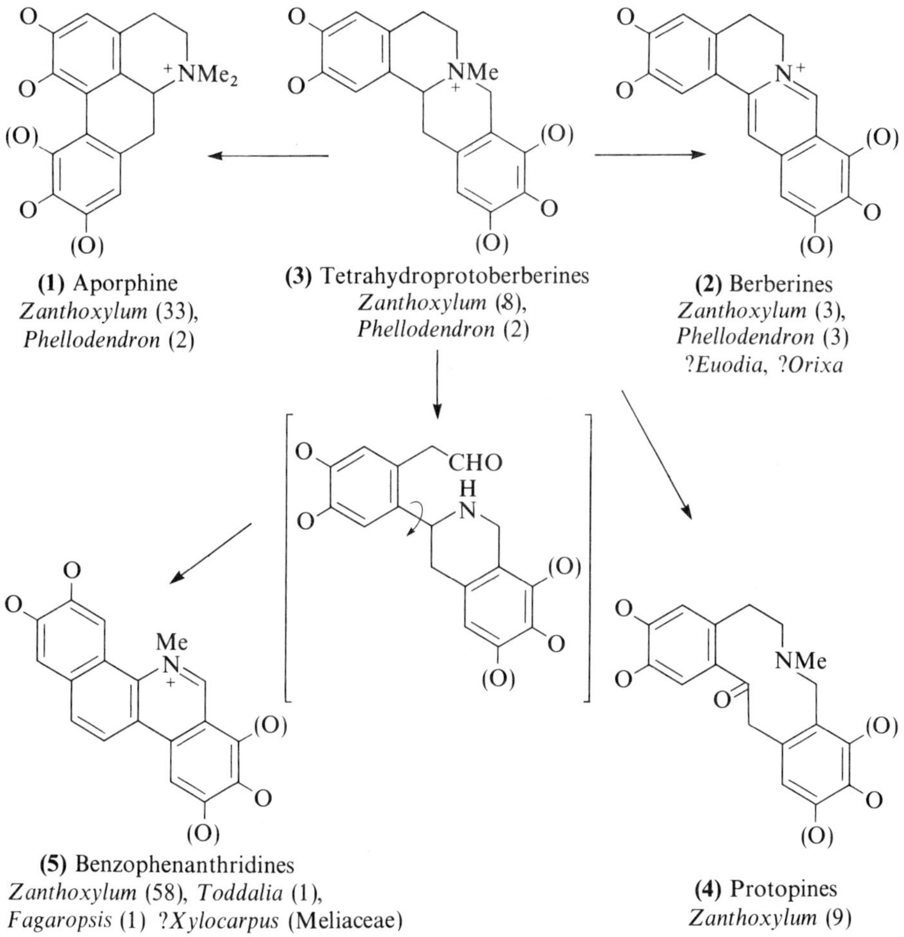

(1) Aporphine
Zanthoxylum (33),
Phellodendron (2)

(3) Tetrahydroprotoberberines
Zanthoxylum (8),
Phellodendron (2)

(2) Berberines
Zanthoxylum (3),
Phellodendron (3)
?*Euodia*, ?*Orixa*

(5) Benzophenanthridines
Zanthoxylum (58), *Toddalia* (1),
Fagaropsis (1) ?*Xylocarpus* (Meliaceae)

(4) Protopines
Zanthoxylum (9)

Fig. 1. The 1-benzyltetrahydroisoquinoline alkaloids of the Rutaceae, shown with normal oxygenation patterns and level of *N*-methylation. Brackets after generic names give the number of species containing that type of alkaloid.

direct information on their biogenesis in the Rutaceae there is no reason to suspect that it differs from that in the Ranales.

Within the four 1-btiq producing genera occurrence of these alkaloids is almost universal; the Malesian species *Z. ovalifolium* is a possible exception inasmuch as several chemical investigations have failed to reveal their presence. The rare protopines are known only from some Australasian and one South American species of *Zanthoxylum* while aporphines are common in *Zanthoxylum*, usually as quaternary derivatives, but have otherwise been recorded only once, in *Phellodendron*. Berberines occur predominantly in *Phellodendron* and also in a few *Zanthoxylum* species, and their tetrahydro derivatives so far only in the latter. Benzophenanthridines are by far the most widespread alkaloids of *Zanthoxylum*, often concentrating in large amounts in the root bark, and are the only 1-btiq alkaloids known from the single species of *Toddalia* and *Fagaropsis* studied. Benzophenanthridines have yet to be found in *Phellodendron*.

The presence of 1-btiq alkaloids in the above four genera is persistent and comparable with their occurrence in typically ranalean families such as the Annonaceae, Magnoliaceae, Berberidaceae, and Papaveraceae. To rationalize their appearance in the Rutaceae it is either necessary to propose an extreme case of chemical convergence through four quite complex alkaloid biosynthetic pathways or to accept that their distribution indicates a direct phylogenetic relationship between Ranales and Rutales through the Rutaceae. While 1-btiq alkaloids offer the most compelling evidence in favour of such a link, other supporting chemical data does exist and has been reviewed elsewhere.[18,23]

Most early systematicists did not favour such a relationship and placed the Rutaceae, and allied families, in the Sapindales[5,16,17,28,37] or Terebinthales[49] far from the Ranales. Hallier[21] also placed the Rutaceae in the Terebinthales but differed in recognizing a direct link to the Berberidaceae. Among modern systematicists there are still many of the most eminent who fail to recognize a close affinity between Ranales and Rutales[10,11,25,41] but a number of others[12,27,42] have been influenced to admit such a relationship. Independent evidence in favour of the relationship comes from cytological studies[14] and the co-evolution between papilionid butterflies and host plants, which centres on the Ranales but spills over into allied groups such as the Rutales.[15]

With the acceptance that the co-occurrence of 1-btiq alkaloids in the Ranales and Rutaceae is phylogenetically significant, then the four 1-btiq producing genera of the Rutaceae must be taken as possessing the most primitive spectrum of secondary metabolism extant within the Order. Accordingly this group, which I term the proto-Rutaceae, can be taken as the base point for an analysis of the chemical systematics of the Rutales.

Finally, is it possible to make any suggestions concerning the closest extant relationship between proto-Rutaceae and Ranales? By far the most striking relationship would appear to be that between the proto-Rutaceae and Papaverales which is the major source of both benzophenanthridine and protopine alkaloids. These two alkaloid types are virtually restricted to these two taxa but both aporphines and berberines are much wider in distribution. While not necessarily advocating a linear relationship between them it does seem likely, on the basis of the alkaloid data, that Papaverales and Rutales have a common source.

III. THE DEVELOPMENT OF TYPICAL SECONDARY METABOLISM WITHIN THE RUTALES

All four of the genera of the proto-Rutaceae do produce at least one type of typical rutalean metabolite as well as 1-btiq alkaloids. However, the distribution of the modern components is not random (Table 1); whilst *Toddalia* and many species of *Zanthoxylum* produce a wide range of anthranilate derived alkaloids (notably furoquinolines) and coumarins that are typical of the Rutaceae, *Phellodendron* and *Fagaropsis* have yielded only limonoids, which are typical of the Rutales as a whole.

TABLE 1
Typical rutaceous metabolites from the proto-Rutaceae

	Anthranilate-alkaloids	Coumarins	Limonoids
Zanthoxylum	+	+	−
Toddalia	+	+	−
Phellodendron	−	−	+
Fagaropsis	−	−	+

This is an interesting dichotomy but one that is difficult to interpret. *Phellodendron* and *Fagaropsis* which produce rutalean limonoids but lack the specifically rutaceous alkaloids and coumarins would appear to be at the earliest biosynthetic level of development extant. With the exception of *Xylocarpus granatum*, which has recently been reported to produce both a benzophenanthridine and a furoquinoline (Chapter 3), these compounds are lacking from the Meliaceae and it seems likely that the Meliaceae split away at a level only slightly more developed than that seen in *Phellodendron* and *Fagaropsis*.

The situation regarding the Simaroubaceae is more complex. This family, which must clearly have developed thè ability to produce quassinoids from a limonoid producing ancestor, shares with *Zanthoxylum*, a non-limonoid producing proto-Rutaceae, the ability to form the unusual canthin-6-one type of alkaloid (Fig. 2). Within the Rutaceae canthin-6-one (6) and related alkaloids are found only in *Zanthoxylum* and in the monotypic Australian genus *Pentaceras*, but in the Simaroubaceae they are more widespread. Similarly the allied β-carbolines (7) occur more widely in Simaroubaceae than in the Rutaceae although β-indoloquinazolines (8) are found only in Rutaceae (Fig. 2). Thus it seems likely that, before the Simaroubaceae and proto-Rutaceae progenitors separated, the ability to synthesize these simple β-carboline type alkaloids must have developed, although the absence of limonoids in *Zanthoxylum* would seem to preclude its direct involvement.

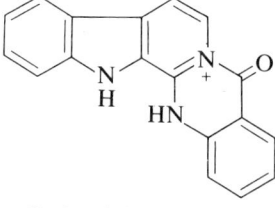

(6) Canthin-6-ones   (7) β-Carbolines

Rutaceae

Zanthoxylum (11),   *Dutaillyea* (1), *Flindersia* (2)   *Araliopsis* (1), *Euodia* (1),
Pentaceras (1)   *Euxylophora* (1), *Hortia* (4),
                                                                        Vepris (1), *Zanthoxylum* (5)

(8) β-Indoloquinazoles

Simaroubaceae

Aeschrion (1), *Ailanthus* (3),   *Aeschrion* (1), *Ailanthus* (2),
Amaroria (1), *Picrasma* (3),   *Perriera* (1), *Picrasma* (4)
Samadera (1), *Simaba* (2),
Simarouba (1), *Soulamea* (2)

FIG. 2. The distribution of canthin-6-ones, β-carbolines, and β-indoloquinazolines among genera of the Rutaceae and Simaroubaceae. Brackets after generic names give the number of species containing that type of alkaloid.

The scenario envisaged therefore is that as the Rutales evolved there was probably an initial development along the pathway of limonoid formation and at this stage the line leading to the Meliaceae began to separate. Further development along the major line of advance of the Order saw the formation of alkaloids based on anthranilic acid and its metabolite tryptophan. At this stage the simaroubaceous stock separated and elaborated the quassinoid pathway with retention of the tryptophan type of alkaloid. Limonoids then persisted into the Rutaceae with development of anthranilic acid and coumarin pathways, and with tryptophan derivatives occurring only rarely. As these developments took place the ability to form 1-btiq alkaloids was rapidly lost. While the above scheme is generally successful in explaining the distribution of the major groups of rutalean metabolites within the Order it does run into problems when other chemical markers are considered. Foremost among these are the hydrolysable tannin constituents gallic acid and ellagic acid which appear quite common in the Simaroubaceae,[3,32] very rare in the Rutaceae,[24] and completely absent from the Ranales.[3] As it is generally considered[3] that the trend is for the loss of these compounds then the suggested phylogeny for the formation of the major rutalean families appears to contradict this.

As pointed out above, all major types of rutaceous secondary metabolites occur together with the 1-btiq alkaloids in the proto-Rutaceae, but so far there is no known species, or for that matter genus, that produces all four. If such a taxon exists it seems probable that it will be among the taxa of the tribe Zanthoxyleae (sub-family Rutoideae), and an extensive search among the numerous uninvestigated genera of this group would be most interesting. One possible candidate must be *Euodia*, a diverse genus found from the temperate regions of China and Japan south to the tropical rain forests of New Caledonia and Australia. Relevant data is available on some 14 species (Table 2) and shows that limonoids are common in northern species but are replaced in more southerly species by furoquinoline and acridone alkaloids and coumarins. The tryptophan derived β-indoloquinazolines are also known from a Japanese species and, most interestingly, there are a number of very old reports of "berberine-like" compounds from two northern species.[24] The substantiation of the latter and the resolution of the relationship between these northern temperate *Euodia* and sympatric *Phellodendron* species could yield information of considerable taxonomic value. The closely related Japanese genus *Orixa* is also reported to yield berberine-like alkaloids[50] as well as anthranilate alkaloids[35,44] and coumarins.[19]

IV. CHEMICAL SYSTEMATICS WITHIN THE RUTACEAE

Traditionally the Rutaceae has been subdivided into seven sub-families[16,17,18] of which three, the Rutoideae, Toddalioideae, and Aurantioideae, are large. All three have been further subdivided by Engler[28] into a number of tribes and sub-tribes, and the Aurantioideae has been similarly treated by Swingle.[40] The Flindersioideae is a smaller sub-family of two genera

TABLE 2
Distribution of secondary metabolites in *Euodia* species

Location	Type of compound*					
Species	A	B	C	D	E	F
India/China/Japan						
E. aubertia	−	−	−	−	−	+
E. danielli	−	−	−	−	+	−
E. fraxinifolia	−	−	−	−	+	−
E. hupehensis	−	−	−	+	+	−
E. meliaefolia	?	−	−	−	+	−
E. rutaecarpa	−	+	−	−	+	−
Madagascar						
E. beleha	−	+	−	+	−	−
E. floribunda	−	−	−	−	−	−
Australisia/Malesiana						
E. alata	−	+	+	−	−	−
E. elleryana	−	+	−	−	−	+
E. hortensis	?	−	−	−	−	+
E. littoralis	−	+	−	+	−	+
E. triphylla†	−	−	+	−	−	−
E. xanthoxyloides	−	+	+	−	−	+

* A = "berberine-like" alkaloids; B = furoquinoline and β-indoloquinazoline alkaloids; C = acridones; D = furo- and pyrano-coumarins; E = limonoids; F = acetophenones and related compounds.
† J. Vaquette, unpublished results.

which has been variously placed in the Rutaceae, Meliaceae, and as a distinct family between these two.[22] In a recent revision of *Flindersia*, Hartley[22] retained the taxon within the Rutaceae, a contributing factor being the presence of many typically rutaceous furoquinolines and coumarins and the absence of true limonoids.

The three remaining sub-families are all small and their positions are contentious. The Dictylomatoideae has been assigned to both Simaroubaceae[29] and Rutaceae,[28] and the only known metabolite, a simple indole, is of little value in resolving this problem. The Rhadbodendroideae is of very doubtful affinity and has been placed near the Centrospermae[34] although there is some morphological evidence favouring retention in the Rutales.[4,36] There is, as yet, no chemical evidence to support either of these placements although our own investigations of the stem bark, wood, and leaves of *Rh. macrophyllum* revealed copious amounts of leucodelphinidin but no typical metabolites of the Rutaceae. The final sub-family, the Spathelioideae, will be discussed in detail in Section V.

A. THE DISTRIBUTION OF FUROQUINOLINES, ACRIDONES, AND COUMARINS
THROUGHOUT THE RUTACEAE (Fig. 3)

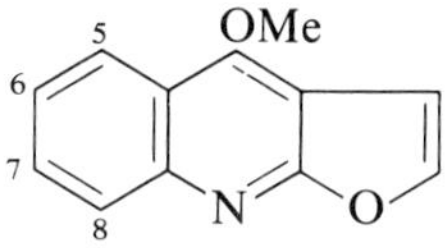

(9) Furoquinolines 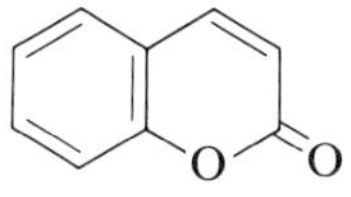   **(10)** Coumarins 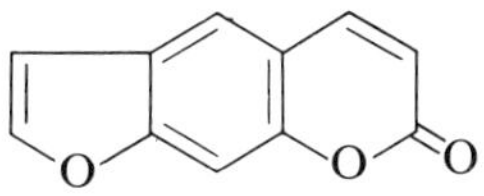   **(11)** Furocoumarins

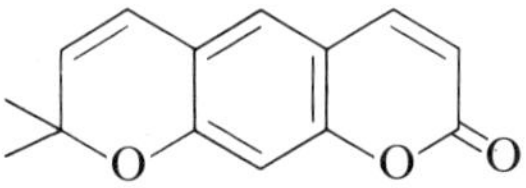

(12) Pyranocoumarins 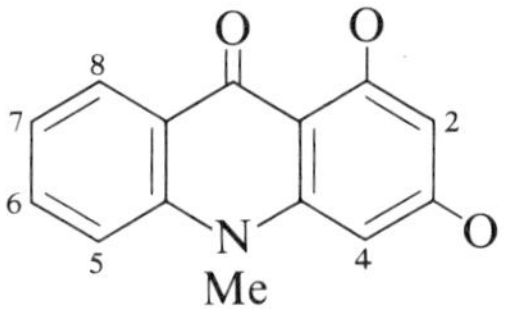   **(13)** Acridones

FIG. 3. The most common types of secondary metabolites produced by the Rutaceae (found in all three major sub-families).

Table 3 gives the occurrence of furoquinolines of different oxygenation patterns in the three major sub-families and the Toddalioideae. They occur widely in the proto-rutaceous genera *Zanthoxylum* (Rutoideae) and *Toddalia* (Toddalioideae) and in more advanced genera of Rutoideae, Toddalioideae, and Flindersioideae. By contrast they are found only to a limited extent among the Aurantioideae. The most interesting feature of this survey is the apparent specialization in production of 6,7- and 6,7,8-oxygenation by genera of the

TABLE 3

Distribution of furoquinoline and pyranoquinolone alkaloids with different substitution patterns among the major sub-families of the Rutaceae (%)

| Sub-family | All | 0* | Substitution patterns | | | | | | |
			8	7	6	7/8	6/8	6/7	6/7/8
	[580]	[126]	[72]	[32]	[23]	[197]	[27]	[75]	[28]
Rutoideae	61.6	66.7	75.0	81.3	78.3	68.5	33.3	37.3	10.7
Flindersioideae	6.7	7.1	2.8	0.0	0.0	4.1	7.4	14.7	25.0
Toddalioideae	28.3	20.6	18.1	15.6	21.7	23.9	59.3	45.3	64.3
Aurantioideae	3.4	5.6	4.2	3.1	0.0	3.6	0.0	2.7	0.0

* No substitution. Figures in square brackets refer to total number of reports of that substitution pattern.

Flindersioideae and Toddalioideae (particularly the African genera *Teclea,* *Vepris,* and *Oricia*). Table 4 gives a comparable breakdown for the distribution of simple, furo- and pyrano-coumarins. In this case the major centres of coumarin elaboration would appear to be among genera of the Rutoideae and Aurantioideae, with smaller numbers, particularly of the more complex coumarins, being produced by Toddalioideae and Flindersioideae.

TABLE 4

Distribution of simple, furo- and pyrano-coumarins within the sub-families of the Rutaceae (%); Data based on Gray and Waterman[19]

Sub-family	All	Simple coumarins	Pyrano-coumarins	Furo-coumarins	Furo-quinolines
	[148]	[92]	[66]	[73]	
Rutoideae	50.0	46.7	45.5	47.9	61.6
Flindersioideae	6.8	7.6	7.6	2.7	6.7
Toddalioideae	13.5	12.0	7.6	19.2	28.3
Aurantioideae	29.7	33.7	39.4	30.1	3.4

Figures in square brackets refer to total number of reports of coumarins from different species.

A third group of compounds that may be capable of throwing some light on the development of the family as a whole are the acridone alkaloids, which have so far been reliably recorded in 19 genera from the three large sub-families, with highest concentrations occurring in the Toddalioideae (Table 5). Major developments from the 1,3-oxygenated precursors are: (i) C-2 oxidation—common, without taxonomic value; (ii) C-4 oxidation—mainly in Australasian species of *Euodia* and *Melicope* (Rutoideae) and *Acronychia* and *Baurella* (Toddalioideae); (iii) C-5 oxidation—*Atalantia* (Aurantioideae) and *Teclea* (Toddalioideae); (iv) C-prenylation followed by pyran ring formation (*Atalantia* and *Acronychia*) or furan ring formation (*Ruta*–Rutoideae); (v) reduction at C-3 and C-1—in *Ruta, Thamnosma, Boenninghausenia* (Rutoideae). Perhaps the most interesting feature to emerge from acridone distribution is the link between Aurantioideae and Toddalioideae in formation of pyranoacridones and in C-5 oxygenation.

From the above it is possible to make a number of speculative points concerning the general development of the Rutaceae. The Aurantioideae does appear to be to some extent discrete from the Rutoideae and Toddalioideae in that it exhibits the strongest overall proliferation of coumarins and limonoids but a relative paucity, in number and diversity, of furoquinoline and acridone alkaloids. It is assumed that the latter must be due to loss of biosynthetic

TABLE 5
Acridone containing genera of the Rutaceae

Genus	0*	Features of substitution patterns				
		2	4	5/6	MD	F/P
Rutoideae						
Zanthoxylum	–	+	–	–	–	–
Euodia	–	+	+	–	+	–
Melicope	–	+	+	–	+	–
Ruta	+	+	–	–	–	f
Boenninghausenia	+	–	–	–	–	p
Thamnosma	+	–	–	–	–	–
Esenbeckia	–	–	–	–	–	–
Monniera	–	+	–	–	–	–
Ravenia	–	+	–	–	–	–
Toddalioideae						
Acronychia	–	+	+	–	+	p
Baurella	–	+	+	–	+	p
Balfourodendron	–	+	–	–	+	–
Diphasia	–	+	–	+	+	–
Oricia	–	+	–	+	+	–
Teclea	–	+	+	+	+	–
Vepris	–	+	–	–	+	–
Aurantioideae						
Glycosmis	–	+	–	+	–	p
Murraya	–	–	–	–	–	p
Atalantia (inc. *Severinia*)	–	–	–	+	–	p

0* = loss of C-1 and/or C-3 oxygenation; 2 = C-2 oxygenation; 4 = C-4 oxygenation; 5/6 = oxygenation at C-5 and/or C-6; MD = methylenedioxy; F/P = furan or pyran ring.

potential. There does not appear to be any comparable distinction that can be drawn between the Rutoideae, Toddalioideae, and Flindersioideae as units other than the previously mentioned distribution of some types of highly oxygenated furoquinolines. That the genera classified as Toddalioideae and Rutoideae do show such overall similarity should not be surprising in view of the fact that taxa implicated in the proto-Rutaceae are found in both the sub-families: *Zanthoxylum* and *Euodia* (Rutoideae); *Phellodendron, Toddalia,* and *Fagaropsis* (Toddalioideae).

It is possible to make some comments regarding the origins of the Aurantioideae. The co-occurrence of pyranoacridones and C-5 oxygenated acridones in *Acronychia* and *Teclea* (Toddalioideae) and *Glycosmis* and *Atalantia* (Aurantioideae) suggests some affinity between these two groups.

This is sustained by the distribution of another minor group of alkaloids, the oxazoles (14), which occur in *Halfordia* and *Amyris* (Toddalioideae) and in *Micromelum*, *Aegle*, and *Aegleopsis* (Aurantioideae).

(14) Oxazole alkaloids
(R^1 = H, Me, or prenyl
R^2 = pyridyl or phenyl)

At first sight the suggested phylogenetic sequence within the Rutaceae appears to conflict with that proposed by Dreyer[13] on the basis of oxidation levels achieved by the C-19 methyl groups of limonoids. This suggests that the simplest limonoids, in which C-19 is unmodified, are typical of the Toddalioideae and that increasing ability to modify around C-19 occurs through the Aurantioideae and into the Rutoideae. The occurrence of C-19 oxidized limonoids in the proto-rutaceous genus *Fagaropsis*[47] is one of a number of exceptions to this generalization (Chapter 7) and suggests that the ability to modify C-19 may have developed early in the phylogeny of the Rutaceae. If this is so then the different oxidation levels that do still to some extent seem to characterize the taxonomic groupings can be seen as either a retention of the pathway evolved in the proto-Rutaceae (Aurantioideae), a further advance in oxidative mechanisms (Rutoideae), or a degradation of the pathway (Toddalioideae).

B. ORIGINS AND DIVISIONS WITHIN THE TODDALIOIDEAE

The distribution of major classes of secondary metabolites among 19 genera of Toddalioideae is shown in Table 6. The most striking feature of this group of genera, arrangement of which takes no account of Engler's proposed[28] subdivision into six sub-tribes, is that while eight genera yield coumarins and six yield acridones in no case are both coumarins and acridones recorded as occurring in the same genus. Furthermore, within the acridone producing taxa there appears to be an emphasis on highly oxygenated furoquinolines, whereas among the coumarin producing taxa most furoquinolines are less substituted. Limonoids are of only sporadic distribution and do not at present appear of any systematic value at this level.

On the basis of Engler's classification[28] the division of the bulk of the Toddalioideae into coumarin and acridone producing groups does not appear significant and no other classification systems are available against which to

TABLE 6

Distribution of secondary metabolites among genera of the Toddalioideae

Genus		Secondary metabolites					Furoquinoline substitution						
		A	B	C	D	E	0	8	7	6	7/8	6/8	6/7	6/7/8
Phellodendron	As	+	+	−	−	−	−	−	−	−	−	−	−	−
Fagaropsis	Af	+	+	−	−	−	−	−	−	−	−	−	−	−
Toddalia	As/Af	+	−	+	−	+	+	+	−	−	+	−	−	−
Skimimia	As	−	−	+	−	+	+	−	−	−	+	−	−	−
Ptelea	Am	−	−	+	−	+	+	−	−	+	+	+	+	−
Hortia	Am	−	−	+	−	+	+	+	−	−	+	−	−	−
Helietta	Am	−	+	+	−	+	+	−	−	+	+	−	+	+
Halfordia	Au	−	−	+	−	+	+	−	−	−	−	−	+	+
Casimiroa	Am	−	+	+	−	+	+	+	−	−	+	−	−	−
Amyris	Am	−	−	+	−	+	+	−	−	−	+	−	−	−
Sargentia	Am	−	+	−	−	+	−	−	−	−	−	−	+	−
Acronychia (Bauerella)	As/Au	−	−	−	+	+	+	−	−	+	−	+	−	+
Balfourodendron	Am	−	−	−	+	+	+	−	−	−	+	+	+	+
Diphasia	Af	−	−	−	+	+	−	−	−	−	+	−	−	+
Teclea	Af	−	+	−	+	+	−	−	+	−	+	−	+	+
Vepris	Af	−	+	−	+	+	−	+	−	−	+	−	+	+
Oricia	Af	−	−	−	+	+	−	−	−	−	+	+	+	+
Oriciopsis	Af	−	−	−	−	+	−	−	−	−	−	−	−	+
Araliopsis	Af	−	+*	−	−	+	+	−	−	−	+	−	+	+

As = Asian; Af = African; Am = American; Au = Australasian.

A = 1-btiq alkaloids; B = limonoids (* = proto-limonoid); C = coumarins; D = acridones; E = furoquinolines.

test its relevance. It is noteworthy, however, that the division does appear to have a geographic component in that most coumarin Toddalioideae are found in the neotropics while most acridone containing genera are African, with both occurring in Asia. It has also been suggested[48] that the secondary compound profiles of African Toddalioideae are influenced by those of sympatric species.

Of the two groups the acridone taxa show less obvious chemical affinity with other Rutaceae, the only clear links being with the acridone producing genera of Aurantioideae (*Glycosmis, Atalantia*) and the acridone and acetophenone producing genera of Rutoideae-Zanthoxyleae (*Euodia, Melicope*). Among the coumarin taxa on the other hand there may well be strong affinity with the rather underworked Rutoideae-Cusparieae, notably the co-occurrence of β-indoloquinazolines in *Hortia* and *Euxylophora* (Cusparieae), and imidazoles in *Pilocarpus* (Cusparieae) and *Casimiroa*. Furthermore *Esenbeckia* (Cusparieae) is probably the most prolific producer of 6,7,8-oxygenated furoquinolines outwith the Toddalioideae and Flindersioideae.

Although the Cusparieae are among the least chemically investigated groups within the Rutaceae the case for a link with the Toddalioideae seems plausible. However, there is nothing to suggest that the Cusparieae is either derived from or the progenitor of the Toddalioideae, and it is more likely that the similarities indicate derivation from a common ancestor. The obvious candidate for the ancestral position would appear to be *Euodia* and the allied Australasian genus *Melicope* (Rutoideae—Zanthoxyleae). Both these taxa exhibit an extremely wide spectrum of metabolites (Table 7) which include coumarins and both furoquinoline and acridone alkaloids. In none of the taxa investigated are coumarins and acridones recorded together.

TABLE 7
Distribution of compounds in *Euodia* and *Melicope* species

| | Number of species in: | |
Compounds	*Euodia*	*Melicope*
Furoquinolines	6	6
Acridones	3	2
Coumarins	3	3
Furoquinolines and coumarins	2	1
Acridones and coumarins	0	0
Limonoids	6	0

It seems a plausible scenario to envisage the development of the Toddalioideae, and indeed the rest of the Rutaceae, from taxa with comparable chemistry to that seen in present-day *Melicope* and *Euodia*. The Aurantioideae and Flindersioideae could then have separately split from the major Rutoideae/Toddalioideae line of advance, presumably from ancestors most nearly comparable to extant Toddalioideae. The Rutoideae and Toddalioideae only then began to separate into groups such as those outlined above for the Toddalioideae and the Rutoideae-Cusparieae.

C. THE RUTOIDEAE-RUTINAE

The Rutinae is a small tribe made up of a number of northern temperate and sub-tropical, shrubby or herbaceous, taxa. Genera for which some data is available include *Ruta, Haplophyllum, Boenninghausenia, Thamnosma,* and

Cneoridium. This group appears able to produce most typically rutaceous types of metabolites although limonoids have yet to be found. Two characteristic groups of metabolites set the Rutinae apart from all other Rutaceae. This is the only group so far reported to synthesize acridones devoid of an oxygen substituent at C-3 and also, in some cases, at C-1. It is also by far the most common source of lignans of aryltetrahydronaphthalene type (15) in the Rutaceae. The latter replace the bicyclooctane lignans (16) that are common in *Zanthoxylum* (Chapter 9). Both acridones and lignans appear to have the characteristics of derived compounds and suggest that the Rutinae represent a relatively modern offshoot either from some other part of the Rutoideae or from the Rutoideae/Toddalioideae conglomerate.

(15) Aryltetrahydronaphthalene lignan   **(16)** Bicyclooctane lignan

D. THE RUTOIDEAE-BORONIEAE AND RUTOIDEAE-DIOSMEAE

The Boronieae consists of some 20 genera occurring in Australia and New Caledonia. The Diosmeae, containing about 10 genera but with a comparable number of species, is confined to South Africa. These two tribes appear to be ecologically similar, having adapted to semi-arid conditions in many cases and appearing to have undergone considerable speciation in the recent past.[38,51] Both groups produce many typically rutaceous compounds but both number and range are less than in many other comparable groups. Both tribes do produce volatile oils which are often characterized by large amounts of phenylpropenes.

Of the two the Boronieae has been the more widely investigated. Typical furoquinolines are common and so are furo- and pyrano-coumarins but both acridones and limonoids are absent. Some genera, notably *Zieria* and *Acradenia*, yield simple acetophenone derivatives similar to those of *Melicope* and *Euodia* species.[24] This complies with the suggestion of Smith-White[38] that the Boronieae is a long isolated group that was initially derived from paleotropical Zanthoxyleae.

Less is known of the chemistry of the Diosmeae. Simple coumarins have been recorded from at least four genera (Chapter 4) but so far no furo- or pyrano-coumarins. Traces of simple furoquinolines have been found in a few species[9] but no other alkaloids are known. The Diosmeae would appear to be the least chemically diverse of the tribes of the Rutoideae and at present it is impossible to recognize any clear affinity with other groups. A recent survey of the flavonoids of *Acmadenia*[46] has shown a taxonomically useful distribution of flavonols and flavones which complies with the hypothesis[51] that there has been a radiation of species from high mountain refuges to lowland and coastal areas. Volatile oils may also be useful chemotaxonomic markers.[6]

E. SUBDIVISIONS WITHIN THE AURANTIOIDEAE

According to Engler[17,28] the Aurantioideae are divided into two tribes, the Hesperathusinae and Citrinae. Swingle[40] also recognized two tribes, each further subdivided into three: the Clauseneae (Clauseninae, Micromelinae, Merrilliinae) and Citreae (Triphasiinae, Citrinae, Balsamocitrinae). As noted previously the Aurantioideae appear to be the most prolific of the Rutaceae in production of coumarins and limonoids but relatively poor in alkaloids. Whilst this is true in terms of numbers of alkaloids there is a considerable variety to be found including, as well as typical quinoline derivatives and acridones, oxazoles, carbazoles, and quinazolines. The distribution of compounds among genera of the Aurantioideae is given in Table 8.

Swingle[40] on morphological grounds and Grieve and Scora[20] on presence of *C*-glycosides rather than *O*-glycosides regarded the Clauseneae as the more primitive tribe. Four genera, *Glycosmis*, *Clausena*, *Murraya* (all Clauseninae), and *Micromelum* (Micromelinae), have been the subject of quite detailed chemical investigations. Furoquinolines occur in *Glycosmis* and *Murraya* and acridones in *Glycosmis* while all four genera have given coumarins. In addition quinazoline alkaloids occur in several *Glycosmis* species and oxazoles have recently been reported from *Micromelum*.[8] Of the alkaloids both acridones and oxazoles could be taken as indicative of an affinity to the Toddalioideae where similar acridones occur in *Acronychia* and oxazoles in *Halfordia* and *Amyris*. Quinazolines have otherwise been reported from only a single *Zanthoxylum* species.

However the most striking taxonomic feature of these four taxa is the presence of 3-methylcarbazoles, an alkaloid type at present unknown from

TABLE 8
Distribution of compounds in the Aurantioideae, arranged after Swingle[40]

Tribe Genus	Alkaloids				Coumarins	Limonoids
	A	B	C	D		
Micromelinae						
Micromelum	+	−	+	+	+	−
Clauseninae						
Glycosmis	+	+	+	−	+	−
Clausena	−	−	+	−	+	+
Murraya	+	+	+	−	+	−
Merrilliinae						
Merrillia	−	−	−	−	−	−
Balsamocitrinae						
Swinglea	−	−	−	−	−	−
Aegle	+	−	−	+	+	−
Aegleopsis	−	−	−	+	+	−
Afraegle	+	−	−	−	+	−
Feronia	−	−	−	−	+	−
Citrinae						
Severinia	−	+	−	−	+	−
Pleiospermum	−	−	−	−	−	−
Hesperethusa	+	−	−	−	+	+
Atalantia	+	+	−	−	+	+
Fortunella	−	−	−	−	−	+
Poncirus	−	−	−	−	+	+
Microcitrus	−	−	−	−	−	+
Citrus	+	−	−	−	+	+
Eremocitrus	−	−	−	−	−	+
Triphasiinae						
Triphasia	−	−	−	−	+	+
Pamburus	−	−	−	−	+	−
Luvunga	−	−	−	−	+	+

A = quinolines (furoquinolines, pyranoquinolines); B = acridones; C = carbazoles; D = oxazoles.

any other source. Furthermore, within the three genera of the Clauseninae there is an interesting progression both in the oxidation level of the 3-methyl moiety and in the complexity of the carbazole which varies from C-13 to C-18 to C-23 through addition of one or two prenyl units (Fig. 4). In both respects an increasing complexity in metabolites produced suggests that *Glycosmis* should be considered the most primitive, *Clausena* intermediate, and *Murraya*

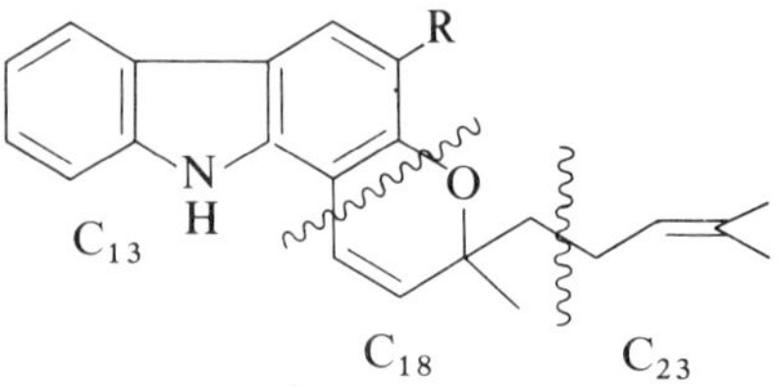

	R = CH$_3$	CHO	COOH	C$_{13}$	C$_{18}$	C$_{23}$
Micromelum (1)	+	−	−	−	+	−
Glycosmis (2)	+	−	−	+	−	−
Clausena (4)	+	+	−	+	+	−
Murraya (2)	+	+	+	+	+	+

FIG. 4. Oxidation levels and degree of prenylation attained by carbazoles from different genera. Numbers in brackets signify the number of species that have yielded carbazoles to date.

most advanced. The single carbazole reported from *Micromelum* shows an oxidation level comparable to that of *Clausena*. Inasmuch as *Murraya* shares with *Citrus* a propensity for producing polymethoxylated flavonoids its placement as the most advanced taxon of the Clauseninae seems most appropriate.

The Balsamocitrinae is usually considered, along with the Citrinae and Triphasiinae, to form the more advanced tribe of the Aurantioideae. However, in three genera, *Aegle*, *Afraegle*, and *Aegleopsis*, oxazoles and a large number of furoquinolines have been found whilst limonoids are so far lacking from the entire sub-tribe. These are features that do not seem to comply with placement of the Balsamocitrinae with the true citroids but which suggest a somewhat earlier separation, possibly from Clauseneae stock. By contrast the Triphasiinae and Citrinae are very rich in limonoids and coumarins but with a relative paucity of alkaloids. The only exceptions are the acridone producing genera *Atalantia* and *Severinia*, the alkaloids of which may be indicative of a development from a Clauseneae stock or from the Toddalioideae. Within the Citrinae it has been suggested[31] that long-chain leaf hydrocarbons may have some systematic value.

V. THE CHROMONE PRODUCING TAXA (PTAEROXYLACEAE, CNEORACEAE, SPATHELIA, HARRISONIA)

After consideration of the major taxa of the Rutales there remain a number of smaller units to be examined. The Ptaeroxylaceae have often been included in the Meliaceae[28] but differ in a number of morphological[22] and chemical[43] features, notably absence of limonoids and presence of coumarins. The Cneoraceae is a small family of two genera and three species which has been the subject of a number of investigations that has yielded both coumarins (Chapter 4) and some extensively modified limonoids (Chapter 6). The Spathelioideae are generally included as a small sub-tribe in the Rutaceae[28] containing the three genera *Spathelia*, *Sohnreyia*, and *Diomma*.[39] It has alternatively been placed in the Simaroubaceae but the isolation of limonoids and the typical rutaceous alkaloid *N*-methylflindersine (17) was taken by Adams *et al.*[1] as strong evidence in favour of retention in the Rutaceae.

In addition to the above all three of these taxa are characterized by the production of a group of 6-prenyl and 6,8-diprenyl derivatives of 5,7-dihydroxy-2-methylchromone (18). In a number of cases the same compounds are known from all three sources. In a previous analysis of the systematic significance of the chromones[44] a loose association was proposed between these three taxa, with the Spathelioideae retained in the Rutaceae, held there by the presence of 17, and the Ptaeroxylaceae and Cneoraceae as distinct taxa whose closest affinities were with each other and with the Spathelioideae sub-family of the Rutaceae.

(17) *N*-Methylflindersine   (18) 5,7-Dihydroxy-2-methylchromone

To this must now be added the recent finding of both simple limonoids and chromones in *Harrisonia abyssinica* (Simaroubaceae) collected in both Nigeria[33] and Ghana.[2] This species has been an undisputed member of the Simaroubaceae although it differs from other taxa of the tribe Simaroubeae in having partly fused carpels.[28] The finding of both chromones and limonoids in *H. abyssinica* has suggested further systematic possibilities in the interpretation of the secondary metabolites of this group.

It is assumed that, as with the 1-btiq alkaloids, the co-occurrence of these identical chromones is systematically meaningful in these taxa and not an accident of convergence. If this is so then the progenitors of this group must have been at an evolutionary stage where coumarin and limonoid synthesis was developed but, quite possibly, when only relatively simple anthranilate derived alkaloids were being produced. Taking into account the morphological relationship of *Harrisonia*, and perhaps *Spathelia*, to the Simaroubaceae it is suggested that they could have developed from proto-Simaroubaceae stock which in turn had derived from limonoid, coumarin, and alkaloid producing proto-Rutaceae. The separation is assumed to have come before the evolution of the quassinoid pathway in the Simaroubaceae.

Subsequent to their separation the precursors of the chromone taxa of today evolved the ability to synthesize chromones. This has then been followed by separation into the four groups seen today (and perhaps more) with the loss of abilities to synthesize alkaloids (in all except *Spathelia*), loss of ability to synthesize coumarins (*Harrisonia*, *Spathelia*), and loss of ability to produce limonoids (Ptaeroxylaceae).

VI. CONCLUSIONS

The above is a highly speculative set of hypotheses in which I have set out to try and rationalize the biochemical evolution of the Rutales in a way which may be comparable to its actual phylogeny. Particular emphasis has been placed on the Rutaceae because at present our knowledge of the chemistry of that family is greater and there is greater chemical diversity. The distributions of a number of types of secondary compounds that I feel may be systematically important, superimposed on a phylogenetic diagram based on the proposals made in this chapter, are shown in Fig. 5.

Despite our considerable knowledge of the chemistry of the Rutales there remain many problems to be resolved. In the course of this chapter I have noted some of the most obvious examples in the Rutaceae such as the chemistry of *Rhabdodendron*, the search for more proto-Rutaceae, and expansion of knowledge on Diosmeae and Cusparieae. Taylor (Chapter 14) has likewise pointed out some of the major shortfalls in the Meliaceae. The Simaroubaceae is also in need of much further work, particularly to resolve the situation of satellite groups such as the Irvingioideae. Finally, some further consideration should be given to the position of the Zygophyllaceae, or at least to the genus *Peganum* which was once included in the Rutaceae by Engler.[16] *Peganum* is the source of many β-carboline alkaloids similar to those of the Simaroubaceae[24] and may offer a clue to the link between Rutales *s.s.* and taxa usually placed in the Sapindales *s.s.*

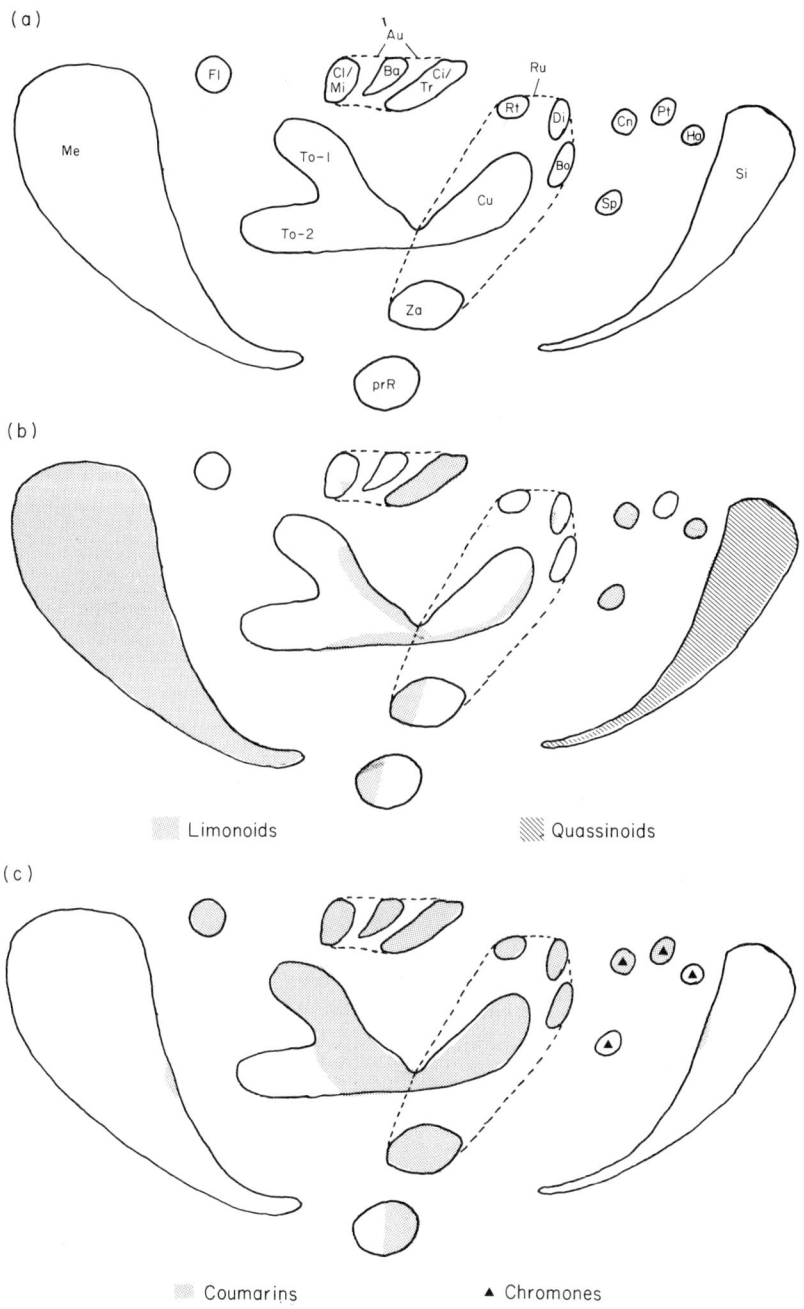

398 P. G. WATERMAN

(d)

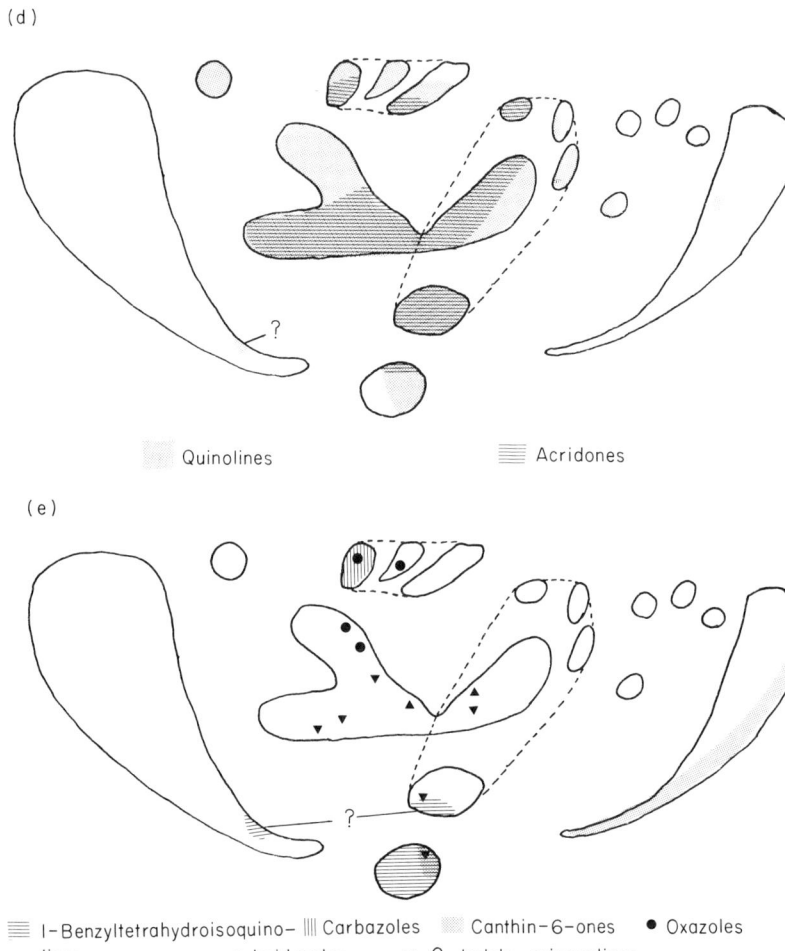

Quinolines   Acridones

(e)

▤ l-Benzyltetrahydroisoquino- ▥ Carbazoles   ▦ Canthin-6-ones   ● Oxazoles
lines   ▲ Imidazoles   ▼ β-Indolo-quinazolines

FIG. 5. A suggested phylogeny of the Rutales, based on the distribution of secondary metabolites
(a) Au = Rutaceae, sub-family Aurantioideae; Ba = tribe Balsamocitrinae; Bo = tribe
Boronieae; Ci = tribe Citrinae; Cl = tribe Clauseninae; Cn = Cneoraceae; Cu = tribe
Cusparieae; Di = tribe Diosmeae; Fl = sub-family Flindersioideae; Ha = *Harrisonia*
(Simaroubaceae); Me = Meliaceae; Mi = tribe Micromelinae; prR = proto-Rutaceae genera; Pt = Ptaeroxylaceae; Rt = tribe Rutinae; Ru = Rutaceae, sub-family Rutoideae;
Si = Simaroubaceae; Sp = sub-family Spathelioideae; To-1 = coumarin containing
Toddalioideae; To-2 = acridone containing Toddalioideae; Tr = tribe Triphasiinae;
Za = tribe Zanthoxyleae.
(b) Distribution of limonoids and quassinoids.
(c) Distribution of coumarins and chromones.
(d) Distribution of quinoline and acridone alkaloids.
(e) Distribution of minor alkaloid groups.

REFERENCES

1. Adams, C. D., Taylor, D. R. and Warner, J. M. (1973). *Phytochemistry* **12**, 1359.
2. Ampofo, S. and Waterman, P. G. (1982), unpublished.
3. Bate-Smith, E. C. (1962). *J. Linn. Soc. Bot.* **58**, 95.
4. Behnke, H.-D. (1977). *Plant Syst. Evol.* **128**, 227.
5. Bessey, C. A. (1915). *Ann. Mo. Bot. Gard.* **2**, 109.
6. Blommaert, K. L. J. and Bartel, E. (1976). *J. S. Afr. Bot.* **42**, 121.
7. Bohm, B. A. (1975). *In* "The Flavonoids" (Harborne, J. B., Mabry, T. J. and Mabry, H. eds.), pp. 560–631. Chapman and Hall, London.
8. Bowen, I. H. and Christopher Perrera, K. P. W. (1982). *Phytochemistry* **21**, 433.
9. Campbell, W. E., Provan, G. J. and Waterman, P. G. (1982). *Phytochemistry* **21**, 1457.
10. Cronquist, A. (1968). "The Evolution and Classification of Flowering Plants". Nelson and Sons Ltd., London.
11. Cronquist, A. (1981). "An Integrated System of Classification of Flowering Plants". Columbia University Press, New York.
12. Dahlgren, R. M. T. (1980). *Bot. J. Linn. Soc.* **80**, 91.
13. Dreyer, D. L., Pickering, M. V. and Cohen, P. (1972). *Phytochemistry* **11**, 705.
14. Ehrendorfer, H. (1974). *In* "Origin and Early Evolution of Angiosperms" (Beck, C. B. ed.), pp. 220–240. Columbia University Press, New York.
15. Ehrlich, P. R. and Raven, P. H. (1964). *Evolution* **18**, 586.
16. Engler, A. (1896). *In* "Die Naturlichen Pflanzenfamilien" (Engler, A. and Prantl, K. eds.), Vol. 3, part 4. Engelmann, Leipzig.
17. Engler, A. (1931). *In* "Die Naturlichen Pflanzenfamilien" (Engler, A. and Prantl, K. eds.), 2nd Edn., Vol. 19a. Engelmann, Leipzig.
18. Fish, F. and Waterman, P. G. (1973). *Taxon* **22**, 177.
19. Gray, A. I. and Waterman, P. G. (1978). *Phytochemistry* **17**, 845.
20. Grieve, C. M. and Scora, R. W. (1980). *Syst. Bot.* **5**, 59.
21. Hallier, H. (1912). *Arch. Neerl. Sci.* Ser. III **1B**, 146.
22. Hartley, T. G. (1969). *J. Arnold Arbor.* **50**, 481.
23. Hegnauer, R. (1963). *In* "Chemical Plant Taxonomy" (Swain, T. ed.), pp. 389–427. Academic Press, London.
24. Hegnauer, R. (1973). "Chemotaxonomie der Pflanzen", Vol. 6. Birkhauser Verlag, Basel.
25. Hutchinson, R. (1969). "Evolution and Phylogeny of Flowering Plants". Academic Press, London.
26. Kubitzki, K. (1969). *Taxon* **18**, 360.
27. Meeuse, A. D. J. (1970). *Acta Bot. Neerl.* **19**, 61 and 133.
28. Melchior, H. (1964). "A. Engler's Syllabus der Pflanzenfamilien", 12th Edn., pp. 262–272. Borntrager Verlag, Berlin.
29. Metcalfe, C. R. and Chalk, L. (1950). "Anatomy of the Dicotyledons", Vol. 1, pp. 305–325. Oxford University Press.
30. Nielsen, B. E. (1971). *In* "The Biology and Chemistry of the Umbelliferae" (Heywood, V. H. ed.), pp. 325–336. Academic Press, London.
31. Nordby, H. E., Hearn, C. J. and Nagy, S. (1975). *Proc. Florida State Hort. Soc.* **88**, 32.
32. Nooteboom, H. P. (1966). *Blumea* **14**, 309.
33. Okorie, D. A. (1982). *Phytochemistry* **21**, 2424.

34. Prance, G. T. (1968). *Bull. Jard. Bot. Nat. Belg.* **38**, 127.
35. Pride, J. R. (1963). *In* "Chemical Plant Taxonomy" (Swain, T. ed.), pp. 428–452. Academic Press, London.
36. Puff, C. and Weber, A. (1976). *Plant Syst. Evol.* **125**, 195.
37. Rendle, A. B. (1925). "The Classification of Flowering Plants", Vol. 2, pp. 283–291. Cambridge University Press, Cambridge.
38. Smith-White, S. (1954). *Aust. J. Bot.* **2**, 287.
39. Stern, W. L. and Brizicky, G. K. (1960). *Mem. N. Y. Bot. Gard.* **10**, 38.
40. Swingle, W. T. (1938). *J. Wash. Acad. Sci.* **28**, 530.
41. Takhtajan, A. L. (1980). *Bot. Rev.* **46**, 225.
42. Thorne, R. F. (1981). *In* "Phytochemistry and Angiosperm Phylogeny" (Young, D. A. and Seigler, D. S. eds.), pp. 233–295. Praeger Publishers, New York.
43. Taylor, D. A. H. (1982) *In* "Flora Neotropica", Monograph 28, Meliaceae, pp. 450–459. New York Botanical Gardens, New York.
44. Waterman, P. G. (1975). *Biochem. Syst. Ecol.* **3**, 149.
45. Waterman, P. G. and Faulkner, D. F. (1981). *Phytochemistry* **20**, 2765.
46. Waterman, P. G. and Hussain, R. A. (1982). *Bot. J. Linn. Soc.*, in the press.
47. Waterman, P. G. and Khalid, S. A. (1981). *Biochem. Syst. Ecol.* **9**, 45.
48. Waterman, P. G., Meshal, I. A., Hall, J. B. and Swaine, M. D. (1978). *Biochem. Syst. Ecol.* **6**, 239.
49. Wettstein, R. (1935). "Handbuch der Systematischen Botanik". F. Deuticke, Wien.
50. Willaman, J. J. and Li, H.-L. (1970). *Lloydia* **33**, supplement.
51. Williams, I. (1982). *J. S. Afr. Bot.*, in press.

CHAPTER 16

Chemical Characters and the Classification of the Rutales

R. HEGNAUER

Laboratorium voor Experimentele Plantensystematiek, Leiden, Netherlands

I. INTRODUCTION

During the past 25 years comparative phytochemistry has been accepted by many plant taxonomists as an auxiliary science valuable to plant classification.* This is well exemplified by papers by Wagenitz,[199] Heywood,[95]

---

* I use the words plant classification, taxonomy, and systematics (and systematic botanists and plant taxonomists) as synonyms like many modern plant taxonomists.[50,57] For other views see Alston,[7] Merxmüller,[126] and Crowson.[52]

Turner,[197] Merxmüller,[126] Kubitzki,[112] Meeuse,[124] Soó,[182] Dahlgren,[54,55] and Cronquist.[50] It is by no means accidental that the publications I use to illustrate my point were all written by botanists more or less actively engaged in plant classification. A large part of recent chemotaxonomic literature has been produced by organic chemists, biochemists, and plant physiologists who became interested in the natural (phylogenetic) classification of plants. It must be stressed, however, that a real improvement of any existing classification can only be achieved by someone who has "the deepest possible understanding of the existing system and its basis".[52] Therefore, as a rule, chemists should rather provide new characters for taxonomists than new classifications and far-reaching taxonomic speculations. The improvement of classification would be committed to the care of the systematic botanists unless the non-taxonomist is ready to spend enough time to become really acquainted with existing proposals for the classification of the taxa under consideration.

My task within the scope of this chapter will be to discuss relationships of the Rutales. This implies examination of the delimitation of the order in the light of chemical evidence and a careful appreciation of the chemical make-up of this taxon. Proposals for the classification of the order will then be examined and evaluated in the light of the chemistry of the dicotyledons as a whole.

II. The Delimitation of Rutales

Thirteen proposals for the contents of an order including the Rutaceae are summarized in Table 1. This shows that there is a rather general agreement between taxonomists that Burseraceae, Cneoraceae, Meliaceae, Rutaceae, and Simaroubaceae should be placed in the same order. Only Hutchinson has two, albeit closely related, orders for the four major families mentioned and includes Cneoraceae in the Celastrales. There are, however, essentially three main areas of disagreement: (i) the sequence of these four families within the order; (ii) the content of the order; (iii) the classification of the Anacardiaceae and Zygophyllaceae. Other points of discussion concern the classification of some small genera or groups of genera and are noted in Table 1 but the three main points require some additional comment.

(1) Most authors place Rutaceae first in the order, but Gundersen and Cronquist start the sequence representing the Rutales *s.s.* with Burseraceae while Cronquist[49,51] puts Rutaceae at the end of the Sapindales *s.l.* and considers Sapindales *s.s.* as more primitive than Rutales *s.s.* Therefore Burseraceae, which connect the rutalean alliance with the sapindalean group of families, have to be classified first in the rutalean group.

TABLE 1

A comparison of taxonomic treatments of the major families of the Rutales

Treatment	Taxa involved	
Bentham and Hooker[15]	Geraniales[a]	Li, Hu, Ma, Zy, Ge, Ru, Si (inc. Cn), Oc, Bu, Me, Ch
Hallier[83]	Terebinthinae (= Rutales[b])	Ru, Cn, Me, Si, Te,[c] Ac, Am, Ur
Rendle[157]	Rutales[d]	Ru, Si, Bu, Me
Wettstein[202]	Terebinthales[e]	(Ru, Si, Bu, Me),[f] Tr,[g] (Po, Xa, Ti, Vo),[h] (An, Sa, Ak, Ae, Ac, Hi),[i] (Co, Cy, Pe, Sb, Ml, Cr, Ba)[j]
Gundersen[81]	Rutales[k]	Bu, An, Ju, Co, Cn, Si, Ru, Me
Emberger[38]	Terebinthales[l]	(Ru, Si, Bu, Me, Bx, Po, Ti, Vo, Ae, An, Sa, Ac, Hi, Ak),[m] (Co, Cy, Pe)[n]
Scholz[169]	Rutales (= Terebinthales pp.[o])	(Ru, Cn, Si (inc. Ir, Ki, Su), Pc, Bu, Me, Ak),[p] (Ma, Ti, Vo),[q] (Tr, Po)[r]
Cronquist[49]	Sapindales[s]	St, Ml, Gr, Ca, Sy (inc. Su), Ak, Sa (inc. Br), Hi, Ac, Bu, An, Ju, Si (inc. Ir, Ki, Bl), Cn, Ru, Me, Zy
Hutchinson[100]	Rutales[t]	Ru (inc. Ph), Si (inc. *Ptaeroxylon*, Su, not inc. Ir), Bu, Av
	Meliales	Me (inc. Cd, Fl)
Thorne[191]	Rutales[u]	(Ru (inc. Cn), Co, Si (inc. Ir, Ki, Su, Bl), Me, Bu, An (inc. Ju)),[v] (Sa, Gy, Sb, Ml, Ak, Ac, Hi, Br),[w] (Rh, Jg)[x]
Takhtajan[189]	Rutales[y]	(Ru, Rb,[z] Cn, Si (inc. Ir, Su), [Zy, Ni, Bl],[aa] Me, Ki,[bb] Pt,[cc] Bu, An, Ju, Pd[dd]),[ee] Co[ff]
Dahlgren[55]	Rutales[gg]	Ru, Cn, Su,[hh] Si, Bu, Me (inc. Ai)
Cronquist[51]	Sapindales[ii]	St, Ml, Br, Ak, Sa, Hi, Ac, Bu, An, Ju, Si,[jj] Cn, Me, Ru,[kk] Zy

Key to taxa: Ac = Aceraceae; Ae = Aextoxicaceae; Ai = Aitoniaceae; Ak = Akaniaceae; Am = Amentaceae; An = Anacardiaceae; Av = Averrhoaceae; Ba = Balsaminaceae; Bl = Balanitaceae; Br = Bretschneideraceae; Bu = Burseraceae; Bx = Buxaceae; Ca = Connaraceae; Cd = *Cedrelopsis*; Ch = Chailletiaceae; Cn = Cneoraceae; Co = Coriariaceae; Cr = Corynocarpaceae; Cy = Cyrillaceae; Fl = Flindersiaceae; Ge = Geraniaceae; Gr = Greyiaceae; Gy = Gyrostemonaceae; Hi = Hippocastanaceae; Hu = Humiriaceae; Ir = Irvingiaceae; Jg = Juglandaceae; Ju = Julianiaceae; Ki = Kirkiaceae; Li = Lineae; Ma = Malpighiaceae; Me = Meliaceae; Ml = Melianthaceae; Ni = Nitrariaceae; Oc = Ochnaceae; Pc = Picrodendraceae; Pd = Podoaceae; Pe = Pentaphylacaceae; Ph = *Phelline*; Po = Polygalaceae; Pt = Ptaeroxylaceae; Rb = Rhabdodendraceae; Rh = Rhoipteleaceae; Ru = Rutaceae; Sa = Sapindaceae; Sb = Sabiaceae; Si = Simaroubaceae; St = Staphyleaceae; Su = Surianaceae; Sy = Stylobasiaceae; Te = Terebinthaceae; Ti = Trigoniaceae; Tr = Tremandraceae; Ur = Urticaceae; Vo = Vochysiaceae; Xa = Xanthophyllaceae; Zy = Zygophyllaceae. (Continued on p.404)

Key to points raised: (*a*) *ex* Willis,[203] An in Sapindales. (*b*) Cn, Me, Si, Te derived from Ru; Zy in Gruinales. (*c*) Inc. Bu, Sb, Ju, An. (*d*) An in Sapindales; Zy in Gruinales. (*e*) Zy and ?Cn in Gruinales but both could also be incorporated in Terebinthales. (*f*) Natural group with ovules epitropous. (*g*) *Incertae sedis*. (*h*) Ovules epitropous. (*i*) Ovules apotropous. (*j*) *Incertae sedis*. (*k*) Zy in Geraniales. (*l*) A further seven families of uncertain position not included; Cn in separate order. (*m*) Terebinthales *s.s.* (*n*) Terebinthales with celastraceous tendencies. (*o*) Zy in Geraniales (= Gruinales) and An in Sapindales (= Acerales = Terebinthales *pp.*). (*p*) Suborder Rutineae (excretory cells, cavities, or canals present). (*q*) Suborder Malpighiineae (no excretory structures, flowers more or less zygomorphic). (*r*) Suborder Polygalineae (flowers usually zygomorphic, excretory structures rare). (*s*) St, Ml, Gr, Ca, Sy, Ak, Zy *incertae sedis*; Zy resemble epitropous orders Geraniales, Linales, Polygalales but best retained in more primitive Sapindales which comprises both apotropous and epitropous taxa. (*t*) Cn mentioned under Celastrales, formerly described;[99] An in Sapindales and Zy in Malpighiales. (*u*) Rutales, Myricales, and Leitneriales form superorder Rutiflorae; Zy in Geraniales. (*v*) Suborder Rutineae. (*w*) Suborder Sapindineae. (*x*) Suborder Juglandineae. (*y*) Rutales, Sapindales, Geraniales, and Polygalales form superorder Rutanae. (*z*) Included in Ru by many authors. (*aa*) Ni and Bl included in Zy by most authors. (*bb*) Usually included in Si. (*cc*) This family comprises *Cedrelopsis* and *Ptaeroxylon* and is often included in Me. (*dd*) Usually included in An. (*ee*) Suborder Rutineae. (*ff*) Suborder Coriariineae. (*gg*) Rutales, Sapindales, Balsaminales, Polygalales, Geraniales, and Tropaeolales form superorder Rutiflorae; An in Sapindales; Zy in Geraniales. (*hh*) Often included in Si. (*ii*) Natural group; St connect to Rosales and Zy remind of Geraniales; Ca and Gr in Rosales; Sy in Su and transferred to Rosales. (*jj*) Includes Ir and Ki but not Su. (*kk*) Rb removed from Ru and transferred to Rosales.

----

(2) Rutales *s.s.* of Rendle correspond more or less with Rutales *s.s.* of Hallier (if doubtfully included amentiferous taxa are excluded), Gundersen, Hutchinson, and Dahlgren, and with the suborder Rutineae of Scholz, Thorne, and Takhtajan. Wettstein also has this natural group (note *f*, Table 1) but without giving it formal recognition. Bentham and Hooker have Rutales *s.s.* in the Geraniales, and Wettstein, Emberger, Cronquist, and Thorne have an order which roughly corresponds to Rutales *s.s.* + Sapindales *s.s.* of other taxonomists, or even includes Polygalales (Wettstein, Emberger). The delimitation of Rutales by Scholz is similar to Wettstein's concept insofar as he includes what he calls Malpighineae (Malpighiaceae is in Geraniales in Wettstein's system) and Polygalineae in Rutales; he differs from Wettstein by accepting a separate order Sapindales which comprises 9 of the families of Wettstein's Terebinthales. Emberger's Terebinthales contain roughly the same families as Wettstein's but include, in addition, the Buxaceae.

(3) The Anacardiaceae and Zygophyllaceae illustrate very well the difficulties encountered in delimiting orders of polypetalous dicotyledons. Their various classifications are listed in Table 2. Hallier has Burseraceae and Anacardiaceae in the same family (Terebinthaceae), and Gundersen, Cronquist, Thorne, and Takhtajan place Anacardiaceae next to Burseraceae. By contrast Bentham and Hooker, Rendle, Scholz, Hutchinson, and Dahlgren have Burseraceae and Anacardiaceae in different orders. It is interesting to note that Gundersen puts Anacardiaceae in Rutales although he

TABLE 2

Classification of Anacardiaceae and Zygophyllaceae by various authors

Author(s)	Anacardiaceae	Zygophyllaceae
Bentham and Hooker[15]	Sapindales	Geraniales*
Hallier[83]	Rutales (Terebinthinae)	Geraniales[a]
Rendle[157]	Sapindales s.s.	Geraniales[a]
Wettstein[202]	Terebinthales*	Geraniales[a,b]
Gundersen[81]	Rutales s.s.	Geraniales
Scholz[169]	Sapindales s.s.	Geraniales
Cronquist[49]	Sapindales s.l.*	Sapindales s.l.*
Hutchinson[100]	Sapindales s.s.	Malpighiales
Thorne[191]	Rutales s.l.	Geraniales
Takhtajan[189]	Rutales s.s.	Rutales s.s.
Dahlgren[55]	Sapindales s.s.	Geraniales
Cronquist[51]	Sapindales s.l.*	Sapindales s.l.*

* Rutales s.s. placed in the same order; (a) = Gruinales; (b) Zygophyllaceae might perhaps better be transferred to Terebinthales.

has a separate Sapindales. The difficulties in classification of Anacardiaceae arise because of its many similarities to the Burseraceae from which it differs mainly in ovule orientation. Whilst in Burseraceae ovules are epitropous (as in Rutales s.s.), in Anacardiaceae they are apotropous (as in Sapindales s.s.).

III. THE CHEMICAL MAKE-UP OF THE RUTALES

For a discussion of the chemical characters of the order I have accepted Dahlgren's delimitation (Table 1). This choice will be commented upon in Section V. Many references to the distribution of compounds in these families will be found in the four relevant volumes of "Chemotaxonomie der Pflanzen".[86,88,89,91] Moreover, the chemistry of all five families of Rutales s.s. has been treated by other contributors. Therefore the following general discussion will only exceptionally be documented by reference to literature.

A. ESSENTIAL OILS

Excretory structures are widespread in Rutales and occur in most plant parts.[45,128,161,181] In many instances these structures contain essential oils or oleo-resins with varying proportions of steam-volatile constituents. Some resin ducts contain mixtures of oleo-resin and mucilage, the so-called oleo-gum-resins like frankincense and myrrh. Some features of essential oil production in Rutales are summarized in Table 3.

With the exception of some Rutaceae, essential oils in the Rutales are poor in phenylpropanoid constituents. Cembranoid diterpenes like mukulol (1) occur in Burseraceae, and Cneoraceae produce azulenogenic C_5-extended sesquiterpenes (e.g. 2) which recall similar diterpenoid constituents of Myoporaceae. In Simaroubaceae mevalonate seems to be totally used for the synthesis of quassinoids and simarubolides with very little steam-volatile material in resin ducts.

B. ARYLPROPANOIDS AND DIARYLPROPANOIDS (LIGNANS AND NEOLIGNANS)
(where possible numbering follows Chapter 9)

The common cinnamic acids occur widely in Rutales (Table 4), and many rutaceous essential oils contain appreciable amounts of phenylpropanoids (Table 3). A number of new arylpropanoids have recently been described from *Zanthoxylum* species;[1,104,105] these are cuspidiol (3), dihydro-*p*-coumaryl alcohol, boninenal (4), methyl boninenalate, and methyl 2,4-dimethoxy-5-hydroxycinnamate (5). A characteristic feature of the order seems to be a tendency for its members to accumulate lignans. Cubebin type lignans (one C–C bond; called type A here), arylnaphthalene type lignans (including dihydro and tetrahydro compounds; type B), and difuranoid lignans (type C) are known from the intensively studied Rutaceae. Examples of type A include phebalarin (6), sventenin (2-6),* and the monofuranoid brassilignan (1-4) and sanshodiol (1-2); of type B justicidin A (5-5), justicidin B (5-2), and austro-bailignan (3-4); and of type C the stereoisomeric compounds asarinin (6-2) and sesamin (6-1).

If Gottlieb's[78] definition of lignans and neolignans is accepted, Rutales produce predominantly lignans. According to Gottlieb, lignans and neolig-nans are biosynthetically analogous (i.e. non-homologous) plant constituents, the former being diarylpropanoids with oxygenated γ-C atoms and the latter diarylpropanoids with the γ-C non-oxygenated. The coumarinolignan pro-pacin (7) belongs to the neolignoid rather than the lignoid group of plant constituents. Some features of the distribution of arylpropanoids in the Rutales are indicated in Table 4.

---

* In numbers of this type the first part refers to the table number and the second to the structure number, in this case in Chapter 9.

TABLE 3

Features of essential oil production by families of Rutales

Family	Types of excretory structure	Main types of excretion	Main types of essential oil constituents
Rutaceae	Lysigenous cavities; oil cells	Essential oils; bitter resins and probably coumarins	Mono- and sesqui-terpenoids; aliphatic ketones; phenylpropanoids
Cneoraceae	Resin cells	Bitter resins with a low content[a] of volatiles	Azulenogenic cneorubins (e.g. **2**)[184,195]
Simaroubaceae	Resin ducts and cells in many genera; oil glands in *Kirkia*; glandular hairs[33]	Bitter resins with low proportions of volatile compounds	Not investigated; yields very low
Burseraceae	Ducts; cells in wood	Essential oils; oleo-gum-resins and oleo-resins	Mono- and sesqui-terpenes predominate; α-camphorenoid and cembranoid diterpenes (e.g. **1**) sometimes present;[72,111,142,153,164] a little methylchavicol in the oleo-resin of *Boswellia serrata*;[142] furanoid sesqui-terpenes in the essential oil of gum myrrh[35]
Meliaceae	Excretory cells	Bitter resins and essential oils	Mono- and sesqui-terpenes; geranyl-geraniol in leaf oil of *Cedrela toona*

(a) 0.13–0.25 % in dry leaves.[195]

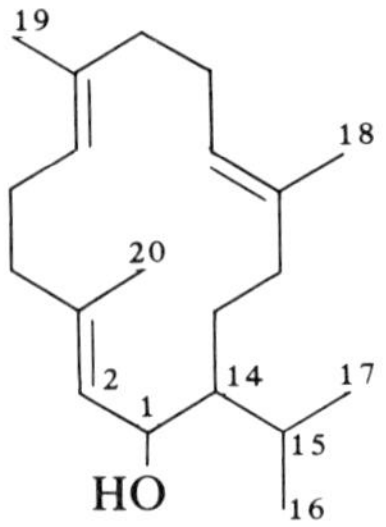

(1) Mukulol

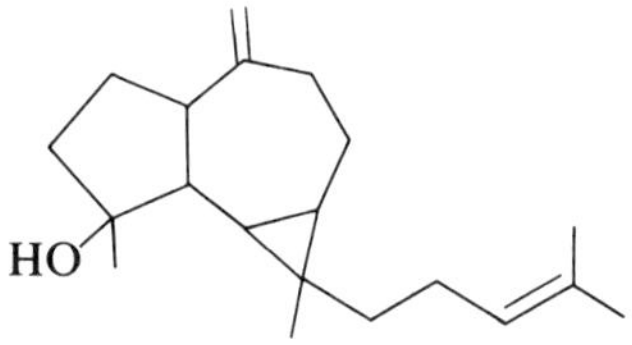

(2)

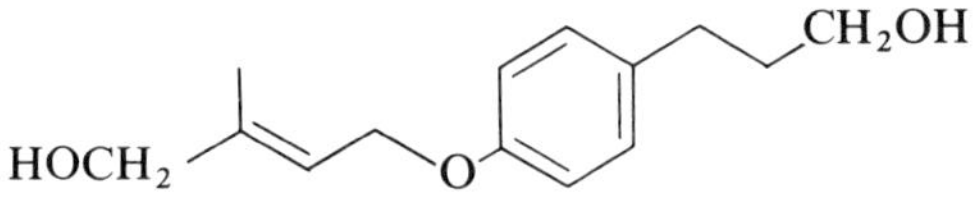

(3) Cuspidiol

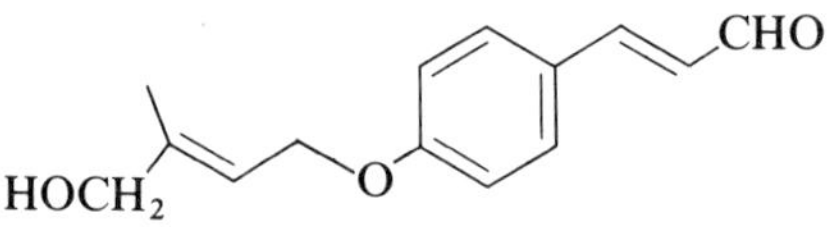

(4) Boninenal

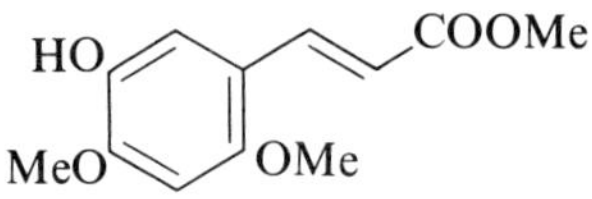

(5)

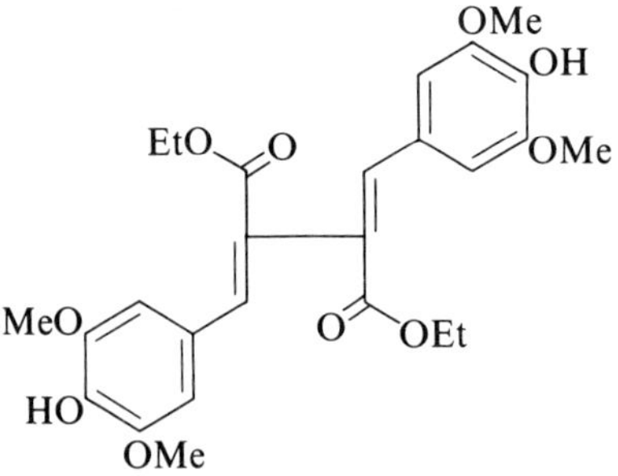

(6) Phebalarin

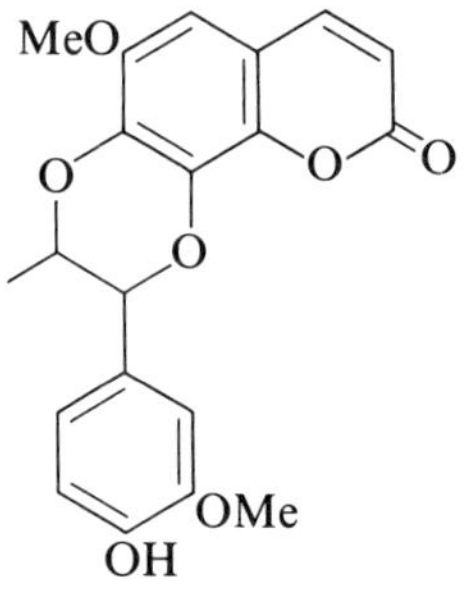

(7) Propacin

TABLE 4

Distribution of some classes of phenolics in the Rutales (supplementing data previously reported)[86,88,89,91]

Classes of constituents	Rutaceae	Cneoraceae	Simaroubaceae	Burseraceae	Meliaceae
Arylpropanoids and diarylpropanoids (lignans)	Common cinnamic acids in leaves.[a] Allyl and propenyl benzenes in essential oils. A-, B-, C-type lignans common[b]	Ferulic acid in leaves; diarylpropanoids not yet detected	Caffeic acid in leaves; syringaresinol (C-type lignan)	Caffeic acid in leaves; A-, B-, and C-type lignans.[c] Propacin[d]	Caffeic acid in leaves; lignans not yet detected
Acetogenins (chromones and phloroglucin derivatives)	2-Methylchromones rare[e] Phloroacetophenone derived volatile and nonvolatile chromenes quite common.[f] Resorcin derived chromenes rare.[g] Prenylated 2,4,6-trihydroxycinnamic acid derivatives in several taxa[h]	2-Methylchromones in both genera[e]	A phloroacetophenone derivative[i]	Not yet detected	2-Methylchromones in Aphanamixis, Cedrelopsis, and Ptaeroxylon.[e,j] Phloroacetophenone derivatives in Cedrelopsis.[k] Acetogenic coumarins[l]
Coumarins	Simple coumarins, prenylated coumarins, and furo- and pyrano-coumarins very common[m]	Cedrelopsin and Cneorum-coumarin B.[e] Obliquin and three new coumarins[n]	Simple coumarins only[o]	Scopeletin from wood of two species of Bursera[p]	Common in Cedrelopsis and Ptaeroxylon.[e,j,k] A few in other genera[q]

TABLE 4—continued

Flavonoids	Flavonols, flavones, and flavanones, and flavanonols, as glycosides and polymethoxylated more or less lipophilic derivatives. Prenylated and C-methylated flavonoids rather common[r]	Kaempferol and quercetin glycosides. Rutin in *Neochamaelea pulverulenta* leaves[129]	Kaempferol, quercetin, and myricetin glycosides	Kaempferol and quercetin glycosides. Amentoflavone[s]	Quercetin and myricetin[186] glycosides in leaves and flowers.[t] Methylated flavones in leaves.[u] Several types in wood.[v]
Tannins and their precursors (cf. ref. 87)	Procyanidins and prodelphinidins rather common. Gallic and ellagic acids rather rare. Tannin content low to considerable	Not detected in leaves or wood[204]	Procyanidins, prodelphinidins, gallic and ellagic acid relatively frequent in leaves and stems.[w] Tannin content low to considerable	Procyanidins in leaves. Ellagic acid in wood.[p] Tannin content often high	Catechins, proanthocyanidins, condensed tannins. Tannin content sometimes considerable[x]

(*a*) Caffeic, ferulic, sinapic, and *p*-coumaric acids. (*b*) Phebalarin (**6**)[36] and see Chapter 9. (*c*) See Chapter 4 and structure **7**. (*e*) See Chapter 4. (*f*) See ref. 91 and **8** from *Acronychia laurifolia*;[13] **9**, franklinol, and franklinene from *Bosistoa euodiformis*;[47] **11** from *Melicope broadbentia*;[12] **10** from *M. octandra*.[73] (*g*) **12** from *Boenninghausenia albiflora*.[187] (*h*) **13** from *Adiscanthus fusciflorus*[198] and *Hortia badinii*.[14] (*i*) From *Harrisonia abyssinica*.[117] (*j*) Review[58] of Ptaeroxylaceae and see Chapter 4. (*k*) A coumarin and four acetophenones, including alloevodionol.[170] (*l*) See Chapter 4. (*m*) See Chapter 4. (*n*) In *Cneorum tricoccon*, Chapter 4. (*o*) *Ailanthus* and *Picrasma*, Chapter 4. (*p*) See Chapter 4. (*q*) *Cedrela*, *Chukrasia*, and *Swietenia*, Chapter 4. (*r*) See ref. 91 and Chapter 5 and, in addition, flavonoids from *Feronia elephantum*,[175] *Flindersia laevicarpa*,[107,148] *Pamburus missionis*,[62] *Pelea barbigera*,[97] and *Phellodendron amurense*.[165] (*s*) (*t*) (*u*) (*v*) See Chapter 5. (*w*) In addition to taxa previously mentioned[91] gallic and ellagic acid have been detected in the leaves of *Kirkia wilmsii* and in species of Irvingiaceae.[138] (*x*) in addition to sources reported by Hegnauer[89] condensed tannins have been recorded in the leaves of two Madagascan species[143] and dimeric and trimeric procyanidins in the heartwood of *Cedrela toona*.[16]

C. CHROMONES, CHROMENES, AND PHLOROGLUCIDES (ACETOGENINS) (where possible numbering follows Chapter 4)

2-Methylchromones, prenylated phloroglucins and resorcins, and phloracetophenone derivatives are known. They represent a rather distinct chemical feature of the taxa concerned. Examples include spatheliabis-chromene (31-16), 6-demethylacronylin (8), franklinone (9), octandrenolone (10), melicopol (11), 7-methoxy-2,2-dimethylchromene (12, R = H), and ageratochromene (12, R = OMe). The distribution of this type of acetogenin is outlined in Table 4.

Dihydrocinnamic acid derivatives with a resorcinol or phloroglucinol type substitution pattern occur in some Amazonian Rutaceae (13, R^1 = COOMe, COOH, or CH_2OH; R^2 = H or OMe). It is not known whether they represent acetogenins or are derived from phenylalanine; for convenience they are included in Table 4. Two 5-methylcoumarins from Meliaceae, ekersenin (1-2) and siderin (1-3), are probably acetogenic in origin.[43] An orange-red bisiso-coumarin, castanaguyone (14), occurs in the fruit of *Zanthoxylum fagara*;[177]

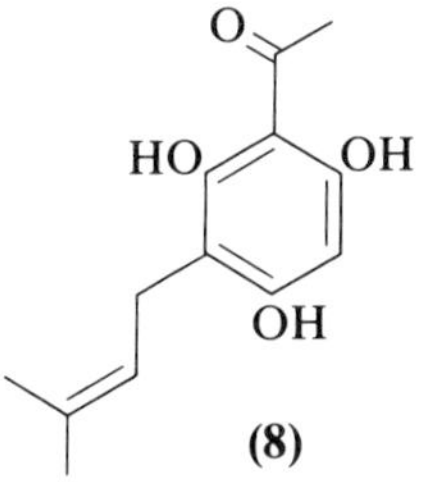

(8)

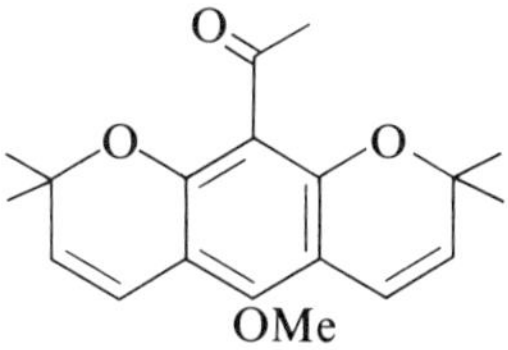

(9) Franklinone

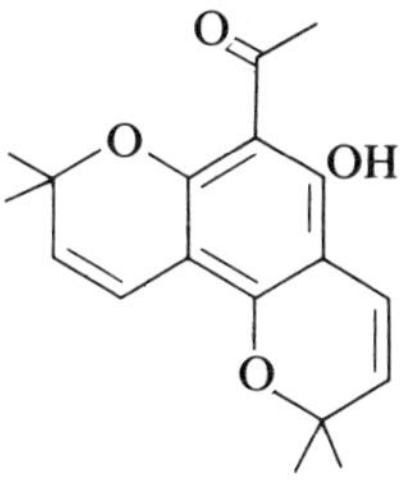

(10) Octandrenolone

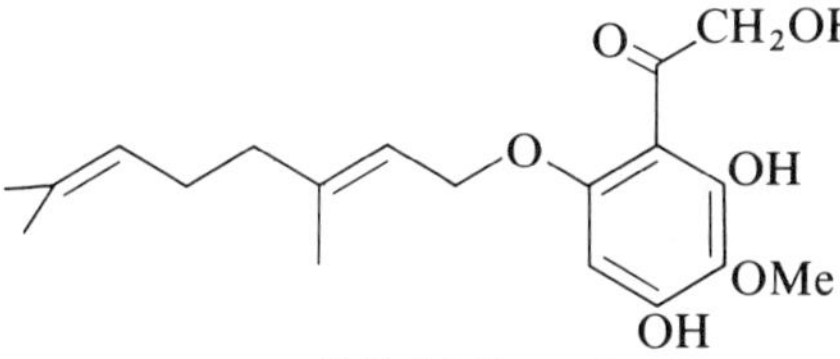

(11) Melicopol

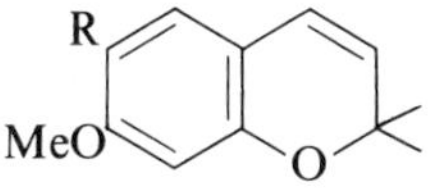

(12)

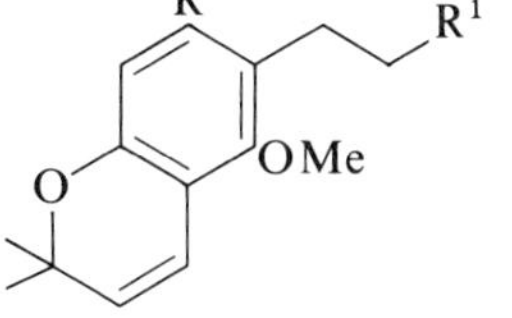

(13)

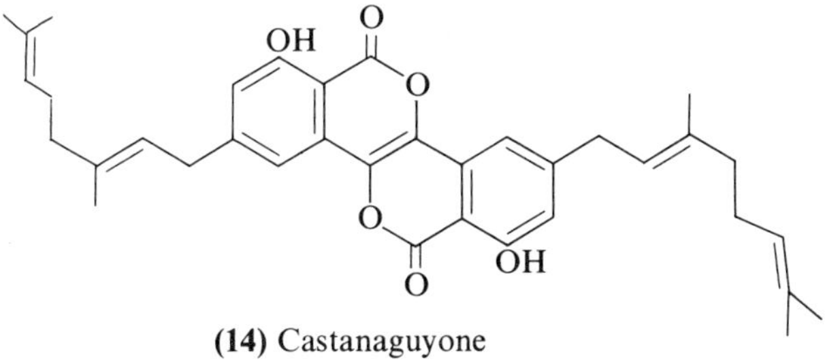

(14) Castanaguyone

412   R. HEGNAUER

its biosynthesis is not known, but it is possibly an isoprenylated acetogenin. Two types of compounds isolated from *Flindersia laevicarpa*,[148] flindersia-chromene (**34**-1) and 1,5-diphenylpentan-1,3-diol, are of uncertain biogenetic origin and are included for convenience.

D. COUMARINS (where possible numbering follows Chapter 4)

Coumarins, furocoumarins, pyranocoumarins, and other types of pre-nylated coumarins are highly characteristic of Rutaceae[79] and Cneoraceae. Isoprenylated coumarins seem to be rare, however, in all other families of the Rutales, except for a small part of Meliaceae, if *Ptaeroxylon* and *Cedrelopsis* (Ptaeroxylaceae) are included in this family. Nevertheless, the isolation of bergapten from the seeds of *Cedrela toona* (Meliaceae) indicated that cinnamic acid metabolism is perhaps more uniform in Rutales than present knowledge of coumarin distribution suggests. Coumarins such as balsamiferone (**28**-1) from *Amyris balsamifera*[37] confirm that the affinities of *Amyris* lay with Rutaceae rather than Burseraceae.

E. FLAVONOIDS

Flavonol glycosides including derivatives of myricetin seem to be most common in the order. Some recently discovered features of flavonoid patterns of Rutales are included in Table 4. Three recent observations might prove, in the future, to have special taxonomic interest; (i) isolation of flavonoids with a trihydroxylated B-ring from *Soymida febrifuga* wood; (ii) isolation of a biflavonoid from *Garuga pinnata* leaves; (iii) occurrence of 5-deoxyflavonoids in leaves and wood of *Amphipterygium adstringens* (Julianiaceae) which led to the classification of Julianiaceae as a subtribe of Anacardiaceae.[204] These findings suggest that Hallier (Table 1) was not far from the truth when he included Anacardiaceae, Burseraceae, and Julianiaceae in one family, Terebinthaceae. Biflavonoids, including amentoflavone, seem to be rather common in the Anacardiaceae.[42,64,106,131,132]

F. TANNINS AND THEIR PRECURSORS

Members of the order produce catechins, proanthocyanidins, gallic acid, and ellagic acid. The tannins themselves have not been investigated thorough-ly, but some statements seem to be possible. Condensed tannins are widespread in roots, stems, and leaves, and are often accompanied by hydrolysable tannins. The latter predominate in some instances, as in chinese tannin which derives from leaf galls of eastern asiatic species of the related Anacardiaceae. High tannin content is rather rare in true Rutales, but rather common in some Anacardiaceae. The main features of tannin distribution in Rutales are summarized in Table 4.

G. AMIDES (where possible numbering follows Chapter 3)

Insecticidal isobutylamides, fagaramide (**31**-3), and several tyramine derived amides are rather characteristic of the Rutaceae.[61,91,127,200] Some recent additions to our knowledge are incorporated in Table 5. These include benzamides (**32**-4, **13**-5, **19**-12, **19**-13, **19**-15, **19**-20, **19**-33, **20**-3), cinnamides

TABLE 5
Some recently detected amides of the Rutales

Taxa	Compounds*
Rutaceae	
Aegle marmelos	Marmeline (**19**-26); its methyl ether (**19**-28) was probably formed during extraction with methanol
Amyris balsamifera	Balsamide (**19**-20) and the corresponding oxazole (**20**-3)
A. plumieri	Two nicotinamides (**19**-16 and **19**-22) and an oxazole (**20**-4)
Atalantia monophylla	Severine palmitate (**19**-33)
Clausena lansium (= *C. wampi*)	Lansamide I (**19**-14)
Fagara rubescens	Rubesamide (**19**-10)
Hesperethusa crenulata	Severine palmitate (**19**-33)
Myrtopsis macroçarpa	Benzamide (**32**-4)
M. myrtoidea	Benzamide (**32**-4), *N*-benzoyltyramine methyl ether (**19**-13), and *N*-benzoyltryptamine (**13**-5)
Pleiospermum alatum (= *Hesperethusa alata*)	Alatamide (**19**-12), dihydroalatamide (**19**-13), and *N*-homoveratroylhomoveratrylamine (**19**-30)
Zanthoxylum inerme	Dihydroalatamide (**19**-13) and tembamide (**19**-15)
Z. tingoassuiba	Tembamide (**19**-15)
Meliaceae	
Aglaia sp.	Tiglamide (**32**-3)
A. odorata	Odorine (**28**-2) and odorinol (**28**-3)
A. roxburghiana	Roxburghilin and hydroxyroxburghilin, identical to **28**-2 and **28**-3, respectively.

* Numbering of compounds follows that of Chapter 3.

(19-26, 19-28, 19-14), and nicotinamides (19-16, 19-22, 20-4). Atypical examples are rubesamide (19-10), which has a cyclopropanecarboxylic acid N-acylating group, and another (19-30) in which the acylating group is a derivative of phenylacetic acid. The amines which participate in the formation of these amides derive mainly from phenylalanine and tyrosine, exceptions being benzamide (32-4) and a tryptamine amide (13-5). Similar amides are at present unknown from other rutalean families, but a few rather different amides (28-2, 28-3, 32-3) are recorded from Meliaceae (Table 5).

H. AMINES (= PROTOALKALOIDS) AND ALKALOIDS (numbering follows Chapter 3)

Most alkaloids of the Rutales have been isolated from the Rutaceae, and these have been reviewed many times.[86,91,127,154,200] They may be grouped according to their assumed biogenetic origin as follows.

(I) Phenylalanine and tyrosine derived alkaloids; phenylethylamines, benzylisoquinolines, aporphines, protopines, berberines, and benzophenanthridines.

(II) Tryptophan derived alkaloids; tryptamines, β-carbolines, canthin-6-ones, indoloquinazolines (mixed origin — tryptamine + anthranilic acid).

(III) Anthranilic acid derived alkaloids; quinazolines and quinazolones, acridones, quinolines and quinolones, furoquinolines and furoquinolones.

(IV) Histidine derived alkaloids, imidazoles.

(V) Uncertain origin (either anthranilic acid or tryptophan), carbazoles.[39]

Most of these alkaloid types seem to be restricted in the order to one family, Rutaceae. This is true of group III which represents an outstanding character of the Rutaceae. Group I, which is highly characteristic of the Magnoliidae sensu Cronquist[51] (= Polycarpicae sensu Wettstein), seems to occur only in the primitive genera Zanthoxylum (incl. Fagara), Phellodendron, Fagaropsis, and Toddalia,[201] and β-carbolines and carbolines occur in Zanthoxylum and Pentaceras (Rutaceae) and seem to be rather common in the Simaroubaceae.[91] Some recent additions to these simaroubaceous alkaloids are the β-carbolines hannine (17-2) from Hannoa klaineana, 17-10 and three canthinones (18-1, 18-10, 18-17) from Lignum Quassiae (Picrasma excelsa), and two new β-carbolines (17-17, 17-23) and a new canthinone (18-15) from the root bark of Ailanthus altissima. New canthinones have also been isolated from Amaroria soulameoides (18-7, 18-13), Simaba cuspidata (18-11, 18-18), and the root bark of Simarouba amara (18-4).

No alkaloids are known from Burseraceae, and Meliaceae has hardly been investigated for this type of constituent. Available evidence suggests that true alkaloids are rare in the latter although a number of basic compounds are known. In addition to the previously mentioned amides, 3-hydroxypyridine (29-1) has been isolated from *Entandrophragma cylindricum* bark, and rohitukine (29-3) from the leaves of *Amoora rohituka* (*Aphanamixis polystachya*) is a combination of noreugenin and *N*-methyl-3-hydroxypiperidine. The basic principles of *E. caudatum* seem to be esters of nicotinic acid with the limonoid demethylphragmalin.[8]

Before leaving alkaloids of the Rutales some recent developments for the Rutaceae need to be mentioned. Plants sometimes store alkaloids as the glucosides, as in gravacridonediol glucoside (9-12) from *Ruta graveolens* roots. Dimeric quinolone alkaloids, such as paraensidimerine D (7-31), have been found in several species. Tecleanone (8-35), a biogenetic precursor of the acridones, has been found in several African species. The β-carboline 7-17 occurs in the leaves of *Dutaillyea oreophila* where it is accompanied by hordenine (19-3) and 5-methoxy-*N*-methyltryptamine (13-4), the latter also occurring in the leaves of *D. drupacea*. The β-carboline 7-17 has previously been recorded as a constituent of the wood of *Nectandra megapotamica*.[60]

Harmalan (17-1) has been isolated from the leaves of *Flindersia laevicarpa*, and the β-carbolines borrerine (17-15) and isoborrerine (17-16), together with indole-monoterpenoid alkaloids like isoborreverine (13-7), from the leaves of *Fl. fournieri*. Both 17-15 and 13-7, and allied derivatives, were first isolated from *Borreria verticillata* (Rubiaceae), a clear-cut example of convergence of alkaloid metabolism. It should be stressed that 13-7 and related alkaloids found in these two taxa do not belong to the characteristic rubiaceous indole-iridoid (secologanin) alkaloid type.

Hordenine (19-3) has recently been encountered in several rutaceous plants, in *Geijera balansae* together with 19-8, in *Zanthoxylum microcarpum* with 19-8, in *Z. coriaceum* with alfileramine (19-35), and in *Z. procerum* with alfileramine-like alkaloids. *Z. culantrillo* has culantraramine, another alfileramine-like alkaloid, in the leaves, and synephrine (19-4) and candicine (19-7) in the stems. Candicine also occurs in the root bark of *Z. simulans* and synephrine in the leaves of *Z. fagara*. Alfileramine was first isolated from the leaves of *Z. punctatum*. A piperazine derivative (19-34) found in the leaves of *Z. arborescens* represents a new type of alkaloid for the Rutaceae. One aporphine alkaloid deserves special mention; liriodenine (22-1) is another typical magnoliaceous alkaloid found in the Rutaceae, in *Z. cuspidatum*.[105]

I. BITTER NORTRITERPENOIDS AND THEIR TETRACYCLIC TRITERPENOID PRECURSORS

Bitter principles are very widespread in Cneoraceae, Meliaceae, Rutaceae, and Simaroubaceae. When it was recognized that those bitter principles with a

C_{26} skeleton (Rutaceae, Meliaceae) or C_{25}, C_{20}, or C_{19} skeleton (Simaroubaceae) biogenetically represent tetra-, penta-, deca-, and undeca-nortriterpenoids, and derive from adequately oxygenated tetracyclic triterpenes like the sapelins (**15** and **16**), bourjotinolone A (**17**), and hispidone (**18**), the taxonomic potential of this group of bitter principles became apparent. Previous chemotaxonomic discussions of these compounds have been made.[87,89,91]

The tetra- and penta-nortriterpenoid bitter principles of Cneoraceae, the cneorins and tricoccins, became known later.[184] The coidentity of tricoccin S3 with obacunone and the correlation of cneorin R with methyl ivorensate[65] show clearly that the triterpene metabolism of the Cneoraceae is very similar to Rutaceae and Meliaceae. In fact chemistry leaves no doubt about the classification of the Cneoraceae; the family must be included in the Rutales. Just as betalains form an outstanding character of the order Centrospermae, the highly characteristic nortriterpenoid bitter principles represent the chemical marker of the Rutales. There is one exception however; bitter nortriterpenoids have not yet been detected with certainty in Burseraceae (see Chapter 10). This is a similar situation to the Caryophyllaceae and Molluginaceae of the Centrospermae, which do not contain betalains.

Some recent developments in this area are illustrated by the isolation of glucosidic quassinoids from *Brucea* spp.,[141] of new C_{25} simarolides from a *Simaba* sp.,[152] and (if no botanical mistake was made) of the limonoids obacunone (**19**) and harrisonin (**20**) from roots of the simaroubaceous plant *Harrisonia abyssinica*.[114,134] Some hydroxylated tirucallan derivatives have been isolated from *Ailanthus excelsa* bark[173] and from *Simarouba amara*,[151] and the stereo structure of the triterpenoid resin constituent malabaricol, from *Ailanthus malabarica*, resolved.[146] Members of the Burseraceae produce mixtures of isomeric tetracyclic tirucalladien-21-oic acids, known as elemolic and elemonic acids.[46,142] Elemolic acid from *Canarium boivinii* contains the 3α-ol and 3β-ol isomers.[21] Cholesterol and three hydroxylated C_{27} sterols (guggulsterols 1, 2, and 3) and two stereoisomeric pregnanes (guggulsterones) occur in the resin of *Commiphora mukul*,[145] and the stereochemistry of guggulsterol 1 has been defined.[10] $3\beta,7\alpha,20$-Trihydroxystigmast-5-ene is a constituent of the leaves of *Neochamaelea pulverulenta*.[129]

J. DITERPENES AND PENTACYCLIC TRITERPENES

Diterpenes occur in the essential oils of some Rutales (Table 3) but appear to be rare as resin constituents. Non-volatile diterpenes have been recorded for only a few members of the order. In the Rutaceae *Phebalium rude* contains two

(15) Sapelin A (R = α-OH, β-H)   **(16)** Sapelin B (R = α-OH, β-H)
(17) Bourjotinolone (R = O)   **(18)** Hispidone (R = O)

(19) Obacunone   **(20)** Harrisonin

kaurane type acids[91] and *Pamburus missionis* has the furanoid daniellic acid and corresponding butenolide in the leaves.[62] The taxonomically most interesting diterpenes have been isolated from the madagascan species *Euodia floribunda*[18–20] which yielded the *cis*-clerodane derivatives floribundic acid **(21)**, floridiolic acid **(22)**, floridiolides A **(23)** and B **(24)**, and a related hydroxy–lactone. All these are present as either glycosides or sugar esters and were isolated after autolysis. Floribundic acid **(21)** is closely related to the menispermaceous bitter principles like columbin and tinophyllone **(25)**. In the Meliaceae sugiol, an aromatic diterpene, and nimbiol, a norditerpene, occur in the Neem tree (*Azadirachta indica*)[89] and eperu-13-en-8β,15-diol in *Aphanamixis polystachya*.[41]

(21) Floribundic acid (R^1 = H; R^2 = H_2)   **(23)** Floridiolide A (R = Me)

(25) Tinophyllone (R^1 = Me; R^2 = O)   **(24)** Floridiolide B (R = CH_2OH)

Pentacyclic triterpenes are common constituents of the Rutales. They occur mainly as alcohols (e.g. lupeol, α- and β-amyrin) and ketones (e.g. friedelin, taraxerone), but pentacyclic and seco-pentacyclic acids are known, notably in Burseraceae. In one instance onocerane derivatives have been recorded. The occurrence of pentacyclic triterpenes has been reviewed by Hegnauer[86,89,91] and can be summarized as follows. *Burseraceae:* oleanane type and ursane type resin acids and alcohols, 3-epilupeol, A-ring seco oleananes, canaric acid, 3,4-secolupeol (see Chapter 10). *Meliaceae:* betulin, betulinic acid, katonic and indicic acids, walsurenol, onoceradienone and its seco derivative lansic acid. *Rutaceae:* arborinol, isoarborinol, arborinone, bauerenol, isobaurenol, friedelin, germanicol, germanicone, lupeol, multiflorenol, isomultiflorenol, myricadiol, myricolal, taraxerol, taraxerone, ursolic and ifflaionic acids; distribution in Aurantioideae reviewed by Dreyer.[61] *Simaroubaceae:* lupeol, taraxerone, 3-epibetulinic acid.

More recent observations are available for Burseraceae, Rutaceae, and Simaroubaceae and illustrate the diversity of pentacyclic triterpenes in the order and the common occurrence of 3-epitriterpenols and A-ring seco compounds. Some examples of recent reports are summarized below.

Amyrins are typical constituents of resins of many Burseraceae (Chapter 10) and of fossil resins of burseraceous origin.[74] *Boswellia carteri* contains α- and β-amyrenones and 3-epiamyrins and their acetates and the acetate of β-boswellic acid,[176] while frankincense from *B. frereana* contains lupeol and 3-epilupeol.[155] Betulonic acid and benulin, a hemiketal derivable from betulonic acid, occur in *Bursera arida.*[103] The oleoresin of *Canarium boivinii* from Madagascar yields α- and β-amyrin as well as elemolic and elemonic acids,[21] and elemi from *C. muelleri* lupeol and canaric acid, a derivative of 3,4-

secolupeol. Indian black dammar resin from *Canarium strictum* contains α- and β-amyrins and their acetates, 11-oxo-α-amyrin, pseudotaraxasterol, a pseudotaraxastane-3,20-diol, and a pseudotaraxastane-3-one-20-ol.[110] α-Amyrin is the major triterpene of wood, bark, and oleoresin of *Canarium zeylanicum*; the bark and resin also contain β-amyrin, 11-oxo-α-amyrin (neoilexonol), 11-oxo-β-amyrin, and the corresponding dioxo compounds, and the oleoresin additionally α- and β-amyrenone and canaric acid (Chapter 10). The commic acids of *Commiphora* sp. (gum myrrh) are ursane type and oleanane type triterpenes.[190] The oleoresin of *Dacryodes edulis* (= *Pachylobus edulis*) contains canaric acid, α-amyrin, 3-epi-α-amyrin, and 3-epilupeol (Chapter 10).

In the Rutaceae *Bosistoa sapindiformis* yielded taraxerol and sawamilletin from the leaves,[48] while the leaves of *Melicope octandra* gave multiflorenol.[73] Four *Myrtopsis* spp. from New Caledonia contained lupeol in both leaves and twigs.[96] In the Simaroubaceae *Simarouba versicolor* (= *S. antisyphilitica*) yielded 3-epilupeol,[77] and the bark of *Picramnia sellowii* betulinic and 3-epibetulinic acids.[116]

K. SAPONINS

Saponins have been reported from some Meliaceae[89] and Rutaceae[91] but seem to be rare in the order. Structural work has been described for a seed saponin of *Aphanamixis polystachya* (= *Amoora rohituka*) which is a diglycoside of stigmastadienol,[17] a rare type of saponin in that it has a C_{29} phytosterol as sapogenin. Phytosterols and compounds with haemolytic activity occur in roots of *Eurycoma longifolia* (Simaroubaceae),[140] but have yet to be analysed.

L. VARIOUS SECONDARY METABOLITES

Tariric acid (18:1,6a) is the main fatty acid of seed oils of *Picramnia* spp. (Simaroubaceae) and the only acetylenic compound known from Rutales. It is accompanied by petroselinic acid (18:1, 6*cis*), an isomer of oleic acid, which also occurs in *Picrasma* seed oils.[91]

Cyanogenesis is infrequent in Rutales. *Boronia* and *Zieria* spp. (Rutaceae) are reported to be cyanogenic, and zierin (*m*-hydroxysambunigrin) has been isolated from *Z. laevigata*,[91] while the leaves of *Z. cytisoides* contain amygdonitrile glucoside (sambunigrin or prunasin) as the main cyanogenic component and zierin or its isomer holocalin as a minor constituent.[69] Co-occurrence of benzaldehyde cyanohydrin glucosides (sambunigrin or prunasin) with their *m*-hydroxy derivatives (zierin or holocalin) is by no means rare, the latter probably deriving biogenetically from the former.

Cyanogenesis has been reported for *Commiphora africana* (Burseraceae)[174] and from the leaves of *Loureira* (= *Glycosmis*) *cochinchinensis*,[75] although in the latter identification of material was uncertain. Gibbs[75] found that cyanogenesis was absent from 17 species of Rutaceae, 3 Simaroubaceae, and *Cneorum tricoccon*, and doubtful in *Dysoxylum fraseranum* shoots. From literature data he[75] enumerated 6 *Citrus* spp. and *Triphasia trifoliata* as positive. These latter observations concern work performed in the Philippines,[156] and it should be stressed[84] that these results need confirmation as the picrate test was used in a very unusual way which was likely to lead to false positives.

2,6-Dimethoxy-*p*-benzoquinone (**26**) occurs in Meliaceae[89] and in many Simaroubaceae.[91] More recently anthraquinones have been isolated from two simaroubaceous spp.: chrysophanol (**27**) and a glucoside from *Alvaradoa amorphoides* roots,[183] and **27**, emodin (**28**), and physcion (**29**) from *Picramnia sellowii* bark.[116] Physcion (**29**) has also been described from *Euodia meliaefolia*, 8-deoxyemodin (**30**) from *Ruta graveolens*, and **29** and barbaloin from *Zanthoxylum acanthopodium*.[162] There are, however, indications that other biogenetic pathways leading to anthraquinones exist in Rutaceae. 2-Methylanthraquinone (tectoquinone, **31**) occurs in *Clausena heptaphylla* and has a C-ring that may well be mevalonate derived.[40] Two quinoid coumarins have also been isolated from Rutaceae: 5,8-dioxopsoralen (**32**) from *Clausena indica*[207] and naphthoherniarin from *Ruta graveolens*.[163]

(**26**)

(**27**) Chrysophanol (R^1 = H; R^2 = OH)

(**28**) Emodin (R^1 = OH; R^2 = OH)

(**29**) Physcion (R^1 = OMe; R^2 = OH)

(**30**) (R^1 = OH; R^2 = H)

(**31**) Tectoquinone   (**32**)

Hydroquinone has been isolated from a sample of the wood of *Afraegle paniculata* from Nigeria,[3] and 2-hydroxymethylbenzoic acid is one of the antisickling principles of *Zanthoxylum (Fagara) xanthoxyloides*.[180] An interesting series of phenylethane derivatives occur in the absolute of *Citrus unshiu* flowers, comprising 2-phenylethanol, phenylacetic acid, phenylacetaldehyde, benzyl cyanide, phenylacetaldoxime, and 1-phenyl-2-nitroethane.[166] The distribution of proline and pipecolinic acid in the seeds and leaves of many populations of *Ptelea trifoliata* has been studied. High concentrations proved to be restricted to populations occupying xeric habitats.[160] *Evodiella muelleri* has yielded 1.6% of muellitol, a triprenylated scyllitol, from dry leaves.[67]

M. MUCILAGES

Many Rutales tend to deposit acidic mucilage in epidermal cells. These membrane mucilages are relatively well known from an anatomical viewpoint, but practically unknown chemically. Gummosis, the production of mucilages after injury, is well known from a number of Meliaceae[89] and Rutaceae,[91] and many Burseraceae yield oleo-gum-resins (Table 3) which contain appreciable amounts of mucilage.[86] These mucilages seem to have 4-*O*-methylglucuronic acid as one of their characteristic constituents. The gum of the oleo-gum-resin of *Commiphora mukul* contains a main chain of galactose and 4-*O*-methylglucuronic acid, and side-chains with arabinose, galactose, and some fucose.[34]

N. STORAGE PRODUCTS OF SEEDS

Proteins and fixed oils seem to be the main storage products.[86,89,91] As far as is known, seed oils are of the usual types containing palmitic, stearic, oleic, and linoleic acids as major components. The composition of seed oils is not generally characteristic at the family level, but at infrafamiliar level only. For example, Simaroubaceae have at least two oil types, the common oleic-linoleic type and the less common petroselinic-tariric acid type. The burseraceous species *Boswellia serrata* yields a seed oil with the common constituents.

Aburano et al.[2] investigated seed oils of 6 species of Rutaceae and found 16:0 (6–37%), 18:0 (0.8–5%), 18:1(9-*cis*) (9–33%), 18:2(9,12-*cis*) (20–49%), and 18:3(9,12,15-*cis*) (12–41.5%) with, as would be expected, highest contents of linolenic acid in the temperate species *Phellodendron amurense* (41.5%) and *Ruta graveolens* (31.6%).

Netolitzki[135] described the seeds of a number of taxa of Rutales. In most instances fatty oil and aleuron bodies (protein) were the major storage products of cotyledons and, where present, endosperm. Starch occurs only in ripe seeds of some Meliaceae (*Carapa, Amoora*), Simaroubaceae (*Simaba, Perrieria*), and Rutaceae (*Amyris, Casimiroa*). Most Meliaceae have "resin" cells in cotyledons, and mucilage cavities occur in cotyledons of Simaroubaceae-Irvingioideae (Irvingiaceae). Czaja[53] confirmed that fatty oil and protein were the major storage products of the seeds of Rutales he investigated. Moreover, he noted reserve celluloses in some members of Burseraceae and Simaroubaceae, and starch in unripe seeds of species of *Citrus, Dictamnus, Poncirus*, and *Ruta*, and in ripe seeds of *Casimiroa*.

O. INORGANIC COMPOUNDS

Calcium is deposited in large amounts in many tissues of Rutales. As far as is known it occurs mainly in the form of crystals of oxalate of different shape and size. Many members of the order deposit large amounts of silica (SiO_2) in leaves and wood. The taxonomic meaning of this feature was thoroughly discussed by Edman[63] and later treated by Hegnauer.[86,89,91] According to Edman heavy silicification is a primitive feature in Rutales.

IV. AFFINITIES PROPOSED IN LITERATURE FOR RUTALES

As illustrated by Tables 1 and 2, many authors assume rather clear-cut affinities between Rutales *s.s.* and Sapindales (illustrated by Anacardiaceae) and Geraniales (illustrated by Zygophyllaceae), but the circumscription of orders varies widely. Table 6 broadly outlines the ideas of some taxonomists concerning rutalean relationships. If my own proposal is compared with those in Table 6, there is more or less good agreement with Hallier and Thorne as far as descent of Rutales is concerned. Umbelliflorae (= Apiales = Araliales) are assumed to have evolved from rutalean stock by Emberger and Cronquist.

There seems to exist little agreement between botanists concerning the origin of Asterales. Cronquist is convinced that Asteridae evolved from proto-Rosales. Wettstein is rather vague with regard to Synandrae (Asterales). In his scheme[202] the taxon is mentioned twice; one possibility is evolution with Cucurbitales from parietalean stock, and the second is derivation of Rubiales and Synandrae from Umbelliflorae. With respect to the latter it should be stressed, however, that Wettstein included Cornaceae and allied iridoid producing families in Umbelliflorae. He primarily had cornalean members in

TABLE 6
Some relationships suggested for Rutales

Authority	Order descended from	Order is ancestral to	Remarks
Hallier[83]	Proto-Berberideae	? Amentiferae	Rosales, Sapindales (= Aesculinae, also comprising Leguminosae), and Geraniales (= Gruinales) belong to the same evolutionary line (Rhodophyles) as Terebinthinae. Umbelliferae and Compositae on another branch (Anonophyles) of the evolutionary tree
Wettstein[202]	Tricoccae (= Euphorbiales)	Part of Sympetalae, e.g. Ligustrales, Contortae, Tubiflorae	Perhaps a line Umbelliflorae to Rubiales to Synandrae (= Asterales) also derives from Tricoccae
Emberger[38]	"Souche encore inconnue"	Garryales, Ombelliflorae, and Rubiales	Rhamnales, Ligustrales, and Celastrales evolved from same "souche", Compositae in a separate line
Hutchinson[100]	Magnoliales, via Dilleniales, Bixales, Theales and Celastrales	Meliales, Sapindales, and Euphorbiales p.p.	Asterales belong to Herbaceae
Thorne[191]	Berberidales (belonging to Annoniflorae)	In the Rutales the suborder Rutineae gave rise to Sapindineae. Araliales placed in a separate branch called Corniflorae	Asteridae, with Asterales and Asteraceae (Compositae) only, possibly evolved from Rutales (in Fig. 2) but text (p. 95) stresses Cornifloreae relationships of Asteriflorae

TABLE 6—continued

Takhtajan[189]	Saxifragales	Sapindales and Geraniales to Polygalales	Proto-Saxifragales also gave rise to Rosales, to Cornales, Araliales, and to Asteridae with two evolutionary lines: (i) asteralean; (ii) gentianalean-scrophularialean-lamialean
Dahlgren[55]	Unknown, a stock of Proto-Angiospermae	Climax group	Rutales, Sapindales, Polygalales, and Geraniales closely related and form together the core of the superorder Rutiflorae which is linked through Sapindales with Fabiflorae (Leguminosae)
Cronquist[51]	Rosales	Geraniales and Apiales (not comprising Cornaceae and allies)	Rosales also yielded Fabales; Asteridae are assumed to have their origin in Proto-Rosales; in Asteridae Asterales believed to originate from Rubiales
Thorne[192]	Near Annoniflorae from Proto-Angiospermae	Climax group	Asterales an offshoot of Corniflorae (with Cornales, Araliales, and Dipsacales)
Hegnauer[90]	Ranalean stock	Araliales (Pittosporaceae, Araliaceae, Umbelliferae)	Asterales assumed to have evolved from aralian stock

mind, when he suggested the link between Umbelliflorae and Rubiales and Synandrae. The same holds for Thorne's proposal of a connection between Corniflorae and Asteriflorae (Table 6). My own proposal on the other hand accepts a direct link between Araliaceae *sensu* Thorne (Aralioideae, Hydrocotyloideae, Saniculoideae, Apioideae) and Asterales. Chemical evidence will be treated in Section V.

With regard to recent proposals for angiosperm classification[55,191] one should always keep in mind that they are explicitly considered to be provisional by their respective authors. In the last version of Dahlgren's classification[56] Torricelliaceae were transferred from Cornales to Araliales, and in Thorne's recent version[192] Leguminosae were transferred from Rosiflorae to Rutiflorae. The latter agrees very well with Dahlgren[55] and with arguments discussed below.

V. Chemical Evidence for the Delimitation and Classification of Rutales

A. DELIMITATION

As already mentioned, the bitter principles derived from tetracyclic triterpenoids form an extremely good marker for the Rutales. Their use in this context would imply exclusion of Simaroubaceae-Irvingioideae (Irvingiaceae) and Burseraceae *s.s.* from Rutales *s.s.* This is by no means a new idea. Irvingiaceae are often treated as a separate family or transferred to the Ixonanthaceae,[6] whose affinities have probably to be sought elsewhere. Burseraceae were included in Anacardiaceae by Hallier,[83] which is in line with the presence of resin ducts and biflavonoids in both families, and with the probable lack of limonoid and quassinoid bitter principles in Burseraceae. Whether Rutales and Sapindales should be treated as separate orders or as suborders of Sapindales (Cronquist), Rutales (Thorne), or Terebinthales (Emberger) is solely a matter of convenience or personal preference. If we restrict ourselves to the main families the following classification seems chemically feasible:

Rutaceae Cneoraceae Meliaceae Simaroubaceae	⎱ Rutales *s.s.*	⎱
(Zygophyllaceae) Burseraceae Anacardiaceae (incl. Julianiaceae) Sapindaceae Hippocastanaceae Aceraceae	⎱ Sapindales *s.s.*	Rutales or Sapindales *s.l.*

B. CLASSIFICATION

Table 7 summarizes some chemical evidence supporting an evolutionary line proposed some years ago by me[85,90] and still considered acceptable today. Cronquist[50] remarked: "I must disagree completely, however, with Hegnauer that the Asterales originated from Apiales. I would not address myself to the matter in a formal public way, except for the fact that some people seem to be taking the suggestion seriously". A glance at Table 7, however, shows clearly that there are too many biochemical similarities between Apiales and Asterales to allow neglect of them. The most convincing similarities include synthesis and accumulation of essential oils, isoprenylated coumarins, germa-cranolide, eudesmanolide, guaianolide, elemanolide, and eremophilanolide types of sesquiterpene lactones,[70] diterpenes, triterpene saponins, acetylenic compounds including falcarinone type C_{17} polyacetylenes, and the total lack of tannins and iridoids (but see note m, Table 7). The recent isolation of 1α-angeloyloxy-6α-hydroxy-9-oxocarot-2-ene (33) from *Inula crithmoides*[122] points in the same direction; carotane type sesquiterpenes were previously only known from Umbelliferae.

(33)

Rutales are biochemically intimately connected with Apiales by essential oils, chromones, identical types of prenylated coumarins, and a complete lack of iridoids (but see notes l and m, Table 7). Chemical links between the ranalean stock and Rutales have already been discussed.[85] To the connecting chemical characters (essential oils, benzylisoquinoline alkaloids, acrid amides, prenylated flavonoids, lignans, accumulation of SiO_2) can now be added the absence of iridoids and the occurrence of tinosporone like diterpenes (21) in Rutaceae.

The best explanation for the chemical facts compiled in Table 7 would be an evolutionary line as illustrated in Fig. 1. With regard to coumarins it is interesting to note that, from the 503 compounds reported by Murray,[130] 459 most probably represent true coumarins in a biogenetic sense. Most first

TABLE 7

Some chemical features of Magnoliidae, Rutales, Araliales, and Asterales (cf. Hegnauer[93])

Classes of compounds	Magnoliidae, *sensu* Cronquist[51]	Rutales	Apiales, *sensu* Cronquist[51]	Asterales, *sensu* Cronquist[51]
Essential oils	Common in woody members	Table 3; common in Ac	Very common	Very common except Lactuceae
Phenylpropanes and lignans	Very common	Table 4	Common	Common
Acetogenins (chromones, phloracetophenones, 5-methylcoumarins, etc.)	Chromones in *Cimicifuga* and *Eranthis*; siderin in *Clematis*; glaupalol and glaupadiol in *Glaucidium*[a]	Table 4 (rather common)	Chromones rather common in Um	2-Methylchromones from some genera; *Marshallia, Mikania, Stevia*; chromenes in *Ageratum*; 5-methylcoumarin derivatives and allied 2-hydroxychromones in some Mutisieae and Vernonieae[b]
Coumarins: isoprenylated, furo- and pyrano-deriv.	Not yet known	Table 4; very common in Ru, Cn, Pt	Very common in Um	Known from many genera[c]
Tannins and their precursors	Mainly oligomeric procyanidins and condensed tannins	Section IIIF; usually condensed; ellagic acid in Si; gallitannins in Si and Ac	Very rare; caffeic acid derivatives in large amounts	Very rare; flavan-3,4-diols and large amounts of caffeic acid derivatives
Amides (often pungent, insecticidal)	Not rare (Ar, Ch, La, and Pi)	Section IIIG; in Ru	Not yet known	In several genera[d]
Alkaloids	Common; mainly benzylisoquinolines and derivatives	Section IIIH; many	Very rare[e]	Rare except pyrrolizidine alkaloids in Eupatorieae and Senecioneae[f]
Sesquiterpene lactones	From Ar, Ch, La, and Ma	Only from Bu[g]	In many Um	Very common

TABLE 7—continued

Diterpenes	Relatively common in An and Mn; Aconitum type pseudoalkaloids in *Aconitum* and *Delphinium*	Section IIL; seem to be rather rare	Occur in Al and Um; seem to be rather rare[h]	Very frequent; many types including Aconitum-type pseudoalkaloids
Triterpenes	Free, mainly minor, tetracyclic; pentacyclic sapogenins	Sections IIIi, j, and k; saponins rare and sapogenins not studied	Free, minor amounts, tetra- and penta-cyclic sapogenins common	Tetracyclic and pentacyclic; free and as sapogenins; common
Acetylenic compounds	Uncommon; An and La only[i]	Section IIL; Si only	Common	Very common
Storage products in seeds	Mainly oil and protein; amyloid in An, and starch in some La, My	Section IIIN; mainly oil and protein	Oil, protein, and mannan[j]	Mainly oil and protein
Free cyclitols	Common in some families	Section IIL; rare	Frequent in small amounts	Common
Accumulation of aluminium	In several taxa	Not known	Not known	Not known
Accumulation of SiO$_2$	Rather common in leaves and wood[k]	Section IIIO; rather common in leaves and wood	Not known	Not uncommon in leaves
Iridoids	Not known	One doubtful report[l]	Not known if *Aralidium* is excluded[m]	Not known

Ac = Anacardiaceae; Al = Araliaceae; An = Annonaceae; Ar = Aristolochiaceae; Bu = Burseraceae; Ch = Chloranthaceae; Cn = Cneoraceae; La = Lauraceae; Ma = Magnoliaceae; Mn = Menispermaceae; My = Myristicaceae; Pi = Piperaceae; Pt = Ptaeroxylaceae; Ru = Rutaceae; Si = Simaroubaceae; Um = Umbelliferae.

(a) Siderin from roots of *Clematis ligusticifolia*.[9] Glaupalol and glaupadiol = prenylated derivatives of 4-hydroxy-5-methylcoumarin (also Compositae-Mutisieae and -Vernonieae). (b) *Marshallia obovata*:[26] spatheliachromene, 9-hydroxyspatheliachromene, peucenin 7-methyl ether, and 3 esters of marshalliachromone from roots; roots and leaves yielded flavan derivatives. *M. grandiflora*:[30] identical flavan-3,4-diol derivatives. *Mikania alvimii*:[23] 2-methyl-5,7-dihydroxychromone. Compositae-Mutisieae:[27,32] prenylated and simple derivatives of 4-hydroxy-5-methylcoumarin in *Gerbera, Jungia, Onoseris,* and *Perezia*. Compositae-Vernonieae:[11,24,28,29,93] prenylated derivatives of 4-hydroxy-5-methylcoumarins and 2-hydroxychromones in *Bothriocline, Erlangea,* and *Ethulia*. (c) Recent examples from *Pterocaulon balansae, P. lanatum,* and *P. virgatum*,[22,59,119] Ptaeroxylon-type coumarins (obliquinhydrate) from *Eupatorium lancifolium*[94] and *Acritopappus prunifolius*.[31] See also ref. 93. (d) New examples: 3 species of *Achillea* and *Leucocyclus formosus*.[80] (e) Anthranilic acid derived alkaloids possibly occur in *Mackinlaya* (Araliaceae), and acetogenic coniine derivatives and some other nitrogen-containing compounds from Umbelliferae.[93] Dimeric *o*-aminotolyl derivative of *Cachrys sicula*[149] reminiscent of *N*-methylanthranilic acid and 6-hydroxykynurenic acid from *Thapsia villosa*[125] suggests a metabolite on the pathway connecting tryptophan and anthranilic acid. Two bisbenzylisoquinoline alkaloids from *Heracleum wallichii* roots[82] but this may be a mistake with regard to the investigated crude drug. (f) Other alkaloids little studied; betains rather common; Aconitum type nitrogenous diterpenes in *Inula*,[92] anthranilic acid derived quinolines in *Echinops*,[92] and see ref. 93. (g) Commiferin: see Chapter 10. (h) New sources: *Margotia gummifera*,[150,159] *Magydaris panacifolia*,[144] and *Peucedanum oreoselinum*;[115] and see ref. 93. (i) Refs. 25 and 76. (j) Reserve cellulose of umbelliferous endosperm mainly β(1,4)-mannan.[98] (k) Magnoliaceae[178] and Lauraceae.[158] (l) A configurationally aberrant secoiridoid, xylomollin, was described from unripe fruits of an east African meliaceous tree, *Xylocarpus molluscensis*.[113,133] The epithet *molluscensis* does not appear to exist in *Xylocarpus* (often included in *Carapa*). Most probably *Carapa (Xylocarpus) moluccensis* was the species meant. There is, however, possibly a more serious botanical error as Connolly *et al.*[44] obtained only limonoids from African *X. moluccensis* and *X. granatum*. Another botanical confusion concerns the report of limonoids in seeds of *Uncaria gambir* (Rubiaceae).[5] These seeds later proved to be derived from *X. granatum*[137] and the species was ascribed as the source of xylomollin (without convincing evidence). It is my feeling that the occurrence of secoiridoids in Meliaceae is not yet established (a view supported by S. Rosendal-Jensen and B. J. Nielsen (personal communication) who looked unsuccessfully for iridoids and secoiridoids). (m) *Aralidium pinnatifidum* contains two iridoid glucosides. The genus was recently transferred from Araliaceae to a monotypic family, Aralidiaceae, which seems to be closer to Cornaceae and its allies than to Araliaceae.[147]

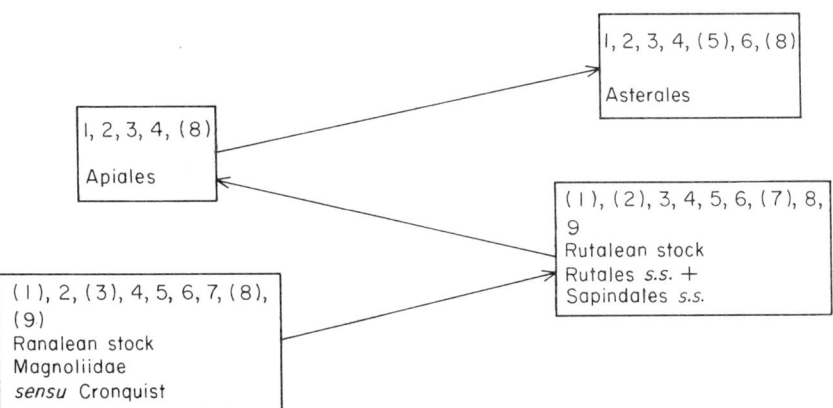

FIG. 1. Chemical evidence for the classification of Rutales in an evolutionary line which also comprises Magnoliidae, Apiales, and Asterales. Major chemical trends of the whole line: essential oils, oily seeds, total lack of iridoids. 1 = polyacetylenic compounds; 2 = sesquiterpene lactones; 3 = prenylated coumarins (and flavonoids); 4 = acetogenic chromones and phloroacetophenones; 5 = silicia bodies and incrustations; 6 = pungent amides; 7 = alkaloids derived from phenylalanine and tyrosine; 8 = alkaloids derived from anthranilic acid; 9 = alkaloids derived from tryptophan. Numbers in brackets refer to those known from one or a few taxa only.

isolations of individual true coumarins concern Umbelliferae (251 compounds, mostly C- or O-prenylated), then Rutaceae (127 compounds, mostly prenylated), Compositae (22, many prenylated), and Meliaceae-Ptaeroxyloideae (9, all prenylated). The other 45 true coumarins (many non-prenylated) were originally isolated from species of 18 different Angiosperm families. Our knowledge of plant coumarins is still fragmentary. Nevertheless this striking picture might indicate tendencies of coumarin accumulation in angiosperms and express phylogenetic connections.

Some remarks concerning classification of Zygophyllaceae and affinities of Leguminosae (Fabales) should be made here. The chemotaxonomy of Zygophyllaceae was discussed by Hegnauer.[91] Arylpropanoid resins (lignans), alkaloids derived from tryptophan (β-carbolines and related non-iridoid indoles) and anthranilic acid (quinazolines), saponins with triterpene or steroid (*Balanites*, *Tribulus*) genins, small amounts of essential oils, seeds without starch but with proteins, fatty oils, and sometimes reserve cellulose are the outstanding features. At present most is known about flavonoids[167] and the distribution of steroidal sapogenins (restricted to *Balanites*, *Kallstroemia*, *Tribulus*).[120,121,179,193,194] The structure of *nitrarine* (**34**) has been revised and new alkaloids described for *Nitraria komarovii* [komarovine (**35**) and its dihydro derivative].[196] *N. schoberi* has yielded nitramine (**36**),[102] nitrarine and the allied alkaloids isonitrarine, nitramidine, and schoberidine,[101] and nitraramine (**37**) and *N*-hydroxynitraramine,[139] whilst *N. siberica* gave isonitramine.[102] *Fagonia cretica* has yielded harmalin, and seedlings of *Peganum harmala* ruine (**38**), dihydroruine, harmalol, and 6-hydroxytrypt-amine.[123,136] Alkaloids of Zygophyllaceae are reminiscent of the canthin-6-ones and β-carbolines of Rutaceae and Simaroubaceae, and the only alkaloid known from Anacardiaceae, **39** from *Dracontomelon mangiferum*.[109] Stem bark of *Balanites aegyptica* has given small amounts of bergapten and marmesin,[171] two furocoumarins widespread in Rutaceae and Umbelliferae. The genus best known from the chemical view at present (lignans, flavonoids, waxes, triterpenes, saponins, essential oil) is *Larrea*[118] which illustrates diversity in flavonoid metabolism (flavonols include derivatives of kaempferol, quercetin, herbacetin, myricetin, and gossypetin, flavones, and *C*-glycosyl-flavones).

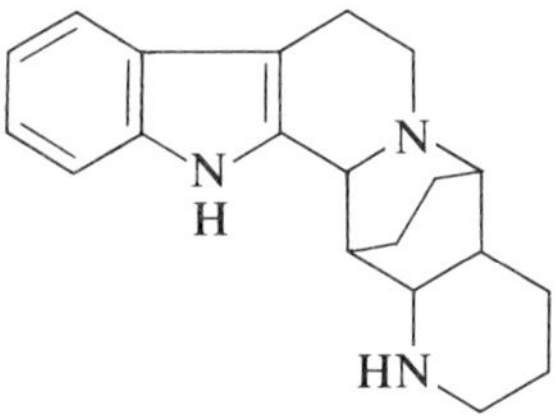

(**34**) Nitrarine

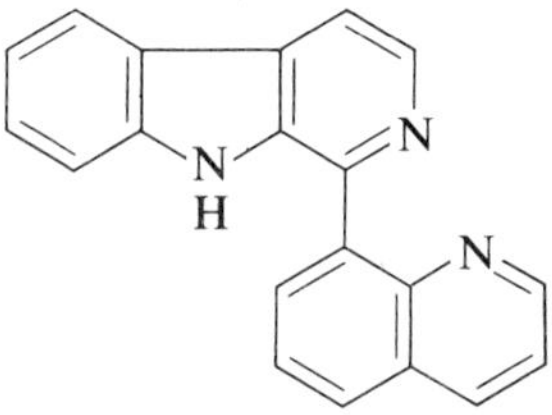

(**35**) Komarovine

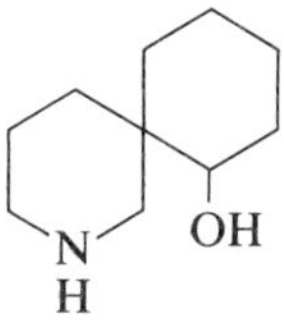

(**36**) Nitramine

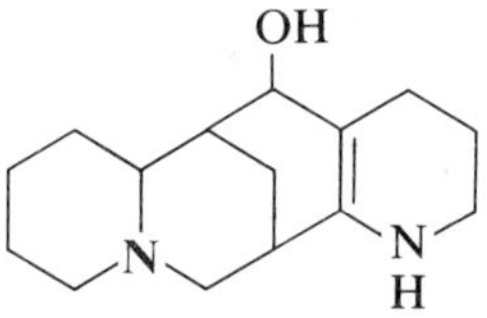

(**37**) Nitraramine

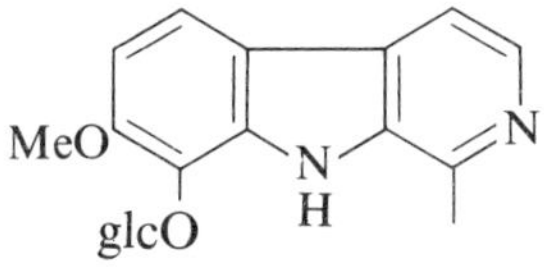

(**38**)

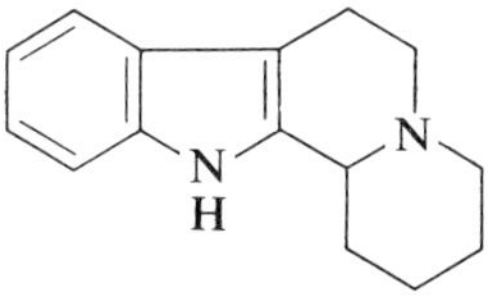

(**39**)

The position ascribed to Zygophyllaceae by Cronquist[51] agrees very well with its chemistry. If the proposal made for the classification of Rutales *s.l.* on p. 425 is accepted, the family should be included in Sapindales *s.s.* but with many links to Rutales *s.s.* Classification of the family in the Geraniales seems to be far less natural in several respects.

Leguminosae (Fabiflorae *sensu* Dahlgren[55]) were included in Sapindales by Hallier.[83] More recently Leguminosae have often been associated with Rosales, but Dahlgren[55] proposed that Leguminosae and Sapindales were closely related. The connecting link between these taxa is believed to be represented by the Connaraceae which are classified in Sapindales by Dahlgren. Thorne[192] also transferred Leguminosae from Rosiflorae to Rutiflorae.

Chemical evidence for such a classification can be found in the condensed and hydrolysable tannins present in many Leguminosae and Anacardiaceae, a good example being the tetraflavanoid condensed tannins from the wood of *Rhus lancea* which derive from leucofisetinidin (5-deoxyleucocyanidin) and (+)-catechin.[68] The isolation of pongapin, fisetin tetramethyl ether, dimethoxykanugin, and ovalitenone, flavonoids known from the leguminous tribe Tephrosieae, from the wood of *Rhus chinensis* (= *R. semialata* = *R. javanica*),[4] and of the isoflavone glycoside sophoricoside from the leaves of *Schinus latifolius*,[172] form other biochemical links between Anacardiaceae and Leguminosae. 5-Deoxyflavonoids are common in Leguminosae and also occur[204,205] in many Anacardiaceae. Young[206] accentuated the systematic potential of 5- and 7-deoxyflavonoids, finding them to occur jointly only in Primulaceae, Fabaceae (Leguminosae), Anacardiaceae, Rutaceae, and Compositae. Besides Primulaceae, these are all taxa belonging to the assumed evolutionary line which is illustrated in Fig. 1.

Other areas of biochemical similarity between Leguminosae and Rutales should also be noted. 4-Methylgallic acid from the wood of *Poupartia axillaris*[188] might be a precursor of bergenin, a compound present in several Leguminosae but not yet found in Anacardiaceae. The erratic occurrence of steroidal C_{27} sapogenins in both Zygophyllaceae and Leguminosae, and of acetogenic anthraquinones in Simaroubaceae, Anacardiaceae,[185] and Leguminosae, also suggests biochemical similarities between Sapindales *s.l.* and Leguminosae. Leucine derived cyanogenic glycosides (proacacipetalin, heterodendrin, cardiospermin) are only known from Sapindaceae, Leguminosae (*Acacia*), and Rosaceae (*Sorbaria*).[168] The corresponding cyanolipids are restricted to Sapindaceae. The convergent occurrence of heterodendrin in two genera of grasses does not merit discussion in the present context.

VI. CONCLUDING REMARKS

Chemical characters strongly link Magnoliidae and Rutales (Table 7) on one side and Rutales and Apiales on the other. Moreover, there are very convincing links between Apiales and Asterales and, to some extent, between Magnoliidae and Asterales (e.g. C_{15} lactones). It is true that Jensen[108] observed few serological relationships between Berberidaceae and Rutaceae, but the behaviour of his two antisera (*Berberis*, *Phellodendron*) were rather unexpected in several respects. It is also true that we should avoid putting too much weight on individual sets of characters in matters of phylogenetic speculation. Biochemical convergencies are many, and two have already been mentioned, the indole alkaloid borrerine (Section IIIH) and heterodendrin (above). Another example is offered by the recent demonstration of the occurrence of the furoquinoline alkaloid skimmianine in *Tylophora asthmatica* (= *T. indica*, Asclepiadaceae, aerial parts) and *Vinca herbacea* (Apocynaceae), and of the allied γ-fagarine in *T. asthmatica*.[66] Chemical convergence is also illustrated by germacranolide, eudesmanolide, and guaianolide type sesquiterpene lactones which occur not only in Magnoliidae, Burseraceae, Umbelliferae, and Compositae but also in Hepaticae, Cupressaceae, and Labiatae.[70]

Nevertheless, chemical characters have proved to be useful tools in plant taxonomy. If all available facts are considered,[56,71,86,89,91,200,201] the pitfalls caused by convergence should become avoidable. We must be prepared, however, to modify our conclusions if new phytochemical evidence tends to show that we are wrong. In this respect the confirmation of presence or definite proof of absence of true secoiridoids or iridoids in Rutales and Araliales (see Table 7, notes *l* and *m*) must be considered as being crucial.

REFERENCES

1. Abe, F., Furukawa, M., Nonaka, G., Okabe, H. and Nishioka, I. (1973). *Yakugaku Zasshi* **93**, 624.
2. Aburano, S., Goichi, K., Moromoto, M. and Nishikawa, Y. (1972). *Yakugaku Zasshi* **92**, 1298.
3. Adesogan, E. K. (1973). *Phytochemistry* **12**, 2310.
4. Ahmad, J., Khan, H. and Shamsuddin, K. M. (1980). *Indian J. Chem.*, **19B**, 420.
5. Ahmed, F. R., Ng, A. S. and Fallis, A. G. (1978). *Can. J. Chem.* **56**, 1020.
6. Airy Shaw, H. K. (1973). "A Dictionary of the Flowering Plants and Ferns" 8th Edn. Cambridge University Press, Cambridge.

7. Alston, R. E. (1966). In "Comparative Phytochemistry" (Swain, T. ed.), pp. 33–56. Academic Press, London.
8. Arndt, R. R. and Baarschers, W. H. (1972). Tetrahedron **28**, 2333.
9. Ayer, W. A. and Brown, L. M. (1975). Phytochemistry **14**, 1457.
10. Bajaj, A. G., Sukh Dev, Arnold, E., Tagle, B. and Clardy, J. (1981). Tetrahedron Lett. **22**, 4623.
11. Balbaa, S. I., Halim, A. F., Halaweish, F. T. and Bohlmann, F. (1980). Phytochemistry **19**, 1519.
12. Balgir, B. S., Mander, L. N. and Mander, S. T. K. (1973). Aust. J. Chem. **26**, 2459.
13. Banerji, J., Rej, R. N. and Chatterjee, A. (1973). Indian J. Chem. **11**, 693.
14. Barros Correa, de D., Gottlieb, O. R. and Pimenta di Padua, A. (1975). Phytochemistry **14**, 2059; (1979). ibid. **18**, 351.
15. Bentham, G. and Hooker, J. D. (1862). "Genera Plantarum", Vol. 1. A. Black, London.
16. Bhatia, V. K., Madhav, R. and Seshadri, T. R. (1969). Indian J. Chem. **7**, 121.
17. Bhatt, S. K., Saxena, W. K. and Nigam, S. S. (1981). Phytochemistry **20**, 1749.
18. Billet, D., Durgeat, M., Heitz, S. and Ahond, A. (1975). Tetrahedron Lett. 3825.
19. Billet, D., Durgeat, M., Heitz, S. and Brouard, J.-P. (1978). J. Chem. Res. (S) 110.
20. Billet, D., Durgeat, M., Heitz, S., Brouard, J.-P., and Ahond, A. (1976). Tetrahedron Lett. 2773, 2777.
21. Billet, D., Heitz, S., Raulais, D. and Matschenko, A. (1971). Phytochemistry **10**, 1681.
22. Bohlmann, F., Abraham, W. R., King, R. M. and Robinson, H. (1981). Phytochemistry **20**, 825.
23. Bohlmann, F., Adler, A., King, R. M. and Robinson, H. (1982). Phytochemistry **21**, 173.
24. Bohlmann, F., Balbaa, S., Halim, A. and Halaweish, F. T. (1981). Phytochemistry **20**, 177.
25. Bohlmann, F., Burkhard, T. and Zdero, C. (1973). "Naturally Occurring Acetylenes." Academic Press, London.
26. Bohlmann, F., Jakupovic, J., King, R. M. and Robinson, H. (1980). Phytochemistry **19**, 1815.
27. Bohlmann, F. and Zdero, C. (1977). Phytochemistry **16**, 239.
28. Bohlmann, F. and Zdero, C. (1977). Phytochemistry **16**, 1092; 1261.
29. Bohlmann, F. and Zdero, C. (1977). Chem. Ber. **110**, 1755.
30. Bohlmann, F., Zdero, C., King, R. M. and Robinson, H. (1979). Phytochemistry **18**, 1246.
31. Bohlmann, F., Zdero, C., King, R. M. and Robinson, H. (1982). Phytochemistry **21**, 147.
32. Bohlmann, F., Zdero, C. and Ngo, LeVan. (1979). Phytochemistry **18**, 99.
33. Bory, G. and Clair-Maczulajtys, D. (1980). Phytomorphology **30**, 67.
34. Bose, S. and Gupta, K. C. (1964). Indian J. Chem. **2**, 57; 156; (1966). ibid. **4**, 87.
35. Brieskorn, C. H. and Noble, P. (1982). Planta Med. **44**, 87.
36. Brown, K. L., Burfitt, A. I. R., Cambie, R. C., Holland, D. and Mathai, K. P. (1975). Aust. J. Chem. **28**, 1327.
37. Burke, B. A. and Parkins, H. (1979). Phytochemistry **18**, 1073.
38. Chadefaud, M. and Emberger, L. (1960). "Traité de Botanique Systématique", Vol. 2, pp. 622–753. Masson et Cie, Paris.
39. Chakraborty, D. P. (1977). Fortschr. Naturst. **34**, 299.

40. Chakraborty, D. P., Islam, A. and Roy, S. (1978). *Phytochemistry* **17**, 2043.
41. Chandrasekharan, S. and Chakrabortty, T. (1968). *J. Indian Chem. Soc.* **45**, 208.
42. Chen, F.-C. and Lin, Y.-M. (1976). *J. Chem. Soc. Perkin Trans. 1* 98.
43. Chexal, K. K., Fouweather, C. and Holker, J. S. (1975). *J. Chem. Soc. Perkin Trans. 1* 554.
44. Connolly, J. D., MacLellan, M., Okorie, D. A. and Taylor, D. A. H. (1976). *J. Chem. Soc. Perkin Trans. 1* 1993.
45. Corner, E. J. H. (1976). "The Seeds of Dicotyledons" (2 volumes). Cambridge University Press, Cambridge.
46. Cotterrell, G. P., Halsall, T. G. and Wrigglesworth, M. J. (1970). *J. Chem. Soc. (C)* 739.
47. Croft, J. A., Ritchie, E. and Taylor, W. C. (1975). *Aust. J. Chem.* **28**, 2019.
48. Croft, J. A., Ritchie, E. and Taylor, W. C. (1975). *Aust. J. Chem.* **28**, 2093.
49. Cronquist, A. (1968). "The Evolution and Classification of Flowering Plants". Houghton, Mifflin, Boston.
50. Cronquist, A. (1980). *In* "Chemosystematics: Principles and Practice" (Bisby, F. A., Vaughan, J. G. and Wright, C. A. eds.), pp. 1–27. Academic Press, London.
51. Cronquist, A. (1981). "An Integrated System of Classification of Flowering Plants". Columbia University Press, New York.
52. Crowson, R. A. (1970). "Classification and Biology". Heinemann, London.
53. Czaja, A. T. (1978). "Stärke und Stärkespeicherung bei Gefässpflanzen". Gustav Fischer Verlag, Stuttgart.
54. Dahlgren, R. (1975). *Bot. Notiser.* **128**, 119.
55. Dahlgren, R. (1980). *Bot. J. Linn. Soc.* **80**, 91.
56. Dahlgren, R., Rosendal-Jensen, S. and Nielsen, B. J. (1981). *In* "Phytochemistry and Phylogeny" (Young, D. A. and Seigler, D. S. eds.), pp. 149–204. Praeger, New York.
57. Davis, P. H. and Heywood, V. H. (1963). "Principles of Angiosperm Taxonomy". Oliver and Boyd, Edinburgh (reprint with supplementary bibliography).
58. Dean, F. M. and Robinson, M. L. (1971). *Phytochemistry* **10**, 3221.
59. Debenedetti, S. L., Ferraro, G. E. and Coussio, J. D. (1981). *Planta Med.* **42**, 97.
60. Dos Santos Filho, D. and Gilbert, B. (1975). *Phytochemistry* **14**, 821.
61. Dreyer, D. L. (1977). *Rev. Latinoam. Quim.* **8**, 11.
62. Dreyer, D. L. and Park, K.-H. (1975). *Phytochemistry* **14**, 1617.
63. Edman, G. (1936). *Svensk. Botan. Tidskr.* **30**, 493.
64. El Sohly, M. A., Craig, J. C., Waller, C. W. and Turner, C. E. (1978). *Phytochemistry* **17**, 2140.
65. Epe, E. and Mondon, A. (1979). *Tetrahedron Lett.* 2015.
66. Etherington, T., Herbert, R. B. and Jackson, F. B. (1977). *Phytochemistry* **16**, 1125.
67. Fazldeen, H., Hegarty, M. P. and Lahey, F. N. (1978). *Phytochemistry* **17**, 1609.
68. Ferreira, D., Hundt, H. K. L. and Roux, D. G. (1971). *J. Chem. Soc. Chem. Commun.* 1257.
69. Fikenscher, L. H. and Hegnauer, R. (1977). *Pharm. Weekblad* **112**, 11.
70. Fischer, N. H., Olivier, E. J. and Fischer, H. D. (1979). *Fortschr. Naturst.* **38**, 47.
71. Fish, F. and Waterman, P. G. (1973). *Taxon* **22**, 177.
72. Forcellese, M. L., Nicoletti, R. and Petrossi, U. (1972). *Tetrahedron* **28**, 325.
73. Free, A. J., Read, R. W., Ritchie, E. and Taylor, W. C. (1976). *Aust. J. Chem.* **29**, 695.
74. Frondel J. W. (1967). *Nature* (London) **215**, 1360; (1969). *Naturwiss.* **56**, 280.

75. Gibbs, R. D. (1974). "Chemotaxonomy of Flowering Plants", Vol. 3, pp. 1669–1674. McGill Queens University Press, Montreal.
76. Gopinath, K. W., Mahanta, P. K., Bohlmann, F. and Zdero, C. (1976). *Tetrahedron* **32**, 737.
77. Gosh, P. C., Larrahondo, J. E., Le Quesne, P. W. and Raffauf, R. F. (1977). *Lloydia* **40**, 364.
78. Gottlieb, O. R. (1974). *Rev. Latinoam. Quim.* **5**, 1.
79. Gray, A. I. and Waterman, P. G. (1978). *Phytochemistry* **17**, 845.
80. Greger, H., Grenz, M. and Bohlmann, F. (1981). *Phytochemistry* **20**, 2579.
81. Gundersen, A. (1950). "Families of Dicotyledons". Chronica Botanica Co., Waltham, Mass.
82. Gupta, B. D., Banerjee, S. K. and Handa, K. L. (1976). *Phytochemistry* **15**, 576.
83. Hallier, H. (1912). *Archiv. Néerl. Sci.* Ser. III B **1**, 146.
84. Hegnauer, R. (1960). *Pharmazeutische Zentralhalle* **99**, 322 (see point 5, p. 324).
85. Hegnauer, R. (1963). *In* "Chemical Plant Taxonomy" (Swain, T. ed.), pp. 389–427. Academic Press, London.
86. Hegnauer, R. (1964). "Chemotaxonomie der Pflanzen", Vol. 3 (Anacardiaceae, Burseraceae, Cneoraceae). Birkhäuser Verlag, Basel.
87. Hegnauer, R. (1965). *Bull. Soc. Bot. Franc. Mémoires*, 103–116 (publ. 1967).
88. Hegnauer, R. (1966). "Chemotaxonomie der Pflanzen", Vol. 4 (Irvingiaceae, Julianiaceae), Birkhäuser, Verlag, Basel.
89. Hegnauer, R. (1969). "Chemotaxonomie der Pflanzen", Vol. 5 (Meliaceae). Birkhäuser Verlag, Basel.
90. Hegnauer, R. (1969). *In* "Perspectives in Phytochemistry" (Harborne, J. B. and Swain, T. eds.), pp. 121–138. Academic Press, London.
91. Hegnauer, R. (1973). "Chemotaxonomie der Pflanzen", Vol. 6 (Rutaceae, Sapindaceae, Simaroubaceae, Zygophyllaceae). Birkhäuser Verlag, Basel.
92. Hegnauer, R. (1977). *In* "The Biology and Chemistry of the Compositae" (Heywood, V. H., Harborne, J. B. and Turner, B. L. eds.), pp. 283–335. Academic Press, London.
93. Hegnauer, R. (1978). *In* "Les Ombellifères. Contribution pluridisciplinaires à la Systématique" (Cauwet-Marc, A. M. and Carbonnier, J. eds.), pp. 335–363. C.N.R.S. Perpignan.
94. Herz, W., Govindan, S. V. and Kumar, N. (1981). *Phytochemistry* **20**, 1343.
95. Heywood, V. H. (1966). *In* "Comparative Phytochemistry" (Swain, T. ed.), pp. 1–20. Academic Press, London.
96. Hifnawy, M. S., Vaquette, J., Sévenet, T., Pousset, J.-L. and Cavé, A. (1977). *Phytochemistry* **16**, 1035.
97. Higa, T. and Scheuer, P. J. (1974). *J. Chem. Soc. Perkin Trans. 1* 1350.
98. Hopf, H. and Kandler, O. (1977). *Phytochemistry* **16**, 1715.
99. Hutchinson, J. (1959). "The Families of Flowering Plants", 2nd Edn, Vol. 2. Clarendon Press, Oxford.
100. Hutchinson, J. (1969). "Evolution and Phylogeny of Flowering Plants". Academic Press, London.
101. Ibragimov, A. A., Nasirov, S. M., Andrianov, V. T., Maekh, S. K., Struchkov, Yu. T. and Yunusov, S. Yu. (1975). *Khim. Prir. Soedin.* 273.
102. Ibragimov, A. A., Osmanov, Z., Tashchodzhaev, B., Abdulaev, N. D., Yagudaev, M. R. and Yunusov, S. Yu. (1981). *Khim. Prir. Soedin.* 623.
103. Ionescu, F., Jolad, S. D., Cole, J. R., Arora, S. K. and Bates, R. B. (1977). *J. Org. Chem.* **42**, 1627.

104. Ishii, H., Ishikawa, T. and Chen, I.-S. (1973). *Tetrahedron Lett.* 4189.
105. Ishii, H., Ishikawa, T., Lu, S.-T. and Chen, I.-S. (1976). *Yakugaku Zasshi* **96**, 1458.
106. Ishratullah, K., Ansari, W. H., Rahman, W., Okigawa, M. and Kawano, N. (1977). *Indian J. Chem.* **15B**, 615.
107. Jain, A. C., Khazanchi, R. and Kumar, A. (1978). *Chemistry Lett.* 991.
108. Jensen, U. (1974). The Serological Museum, *Bull.* No. 50, pp. 4–7.
109. Johns, S. R., Lamberton, J. A. and Occolowitz, J. L. (1966). *Aust. J. Chem.* **19**, 1951.
110. Kamala Devi, A., Bhat, S. V. and Bhattacharyya, S. C. (1969). *Indian J. Chem.* **7**, 1279.
111. Klein, E. and Obermann, H. (1978). *Tetrahedron Lett.* 349.
112. Kubitzki, K. (1969). *Taxon* **18**, 360.
113. Kubo, I., Miura, I. and Nakanishi, K. (1976). *J. Am. Chem. Soc.* **98**, 6704.
114. Kubo, I., Tanis, S. P., Lee, Y.-W., Miura, K., Nakanishi, K. and Chapya, A. (1976). *Heterocycles* **5**, 485.
115. Lemmich, E. (1979). *Phytochemistry* **18**, 1195.
116. Leon, C. J. J. (1975). *Boll. Soc. Quim. Peru* **41**, 14.
117. Liu, W. H., Kusumi, T. and Nakanishi, K. (1981). *J. Chem. Soc. Chem. Commun.* 1271.
118. Mabry, T. J., Hunziker, J. H. and Difeo, D. R. (eds.) (1977). "Creosote Bush. Biology and Chemistry of Larrea in New World Deserts". Dowden, Hutchinson and Ross, Pennsylvania.
119. Magalhaes, A. F., Leitao Filho, H. F., Frighetto, R. T. S. and Barros, S. M. G. (1981). *Phytochemistry* **20**, 1369.
120. Mahato, S. B., Sahu, N. P. and Chakravarti, R. N. (1970). *Indian J. Chem.* **15B**, 445.
121. Mahato, S. B., Sahu, N. P., Ganguly, A. N., Miyahara, K. and Kawasaki, T. (1981). *J. Chem. Soc. Perkin Trans. 1* 2405.
122. Mahmoud, Z. F., Abdel Salem, N. A., Sarg, T. M. and Bohlmann, F. (1981). *Phytochemistry* **20**, 735.
123. McKenzie, E., Nettleship, L. and Slaytor, M. (1975). *Phytochemistry* **14**, 273.
124. Meeuse, A. D. J. (1970) *Acta Bot. Neerl.* **19**, 61; 133.
125. Méndéz, J. and Masa, A. (1975). *Phytochemistry* **14**, 1136.
126. Merxmüller, H. (1967). *Ber. Deutsch. Bot. Ges.* **80**, 608.
127. Mester, I. (1973). *Fitoterapia* **44**, 123; (1977). *ibid.* **48**, 268.
128. Metcalfe, C. R. and Chalk, L. (1950). "Anatomy of the Dicotyledons" (2 volumes). Clarendon Press, Oxford.
129. Mondon, A., Callsen, H. and Hartmann, P. (1975). *Chem. Ber.* **108**, 1989.
130. Murray, R. D. H. (1978). *Fortschr. Naturst.* **35**, 199.
131. Murthy, S. S. N., Rao, N. S. P., Anjaneyulu, A. S. R. and Row, L. R. (1981). *Planta Med.* **43**, 46.
132. Murthy, S. S. N., Anjaneyulu, A. S. R. and Row, L. R. (1981). *Indian J. Chem.* **20B**, 150.
133. Nakane, M., Hutchinson, C. R., Van Engen, D. and Clardy, J. (1978). *J. Am. Chem. Soc.* **100**, 7079.
134. Nakanishi, K. (1977). *Pontificiae Academiae Scientiarum Scripta Varia* **41**, 185.
135. Netolitzki, F. (1926). "Anatomie der Angiospermen-Samen"; Handbuch der Pflanzenanatomie (ed. K. Linsbauer), Vol. 2(X). Borntraeger, Berlin.
136. Nettleship, L. and Slaytor, M. (1971). *Phytochemistry* **10**, 231.
137. Ng, A. S. and Fallis, A. G. (1979). *Can. J. Chem.* **57**, 3088.
138. Nooteboom, H. P. (1967). *Adansonia* Ser. II **7**, 161.

139. Novgorodova, N. Yu., Maekh, S. K. and Yunusov, S. Yu. (1975). *Khim. Prir. Soedin.* 435; 529.
140. Oei-Koch, A. and Kraus, L. (1978). *Planta Med.* **34**, 339.
141. Okano, M., Lee, H.-K. and Hall, I. H. (1981). *J. Nat. Prod.* **44**, 470.
142. Pardhy, R. S. and Bhattacharyya, S. C. (1978) *Indian J. Chem.* **16B**, 171; 174; 176.
143. Paris, P. R. and Debray, M. (1972). *Plant. Méd. Phytothér.* **6**, 311.
144. Pascual, T. de J., Grande, C. and Grande, M. (1978). *Tetrahedron Lett.* 4563.
145. Patil, V. D., Nayak, U. R. and Sukh Dev. (1972). *Tetrahedron* **28**, 2341.
146. Paton, W. F., Paul, I. C., Bajaj, A. G. and Sukh Dev. (1979). *Tetrahedron Lett.* 4153.
147. Philipson, W. R. and Stone, B. C. (1980). *Taxon* **29**, 391.
148. Picker, K., Ritchie, E. and Taylor, W. C. (1976). *Aust. J. Chem.* **29**, 2023.
149. Pinar, M. and Alemany, A. (1975). *Phytochemistry* **14**, 313.
150. Pinar, M., Rodriguez, B. and Alemany, A. (1978). *Phytochemistry* **17**, 1636.
151. Polonsky, J., Baskevitch-Varon, Z. and Das, B. C. (1976). *Phytochemistry* **15**, 357.
152. Polonsky, J., Varon, Z., Prange, T., Pascard, C. and Moretti, C. (1981). *Tetrahedron Lett.* **22**, 3605.
153. Prasad, R. S. and Sukh Dev. (1976). *Tetrahedron* **32**, 1437.
154. Price, J. R. (1963). *In* "Chemical Plant Taxonomy" (Swain, T. ed.), pp. 429–452. Academic Press, London.
155. Proietti, G., Strappaghetti, G. and Corsano, S. (1981). *Planta Med.* **41**, 417.
156. Quisumbing, E. (1951). "Medicinal Plants of the Philippines", Tech. Bull. No. 16, Dept. Agric. and Nat. Resources, Philippines, Manila.
157. Rendle, A. B. (1925). "The Classification of Flowering Plants", Vol. 2 (1956 reprint). Cambridge University Press, Cambridge.
158. Richter, H. G. (1980). *Wood Sci. Technol.* **14**, 34.
159. Rodriguez, B. and Pinar, M. (1979). *Phytochemistry* **18**, 891.
160. Romeo, J. T. and Prass, W. R. (1980). *Biochem. Syst. Ecol.* **8**, 69.
161. Roth, I. (1977). "Fruits of angiosperms"; "Handbuch der Pflanzenanatomie" (Zimmermann, W., Carlquist, S., Ozenda, P. and Wulff, H. D. eds.), Vol. 1/X. Borntraeger, Berlin.
162. Rozsa, Z., Reisch, J., Lingyel, E., Gellert, M., Szendrei, K., Novak, I. and Minker. E. (1977). *Planta Med.* **32A**, 57.
163. Rozsa, Z., Mester, I., Reisch, J. and Szendrei, K. (1980). *Planta Med.* **39**, 219.
164. Rücker, G. (1972). *Arch. Pharm.* **305**, 486.
165. Sakai, S. and Hasegawa, M. (1974). *Phytochemistry* **13**, 303.
166. Sakurai, K., Toyoda, T., Muraki, S. and Yoshida, T. (1979). *Agric. Biol. Chem.* **43**, 195.
167. Saleh, N. A. M. and El Hadidi, M. N. (1977). *Biochem. Syst. Ecol.* **5**, 121.
168. Saupe, S. G. (1981). *In* "Phytochemistry and Angiosperm Phylogeny" (Young, D. A. and Seigler, D. S. ed.), pp. 80–116. Praeger, New York.
169. Scholz, H. (1964). *In* "A. Engler's Syllabus der Pflanzenfamilien", 12th Edn, Vol. 2, pp. 262–277. Borntraeger, Berlin.
170. Schulte, K., Rücker, G. and Klewe, U. (1973). *Archiv. Pharm.* **306**, 857.
171. Seida, A. M., Kinghorn, A. D., Cordell, G. A. and Farnsworth, N. R. (1981). *Planta Med.* **43**, 92.
172. Sepuldeva, S. and Cassels, B. K. (1979). *Rev. Latinoam. Quim.* **10**, 136.
173. Sherman, M. M., Boris, R. P., Ogura, M., Cordell, G. A. and Farnsworth, N. R. (1980). *Phytochemistry* **19**, 1499.
174. Shone, D. K. and Drummond, R. B. (1965). *Rhodesia Agric. J.* **62**, 59.

175. Shukla, S. and Tiwari, R. D. (1971). *Indian J. Chem.* **9**, 287.
176. Snatzke, G. and Vertésy, L. (1967). *Monatsh. Chem.* **98**, 121.
177. Snyder, J., Nakanishi, K., Chaverria, G., Leal, Y., Ochoa, C. C. and Dominguez, X. A. (1981). *Tetrahedron Lett.* **22**, 5015.
178. Söderberg, E. (1936). *Svensk. Botan. Tidskr.* **30**, 537.
179. Sofowora, E. A. and Hardman, R. (1974). *Planta Med.* **25**, 22.
180. Sofowora, E. A., Isaac-Sodeye, W. A. and Ogunkoya, L. O. (1975). *Lloydia* **38**, 169.
181. Solereder, H. (1908). "Systematic Anatomy of the Dicotyledons", Vols. I and II. Clarendon Press, Oxford.
182. Soó, C. R. (1975). *Taxon* **24**, 585.
183. Soto de Villatoro, B., Giral Gonzalez, F., Polonsky, J. and Baskevitch-Varon, Z. (1974). *Phytochemistry* **13**, 2018.
184. Straka, H., Albers, F. and Mondon, A. (1976). *Beitr. Biol. Pflanz.* **52**, 267.
185. Subramanian, S. S. and Nair, A. G. R. (1971). *Phytochemistry* **10**, 1939.
186. Subramanian, S. S. and Nair, A. G. R. (1972). *Indian J. Chem.* **10**, 452.
187. Suga, T., Shishibori, T., Kosela, S. and Sood, V. K. (1975). *Phytochemistry* **14**, 308.
188. Takahashi, T. and Imamura, H. (1966). *Agric. Biol. Chem.* **30**, 617.
189. Takhtajan, A. L. (1980). *Bot. Rev.* **46**, 225.
190. Thomas, A. F. and Willhalm, B. (1964). *Tetrahedron Lett.* 3177.
191. Thorne, R. F. (1976). "Evolutionary Biology", Vol. 9, pp. 35–106. Plenum Press, New York.
192. Thorne, R. F. (1981). *In* "Phytochemistry and Angiosperm Phylogeny" (Young, D. A. and Seigler, D. S. eds.), pp. 233–295. Praeger Publishers, New York.
193. Tomowa, M. P. and Gjulemetowa, R. (1978). *Planta Med.* **34**, 188.
194. Tomowa, M. P., Panowa, D. and Wulfson, N. S. (1974). *Planta Med.* **25**, 231.
195. Trautmann, D., Epe, B., Oelbermann, U. and Mondon, A. (1980). *Chem. Ber.* **113**, 3848.
196. Tulyaganov, T. S., Ibragimov, A. A. and Yunusov, S. Yu. (1980). *Khim. Prir. Soedin.* 732.
197. Turner, B. L. (1967). *Pure Appl. Chem.* **14**, 189.
198. Vieira, P. C., Alvarenga, de, M. A., Gottlieb, O. R., Nazare, de, M., McDougall, V. and Reis, de, A. M. F. (1980). *Phytochemistry* **19**, 472.
199. Wagenitz, G. (1959). *Bot. Jahrb.* **79**, 17.
200. Waterman, P. G. (1975). *Biochem. Syst. Ecol.* **3**, 149.
201. Waterman, P. G. and Khalid, S. A. (1981). *Biochem. Syst. Ecol.* **9**, 45.
202. Wettstein, R. (1935). "Handbuch der Systematischen Botanik", 4th Edn. Franz Deuticke, Leipzig.
203. Willis, J. C. (1931). "A Dictionary of the Flowering Plants and Ferns", 6th Edn. University Press, Cambridge. Reprint 1955.
204. Young, D. A. (1976). *Syst. Bot.* **1**, 149.
205. Young, D. A. (1979). *Am. J. Bot.* **66**, 502.
206. Young, D. A. (1981). *In* "Phytochemistry and Angiosperm Phylogeny" (Young, D. A. and Seigler, D. S. eds.), pp. 205–232. Praeger Publishers, New York.
207. Joshi, B. S., Kamat, V. N. and Gawad, D. H. (1974). *J. Chem. Soc. Perkin Trans. 1* 1561.

For addendum see following page.

From a chemotaxonomic viewpoint the following recent contributions to the phytochemistry of the Rutales are noteworthy. *Tannins*—Ellagic acid occurs in the flowers of *Commiphora mukul* and the ellagitannins geraniin and isogeraniin in the leaves of three *Ailanthus* spp. [Kakrani, H. K. *et al.* (1982). *Fitoterapia* **52**, 221; Haddock, E. H. *et al.* (1982). *Phytochemistry* **21**, 1049]. *Sesquiterpenes*— Lindestrene (known from Lauraceae) and similar furanoid eudesmadienes, furanoid germacradienes and a furanoid guaiadiene occur in *Commiphora molmol* (myrrh) [Brieskorn, C. H. and Pia Noble (1983). *Phytochemistry* **22**, 187, 1207]; chemical relationships between Compositae and Umbelliferae are demonstrated once more by the occurrence of fastigiolide, the first daucane-based sesquiterpenelactone in *Ageratum fastigiatum* [Bohlmann, F. *et al.* (1983). *Phytochemistry* **22**, 983]. *2-Methylchromones*—Peucenin, alloptaerox-ylin and its methyl ether occur together with rutaceous limonoids in the roots of *Harrisonia abyssinica* (Simaroubaceae), a species that is clearly chemically atypical of the family [Okorie, D. A. (1982). *Phytochemistry* **21**, 2424]. *Alkaloids and alkaloid-like compounds*—*Aglaia elliptifolia* (Meliaceae) contains a lignanoid amide, rocaglamide [Ming Lu King *et al.* (1982). *J. Chem. Soc. Chem. Comm.* 1150]. Leaves of *Dysoxylum lenticellare* (Meliaceae) contain phenethylisoquinoline-type alkaloids such as homolaudanosine and the derived epischelhammericine [Aladesanmi, A. J. *et al.* (1983). *J. Nat. Prod.*, **46**, 127]. The latex of the fruit peel of *Lansium domesticum* (Meliaceae) contains large amounts of a glycoside of lansic acid, the sugar being *N*-acetylglucosamine [Nishizawa, M. *et al.* (1982). *Tetrahedron Letters* **23**, 1349].

Index

441

Borreverine, 35, 62
Bosistoa euodiformis, 409–10
Bosistoa sapindiformis, 419
Boswellia carteri, 284–5, 287, 292–3, 418
Boswellia dalzielli, 292–3
Boswellia frereana, 283, 291–3, 418
Boswellia serrata, 291, 296, 407, 421
α-Boswellic acid, 293
β-Boswellic acid, 292, 418
   3-acetoxy-11-hydroxy, 291
Bothriocline, 429
Bourjotinolone A, 289, 416–7
Brassilignan, 268, 270, 406
Brayleanin, 102, 114
Braylin, 125
Brein, 292
Bretschneideraceae, 403
Brosiparin, 114
Brucea spp., 252, 416
   antileukemic principles, 255
Br. amarissima, 255
Br. antidysenterica, 255
Bruceantin, 255, 259–61
Bruceantinol, 255, 259–61
Bruceantinoside A and B, 255
Bruceine A, B, and C, 255, 260–1
Bruceol, 127
   deoxy, 127
Bruceolide, 255–9
Bruceoside A and B, 255
Bucharaine, 20, 43
Bucharamine, 20
Bucharaminol, 33, 46
Bucharidine, 20, 43
Burkillanthus malaccensis, 307
Bursera spp., 409–10
B. aptera, 135
B. arida, 267, 276, 296, 418
B. balsamifera, 141
B. delpechinana, 284, 291–3
B. fagaroides, 276, 296
B. galleotiana, 291–2
B. gorullensis, 291–2
B. graveolens, 293
B. klugii, 289
B. leptophloeos, 141
B. microphylla, 276, 291, 296
B. morelensis, 135, 276, 296

B. schlechtendalii, 276, 292, 296
B. sessiliformis, 291–2
Burseraceae, 5, 31, 281ff., 308, 353
   coumarins in, 97ff.
   lignans in, 267ff.
   links to Rutales *s.s.*, 297, 378, 402ff.
Burseran, 270
Bussein A and B, 201
Buxaceae, 403–4
Byakangelicin, 124, 313
Byakangelicol, 124

C

Cabralea eichleriana, 200, 366
Cachrys sicula, 429
δ-Cadinene, 285
Caesalpiniaceae, 378
Caffeic acid, 409, 427
Calamin, 225–6
Calamondin, 225–6
Calodendrolide, 232, 235
Calodendron, 241
Campanulales, 6
Camphene, 284
α-Camphorene, 287
α-Canadine, 73
   N-methyl, 37, 73
Canarium album, 285, 296
Canarium boivinii, 416, 418
Canarium luzonicum, 292–3
Canarium muelleri, 291, 418
Canarium samoense, 285
Canarium schweinfurthii, 289, 292–3
Canarium strictum, 285, 291, 419
Canarium zeylanicum, 291–3, 419
Canarone, 285
Candicine, 36, 68, 415
Candollein, 201
Cannaric acid, 290, 418–9
Canthinone alkaloids, 36, 67, 382, 414, 431
Canthin-2,6-dione, 67
Canthin-6-one, 304
   5-methoxy, 308–9
   4-thiomethyl, 308–9
Capensin, 115
Capuronianthus, 373

coumarins in, 139–40
lignans in, 276
R. angustifolia, 157
R. chalepensis, 157
R. graveolens, 10–1, 13, 17, 147, 153,
  157, 301, 415, 420–1
R. montana, 157, 314
R. pinnata, 17, 21, 23
Rutaceae, 4ff., 289ff., 308, 377ff., 406ff.
  alkaloids in, 31ff.
  chromones in, 97ff.
  coumarins in, 97ff.
  flavonoids in, 158ff.
  lignans in, 269ff.
  limonoids in, 215ff., 319ff., 353,
    368–9, 373
Rutacridone, 34, 58
Rutacridone epoxide, 58
  hydroxy, 58
Rutacultin, 128
Rutaecarpine, 35, 61, 302
  dihydro, 61
Rutaevin, 217, 235, 240–1
  acetate, 224–5
Rutales, 4ff., 401ff.
  co-evolution with Papilionidae, 6,
    380
  delimitation of, 402–4, 425
  distribution of alkaloids in, 31ff., 376
  distribution of chromones in, 378
  distribution of coumarins in, 378
  distribution of flavonoids in, 169–70
  distribution of limonoids in, 378
  distribution of quassinoids in, 378
  origins in the Ranales, 5ff.
  possible phylogenetic affinities,
    422–4
Rutalinium cation, 34, 53
Rutamarin, 16–7, 128, 313
  alcohol, 128
Rutaretin, 122
  methyl ether, 122
Rutarin, 122
Rutaverine, 45
Rutin, 147–8, 152ff.
Rutinae, 390–1
Rutoideae, 39, 164–6, 383ff.
  limonoids in, 240–1
Rutolide, 127

S

Sabandin, 116, 313
Sabandinin, 115, 313
Sabandinol, 115
Sabandinone, 115
Sabiaceae, 403
Sabinene, 345, 347–9
Saccharomyces cerevisiae, 315
Salannin, 236–8
Samadera indica, 87, 382
Samaderine A and C, 257
Sandoricum, 369
Sanguinarine, 74
Sanshoamide, 80
Sanshodiol, 270, 406
γ-Sanshool, 38, 80
  hydroxy, 80
β-Santalene, 285
Sapelin A and B, 177, 193, 289, 416–7
Sapelin C–F, 194
Sapindaceae, 5, 308, 403, 425, 432
Sapindales, 4ff., 169–70, 380, 396,
  402–5, 422–5, 430–2
Saponins, 419
Sargentia greggii, 87, 162–3, 171, 240–1,
  389
Savinin, 271
Sawamilletin, 419
Schinus latifolius, 432
Schoberidine, 430
Scoparon, 112
Scopeletin, 101, 112, 409–10
Scopelin, 113
Seed oils, 421, 428–9
Seed proteins, 422, 428–9
Semecarpus, 169
Sendanal, 196
Sendanin, 195
  desacetyl, 195
Sendanolactone, 177, 193
Sergeolide, 252, 255
Sesamin, 274, 406
Seselin, 21, 125
Sesquiterpenes
  in Burseraceae, 285–7, 440
  lactones, 426–7, 429, 433, 440
Severine palmitate, 413
Severinia buxifolia, 88, 140, 387, 393–4